Preface

The control of engineering operations, and indeed of many others, requires first the measurement of the variables concerned, to a suitable degree of accuracy and precision. These measurements may then be acted upon by a human operator, but in many cases, and to an increasing extent, control is carried out by automatic systems. The engineer of the future will have an ever increasing need for knowledge of measurement techniques and their incorporation into systems of various kinds. This need is reflected by the inclusion in courses at both technician and technologist level of syllabuses such as 'Engineering Measurements', 'Automatic Control', 'Systems', etc. Such topics are included in the courses OND Technology (Engineering), HNC Engineering, HND in Measurement and Control, as well as other HND and City and Guilds courses.

This text seeks to cover the wide range of basic measurement topics which the aeronautical, mechanical, and production engineer needs, placing emphasis on the fundamental principles involved, without introducing excessive detail of hardware. A course in measurements should have a large proportion of time in the measurements laboratory, or engaged in measurements in other laboratories and workshops. Lectures and tutorial periods, occupying less time, should be designed to stimulate the students' interest. The role of a textbook of this nature should be to provide the basic framework of theory to which student and teacher can refer, without the need for hours spent in copying notes and redrawing diagrams.

The aim has been to use an across-the-board approach, avoiding the narrow specialism that has been characteristic of many past engineering syllabuses. This wider concept was used in the HNC Engineering syllabus 'Engineering Measurements', on which this book is based, and has since been extended to other courses at comparable level. The use of measurement in a systems context has been emphasised, and the reader is introduced to transfer operators and responses. A knowledge of the basic theory of mechanics, electrics, and heat is assumed in the reader, and some of the mathematical coverage is included in appendices. Some suggestions for tutorial work are included in each chapter, and it is hoped that these will help to stimulate discussion in group tutorial periods.

I readily acknowledge the debt owed to the many workers and writers in the field of measurement on whose work I have drawn, and to the many manufacturers of instruments who readily provided information. Errors will undoubtedly have occurred in the book, and I will be grateful if readers will notify me of these, via the publishers.

L. F. Adams

Acknowledgements
The author and publisher would like to thank the following organisations for permission to reproduce photographs or illustrations. The British Standards Institution, 2 Park St, London WIA 2BS (fig. 1.8, from BS 1780: 1960; fig. 7.28(b), from BS 188: 1957; fig. 10.32, from BS 1134: part 1: 1972); A. Maclow-Smith Ltd (fig. 5.3); Kistler Instruments Ltd (figs 5.9, 5.10, and 5.34); Torquemeters Ltd (fig. 5.16(b)); Venner Ltd (fig. 6.3); Veeder-Root Ltd (fig. 6.4(a) and (c)); SE Laboratories (Engineering) Ltd (fig. 6.6); Ealing Scientific Ltd (fig. 6.9(c)); James Scott (Precision Engineers) Ltd (fig. 6.11); Gilbarco Ltd (fig. 7.4); Ricardo and Co. (Engineers) Ltd (fig. 7.18); The National Physical Laboratory (figs 10.7, 13.31, and 13.32 (all Crown copyright)); Herbert Control and Instruments Ltd (figs 10.10, 10.15, 10.16, and 11.27); Solex (Gauges) Ltd (figs 10.13 and 12.7); Rank Precision Industrics Ltd (fig. 10.31, first used in *Modern Workshop Technology* by H. W. Baker (Macmillan); fig. 11.10); Wayne Kerr Co. Ltd (fig. 11.13); Jackson Bros (London) Ltd (fig. 11.28); G. V. Planar Ltd (fig. 12.15).

Engineering Measurements and Instrumentation

L. F. Adams BSc(Eng), CEng, MIMechE

Senior Lecturer in Engineering
Grantham College for Further Education

The English Universities Press Ltd

This book is dedicated to my wife, Joan, who patiently typed the script, and without whose help and encouragement it would not have been written.

The Higher Technician Series

General editor

M. G. Page BSc, CEng, MIMechE, MIProdE, MBIM, FSS

Head of the Department of Production Engineering

The Polytechnic, Wolverhampton

ISBN 0 340 08342 5 (boards edition)

0 340 08354 9 (paperback edition)

First printed 1975

The English Universities Press Ltd

St. Paul's House, Warwick Lane, London EC4P 4AH

Printed and bound in Great Britain by

The Pitman Press, Bath

Contents

Physical and Mathematical Symbols Used in the Text

Physical quantity symbols used in the text

A		amplitude
A	a	area
B		magnetic flux density
C		electrical capacitance
	c	specific heat capacity
C_D		coefficient of discharge
D	d	diameter
E		electromotive force, amplitude value
		modulus of elasticity
		velocity of approach factor
	e	electromotive force – steady or instantaneous value
F		force
F_g		force due to gravity
F_f		force due to friction
	f	damping force or torque
		frequency
G		modulus of rigidity
	g	galvanometer resistance ratio
		gravitational acceleration
H		magnetic field strength
H	h	head of fluid
I		electric current – steady, r.m.s., or amplitude value
		moment of inertia
		second moment of area
	i	electric current – instantaneous value
K		constant
	k	spring stiffness

L		inductance
L	l	length
M		bending moment
	m	bridge resistance ratio
		mass
N		angular speed (rev/min or rev/s)
N	n	number of items, turns, etc
	n	bridge resistance ratio
P		power
		pressure
	p	circular frequency (forcing)
Q		electric charge
		volume of fluid
Q	q	quantity of heat
	q	energy radiated per unit wavelength
		fractional change of resistance
R		electrical resistance
		gas constant
		displacement along scale
(Re)		Reynolds number
R	r	radius
S		reluctance
	s	distance
T		temperature (thermodynamic)
		torque
T_p		periodic time of oscillation
	t	thickness
		time
V		linear velocity
		potential difference – steady, r.m.s., or amplitude value
	v	potential difference – instantaneous value
W		force due to gravity
	w	specific weight
X		displacement, linear
		reactance
	x, y, z	cartesian coordinates, displacements
Z		impedance
	α *alpha*	angle, phase-angle
		angular acceleration
		temperature coefficient of linear expansion
		temperature coefficient of resistance

	β	*beta*	angle, phase-angle
			temperature coefficient of volume expansion
	γ	*gamma*	shear strain
Δ		*delta*	a change of (e.g. ΔT, a change of temperature)
	δ		a small change of (e.g. δR, a small change of resistance)
	ϵ	*epsilon*	direct (tensile or compressive) strain
	ζ	*zeta*	damping ratio
	η	*eta*	dynamic viscosity
			efficiency
	θ	*theta*	angle, angular displacement
			general symbol for input or output signal
			temperature (Celsius)
	λ	*lambda*	thermal conductivity
			wavelength
	μ	*mu*	friction coefficient
	ν	*nu*	kinematic viscosity
			Poisson's ratio
	ρ	*rho*	mass density
			electrical resistivity
	σ	*sigma*	direct stress
	τ	*tau*	shear stress
			time-constant
	ϕ	*phi*	angle, phase-angle
Ω		*omega*	ramp-input constant
	ω		angular speed or velocity
			circular frequency

Mathematical symbols used in the text

$\propto$	directly proportional to		
f( )	a function of variable in bracket		
$\sum$	the sum of		
$\bar{\ }$	mean value. (e.g. $\bar{V}$ is mean velocity)		
$\hat{\ }$	maximum (amplitude) value. (e.g. $\hat{a}$ is maximum acceleration)		
$\neq$	not equal to		
$\equiv$	identical with		
$\approx$	approximately equal to		
$>$	greater than	$\geqslant$	equal to or greater than
$<$	less than	$\leqslant$	equal to or less than
d	differential operator	∂	partial differential operator
D	differential operator with respect to time ($\mathrm{D}x \equiv \mathrm{d}x/\mathrm{d}t$)		
	differential operator with respect to time ($\dot{m} \equiv \mathrm{d}m/\mathrm{d}t$)		

1

Instrument Selection, Errors, and Calibration

1.1 Types of instrument

It would be difficult to think of any man-made article whose manufacture did not at some stage involve measurement; also, the operation of machines of all kinds has to be controlled, either manually or automatically, and here again the first requirement is to measure the variables concerned. Taking the example of a road vehicle, for safety the driver must know the speed at which the vehicle is travelling. He may guess this fairly accurately from his experience, but large errors occur in some circumstances, for example when slowing down after long periods at high speed; hence the law requires that the vehicle has a speedometer. His odometer will add up the kilometres the vehicle has travelled, and in some cases, such as with commercial vehicles, a speed log will draw a graph of speed against time or distance. Instruments will indicate fuel level, engine cooling-water temperature, battery current, etc. Measuring and controlling systems, e.g. a carburettor or fuel-injection system, will measure the fuel requirements of the engine for different loads, speeds, and accelerator-pedal positions, and will supply the necessary air and fuel to the engine.

Taking the term 'instrument' in its wider sense, the following types may be categorised:

a) indicating instruments – e.g. thermometer, pressure-gauge;
b) recording instruments – e.g. temperature and pressure recorders;
c) controlling instruments – e.g. thermostat, float-type level control, machine-tool carriage-position control.

1.2 Factors affecting instrument selection

1.2.1 Accuracy

Accuracy may be defined as conformity with or nearness to the true value of the quantity being measured. The only time a measurement can be exactly correct is when it is a count of a number of separate items, e.g. a number of components or a number of electrical impulses. In all other cases there will be a difference between the true value and the value the instrument indicates, records, or controls to–i.e. there is a measurement error. The extent of this error, or the accuracy of the instrument, may be specified in several different ways.

a) Point accuracy. Here the accuracy of an instrument is stated for only one or more points in its range. This is particularly applicable to temperature-measuring devices, where points are obtained at the melting- and vapourising-temperatures of pure solids and liquids.

b) Percentage of true value. If the accuracy of an instrument is expressed in this way, then the error is calculated thus:

$$\text{error} = \{(\text{measured value} - \text{true value})/\text{true value}\} \times 100\,\%$$

The percentage error stated is the maximum for any point in the range of the instrument.

c) Percentage of full-scale deflection (f.s.d.). Here the error is calculated on the basis of the maximum value of the scale, thus:

$$\text{error} = \{(\text{measured value} - \text{true value})/\text{maximum scale value}\} \times 100\,\%$$

d) Complete accuracy statement. In some cases, pyrometers for example, it may not be sufficient to specify accuracy at a limited number of points, and the accuracy at a larger number of points is specified in tabular or graphical form. As a further example, the error of each individual gauge in a set of slip gauges is specified; it would not be adequate to specify a percentage error for the set based on the nominal gauge sizes.

It will be seen that an accuracy specified as a percentage of f.s.d. implies a less accurate instrument than one having the same accuracy percentage of true value. For example, an error of $\pm 1\,\%$ of f.s.d. on a pressure-gauge having a range of 1000 kN/m^2 would mean that a true pressure of 100 kN/m^2 could read from 90 to 110 kN/m^2; as a percentage of true value it would read from 99 to 101 kN/m^2 (see example 1.5.1).

1.2.2 Precision or repeatability

This is the repeatability of the readings taken of the same value by the same instrument. Precision is often confused with accuracy, and the difference should be clearly understood. For example, a micrometer may read to 0·002 mm, and the same indicated value to within ±0·002 mm may be obtained every time the same diameter is measured; i.e. the instrument is precise. However, if the barrel or the anvil has been moved out of the correct position, the readings may all be inaccurate, though precision is not altered. In this case the error will be obvious, since the instrument will not read zero when the spindle and anvil are in contact. In other cases the zero may still be correct and the inaccuracy not obvious. 'Drift' of instruments, i.e. change of accuracy, may occur after a time, due to various causes such as wear, change of friction values, oxidisation of pyrometer elements, etc., though precision may still be maintained.

1.2.3 *Resolution or discrimination*

This is the smallest change in the input signal which can be detected by the instrument. It may be expressed as an actual value or as a fraction or percentage of the full-scale value. (In some texts this quantity is referred to as 'sensitivity'.)

1.2.4 *Sensitivity and range*

The sensitivity is taken to mean the relationship between the input signal to an instrument or a part of an instrument system and the output,

i.e. sensitivity = (change of output signal)/(change of input signal)

The sensitivity will therefore be a constant in a linear instrument or element, i.e. where equal changes of the input signal cause equal changes of output.

Sensitivity is usually required to be high, and an instrument should therefore not have a range greatly exceeding the values to be measured, although some margin of excess over the expected values should be allowed, to prevent accidental overload. The values to be measured should not lie in the lower third of the range. This is particularly important if the instrument accuracy is specified as percentage of f.s.d., since considerable error as a percentage of actual value may occur.

1.2.5 *Reliability*

In general, a high degree of reliability is required of all instruments, since control of variables depends on their readings. However, the consequences of complete or partial failure may vary greatly, as for example in the failure of a domestic oven-thermometer and of the jet-pipe temperature-measuring system of an aircraft engine.

1.2.6 Cost

The cost of an instrument increases with the accuracy required, and the rate of increase is very steep as the accuracy increases. It follows that an accuracy should not be specified that is in excess of the accuracy of measurement which is adequate and suitable for the purpose. The precision of an instrument is connected with the accuracy, and similarly affects the cost. It should also be noted that, where several instruments are to be used and their readings combined, the accuracy of the resulting value will be less than that of the least accurate instrument (see appendix C); hence, for economy, all should have similar accuracy. Reliability will also affect cost, and in extreme cases (e.g. in nuclear reactors and space vehicles) systems may have to be duplicated to obtain virtual freedom from failure.

1.2.7 Static and dynamic response

The measured variable may be steady or it may vary slowly with time, e.g. the temperature of a furnace. On the other hand, it may be subject to sudden changes or steps, as when a thermometer is thrust into a hot liquid; or the signal may fluctuate rapidly, e.g. the temperature of the face of a piston, or the displacement of a vibrating component. These modes of input are illustrated in fig. 1.1.

If an instrument output signal is to follow a rapidly fluctuating input signal closely, i.e. it is to have a good 'dynamic response', special attention has to be given to this in designing the instrument or system, and generally the cost is higher than if only static values are to be measured. In at least one case, the piezoelectric crystal for force and pressure measurement, the instrument is suitable for dynamic measurement but not for static, since the signal (an electric charge) leaks more or less rapidly away.

1.2.8 Environment

The particular conditions in which an instrument has to operate may affect its accuracy, precision, and reliability. To counter these effects by design and manufacturing changes usually increases the cost. The following are examples of environmental conditions which may affect instruments.

a) High or low ambient temperatures. These, for example, cause expansion or contraction of gauges and rules, or of links and other members of mechanisms, or change of force, stress, or strain in them, which may affect the output signal.

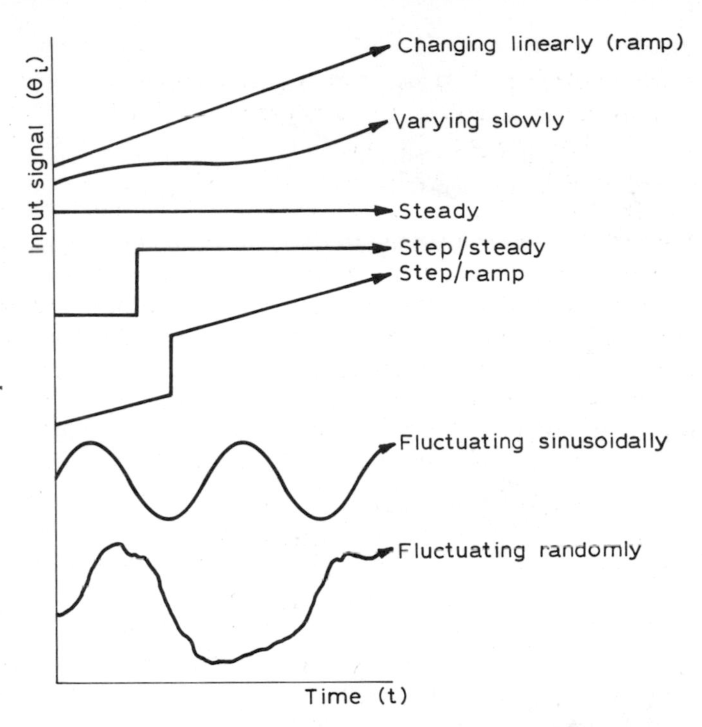

Fig. 1.1 Types of input signal (measured variable)

b) High acceleration. Instruments in moving vehicles are subjected to accelerations other than the standard value in which they were initially calibrated; this may also happen in vibrating machinery. The inertia of links, spring members, or other masses may cause forces, stresses, or strains affecting the output. For example, a vehicle fuel-tank-level float rises and falls rapidly due to motion of the fuel and the inertia of the float as the vehicle is subjected to random accelerations when travelling over rough ground. Rapid fluctuation of the fuel-level pointer results if no design action is taken to give a steady average-level reading.

c) Corrosive media. Measurements of the pressure, temperature, flow rate, etc. of corrosive liquids and gases require special materials in the parts of instruments in contact with them. In other cases, for example in salt-laden atmospheres, the whole of the instrument may need specially resistant materials.

d) Nuclear radiation. Atomic radiation affects the structure of many materials, and the design of instruments for use in radiation environments is highly specialised.

e) Explosive atmospheres. Instruments for use in mines and similarly explosive atmospheres must be specially designed to minimise explosion risk.

1.2.9 Type of output

It has to be decided whether the output signal is to be visible. This may not be necessary in a controlling system, though a visible instantaneous signal or a recorded signal may be considered necessary as a check on the process.

If the output is to be visible, how visible? If the output is important, such as in the machining of parts of an expensive component, or the pressure in a boiler, then the instrument output should be very clearly visible, possibly with bold markings at the limits of size or pressure. An engineer may even consider that an aural warning is desirable, and arrange alarm bells at the acceptable limits.

If the output is to be a continuous trace, it has to be decided whether a short-time or long-time chart, e.g. 24 hour or 7 day, is necessary. (Various British Standards give recommendations regarding dials, scales, charts, etc.)

1.3 Sources of error in measurement systems

Errors may be due to imperfections of the instrument or system, to operator error, to environmental effects, or because the introduction of the instrument has altered the variable it is measuring (application error).

1.3.1 Manufacturing errors

Since tolerances must exist on the dimensions and on the physical quality of components, some departure from the ideal performance must exist. This may result from the backlash of gears, less than ideal fit of bearings, incorrect graduation of scales, inaccuracy of potentiometers, etc. Some of these errors may not be important, since, if the instrument is calibrated, their effect may be removed by adjustment.

1.3.2 Design inadequacy

Many instruments rely on the linear relationship between stress and strain by using the deflection and strain of elements as a measure of force or pressure, e.g. the spring-balance. The general form of stress–strain curves is indicated in fig. 1.2.

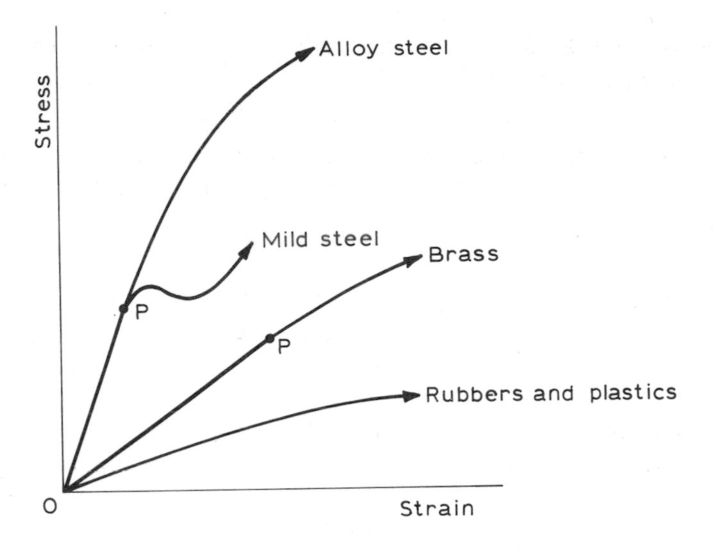

Fig. 1.2

Some materials show a linear range OP, where P is the 'limit of proportionality'; others such as rubbers, and plastics under some conditions, show no linear range at all. Higher-strength steels produced by alloying may not have a much higher proportional limit than mild steel, though their elastic limit, i.e. the stress from which they will return to point O, will be much more.

Many materials show a different line during unloading from that on loading, as shown in fig. 1.3, though they eventually return to zero. Energy absorbed during the cycle is represented by the enclosed loop. The effect, which is known as 'hysteresis', is pronounced in rubbers and plastics, and occurs to a lesser extent in metals. (Similar effects occur in other relationships, for example between flux density and magnetic field strength in an electromagnetic circuit.) The ideal material would have a very high elastic limit and limit of proportionality, and be free from mechanical hysteresis. This would mean that the output signal θ_0 in a form such as linear or angular displacement, or strain, corresponding to an input signal θ_i, such as pressure, force, or torque, would be proportional (linear) as shown in fig. 1.4, and the output would have the same value for a given input value whether the input θ_i had increased or decreased to the particular value. One advantage of this would be equal division of scales, which is easier for reading, and also reduced cost, since scales are more easily printed before assembly and the maximum and minimum values adjusted during calibration.

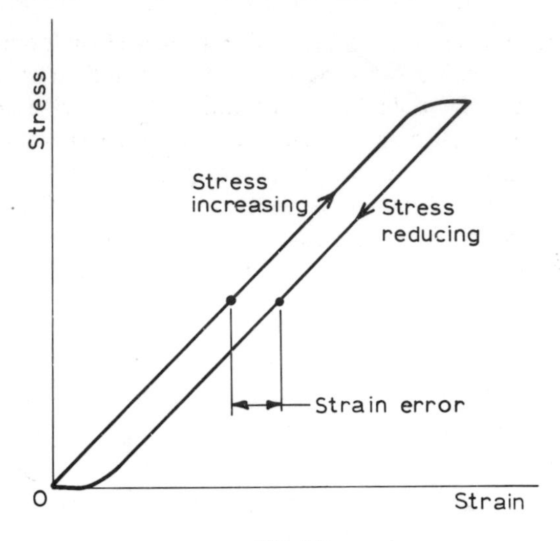

Fig. 1.3

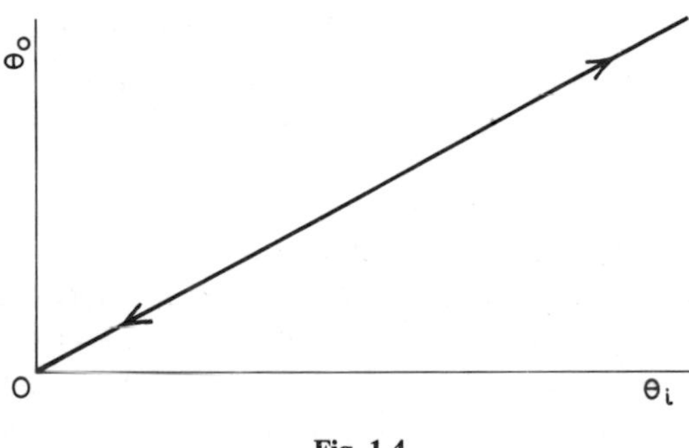

Fig. 1.4

However, if the stress–strain relationship is not linear, or if hysteresis occurs, there is error.

Another important factor is friction, and one of the difficulties is its uncertainty. In mechanical systems it is usually boundary friction, which (on a sub-microscopic scale) consists mainly of making and breaking of bonds in the parent materials, but also in any lubricant present, or in other contaminants invariably found adhering to the surfaces. The force necessary to start motion against boundary friction is more than that required to sustain it. In some conditions in bearings and in hydraulic and fluid systems, the solid components are completely separated by a fluid film, and the friction drag depends on the viscosity of the fluid. Figure 1.5 shows the static friction force or torque over a range AA′, with a reducing but still uncertain value as motion takes place. Once the fluid film is established, the friction force or torque is proportional to relative velocity for given values of bearing pressure, clearance, and lubricant viscosity. (These latter quantities are usually combined into a dimensionless group parameter which is plotted against the friction coefficient to give a graph similar in form to fig. 1.5.)

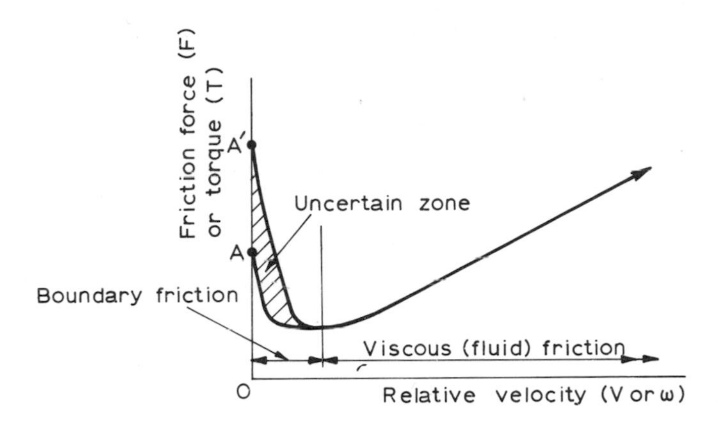

Fig. 1.5

The static friction value means that a certain force or torque has to be applied before motion will commence in either direction, and this results in what is known as a 'dead-zone', illustrated in fig. 1.6. If

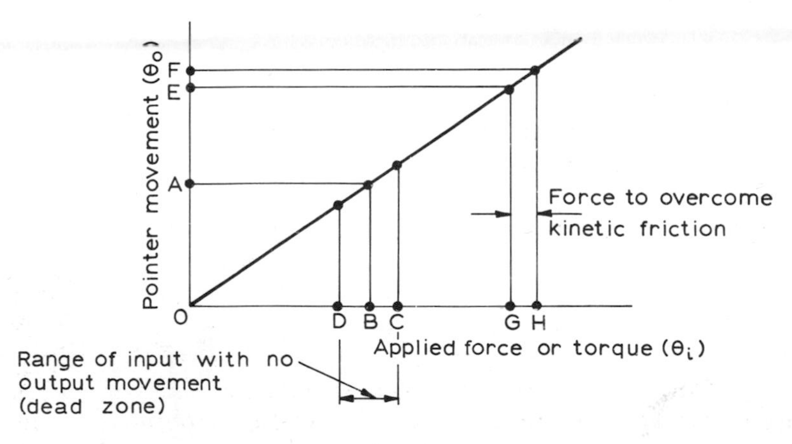

Fig. 1.6

there is an input force or torque OB giving a pointer movement OA, then the input may be reduced to OD or increased to OC without pointer movement, if DB (= BC) is the force or torque to overcome static friction. Once motion starts, the friction opposes the direction of motion, but is less than the static value. Hence the pointer comes to rest at E, which is less than the value F corresponding to the final value H of the force or torque. Vibration may cause the pointer to move nearer to F, and in some cases is deliberately introduced for this purpose.

These difficulties with friction have led to improvements in bearings, such as the use of jewels, but also to the elimination of bearings in cases where only reciprocating motion occurs. For example, the pointer in fig. 1.7(a) may be mounted on a thin rod secured at the ends. The rod acts as a torsion spring, and the pointer rotates with no bearing friction.

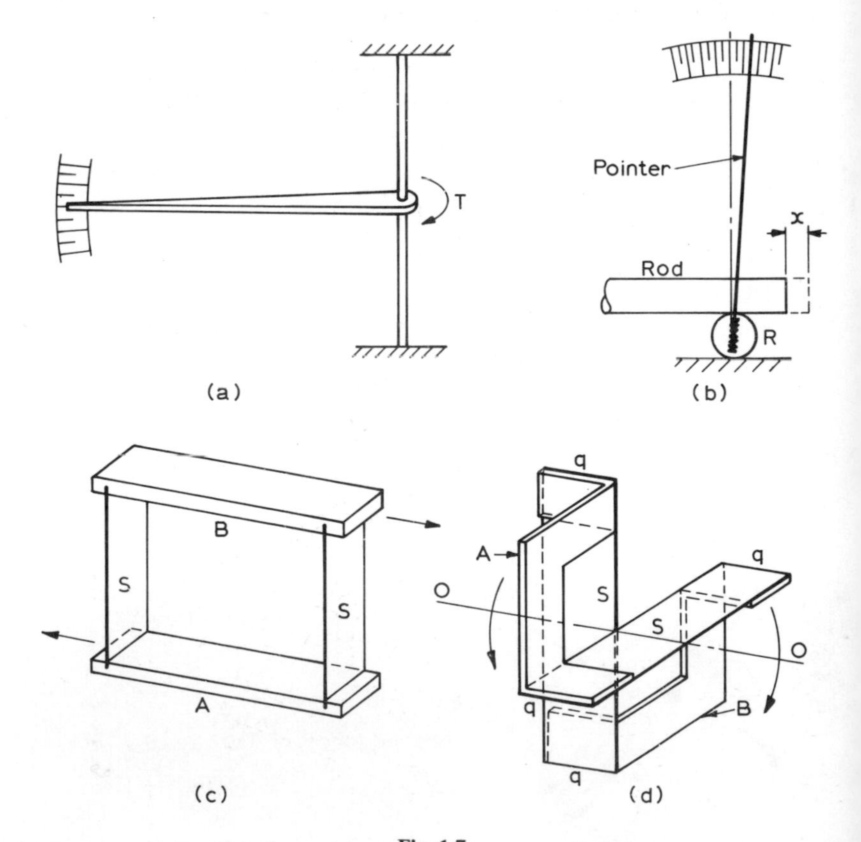

Fig. 1.7

In fig. 1.7(b), the horizontal movement x of the rod resting on the roller R rotates the roller, causing deflection of the pointer on the scale. Rolling friction has been substituted for sliding friction, and is much less.

The mechanical systems of comparators used in metrology employ flexible strip 'hinges'. A parallel movement is shown in fig. 1.7(c), whilst (d) shows a cross-strip hinge for relative rotation of A and B, which takes place approximately about the axis of intersection OO. Two cross-strip hinges are usually used together, being connected by strips parallel to OO secured at the points q. Strip hinges are suitable for only small relative movements, but they completely eliminate friction, and the hysteresis of the hard steel strip used is negligible.

Viscous (fluid) friction may occur naturally in more rapidly rotating bearings and in hydraulic or pneumatic systems, but is more likely to be intentionally introduced in the form of damping elements. The relationships between drag and velocity are:

force = constant × linear velocity, i.e. $F = fV$

or torque = constant × angular velocity, i.e. $T = f\omega$

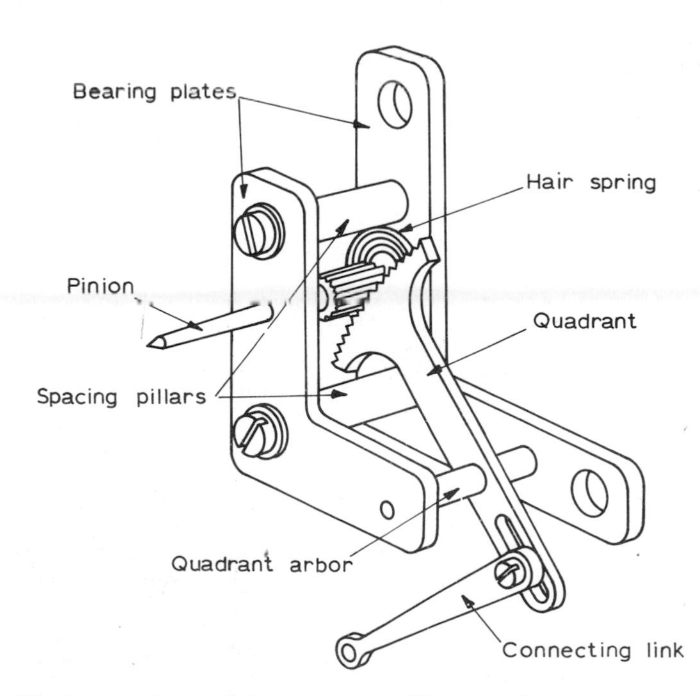

Fig. 1.8 Quadrant and pinion movement of Bourdon-tube pressure-gauge

Backlash in gears may also produce a difference of indicated value according to whether the value is approached by increasing or decreasing values. However, in some mechanisms, e.g. pressure-gauge and dial-gauge movements, backlash is taken up by a torsion spring, as shown in fig. 1.8.

1.3.3 Operating errors

These may arise from a variety of causes, such as failure to read the indicated value correctly, or, in size measurement, to apply the correct pressure between measuring instrument and component, or to apply the instrument squarely to the component, etc.

Figure 1.9 illustrates possible ways of arranging the scale and pointer. The more common way is shown in (a), i.e. with the pointer over the scale, but this is not always easy to read accurately, particularly for parts of divisions. The arrangement in (b) is better, where the pointer does not project over the scale. The method of (c) is best where it is possible, such as in some optical instruments, using two fine lines which have to be positioned either side of a line.

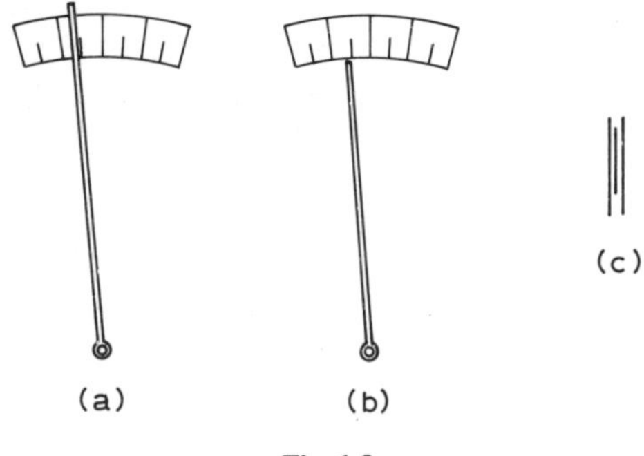

Fig. 1.9

Parallax error may result from the scale and pointer not being contained in the same plane. One method of obviating this is to use a mirror behind the scale. If the operator aligns his eye so that no reflection of the pointer is visible, he is viewing normal to the scale, and parallax is eliminated. A further way is to put the scale and the pointer in the same plane, which is possible with the arrangement shown in fig. 1.9(b). (A more sophisticated way of eliminating it is the digital-reading type of instrument, which is coming into more general use.)

Many size-measuring instruments have to be brought into contact with the measured component. A force is necessary to do this, but always causes deformation, albeit small, of the component and the instrument. This is illustrated in fig. 1.10(a), where slight bending of the vernier members will give a too low reading of diameter d, and in fig. 1.10(b), where the reading of height h will be low due to deformation of both stylus and workpiece, in both cases due to contact forces F. In many instruments, means are designed in to provide a standard contact force, either constant or varying only slightly. The ratchet mechanism on a micrometer and the stylus force on a dial gauge or on a comparator are examples. These are referred to as 'fiducial' devices.

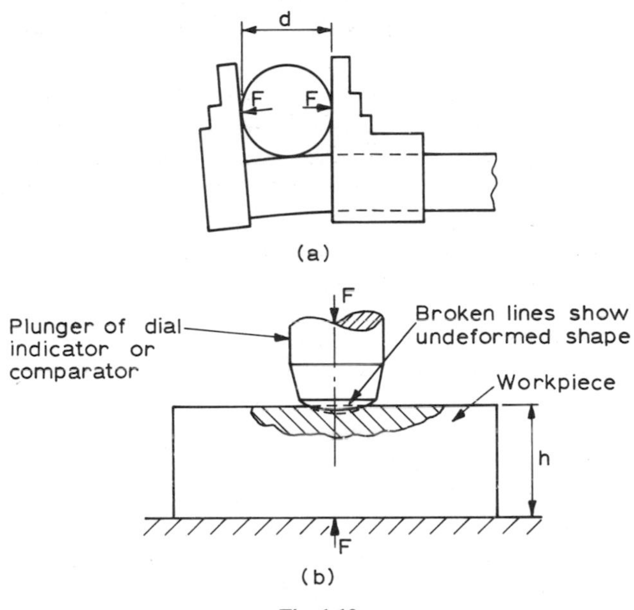

Fig. 1.10

Incorrect measurements may result from misalignment, such as when using a micrometer to measure a diameter, fig. 1.11(a) and (b), or a strain-gauge to measure the axial strain in a bar, (c). The misplacements x or θ give rise to error.

1.3.4 Environmental errors

The effects of variation of local values of pressure, temperature, acceleration, etc. on an instrument have been discussed in section 1.2.8. It is noted here that these effects will lead to errors unless the conditions are brought to a standard, such as temperature control in a laboratory, or unless the effects are compensated for in some way, such as in the use of a device to compensate for the linear expansion of the arm in a pendulum clock.

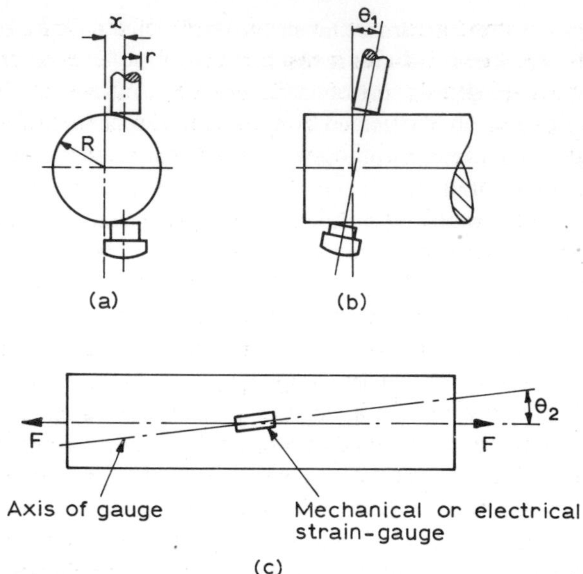

Fig. 1.11

1.3.5 Application errors

The connecting of an ammeter into an electric circuit increases the resistance slightly, and thus causes a small reduction of the current flowing. Hence, even if the ammeter is perfectly accurate, the value it indicates is not the true value in the original circuit.

A thermometer and pocket inserted into a pipe containing a hot flowing fluid will cause a heat flow Q to the surroundings different from, and probably greater than, the heat flow from the pipe without them (fig. 1.12); hence the temperature indicated will not be the same as the fluid temperature would be without the pocket.

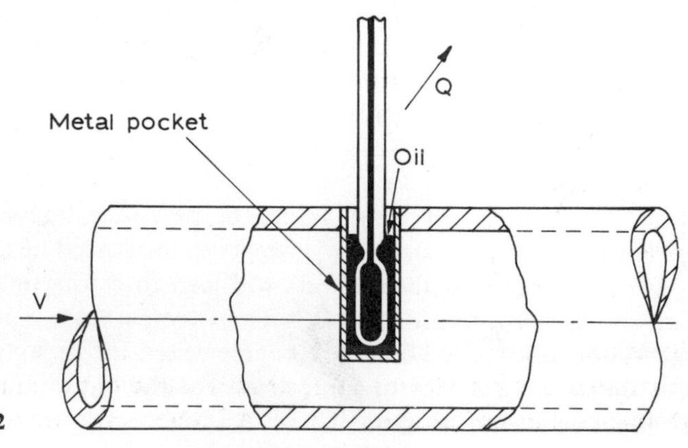

Fig. 1.12

The engineer concerned with measurement of any kind (and how few are not!) must be critical of measurements. Whenever any measurement is made, he should try to estimate the accuracy of the values obtained. There is no virtue at all in quoting the power of an engine to four significant figures if the force on the torque arm is measured to an accuracy of only $\pm 1\%$.

In other cases, it may be difficult to separate the variable being measured from other effects. For example, it may be intended to measure the force (F_1) in a steel-reinforced concrete column by measuring with an extensometer the change (δl) of a gauge-length (l) between points on steel buttons bonded to the surface, and then calculating F_1 from the stress–strain relationships [fig. 1.13(a)]. Unfortunately, concrete shrinks during the hydrating process over a long period; it also suffers from creep, i.e. steady change of length due to stress applied over a long period. Hence the change of length (δl) may be due to all these effects. The effects of shrinkage and creep *alone* may be found by subjecting a concrete block of the same mix as the column to similar stress, temperature, and humidity as the column. The change of height (δx) of the test block *when F_2 is removed* will enable corrections to be made in the calculation to find F_1. The test block is referred to as a 'control', and provides a means of comparison. A similar case is illustrated in the use of brittle coatings in Chapter 12.

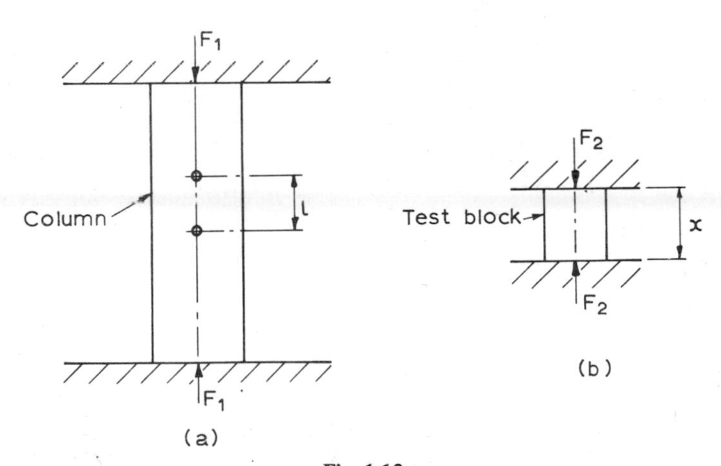

Fig. 1.13

1.4 Calibration

The International System of Units (SI) defines seven basic units and two supplementary units (see appendix A). All other units are derived from these.* The unit of mass is defined as being equal to the mass of the international prototype of the kilogram, kept at the Bureau International des Poids et Mesures (BIPM) at Sèvres, Paris. Copies of this are kept at national standards laboratories throughout the world. The mass of each copy is not the same as that of the prototype, but the difference in each case is known to great precision. These secondary standards may then be used to determine the value of other masses, and the measured mass of an object is thus traced back to comparison with the prototype kilogram.

The standards of length, time, luminous intensity, and temperature are each based on natural phenomena, and with suitable apparatus they may be reproduced to a high degree of accuracy by a laboratory anywhere in the world. The base unit of current, the ampere, is defined with reference to force and length units; but force is a unit derived from mass, length, and time, and hence the calibration of an instrument in amperes should be traceable back to the prototype kilogram and the primary standards of length and time.

Working instruments, i.e. those actually carrying out the operational measurement, may be calibrated against reference instruments of a higher accuracy. These in turn may be calibrated against instruments of still higher grade, or against natural primary standards, or against other accurate standards. It is essential that any measurement made is traceable ultimately to the primary standards involved.

The National Physical Laboratory (NPL) has for many years played a leading part in measurement. It has co-operated with international organisations in the development of standards, and has made many contributions to the science of measurement. It has provided a service to industry in Britain, in that it will carry out certain verifications of reference instruments and gauges, and it also acts in an advisory capacity on measurement in general.

The facilities offered to industry by the NPL have been supplemented since 1966 by the British Calibration Service (BCS). The BCS is Government sponsored (its headquarters are currently in the Department of Trade and Industry), and its aim is to facilitate and encourage the maintenance of high standards of measurement. The headquarters draws on the services of a number of measurement experts and, on the basis of their advice, gives approval for specific measurements to laboratories in industry and in educational and Government establishments, provided that they meet appropriate technical requirements.

* It can be shown that all units may be *expressed* in *dimensions* of mass, length, and time; but it is more convenient to define temperature and luminous intensity from readily reproduced physical phenomena.

Its aim is to provide for British industry a comprehensive service for the accurate calibration of instruments against recognised standards. The address of the headquarters is British Calibration Service, Department of Trade and Industry, 26 Chapter Street, London SW1P 4NS.

1.5 Worked examples

1.5.1 The calibration of a wattmeter was checked using the circuit shown in fig. 1.14. The voltage V_1 applied to the current-coil circuit was varied whilst the voltage across the voltage coil was kept as constant as possible. The ammeter and voltmeter in the circuit were used only for preliminary settings, and the values of V_r across the standard 1 Ω resistor and V_2 across the voltage coil were measured using a precision vernier potentiometer and a precision voltage-divider. From the readings tabulated, determine whether the wattmeter conforms to the maker's specification of accuracy of within ±0·5% of full-scale deflection.

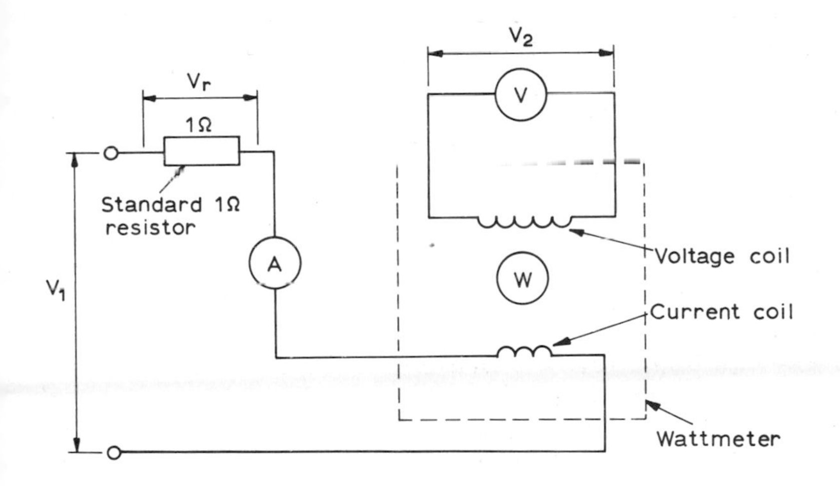

Fig. 1.14

The readings and calculated values are shown in the table. The error in watts and the percentage error for each reading are shown in fig. 1.15. It is seen that the error expressed as percentage of f.s.d. falls slightly outside the ±0·5% range at the 32 watts value. However, this could be obviated by a slight adjustment of the zero, hence the maker's claim is substantiated. The error as a percentage of the actual value is seen to increase at the lower end of the scale.

Readings			Calculated values		
Indicated power P_1 (watts)	Potential difference V_r (volts) $= I$ (amperes)	Potential difference V_2 (volts)	Wattmeter power $P_2 = V_2 I$ (watts)	(a) $\frac{P_1 - P_2}{P_1}$ %	(b) $\frac{P_1 - P_2}{P_{2\,max}}$ %
44	0·58545	75·28	44·08	−0·18	−0·18
40	0·53297	75·18	40·07	−0·18	−0·16
36	0·48051	74·74	35·91	+0·25	+0·20
32	0·42483	74·77	31·76	+0·75	+0·55
28	0·36600	76·20	27·89	+0·39	+0·25
24	0·31574	75·95	23·98	+0·08	+0·05
20	0·26110	75·90	19·82	+0·90	+0·41
16	0·21114	75·65	15·97	+0·19	+0·07
12	0·15786	75·60	11·93	+0·58	+0·16
8	0·10382	75·60	7·85	+1·88	+0·34
4	0·05077	75·60	3·84	+4·00	+0·36

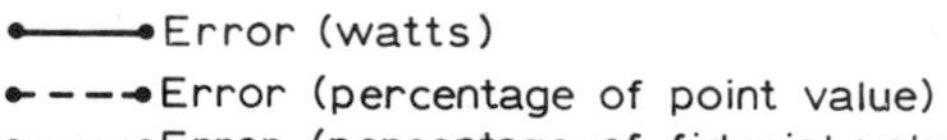

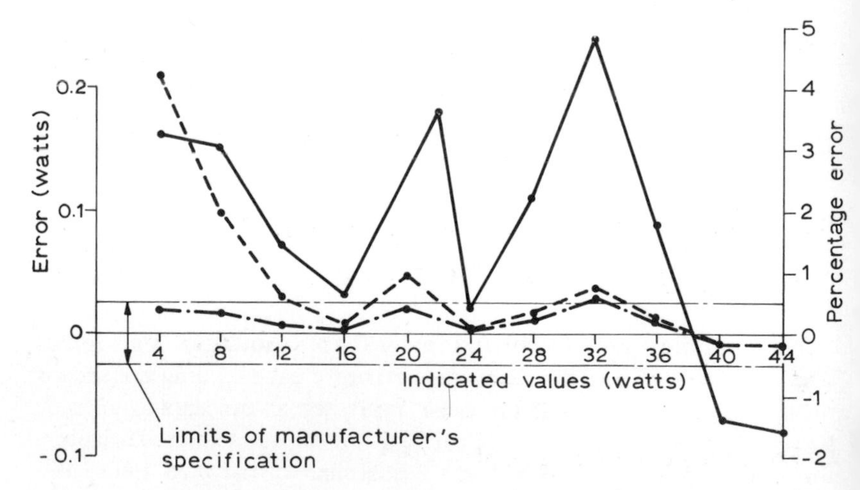

Fig. 1.15 Wattmeter errors

1.5.2 The stiffness, i.e. the rate, of a helical-coil spring is given by $k = Gd^4/8D^3n$ (see appendix D). Using binomial approximation (appendix C), estimate the change $\delta k/k$ of the rate of a steel spring due to change of linear dimensions and of modulus of rigidity, due to temperature change (a) from 20 °C to 50 °C, (b) from 20 °C to −50 °C.

Coefficient of linear expansion $= \delta l/l\,\delta T = 11{\cdot}8 \times 10^{-6}$ per °C

Coefficient of modulus of rigidity change with temperature $= \delta G/G\,\delta T = -24 \times 10^{-5}$ per °C

What will be the effect on the readings if the spring is the sensing element of a spring-balance?

$$k = (1/8n)Gd^4D^{-3}$$

Let k, G, d, and D increase by small quantities, then

$$k + \delta k = (1/8n)(G + \delta G)(d + \delta d)^4(D + \delta D)^{-3}$$

or
$$k\left(1 + \frac{\delta k}{k}\right) = (1/8n)G\left(1 + \frac{\delta G}{G}\right)d^4\left(1 + \frac{\delta d}{d}\right)^4 D^{-3}\left(1 + \frac{\delta D}{D}\right)^{-3}$$

Dividing by $k = (1/8n)Gd^4D^{-3}$,

$$1 + \frac{\delta k}{k} = \left(1 + \frac{\delta G}{G}\right)\left(1 + \frac{\delta d}{d}\right)^4\left(1 + \frac{\delta D}{D}\right)^{-3}$$

Since $\delta G/G$, $\delta d/d$ and $\delta D/D$ are small quantities, binomial approximation gives

$$1 + \frac{\delta k}{k} \approx 1 + \frac{\delta G}{G} + \frac{4\,\delta d}{d} - \frac{3\,\delta D}{D}$$

or
$$\frac{\delta k}{k} = \frac{\delta G}{G} + \frac{4\,\delta d}{d} - \frac{3\,\delta D}{D}$$

But
$$\delta d/d = \delta D/D = \alpha\,\delta T = 11{\cdot}8 \times 10^{-6}\,\delta T,$$

and
$$\delta G/G = -240 \times 10^{-6} \times \delta T,$$

$$\therefore \quad \frac{\delta k}{k} = (-240 \times 10^{-6} + 11{\cdot}8 \times 10^{-6})\,\delta T$$

$$= -228{\cdot}2 \times 10^{-6}\,\delta T$$

a) For a temperature change of +30 °C,

$$\frac{\delta k}{k} = -228{\cdot}2 \times 10^{-6} \times 30$$

$$= -0{\cdot}006\,846$$

$$= -0{\cdot}68\,\%$$

b) For a temperature change of $-70\,°C$,

$$\frac{\delta k}{k} = -228{\cdot}2 \times 10^{-6} \times (-70)$$

$$= +0{\cdot}015\,974$$

$$= +1{\cdot}60\,\%$$

At 50 °C the stiffness is reduced by 0·68 %, hence the balance will read 0·68 % high at all values. At −50 °C the stiffness is increased by 1·60 %, and the balance will read 1·6 % low at all values.

1.5.3 The Hartnell speed-governor mechanism shown in fig. 1.16 uses the centripetal acceleration of the two masses (m) to sense the angular velocity (ω) of a shaft. The centripetal acceleration forces F_c are provided by the force F_s from the spring and the force F_g due to the connecting link, and by sleeve friction. These forces are transmitted from the sleeve to the masses through bell-crank levers AOB pivoted about points O fixed relative to the shaft. In a test to find the effect of friction due to movement of the sleeve and linkage, the spring was removed and a spring-balance was attached to the sleeve. The force to just raise the sleeve was found to be 16 newtons, and that to just lower the sleeve 12 newtons. Calculate the dead-zone of speed corresponding to (a) $\omega = 30$ rad/s, (b) $\omega = 10$ rad/s.

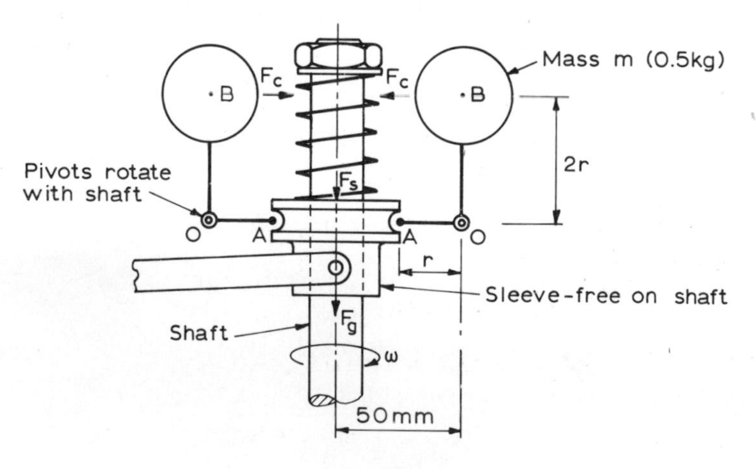

Fig. 1.16

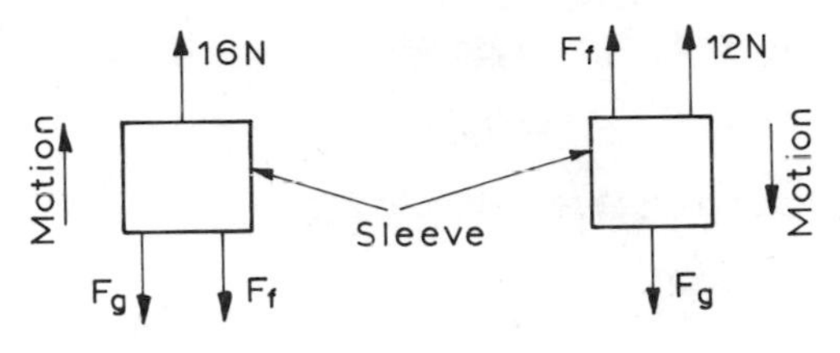

Fig. 1.17

Since the friction force F_f opposes the direction of motion, the just-slipping forces are as shown in fig. 1.17:

$$16 = F_g + F_f$$
$$12 = F_g - F_f$$

Hence $F_f = 2$ newtons. The force to be supplied by each arm is 1 newton, and by each mass 0·5 newton, due to the lever ratio.

Let δF_c be the force at the mass needed to overcome friction, and $\delta\omega$ the change of speed corresponding to this. The initial condition is $F_c = m\omega^2 r$. With friction, this becomes $F_c + \delta F_c = m(\omega + \delta\omega)^2 r$.

$$(\omega + \delta\omega)^2 = (F_c + \delta F_c)/mr$$
$$= \omega^2 + \delta F_c/mr$$
$$\omega + \delta\omega = \omega(1 + \delta F_c/m\omega^2 r)^{1/2}$$
$$= \omega(1 + \delta F_c/F_c)^{1/2}$$

If $\delta F_c/F_c$ is small compared with 1, then

$$\omega + \delta\omega = \omega(1 + \delta F_c/2F_c) \quad \text{by binomial approximation}$$

and $$\delta\omega = \omega\, \delta F_c/2F_c$$

In (a),

$$F_c = 0{\cdot}5 \times 30^2 \times 0{\cdot}050$$
$$= 22{\cdot}5 \text{ N}$$
$$\delta F_c/F_c = 0{\cdot}5/22{\cdot}5 = 0{\cdot}0222$$
$$\delta\omega = 30 \times 0{\cdot}0222/2$$
$$= 0{\cdot}333 \text{ rad/s}$$

The dead-zone of speed is $\pm 0{\cdot}333$ rad/s or $\pm 1{\cdot}11\%$.

In (b),

$$F_c = 0{\cdot}5 \times 10^2 \times 0{\cdot}050$$
$$= 2{\cdot}5 \text{ N}$$

$$\delta F_c/F_c = 0{\cdot}5/2{\cdot}5$$
$$= 0{\cdot}2$$
$$\delta\omega = 10 \times 0{\cdot}2/2$$
$$= 1{\cdot}0 \text{ rad/s}$$

The dead-zone of speed is $\pm 1{\cdot}0$ rad/s or $\pm 10\%$.

The use of binomial approximation in (b) is justified – the initial measurement of the friction force was not very accurate, and the resulting dead-zones of speed should not be regarded as being accurate.

1.5.4 If in the circuit of fig. 1.18 $R_1 = 100\,\Omega$ and $R_2 = 300\,\Omega$, calculate the difference in potential across the points AB with and without the voltmeter connected, if (a) the voltmeter has a resistance of $1000\,\Omega$, (b) it has a resistance of $10000\,\Omega$.

What is the percentage error due to application of the instrument in each case?

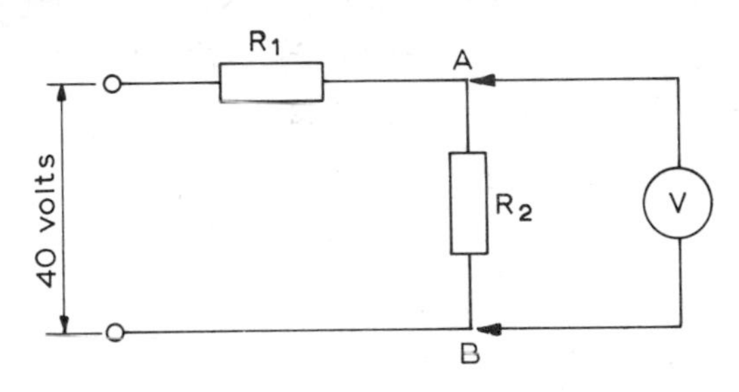

Fig. 1.18

a) Without voltmeter connected,

$$\text{current in circuit } I_1 = E/(R_1 + R_2)$$
$$= 40/(100 + 300)$$
$$= 0{\cdot}1000 \text{ A}$$

p.d. across AB

$$V_{AB} = IR_2$$
$$= 0{\cdot}100 \times 300$$
$$= 30{\cdot}0 \text{ V}$$

With voltmeter connected,

$$\text{resistance across AB} = R_t = R_2R_v/(R_1 + R_v).$$
$$= 300 \times 1000/(300 + 1000)$$
$$= 230{\cdot}8\ \Omega$$

current in circuit $I_1 = E/(R_1 + R_t)$
$$= 40/(100 + 230{\cdot}8)$$
$$= 0{\cdot}1210\ \text{A}$$

p.d. across AB $V_{AB} = I_1R_t$
$$= 0{\cdot}1209 \times 230{\cdot}8 = 27{\cdot}91\ \text{V}$$

b) Without voltmeter connected, $I = 0{\cdot}1000$ A, $V_{AB} = 30{\cdot}0$ volts.
With voltmeter connected,

$$\text{resistance across AB} = 300 \times 10000/(10000 + 300)$$
$$= 291{\cdot}3\ \Omega$$

$$\text{current in circuit } I_1 = E/(R_1 + R_t)$$
$$= 40/(100 + 291{\cdot}3)$$
$$= 0{\cdot}1022\ \text{A}$$

$$\text{p.d. across AB} = I_1R_t$$
$$= 0{\cdot}1022 \times 291{\cdot}3$$
$$= 29{\cdot}78\ \text{V}$$

a) $\%\text{ application error is} = \{(28{\cdot}08 - 30)/30\} \times 100\%$
$$= -6{\cdot}4\%$$

b) $\%\text{ application error is} = \{(29{\cdot}78 - 30)/30\} \times 100\%$
$$= -0{\cdot}22/0{\cdot}3\%$$
$$= -0{\cdot}73\%$$

Hence for minimum application error the voltmeter should have a high resistance. This may be seen more generally by writing down the error thus:

$$\text{error} = ER_2\left(\frac{1}{R_1 + R_2R_v/(R_2 + R_v)} - \frac{1}{R_1 + R_2}\right)$$

As R_v becomes large, $R_2R_v/(R_2 + R_v) \rightarrow R_2$, and the error approaches zero.

1.5.5 In an experiment to measure the stiffness of a cantilever beam, a dial indicator is used to measure the deflection (x) of the end due to the gravity force on a 1 kg mass, as shown in fig. 1.19. The spring force of the plunger of the indicator is $F_I = 0{\cdot}10R + 0{\cdot}30$ newtons, where R is the reading in mm.

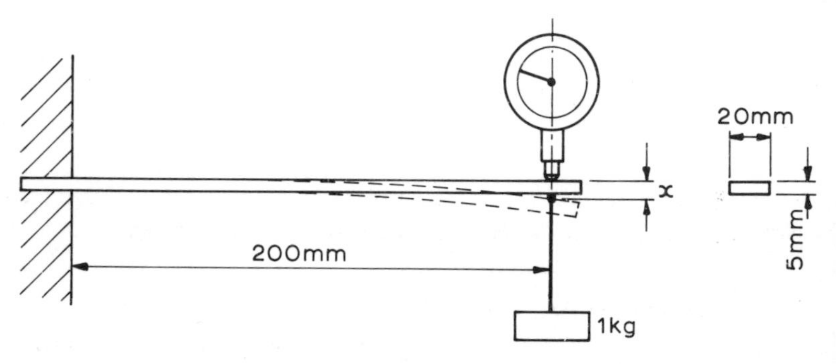

Fig. 1.19

Calculate the error and the percentage error in the measurement of x if the reading of the dial gauge before the 1 kg mass is applied is 2 mm. The modulus of elasticity of the beam material is 200 GN/m². Assume that the indicator reading and the dimensions of the beam are accurate.

From appendix D, deflection $x = Fl^3/3EI$

$$\frac{l^3}{3EI} = \frac{200^3 \times 12}{3 \times 200 \times 10^3 \times 20 \times 5^3}\,\text{mm}^3 \times \frac{\text{mm}^2}{\text{N}} \times \frac{1}{\text{mm}^4}$$

$$= 0{\cdot}0640\ \text{mm/N}$$

Neglecting the spring force of the dial indicator,

$$F = F_g = \text{mg}$$
$$= 1 \times 9{\cdot}807\ \text{N}$$

hence

$$x = 9{\cdot}807 \times 0{\cdot}0640$$
$$= 0{\cdot}628\ \text{mm}$$

Allowing for the spring force,

$$F = F_g + F_I$$
$$= 9{\cdot}807 + 2{\cdot}000 \times 0{\cdot}100 + 0{\cdot}300 - 0{\cdot}100x'$$
$$= 10{\cdot}307 - 0{\cdot}100x'$$

Hence $$x' = (10{\cdot}307 - 0{\cdot}100x') \times 0{\cdot}0640$$

and $$(15{\cdot}625 - 0{\cdot}100)x' = 10{\cdot}307$$

$$x' = 0{\cdot}655 \text{ mm}$$

The measuring error is $x' - x = 0{\cdot}655 - 0{\cdot}628$

$$= +0{\cdot}027 \text{ mm}$$

The percentage error $= \dfrac{0{\cdot}027}{0{\cdot}628} \times 100$

$$= +4{\cdot}3\%$$

It is seen that, for accurate measurement of the spring rate, the deflection-measuring device should apply no force to the member.

1.6 Tutorial and practical work

1.6.1 Errors of measurement may be described as (a) systematic errors, i.e. those which are constant and of the same sign for given conditions, and (b) random errors, i.e. those occurring accidentally. Give several examples of each kind of error, and discuss ways of detecting errors of each kind in an instrument or system.

1.6.2 Give several examples of 'direct measurement', i.e. where direct comparison is made, as in the 'weighing' of masses, and 'indirect measurement', such as when using the deflection of a spring to measure force. Is direct measurement always more accurate than indirect measurement?

1.6.3 The periodic time of small oscillations of a simple pendulum, i.e. a point mass on an inextensible massless cord, is given by $T = 2\pi\sqrt{(l/g)}$. Discuss in detail the accuracy of the determination of the value of g using a 20 mm diameter lead ball on a steel wire 200 mm long and 0·25 mm in diameter supported at the upper end on steel knife-edges, the oscillations being timed by stopwatch. (Refer to mechanics textbooks for derivation of the formula.)

1.6.4 Discuss how the heating effect of the current passing through the coil of an ammeter or voltmeter can affect the accuracy of the measurement.

1.6.5 Give examples of 'summing' instruments, i.e. those which add up or count, and 'integrating' instruments, i.e. those which integrate a quantity with respect to time.

1.6.6 Examine the mechanism of a Bourdon-tube pressure-gauge, and describe the method used to (a) take up the backlash between gear and quadrant, (b) adjust the range or sensitivity, (c) adjust the zero position.

1.6.7 The speed of a shaft may be measured by using a magnetic pick-up to operate a frequency-measuring device giving a read-out of the number of pulses counted in 1 second. Discuss the resolution of the system when using (a) a single steel bolthead, (b) a 100 tooth steel gear-wheel mounted on the shaft to operate the counter.

1.6.8 Give examples of instruments where the input signal is likely to be of the types shown in fig. 1.1.

1.6.9 Discuss ways in which the dead-zone of a moving-coil voltmeter could be measured.

1.6.10 Examine the mechanism of a height comparator working on mechanical amplification. Discuss the methods used in the design to minimise or prevent friction. Take increasing and reducing readings, and attempt to determine the extent of any dead-zone.

1.7 Exercises

1.7.1 A wire-wound resistance-type linear-displacement transducer was tested in conjunction with a 0–50 mm barrel-type precision micrometer head as shown in fig. 1.20. The input voltage (v_i) was maintained constant, and the output voltage (v_0) was measured with a digital voltmeter for different values of spindle displacement (x)

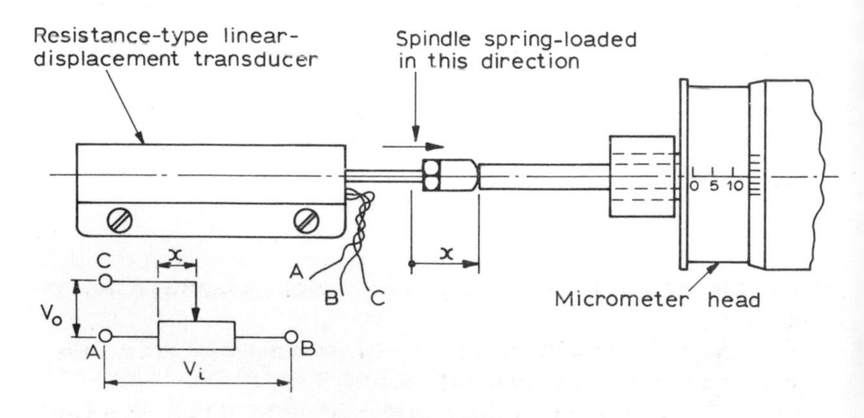

Fig. 1.20

measured by the micrometer. Draw the error/displacement graph, and state the sensitivity of the arrangement. Specify the accuracy of the system (a) as a percentage of true value, (b) as a percentage of full-scale (or fiducial) value.

Discuss the suitability of the transducer for use in a machine-tool position-control system to control dimensions to an accuracy of 0·10 mm.

Readings

x (mm)	0·000	2·000	4·000	6·000	8·000	10·000	12·000	14·000
v_0 (volts)	0·00	2·01	4·00	6·01	8·03	10·01	12·00	13·99

x (mm)	16·000	18·000	20·000	22·000	24·000	26·000
v_0 (volts)	15·98	17·99	20·00	22·02	24·01	26·02
x (mm)	28·000	30·000	32·000	34·000	36·000	38·000
v_0 (volts)	28·03	30·01	32·00	33·99	36·00	37·99
x (mm)	40·000	42·000	44·000	46·000	48·000	50·000
v_0 (volts)	39·97	41·96	43·98	46·01	48·01	50·00

[Sensitivity is 1 V/mm. Point accuracy is within $\pm 0{\cdot}37$ %. Accuracy as percentage of f.s.d. is within $\pm 0{\cdot}08$ %.]

1.7.2 A Bourdon-tube pressure-gauge has a scale from zero to 50 bar over an arc of 270°. The radius of the scale line is 100 mm. During a deadweight calibration test (see Chapter 5), the following values were observed:

Calibration pressure	*Scale value*	*Calibration pressure*	*Scale value*
0 bar	0·0 bar	30 bar	30·1 bar
5	5·0	35	35·3
10	9·8	40	40·2
15	14·8	45	45·1
20	19·9	50	50·0
25	25·1		

a) Determine the sensitivity (i) as a ratio of pressures, (ii) as a ratio of length/pressure.
b) Determine the maximum error as (i) percentage of scale value, (ii) percentage of full-scale (fiducial) value.
c) Could the gauge be adjusted to comply with $\pm 0{\cdot}5$ % f.s.d. accuracy?

[1, 3π mm/bar, -2% at 10 bar, $+0{\cdot}6$% at 35 bar]

1.7.3 A moving-coil electrical element has two flat spiral springs which provide the balancing torque. If the instrument is calibrated at 20 °C, calculate the error due to spring rate when the instrument is used at a temperature of 50 °C. (Refer to appendices C and D.)

Without further calculation, consider whether any change of torque will occur due to the change of electrical conductivity of the circuit through the element, and whether this will tend to balance out or to increase the change of spring rate. The coil and springs are of copper having the following characteristics:

$$\text{coefficient of thermal expansion } (\alpha) = \delta l/(l\ \delta T) = 16{\cdot}6 \times 10^{-6} \text{ per °C}$$

$$\text{temperature coefficient of elasticity} = \delta E/(E\ \delta T) = -300 \times 10^{-6} \text{ per °C}$$

$$\text{temperature coefficient of resistance } (\alpha) = \delta R/(R\ \delta T) = 43 \times 10^{-4} \text{ per °C}$$

[≈ −0·5%]

1.7.4 a) Refering to fig. 1.11(a) and (b), show that if $x > r$ and $(x - r)/R$ is small, the error in measuring the diameter of the bar in (a) is $\{-(x - r)^2/(2R^2)\} \times 100\,\%$, and in (b) is

$$\{100(R + r \sin \theta)/(R \cos \theta) - 100\}\,\%.$$

b) For the cases shown in fig. 1.11(a), (b), and (c), calculate the percentage error of the measurement in each case, if $r = 3$ mm, $R = 10$ mm, $x = 4$ mm, $\theta_1 = 0°30'$, and $\theta_2 = 2°$.

[−0·50%, +0·26%, −0·06%]

1.7.5 For the circuit shown in fig. 1.21, calculate the potential difference over the voltmeter and the current through the ammeter for

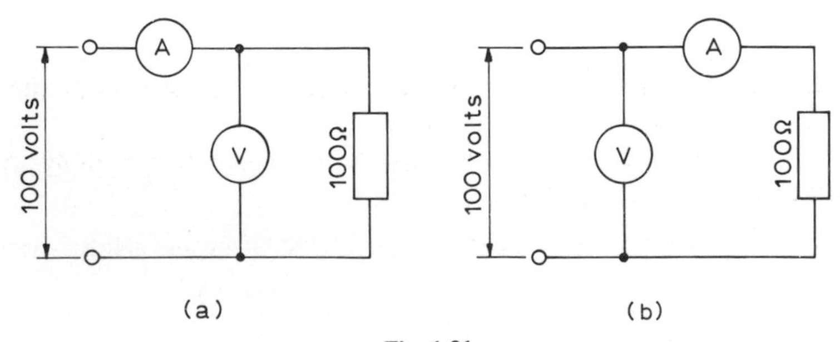

Fig. 1.21

each method of connection, if the voltmeter resistance is 900 Ω and the ammeter resistance is 0·08 Ω.

Calculate the error in the power value obtained by multiplying together the voltage and current readings in (a) and (b),

(i) compared with the power actually dissipated in the 100 Ω resistor,
(ii) compared with the power dissipated without the instruments in circuit.

[99·90 V, 1·110 A, +11 %, +11 %,
100·00 V, 0·992 A, <0·1 %, < −0·1 %]

1.7.6 a) The rate of a small helical-coil spring is to be measured as shown in fig. 1.22. If the true rate of the spring is 1 N/mm, calculate the reading on the dial due to the mass $(M + m)$ if the reading due to the mass m alone is zero. The force exerted by the indicator plunger is $F_I = 0{\cdot}10R + 0{\cdot}20$ newtons, where R is the reading in millimetres, $M = 1{\cdot}000$ kg, and $m = 0{\cdot}100$ kg.

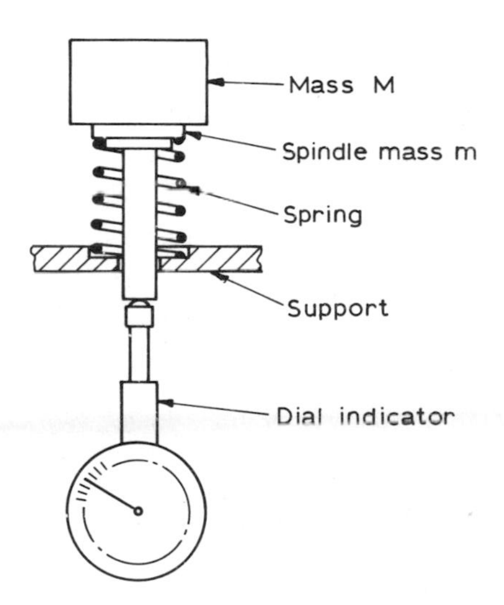

Fig. 1.22

b) Calculate the spring rate from the reading calculated in (a), and the error and percentage error in this rate.
c) Would an increased initial reading of the indicator (with m alone) give a different error? Explain why.

[9·625 mm, +0·019 N/mm, +1·9 %, larger positive error]

2

Elements of an Instrument System

2.1 Elements of an instrument system

An instrument or an instrument system is required to sense some variable quantity, and to indicate or record its value, or to control its value. In between the sensing end of the system and the indicating, recording, or controlling end, the signal from the sensing element may be altered to another form, or it may be amplified, or transmitted from one point to another.

Taking the simple mercury thermometer as an example, the sensing element is the relative change of volume of the mercury and its glass container due to a change of their temperature. This small volume change is amplified by making it occupy the very small-diameter capillary tube of the stem, giving a large movement of the mercury meniscus. The indicating element is the position of the meniscus against a graduated scale.

In this chapter, a number of the typical elements of instrument systems are shown, together with a method of illustrating the combination of elements, by means of block diagrams and transfer functions, and of calculating the effect of several elements connected together.

2.2 Sensing elements

A very large variety of physical principles are used in many different ways to provide elements which will sense the change of some quantity and to give an output signal which depends on that change. Some of these are shown in figs 2.1 to 2.28. Elements which sense an input in one physical quantity and give an output in a different physical quantity, e.g. an element giving a displacement change representing a force change, are known as 'transducers'; however, the term is used more often in the case of elements giving an electrical output.

Sensing elements

Principle of operation	*Examples of application*
Displacement of a rigid element	**Fig. 2.1(a) Length comparator**; **Fig. 2.1(b) Surface-texture measurement**
Deflection of an elastic element The deflection of elements of many different shapes is exactly or nearly proportional to the force or torque producing it, within the elastic range of the material.	**Fig. 2.2(a) Helical spring in spring-balance**; **Fig. 2.2(b) Simply supported beam in Hounsfield Tensometer**; **Fig. 2.2(c) Twist of shaft used to measure torque**

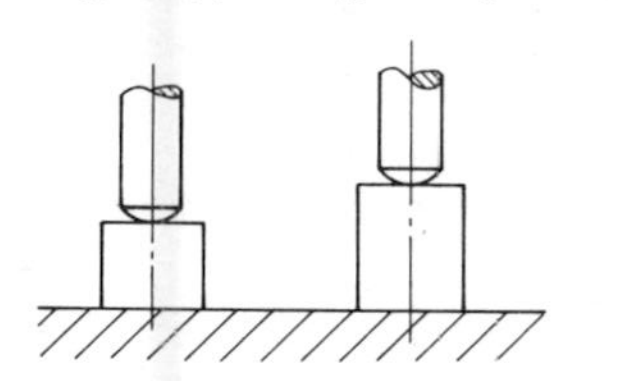

Fig. 2.1(a) Length comparator

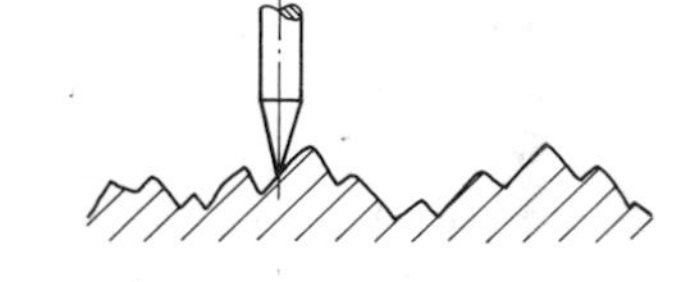

Fig. 2.1(b) Surface-texture measurement

$x \propto F$

Fig. 2.2(a) Helical spring in spring-balance

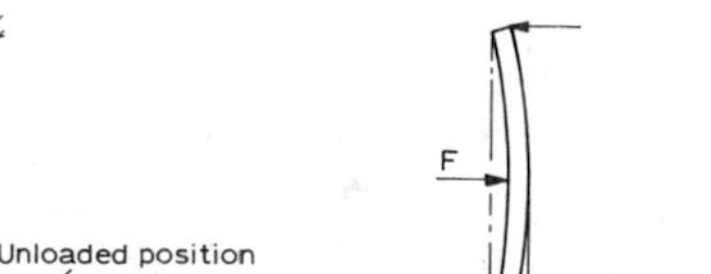

Fig. 2.2(b) Simply supported beam in Hounsfield Tensometer

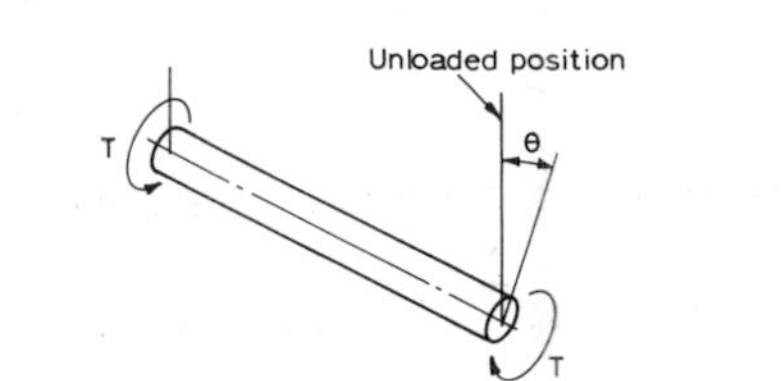

$\theta \propto T$

Fig. 2.2(c) Twist of shaft used to measure torque

Sensing elements

Principle of operation	*Examples of application*
Strain of an elastic element The surface strain of elements of many different shapes is exactly or nearly proportional to the force or torque producing it, within the elastic limit.	**Fig. 2.3(a) Cantilever** $\epsilon \propto F$ **Fig. 2.3(b) Load cell** $\epsilon \propto F$ **Fig. 2.3(c) Torque-sensing by strain-gauge** $\epsilon \propto T$
Balance of gravitational force The force exerted by gravity in a constant gravitational field is proportional to the mass of the object.	**Fig. 2.4(a) Force balance** $(b/a) \propto F$ (m and g constant) **Fig. 2.4(b) Torque balance** $x \propto T$ (m and g constant) $m \propto T$ (x and g constant)

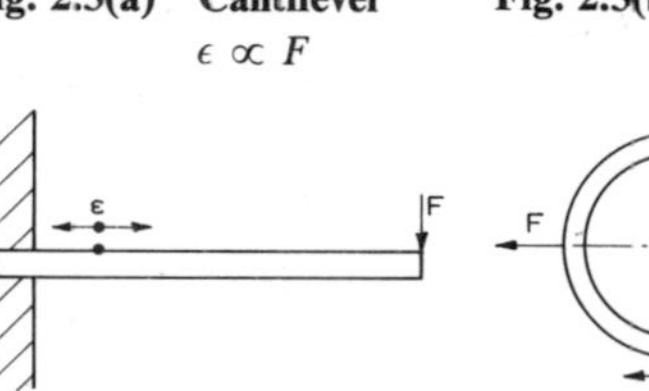

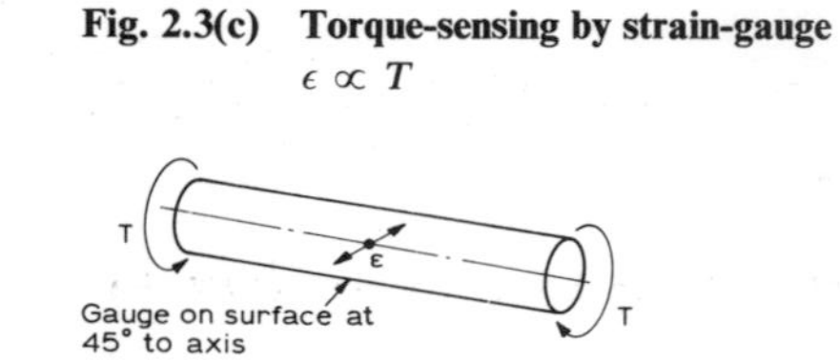

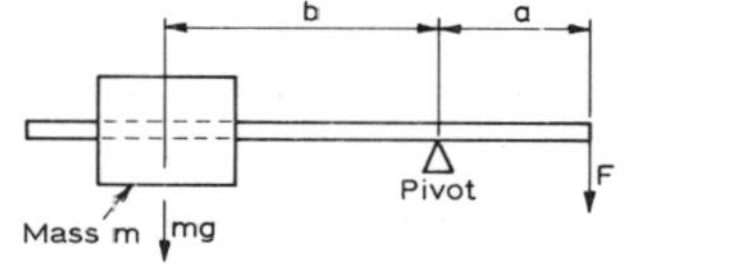

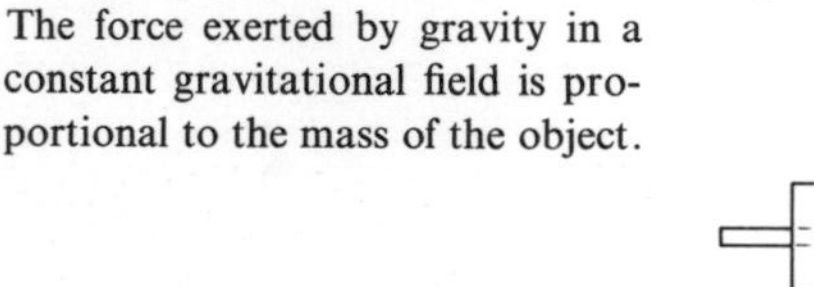

Fluid pressure due to gravity

The pressure due to a column of liquid of uniform density is proportional to the height of the column: $P_{xx} \propto h$, when the density is constant and g is constant.

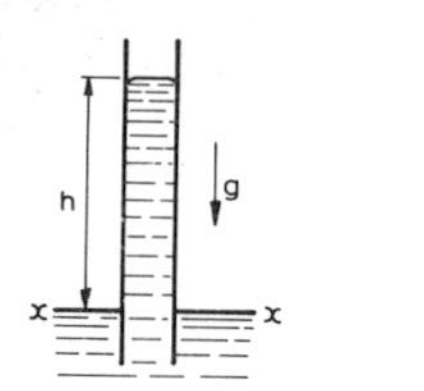

Fig. 2.5(a) Barometer

$h \propto P_a$ (density and g constant)

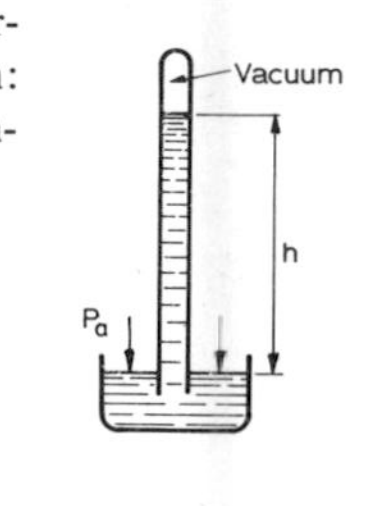

Fig. 2.5(b) Liquid-level measurement

$P \propto h$ (density and g constant)

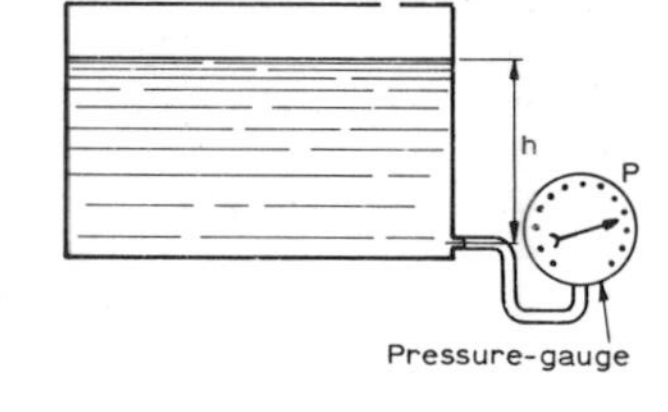

Pressure changes due to restriction of a fluid stream

Fig. 2.6(a) Air-jet sensing

$\delta P_b \propto \delta x$ (P_s constant)

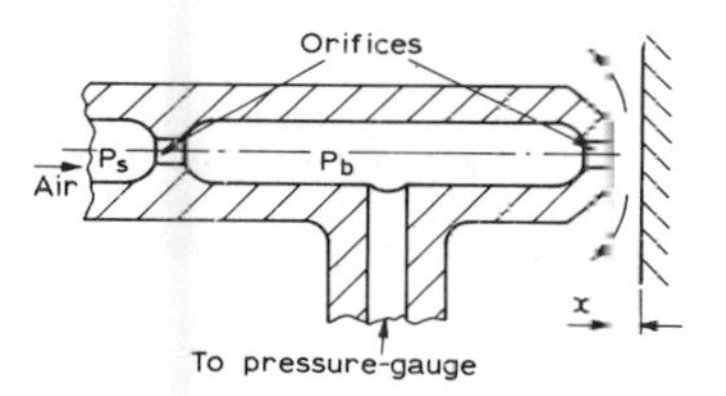

Fig. 2.6(b) Orifice plate for liquid-flow-rate measurement

$h \propto (\dot{Q})^2$

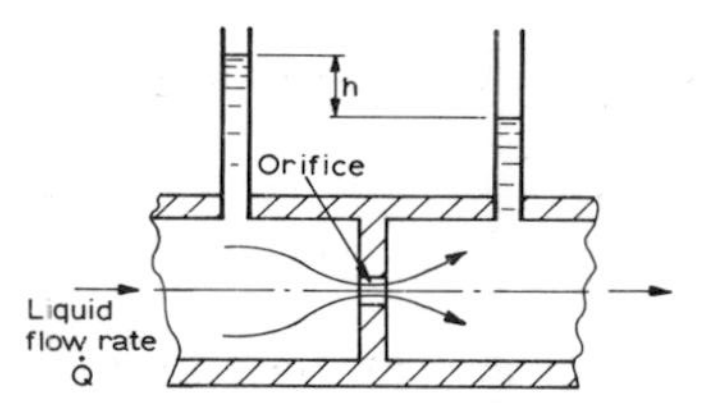

Principle of operation	Examples of application
Force due to or causing acceleration Force = mass × acceleration i.e. $F = ma$	**Fig. 2.7(a) Accelerometer (no gravity force)** F = net force due to springs $F \propto a, F \propto x, \therefore a \propto x$ 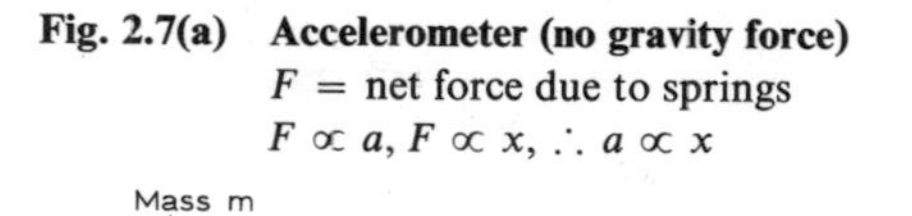 **Fig. 2.7(b) Accelerometer (in gravity field)** $\theta = \mathrm{f}(a)$ **Fig. 2.7(c) Angular-velocity measurement** $F \propto \omega^2$
Torque due to or causing angular acceleration Torque = moment of inertia × angular acceleration i.e. $T = I\alpha$	**Fig. 2.8 Angular-acceleration measurement** $T \propto \alpha, T \propto \theta, \therefore \alpha \propto \theta$ 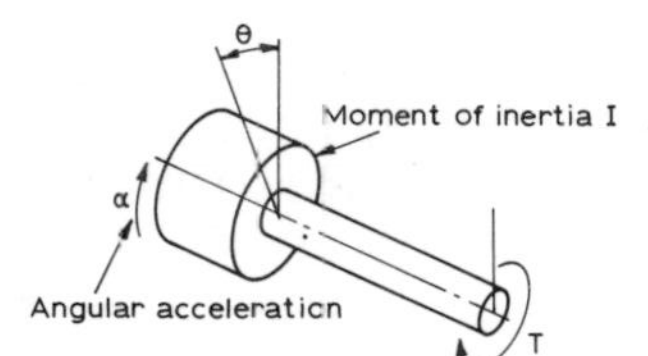

Linear expansion due to temperature change

Change of length is proportional to change of temperature.

i.e. $\delta l \propto \delta T$

Fig. 2.9(a) Rod thermostat

$x \propto (\alpha_A - \alpha_B)\,\delta T$

Fig. 2.9(b) Bimetallic-strip thermostat

x is a function of $\delta T, \alpha_A, \alpha_B, l$

i.e. $x = f(\delta T, \alpha_A, \alpha_B, l)$

Change of volume of a liquid due to change of temperature

$\delta V \propto \delta T$

Fig. 2.10(a) Liquid-in-glass thermometer

δl is proportional to relative volume change of liquid and glass, δV.

$\delta V \propto \delta T, \delta l \propto \delta V, \therefore \delta l \propto \delta T$

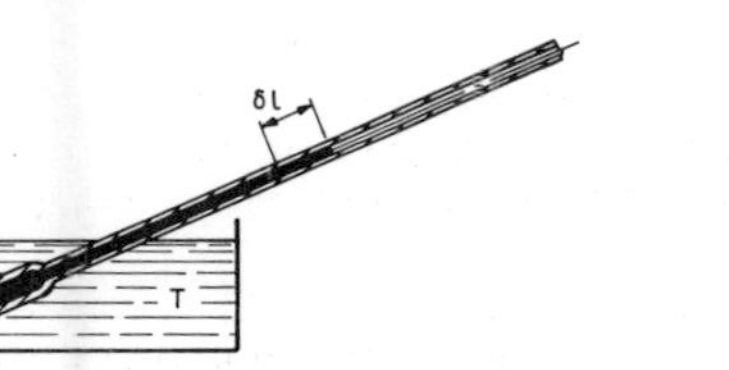

Fig. 2.10(b) Mercury-in-steel thermometer

δR is proportional to the relative volume change of mercury and steel, δV.

$\delta V \propto \delta T, \delta R \propto \delta V, \therefore \delta R \propto \delta T$

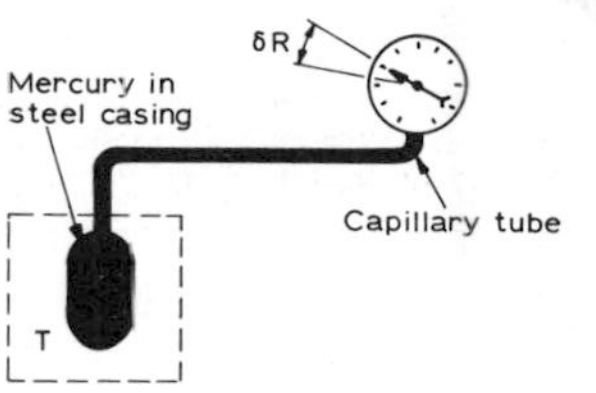

Sensing elements

Principle of operation	*Examples of application*
Change of pressure of a gas or vapour due to temperature change	**Fig. 2.11(a) Constant-volume gas thermometer** $P \propto h, P \propto T, \therefore h \propto T$

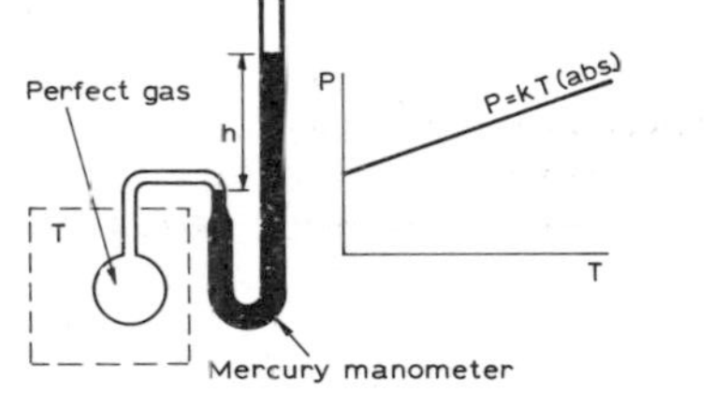

Fig. 2.11(b) Vapour-pressure thermometer
$P = f(T)$ (non-linear)

Principle of operation	*Examples of application*
Change of phase of a substance at a known temperature	Calibration of temperature-measuring instruments at the phase-change points of pure substances.

Fig. 2.12(b) Pyrometric cones
Melting of cones of silicate materials occurs at known temperatures.

Fig. 2.12(a) Automatic sprinkler system
Melting of a metal plug at a known temperature breaks an electrical circuit.

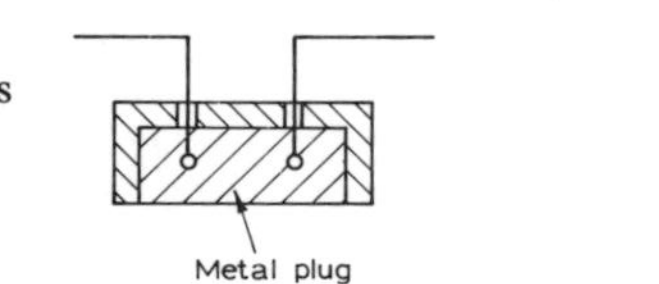

Temperature change due to vapourisation

Fig. 2.13 Wet-and-dry-bulb hygrometer

δT is a function of the moisture content of the surrounding gas.

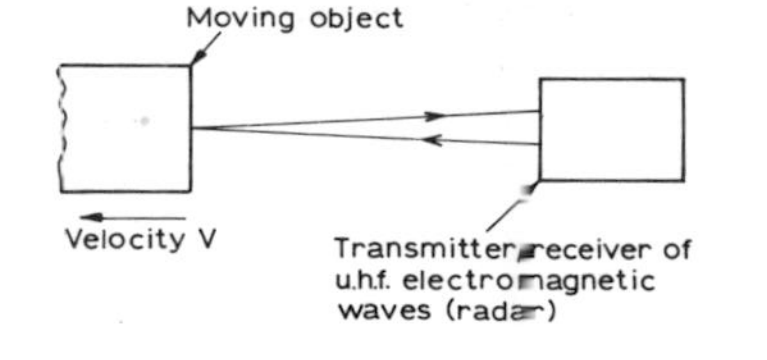

Reflection of waves

Fig. 2.14(a) Ultrasonic level measurement

The time t between transmitting and receiving a pressure pulse is proportional to h'.

i.e. $t \propto h'$

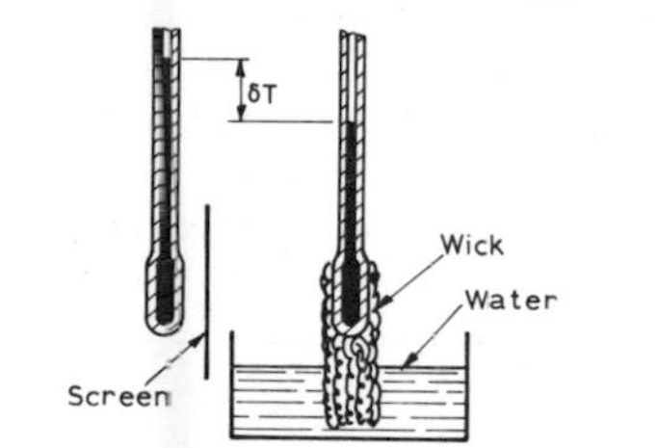

Fig. 2.14(b) Velocity measurement by use of the Doppler effect

The change of frequency between the reflected and the transmitted wave is proportional to velocity V.

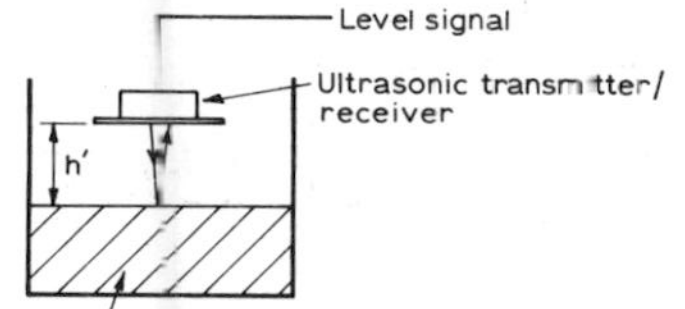

Principle of operation	*Examples of application*	
Interference of waves	Use of Moiré fringes in displacement measurement.	Light interference methods of measuring length, flatness, etc.
Refraction of waves	Photoelastic methods of measuring stress in translucent materials: the degree of refraction is proportional to stress.	
Change of potential difference in a circuit due to resistance	**Fig. 2.15** $\frac{V_0}{V_i} = \frac{x}{l}$ 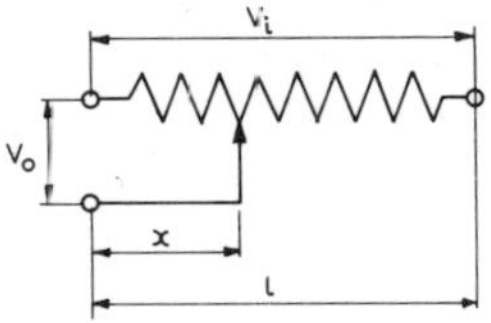	(i) Displacement measurement – $x \propto V_0$ (V_i and l constant) (ii) Measurement of p.d. – $V_0 \propto x$ (V_i and l constant)

Change of contact resistance due to pressure

Fig. 2.16(a) Force measurement by resistance change
$\delta R_{AB} \propto F$

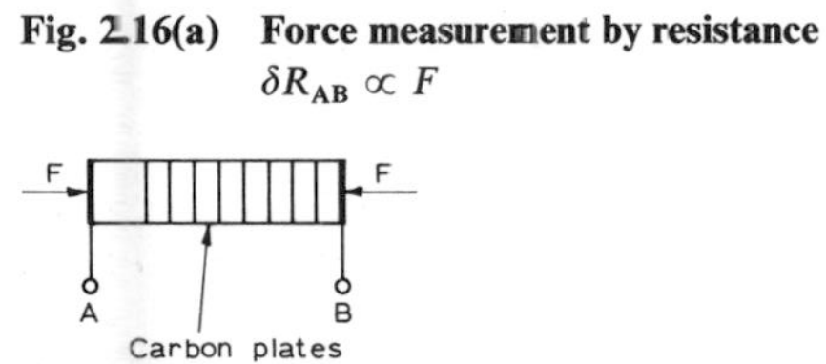

Fig. 2.16(b) Resistance strain-gauge
$\delta R_{AB} \propto \epsilon$

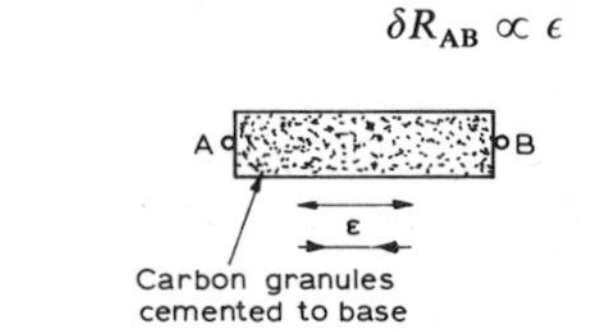

Change of resistance of an electrical conductor due to strain

Fig. 2 17(a) Unbonded strain-gauge
$\delta R_{AB} \propto \delta\epsilon$

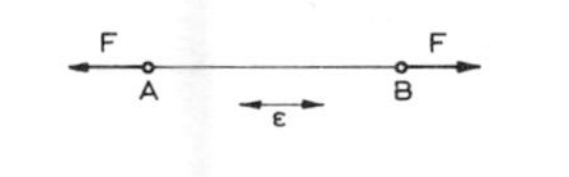

Fig. 2.17(b) Bonded strain-gauge
$\delta R_{AB} \propto \delta\epsilon$

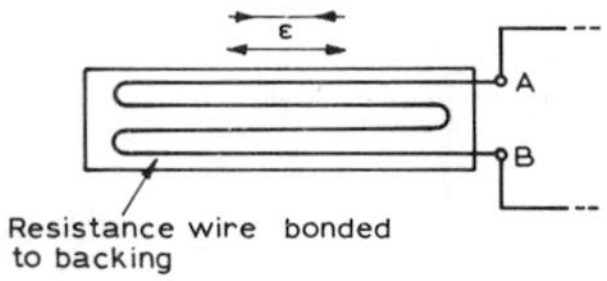

Charge on a crystal due to force (piezoelectric effect)

Fig. 2.18 Force and pressure measurement
$Q_{AB} \propto F$

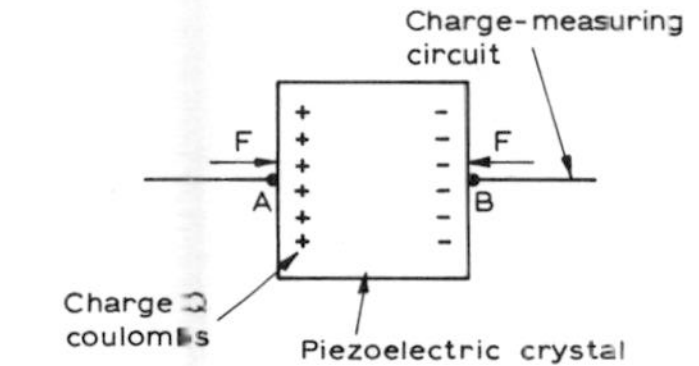

Sensing elements

Principle of operation	*Examples of application*	
Change of resistance of an electrical conductor due to change of temperature	**Fig. 2.19(a) Temperature measurement** $\delta R = f(\delta T)$	**Fig. 2.19(b) Temperature-control systems** $\delta R = f(\delta T)$
Change of capacitance due to proximity	**Fig. 2.20(a) Measurement of distance, displacement, etc.** $C \propto 1/d$ (A constant) $C \propto A$ (d constant) $\therefore C \propto x$	**Fig. 2.20(b) Counting** Change of capacitance due to an intermediate object

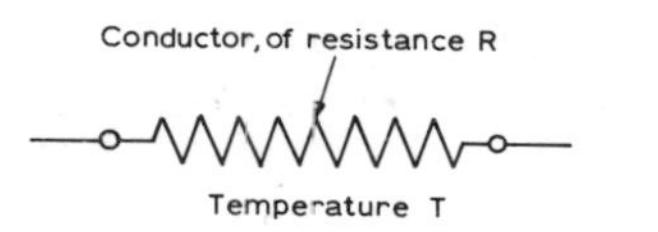

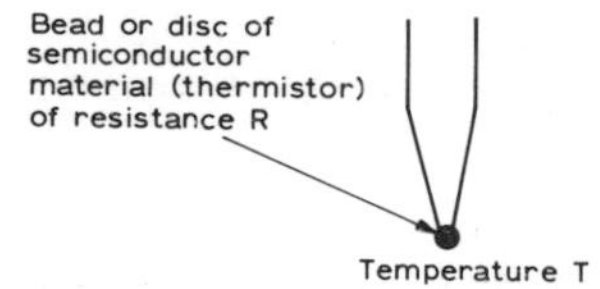

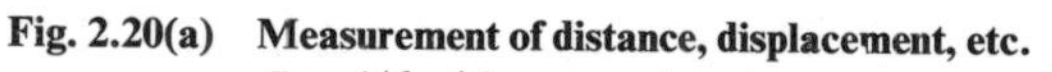

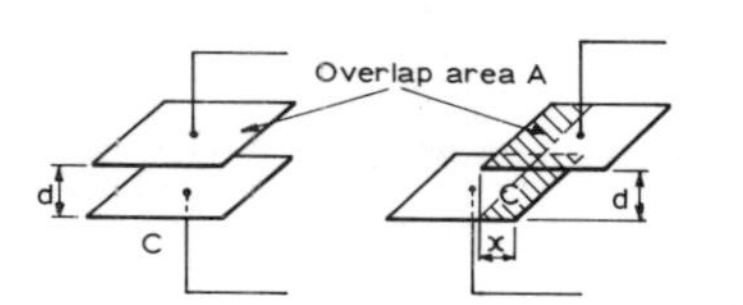

Induced e.m.f. due to relative velocity of a conductor and a surrounding magnetic field

Fig. 2.21(a) Vibration pick-up sensing velocity

Induced e.m.f. $\propto V$

Fig. 2.21(b) Tachogenerator for angular-velocity measurement

$e_{AB} \propto \omega$

Force exerted on a stream of electrons in a magnetic field

Fig. 2.22(a) Cathode-ray oscilloscope (magnetic-deflection type)

$F \propto vB$

$\therefore \theta = f(v, B, l)$

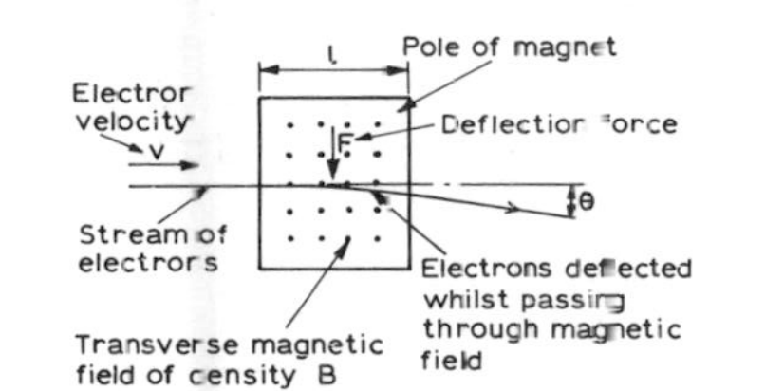

Fig. 2.22(b) Electromagnetic instruments

$F \propto i_{AB}, \therefore T \propto i_{AB}$

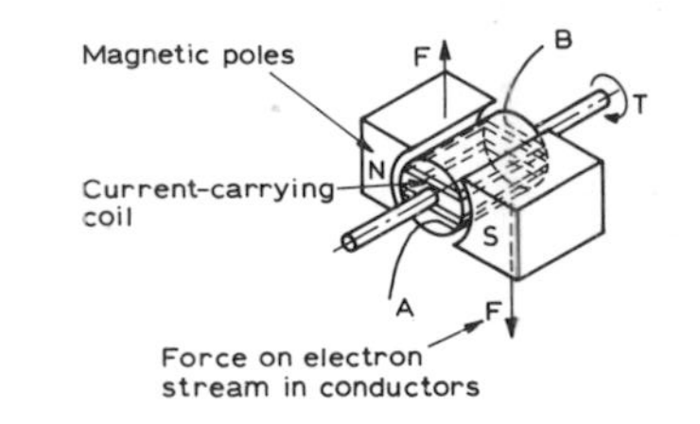

Principle of operation	*Examples of application*

Change of mutual inductance due to the change of reluctance of a magnetic circuit

Fig. 2.23 Displacement measurement

$v_i = V_i \sin \omega t \quad (V_i \text{ constant})$

$v_0 = V_0 \sin(\omega t + \phi)$

$\delta V_0 \propto \delta x$

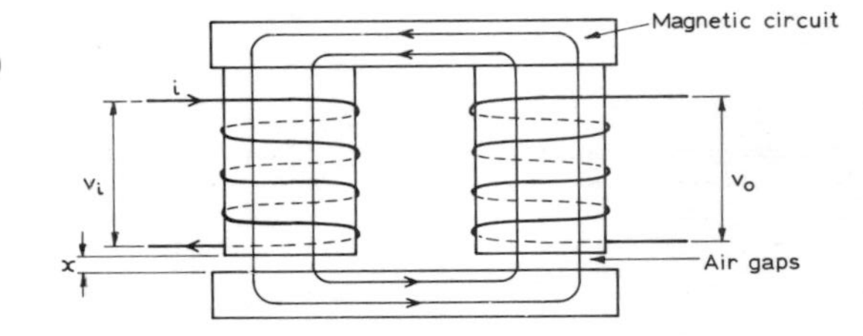

Magnetic fields due to eddy currents caused by relative motion of a conductor and a surrounding magnetic field

The induced field interacts with the original field, and a force is exerted on both members.

Fig. 2.24(a) Angular-speed measurement

Drag torque $T \propto \omega$ (same direction)

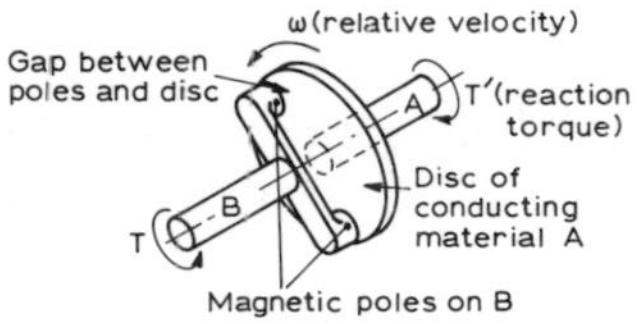

ω (relative velocity)
Gap between poles and disc
T′ (reaction torque)
Disc of conducting material A
Magnetic poles on B
A
B
T

Fig. 2.24(b) Damping of instrument movements

$T' \propto \omega$ (opposite direction)

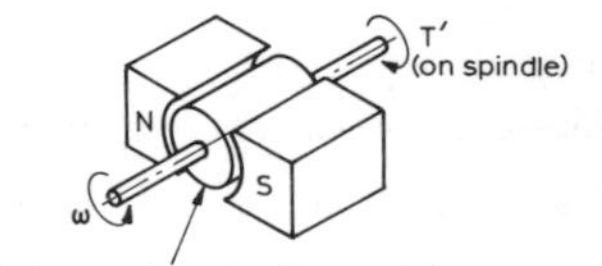

Thermocouple effect
An e.m.f. is generated in a circuit containing two junctions of dissimilar metals at different temperatures.

Fig. 2.25(a) Thermocouple pyrometer
$e = f(T_1 - T_2)$

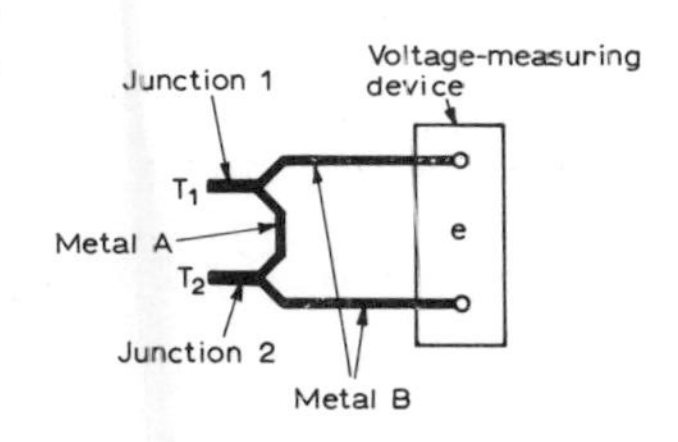

Fig. 2.25(b) Thermocouple ammeter
$e = f(T_1),\ T_1 = f(i_{AB}),\ \therefore e = f(i_{AB})$

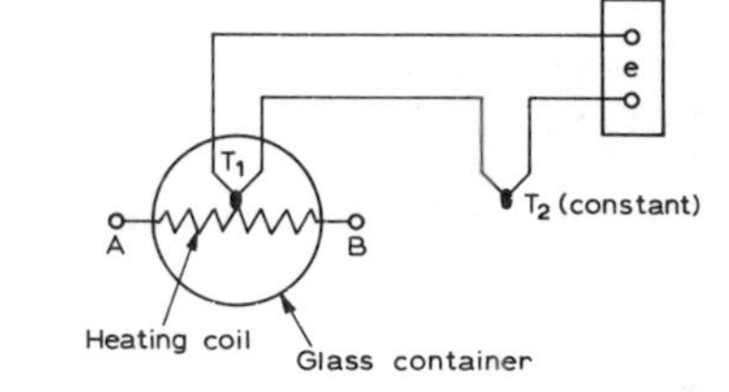

Incident light causes e.m.f. (photo-voltaic cell)

Fig. 2.26 Selenium cell in light-meter
$e = f(\text{light intensity})$

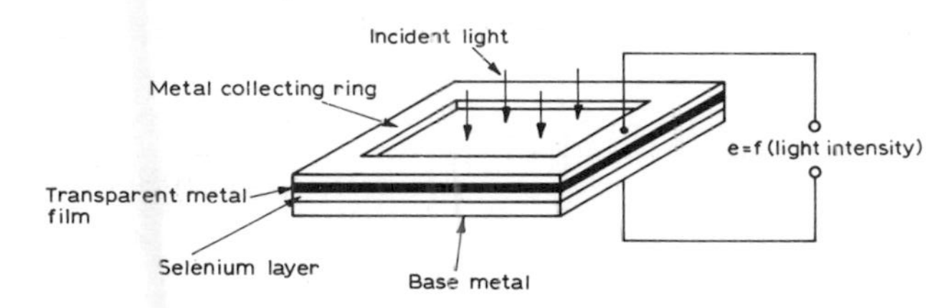

Principle of operation	*Examples of application*
Incident light causes change of resistance (photo-conductive cell)	**Fig. 2.27 Light-operated switch** $\delta R_{AB} = f(\text{light intensity})$
Incident light causes electric-current change	**Fig. 2.28(a) Photo-emissive cell** $I_a = f(\text{light intensity})$; **Fig. 2.28(b) Phototransistor** $I_E = f(\text{light intensity})$

Fig. 2.27 Light-operated switch
$\delta R_{AB} = f(\text{light intensity})$

To resistance-measuring circuit
Incident light
A
B
Interlocking metal fingers sprayed with semiconductive material (selenium or thallous sulphide)
Base of insulating material

Fig. 2.28(a) Photo-emissive cell
$I_a = f(\text{light intensity})$

May be contained in a gas-filled or evacuated tube
Surface coated with photo-emissive material (e.g. caesium)
Cathode
Anode
I_a
Incident light
R

Fig. 2.28(b) Phototransistor
$I_E = f(\text{light intensity})$

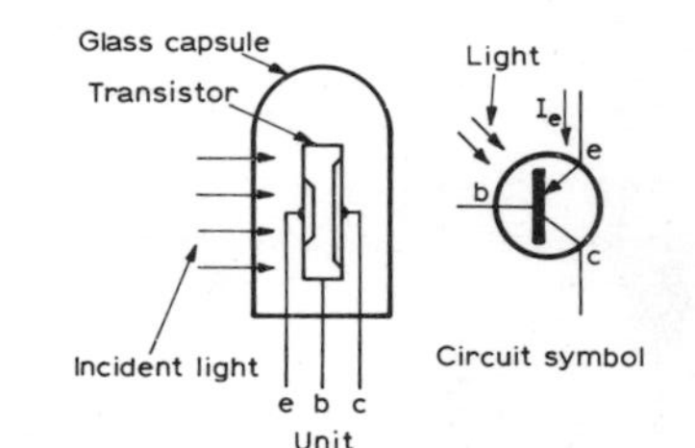

Sensing elements may be described as primary, when they react directly to changes in the quantity to be measured, or secondary, when they occur in the system between the initial sensing point and the indicating or recording points.

2.3 Amplifying elements

The signal from a sensing element is frequently very small and has to be amplified before it is large enough to operate an indicating element or to enable it to be transmitted over some distance. Amplifying elements may operate on mechanical, optical, fluid, or electrical principles, or on a combination of these.

2.3.1 Mechanical amplification

The lever and the simple gear-train, shown in fig. 2.29(a) and (b), are typical of mechanical amplification devices found in instruments.

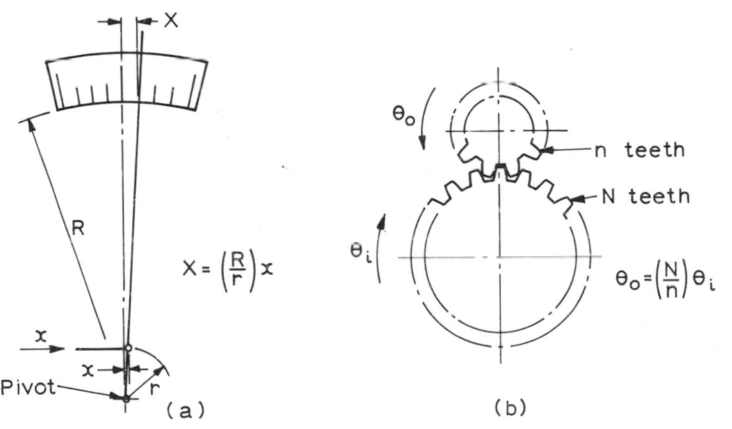

Fig. 2.29

2.3.2 Fluid amplification

a) Liquid. The liquid-in-glass thermometer has already been

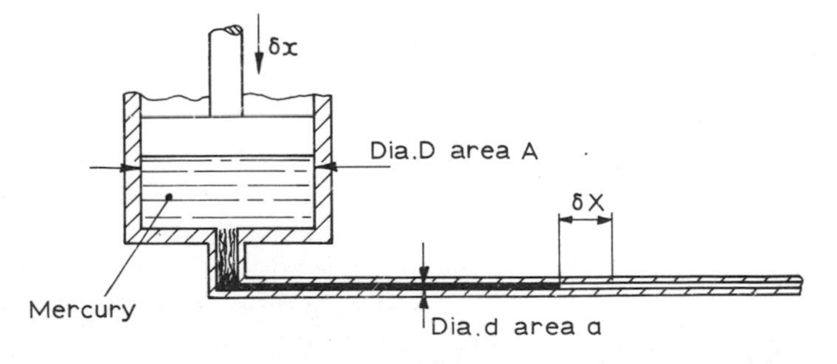

Fig. 2.30

mentioned as an example of an amplifying device. Figure 2.30 shows the method used on the Hounsfield Tensometer to amplify the small deflection δx of the beam which senses the force applied to a test-piece.

Volume of mercury displaced due to movement $\delta x = \delta x \times A = \delta X \times a$

Hence $$\delta X = \left(\frac{A}{a}\right) \times \delta x = \frac{\pi}{4} D^2 \times \frac{4}{\pi d^2} \times \delta x$$

$$\therefore \quad \delta X = \left(\frac{D}{d}\right)^2 \delta x$$

b) Pneumatic. A small change of gas pressure δP_1 may be made to give a larger change of pressure δP_2 if a supply at a higher pressure P_s is available; a rather simplified sketch of such an amplifier is shown in fig. 2.31. Increase of pressure P_1 causes movement of the valve spindle AV to the right, increasing the gap between the valve V and its seat, and increasing the output pressure P_2. A linear relationship may exist over part of the range of movement such that $P_2 = KP_1$, where K is a constant.

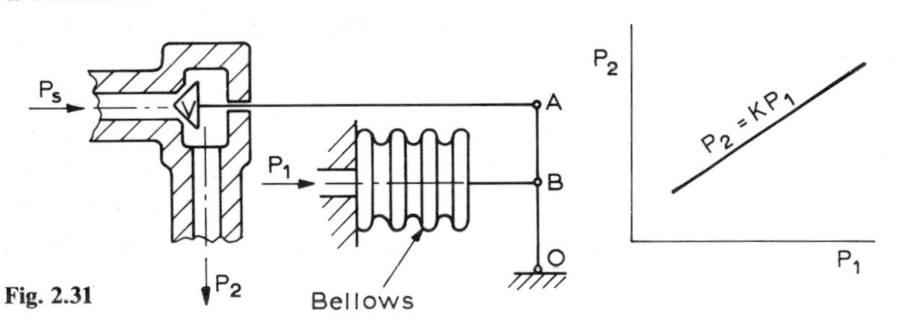

Fig. 2.31

Other pneumatic devices for amplification use the principle illustrated in fig. 2.6(a).

2.3.3 *Optical amplification*

Optical amplification provides a very useful means of amplifying small linear and rotational displacements without applying any measurable force to the displaced element. Figure 2.32 shows a cantilever sensing the magnitude of the force F, giving as an output the slope $(\theta_i) = KF$. Initially the collimated light ray is reflected back along the same path to the scale zero. When F causes an end slope θ_i, the out-going ray OAB is deflected through an angle $2\theta_i$, since both the angle of incidence and the angle of reflection are equal to θ_i. Several mirror surfaces may be used in series, to give a greater amplification of the angle. The radius R may be increased to give a greater scale movement X at the output.

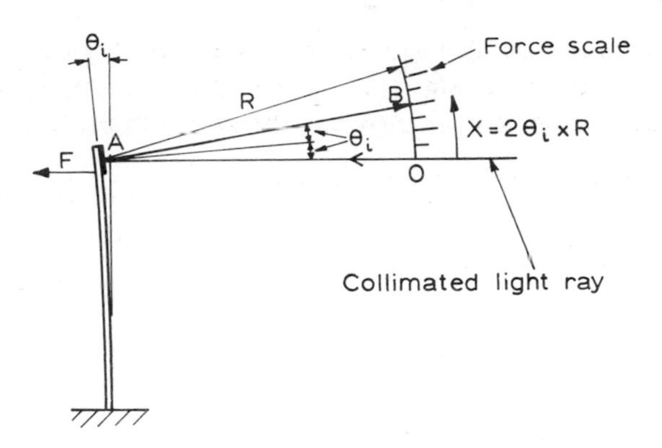

Fig. 2.32

2.3.4 *Electrical amplification*

The output from electrical transducers may be in the form of a small e.m.f., or a change of resistance, inductance, or capacitance. Small static e.m.f.'s may be measured accurately using potentiometer circuits; and static or varying resistances, inductances, or capacitances may be measured using bridge circuits. These devices are discussed in some detail in Chapter 4.

Electronic amplifiers, using transistor circuits, are widely used to take a very small power signal at the input and give a much increased or amplified power at the output, such that a suitable indicating, recording, or controlling device may be driven. An external source of energy is required, and the amplifier may be represented as shown in fig. 2.33.

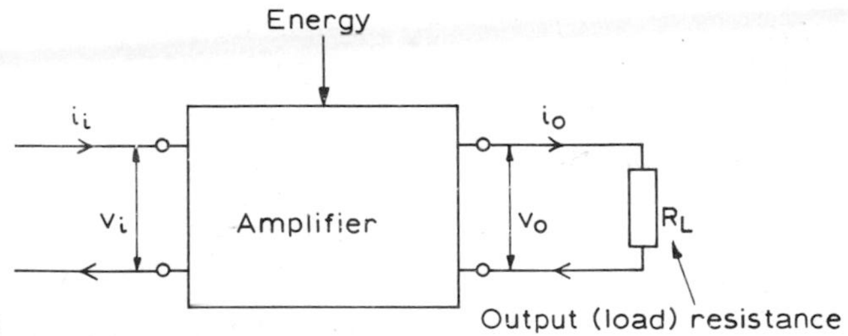

Fig. 2.33

The following definitions are applied:

gain = the ratio, power output/power input

$= v_0 i_0 / v_i i_i$

voltage amplification $= v_0/v_i$

current amplification $= i_0/i_i$

A distinction is made between amplifiers capable of amplifying a constant or slowly varying input signal, i.e. d.c. amplifiers, and those designed to amplify only rapidly fluctuating signals, i.e. a.c. amplifiers (see fig. 1.1).

D.C. amplifiers are used in some instrument systems, and are the essential components of analogue computers. The difficulty is in preventing 'drift', i.e. a slow change of the output signal when the input signal is static, or nearly so. D.C. amplifiers are relatively expensive, and a.c. systems are usually preferred, where they can be used.

A.C. amplifiers are designed to reproduce the fluctuating input signal accurately at the output, but with voltage and/or current increased in a known ratio. A particular amplifier will do this only within a particular range of frequencies of the input signal, as indicated in fig. 2.34. The range of frequencies over which a flat response is obtained is influenced by the load-resistance value, R_L.

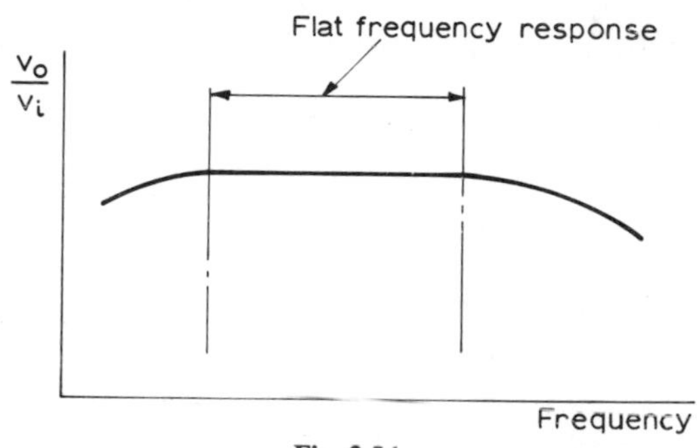

Fig. 2.34

Charge amplifiers are a type of d.c. amplifier used to give an output voltage proportional to the charge Q coulombs at the input. They are used particularly in the measurement of the charge on a piezoelectric crystal due to an applied force or pressure.

2.4 Indicating and recording elements

After the input signal to a system has been sensed and amplified, the increased power then available may be used to drive an indicator or recorder, which should be capable of following the signal from the amplifier without excessive inaccuracy. The indicator or recorder should be able to respond quickly to step changes or to fluctuations or oscillations of the signal. To impose as little 'drag' on the system as possible, the 'pointer' and associated moving parts should have a mass or rotational inertia as small as possible, and, in the case of recording instruments, any friction drag from the 'writing' process should be as small as possible. To achieve this, pointers and other moving parts are made as light as possible (but see Chapter 3), and light rays and electron beams are used in many indicators, these virtually removing inertia from the 'pointer' or 'writer.'

'Writers' may consist of ink systems using capillary action or sometimes pressure feed, or they may use electrical writing on special paper, as shown in fig. 2.35, where current passing through the indicating paper causes a change of colour. Another type uses a heated stylus to mark normal or heat-sensitive paper. A further type of paper

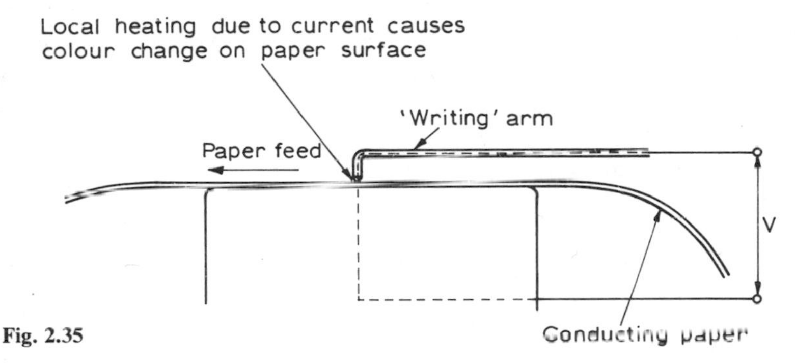

Fig. 2.35

is sensitive to ultra-violet light, and a trace may be produced on this without any pen contact. The signal shown on a cathode-ray oscilloscope may be photographed to obtain a permanent record, either by a single-shot or a movie camera.

Indicators may also be classified broadly as analogue, giving a pointer or meniscus etc. position on a scale, and digital, where an exact value is indicated, such as in a motor-vehicle odometer or a counter.

2.4.1 Mechanical movements

Many instruments have purely mechanical movements, e.g. Bourdon-tube, bellows, or diaphragm pressure-gauges, extensometers, and mechanical strain-gauges, etc.

Figure 2.36 shows the arrangement of a simple gas-pressure recorder. The gas pressure deflects a bellows (elastic sensing element), and the movement of the free end is amplified by a lever system giving a movement, on a circular arc, proportional to the pressure. The chart is rotated at a constant rate, e.g. 24 hours or seven days for one rotation, and a pen on the arm traces a continuous pressure–time record. The disadvantage of the circular-arc output is overcome by using charts

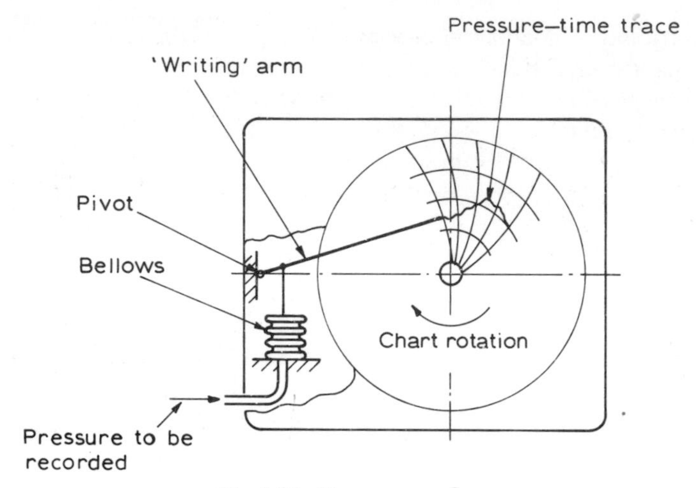

Fig. 2.36 Pressure recorder

having circular-arc scales, but a parallel motion giving straight-line amplified output is also available on some instruments.

2.4.2 *Electromechanical movements*

The basic electromechanical movement is the d'Arsonval galvanometer, the elements of which are shown in fig. 2.22(b). The current (i_{AB}) flowing in the coil suspended in a magnetic field, either from a permanent magnet or from an electromagnet, causes a tangential force to be exerted on the axial conductors in the coil, causing a torque (T). The torque is balanced after a movement (θ) against the restoring torque of a spring, which is in the form of a flat spiral in many instrument movements or a torsion wire in sensitive galvanometers. The movement (θ) may be indicated on a scale by a pointer attached to the shaft, or by a light beam reflected from a mirror attached to the shaft, or by a pen attached to an arm secured to the shaft. These devices are indicated in fig. 2.37. The deflection in each case is proportional to the current flowing in the coil, if the magnetic field intensity is constant.

The pen-recorder movement of fig. 2.37(c) has friction drag from the pen, and hence the springs must have a higher rate (i.e. higher torque per unit angular deflection) than the pointer instrument of 2.37(a), so that the error due to friction represents a lower percentage. The spot galvanometer of 2.37(b) is used to measure very small currents, and sometimes to indicate a null (i.e. no current flow) condition. The UV-recorder movement of fig. 2.37(d) has extremely low inertia for a mechanical system, and may be used for recording rapidly fluctuating signals such as the pressure in a high-speed engine cylinder.

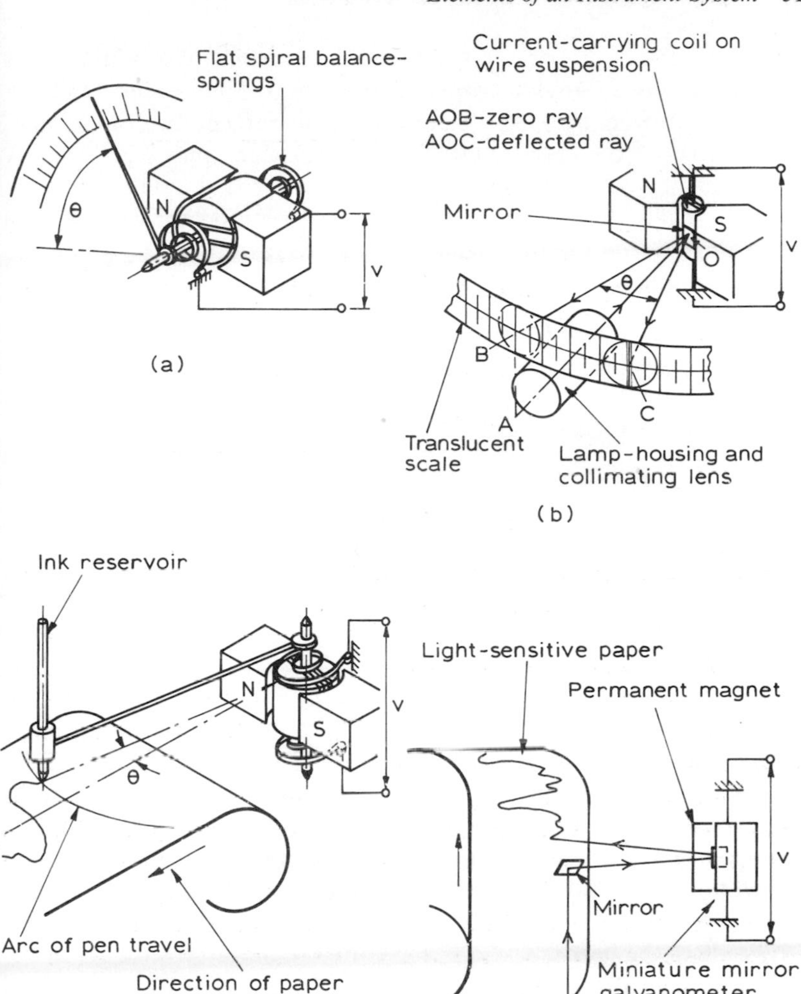

Fig. 2.37 (a) Voltmeter or ammeter movement
(b) Spot galvanometer
(c) Pen recorder
(d) Ultra-violet recorder

The output of the pen and UV recorders is on a base of time, but occasionally one signal has to be plotted relative to another. The *XY* recorder shown in fig. 2.38 is a convenient instrument for this purpose, for relatively slow speeds. Ordinary graph-paper sheets are usually used, and the *X* and *Y* incoming signals may be amplified to give suitable scales. Automatic control systems position the bridge (AB) according to the *X*-input signal and the ink-pen head (C) along the arm according to the *Y*-input.

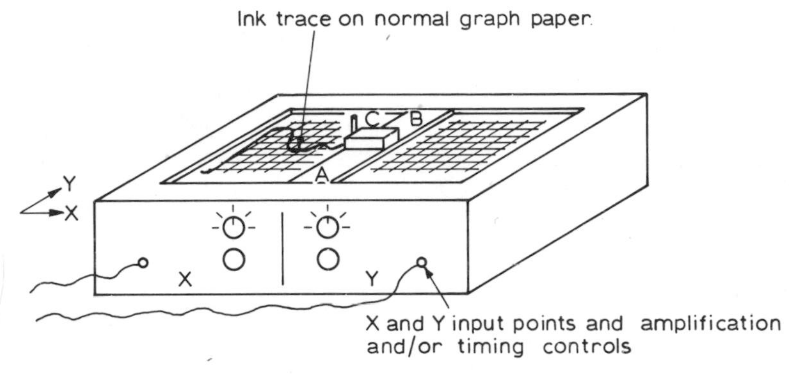

Fig. 2.38 *XY* **recorder**

2.4.3 Electronic indication

The cathode-ray oscilloscope (CRO) is undoubtedly the most useful and versatile indicating instrument available to the engineer. The principle of the deflection of an electron beam in a magnetic field is shown in fig. 2.22(a). More usually in the CRO, the deflection is done by an electrostatic field caused by applying a high voltage to plates on each side of the beam.

Figure 2.39(a) shows the tube of a CRO. The angular deflections θ and ϕ, and the scale deflections x and y of the spot on the fluorescent screen produced by the electron beam, are proportional to the p.d.'s v_x and v_y applied to the plates. The x- and y-input voltages to the CRO are amplified, and the amplifier outputs provide these deflection voltages v_x and v_y. The degree of amplification is controlled by the operator, to give a suitable amount of deflection of the spot on the screen.

Since in the majority of applications the variable is to be measured against time, special arrangements are made for moving the beam in the x-direction. A resistance–capacitance circuit changes v_x, and hence the x-position of the spot, linearly with respect to time as in fig. 2.39(b). This moves the spot at a predetermined rate to the right, and then almost instantaneously returns it to the left of the screen. Hence, if y is

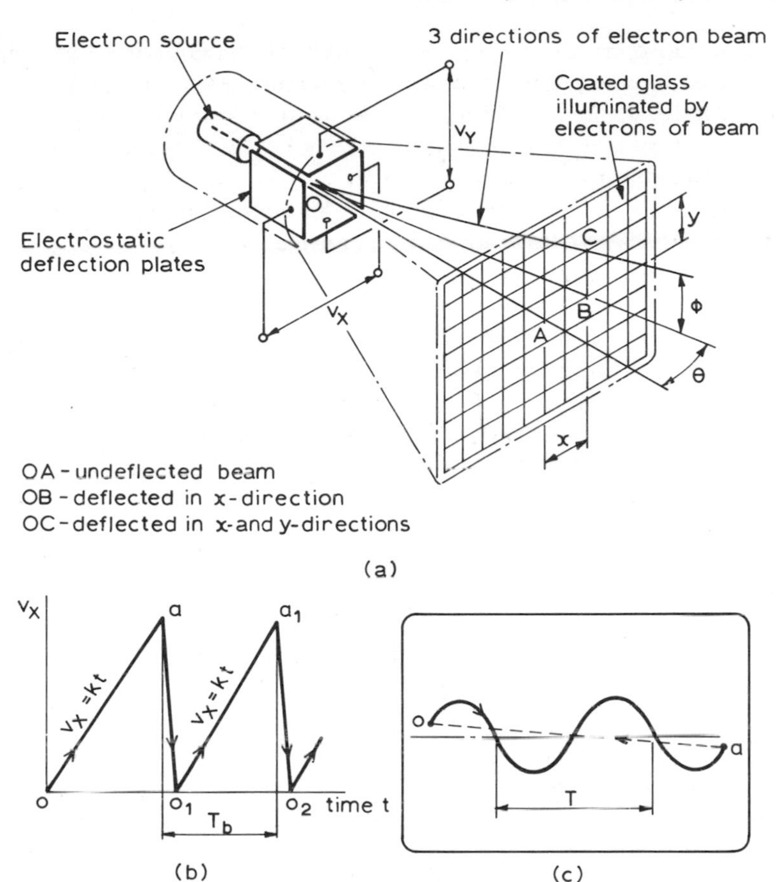

Fig. 2.39 Cathode-ray oscilloscope: (a) Cathode-ray tube (b) 'Saw-tooth' voltage wave (c) Trace of continuous wave

a value varying say sinusoidally with time, then a trace as in fig. 2.39(c) is obtained if the periodic time of the x-motion (T_b) coincides with or is a multiple of the periodic time T_p of the y-variable. The controls are adjusted by trial and error until the static trace is obtained. Alternatively, the x-motion may be triggered off by an external signal, such as that obtained from a pick-up once every revolution of a shaft.

A *double-beam* oscilloscope is a useful tool for comparing two signals on the same base of time, for example the input to and the output from an instrument or control system. Some CRO's have a long-persistence tube which enables the whole of a slow-moving signal trace to be viewed.

A useful accessory is the Polaroid camera, which may be attached to the CRO screen to produce a printed picture of the trace in a matter of seconds.

2.5 Signal transmission

A signal may be transmitted locally in an instrument by means of the movement of a rod, lever, or fluid column, etc., or by a change of various electrical quantities. However, in applications such as industrial process-control, remote indication is becoming increasingly used, where a central control room can observe the complete functioning of a plant from one point. Usually either pneumatic or electric transmission is adopted.

Pneumatic transmissions frequently use the 'flapper valve' shown in fig. 2.40, where the movement x of the end of the 'flapper' causes a proportional change in the back pressure P_b if the supply pressure P_s is constant. P_b may be amplified before being piped to the remote indicator.

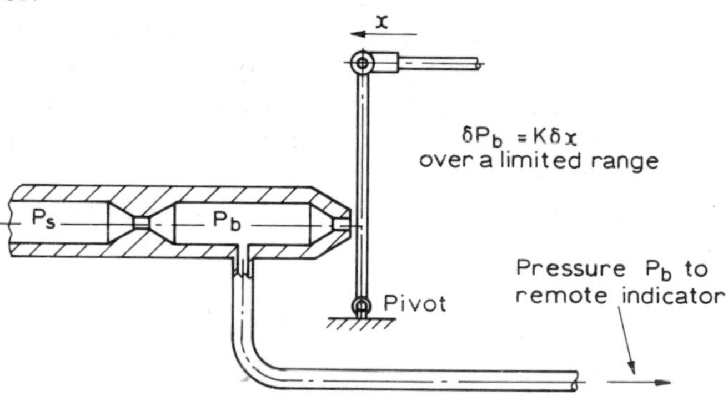

Fig. 2.40 Pneumatic transmission of signal

In electrical transmitters, resistance and inductance values are commonly varied to cause a change of current in the transmitting circuit, as indicated in fig. 2.41. Variation of capacitance is also used.

In other electrical systems, the making or breaking of a circuit may indicate a quantity. The number of pulses in a given time, or the time during which a circuit is open or closed, may indicate a value. A summary of these methods is given in ref. 38.

2.6 Summing, multiplying, integrating, and differentiating elements

Mathematical operations frequently represent or are represented by physical operations, and may be carried out either by a human operator or by a mechanical, electrical, etc. device.

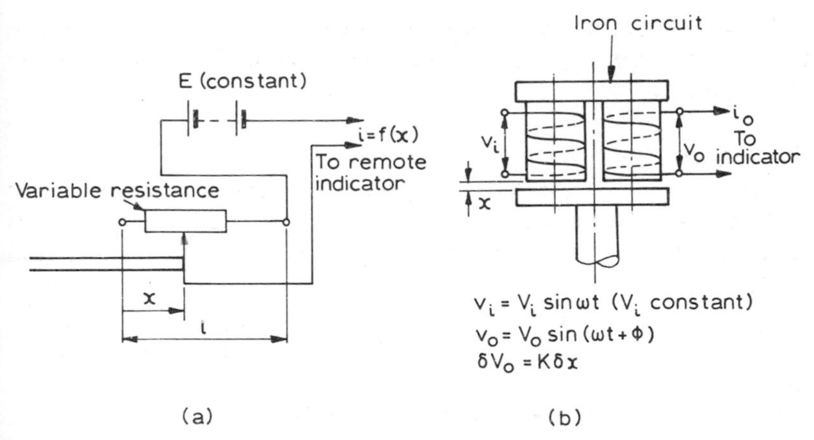

Fig. 2.41 Electrical transmitters: (a) Variable-resistance transmitter (b) Variable-inductance transmitter

2.6.1 *Summation*

This is the operation of adding together like quantities. For example, a sensing device as shown in fig. 2.20(b) may be placed at the output end of a conveyor belt, so that a voltage pulse is given out for every item passing. The pulses may be counted by an electronic counter, to give the total number of items.

In instrument systems, the algebraic summing of signal quantities such as linear displacement, voltage, and current may be carried out as indicated in fig. 2.42.

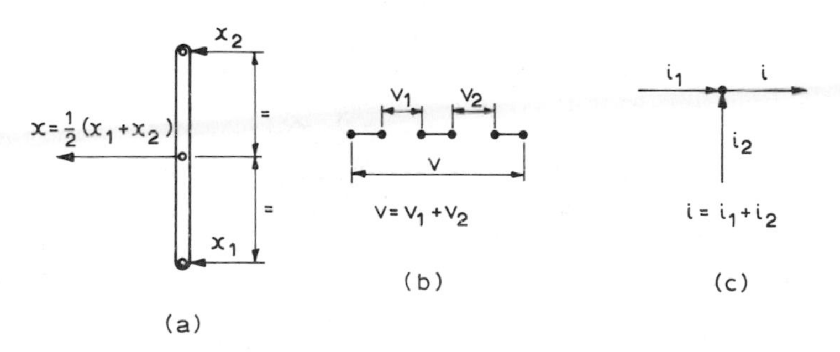

Fig. 2.42 Algebraic summing of signals: (a) Displacement (b) Voltage (c) Current

2.6.2 *Multiplication*

If the electromagnetic movement of figs 2.22(b) and 2.37(a) has the magnetic poles produced by a current-carrying coil, then the torque, and hence the deflection, is proportional to the current in both the fixed coil and the moving coil, and the following relationship applies:

torque $\propto$ (magnetic flux density in gap) $\times$ (current in moving coil)

or torque $= Kiv$ if the magnetic flux density is proportional to the current (i) in part of a circuit, and the current in the moving coil is proportional to the potential difference (v) in that part of the circuit, as shown in fig. 2.43. The device has effectively multiplied together the current and voltage values, to give an output torque proportional to power (see section 8.4).

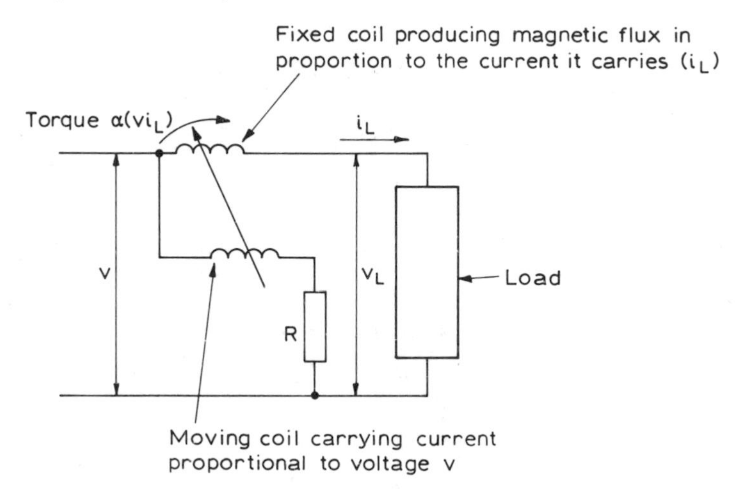

Fig. 2.43 Wattmeter circuit

2.6.3 *Integration*

If the total flow $Q(\text{m}^3)$ through a pipe in a given time (t) is required, then the instantaneous flow rate V (m/s) may be sensed by a transducer giving an output voltage $v = K_1V$ which is fed into a chart recorder giving deflections $y = K_1K_2V$ and $x = K_3t$ on the chart, as shown in fig. 2.44. Q may then be found by integration from the chart trace, or alternatively v may be fed into a suitable integrating device.

If A is the cross-sectional area of flow, then

volume passing in time $\mathrm{d}t = \mathrm{d}Q = AV\,\mathrm{d}t$

Total volume in the time period $(t_2 - t_1)$ is given by

$$Q = \int_{t_1}^{t_2} AV\,\mathrm{d}t = \frac{A}{K_1}\int_{t_1}^{t_2} v\,\mathrm{d}t$$

which would represent the operation of an integrating device, A/K_1 being set as a gain value.

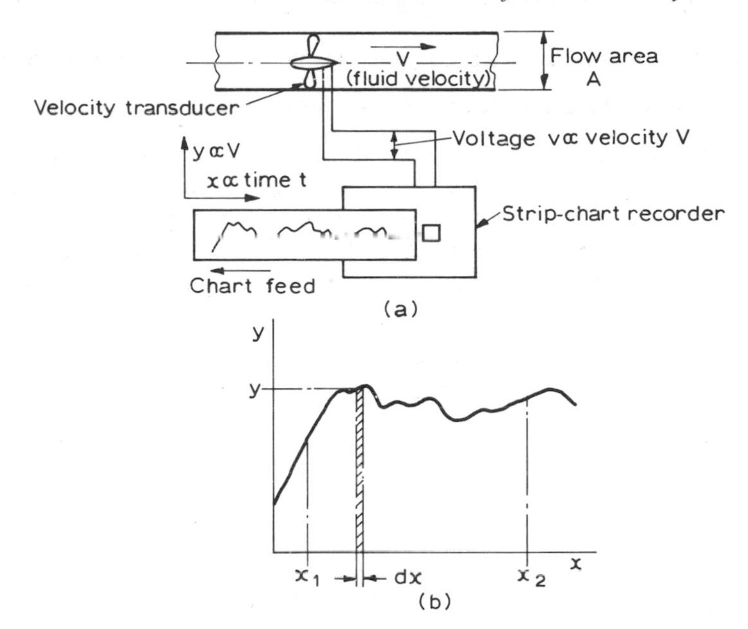

Fig. 2.44 Fluid-flow measurement
$v = K_1V, y = K_2v, \therefore y = K_1K_2V; x = K_3t$

To find the total flow from the chart trace,

$$Q = \frac{A}{K_1}\int_{x_1}^{x_2}\frac{y}{K_2}\times\frac{\mathrm{d}x}{K_3} = \frac{A}{K_1K_2K_3}\int_{x_1}^{x_2} y\,\mathrm{d}x$$

$$= K \times (\text{area under curve between } x_1 \text{ and } x_2)$$

where $K = A/(K_1K_2K_3)$. The area under the curve may be found by using Simpson's method or by using a planimeter, which itself is an example of an integrating instrument.

In an industrial environment, however, an electronic instrument is preferred to carry out the integration automatically, and these are available with either pointer (analogue) or number (digital) output. (Also, many flow-measuring instruments are purely mechanical, see Chapter 7.)

2.6.4 *Differentiation*

Since differentiation is the determining of the rate of change of one quantity with respect to another, many devices measuring 'rates' are in fact carrying out this process. For example, the tachogenerator of fig. 2.21(b) gives an output voltage proportional to angular velocity.

i.e. $$v_0 = K\omega = K\frac{d\theta}{dt}$$

or $$\frac{d\theta}{dt} = \frac{v_0}{K}$$

Hence the device may be regarded as carrying out a differentiation of angle θ with respect to time. Electronic circuits can carry out differentiation and integration; hence, if the output voltage (v_0) is differentiated, the result is a quantity representing angular acceleration.

i.e. $$\text{acceleration} = \frac{d^2\theta}{dt^2} = \frac{1}{K}\frac{dv_0}{dt}$$

Similarly, v_0 could be integrated to give the angle θ through which the shaft had turned in a given time interval.

i.e. $$d\theta = \frac{v_0}{K}dt$$

$$\theta = \frac{1}{K}\int_{t_1}^{t_2} v_0 dt$$

2.7 System representation

A number of elements such as described in this chapter may be connected together to form a system; it will then be necessary to know the overall effect, i.e. the relationship between the signal coming out of the system and that going in. It is convenient to represent the system by a block diagram, each block representing some *function* of the apparatus, rather than some component. Hence the electromagnetic movement shown in fig. 2.37(a) may be broken down into the following functions:

a) converting input current into torque,
b) converting torque into angular movement,
c) converting angular movement into scale displacement.

Function (a) is carried out by the coil and permanent-magnet arrangement, (b) by the balance springs, and (c) by the pointer; hence the system may be shown as in fig. 2.45.

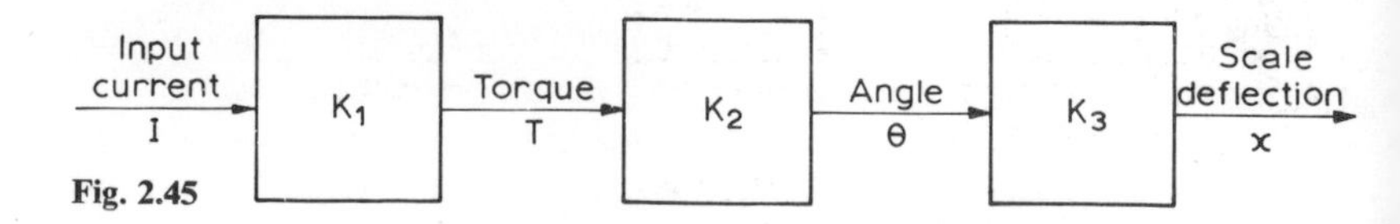

Fig. 2.45

The constants K_1, K_2, and K_3 are arranged so that

$$\begin{aligned} \text{input current} \times K_1 &= \text{torque} \\ \text{torque} \times K_2 &= \text{angle} \\ \text{angle} \times K_3 &= \text{scale deflection} \end{aligned}$$

Hence input current $\times K_1 \times K_2 \times K_3 =$ scale deflection

and
$$\frac{\delta x}{\delta I} = K_1 \times K_2 \times K_3$$

K_1 must have the units N m/A.
K_2 must have the units rad/N m.
K_3 must have the units m/rad.

Hence K_1 N m/A $\times$ K_2 rad/N m $\times$ K_3 m/rad $= K_1K_2K_3$ m/A

and
$$\frac{\delta x}{\delta I} = K_1K_2K_3 \text{ m/A}$$

K_1, K_2, and K_3 are constant values only under *steady conditions*, i.e. where the input and output are not changing with time. The K values for the functional parts, and those for the system, are referred to as *sensitivities*, as defined in Chapter 1, or as gains or amplifications. When the input or the output is changing with time, differential and/or integral terms appear, and the terms *transfer function* or *transfer operator* are used rather than sensitivity, etc. The effects of such terms are discussed in Chapter 3.

2.8 Worked examples

2.8.1 A piezoelectric pressure transducer has a sensitivity of 9·00 picocoulomb/bar. It is connected to a charge amplifier, the gain being set to 0·005 volts/picocoulomb. The amplifier output is connected to an ultra-violet chart recorder, whose sensitivity is set to 20 mm/V. Draw the block diagram for the system, and calculate the overall sensitivity. Calculate the deflection on the chart due to a pressure change of 35 bar.

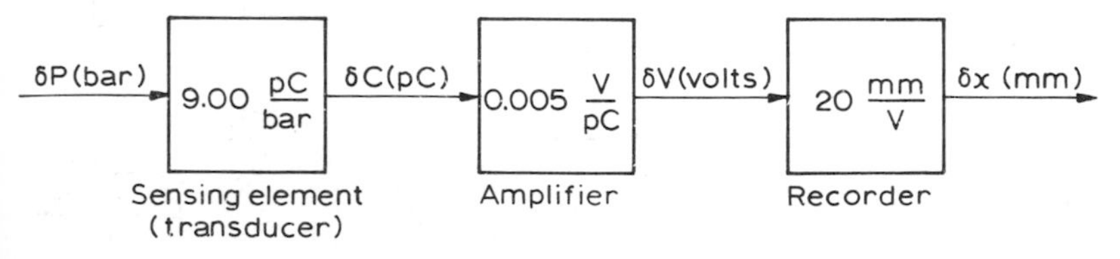

Fig. 2.46

The system is represented by the block diagram of fig. 2.46. The change in the measured variable (δP) is multiplied by the sensitivity of the sensing element, to give the charge change (δC), which is multiplied by the amplifier gain, etc. Each input signal is multiplied by a constant, to give the output signal for that element. Thus

$$\delta P \times 9{\cdot}00 \text{ pC/bar} \times 0{\cdot}005 \text{ V/pC} \times 20 \text{ mm/V} = \delta x$$

The overall sensitivity is $\dfrac{\delta x}{\delta P} = 9{\cdot}00 \times 0{\cdot}005 \times 20$ mm/bar

$$= 0{\cdot}90 \text{ mm/bar}$$

For $\delta P = 35$ bar,

$$\delta x = 0{\cdot}90 \times \delta P$$
$$= 0{\cdot}90 \times 35 = 31{\cdot}5 \text{ mm}$$

2.8.2 A spot galvanometer (fig. 2.37b) has a sensitivity of 0·25 rad/μA. Calculate the overall sensitivity when using a scale at a distance of 200 mm from the galvanometer mirror.

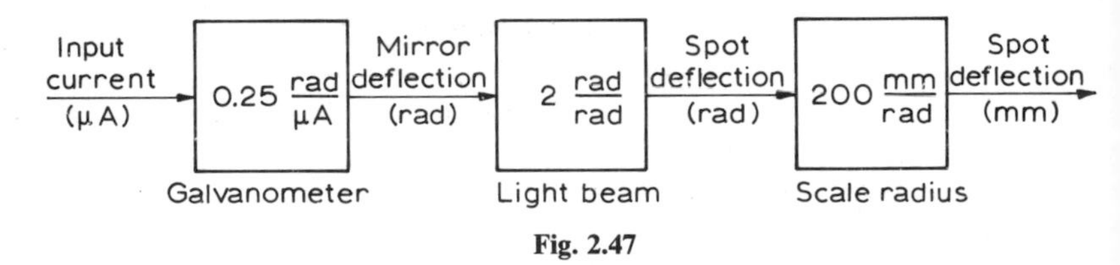

Fig. 2.47

$$\delta I \times 0{\cdot}25 \text{ rad/}\mu\text{A} \times 2 \text{ rad/rad} \times 200 \text{ mm/rad} = \delta x$$

Overall sensitivity $= \dfrac{\delta x}{\delta I} = 0{\cdot}25 \times 2 \times 200$ mm/μA

$$= 100 \text{ mm/}\mu\text{A}$$

(In basic units this should be expressed as 100 km/A, but it is more usefully left as above.)

2.8.3 A piezoelectric force transducer on an actuating ram in a linkage has a sensitivity with its associated amplifier of 1 V/kN, and is connected to the *Y*-axis of an *XY* recorder, the sensitivity setting of the recorder being 20 cm/V. The displacement of the ram is sensed by a variable-resistance transducer, the output from which is 25 V/m. Its output is connected to the *X*-axis of the recorder, the setting being 2·00 cm/V. If the area under the curve traced out during a stroke of the ram is 21 300 mm^2, and the length on the Y-axis under the curve is 225 mm, calculate the work done and the mean effort.

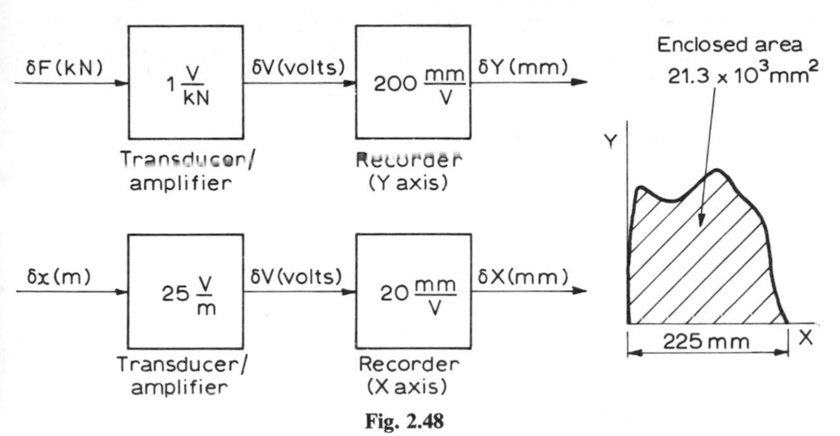

Fig. 2.48

Referring to fig. 2.48,

Force sensitivity

$$\delta F \times 1 \text{ V/kN} \times 200 \text{ mm/V} = \delta Y$$

$$\therefore \quad \frac{\delta Y}{\delta F} = 200 \text{ mm/kN}$$

Distance sensitivity

$$\delta x = 25 \text{ V/m} \times 20 \text{ mm/V} = \delta X$$

$$\therefore \quad \frac{\delta X}{\delta x} = 500 \text{ mm/m}$$

$$\therefore \quad 1 \text{ mm}^2 \text{ represents } \frac{1}{500} \text{ m} \times \frac{1}{200} \text{ kN} = \frac{1}{100} \text{ N m}$$

$$\text{Hence work done} = \frac{21\,300}{100} \text{ N m}$$

$$= 213 \text{ J}$$

$$\text{Mean height of } XY \text{ diagram} = \frac{21\,300}{225} \text{ mm}$$

$$\text{Mean effort} = \frac{21\,300}{225} \times \frac{1}{200}$$

$$= 0{\cdot}472 \text{ kN} = 472 \text{ N}$$

2.8.4 The calibrated sensitivity of a piezoelectric force washer is 42·9 pC/kgf, and this value is set on the charge-amplifier dial to give an overall sensitivity of 10·0 mV/kgf (see fig. 5.10). Calculate the setting required on the amplifier to give an overall sensitivity of 1·0 mV/N, and the resulting change of the amplifier gain.

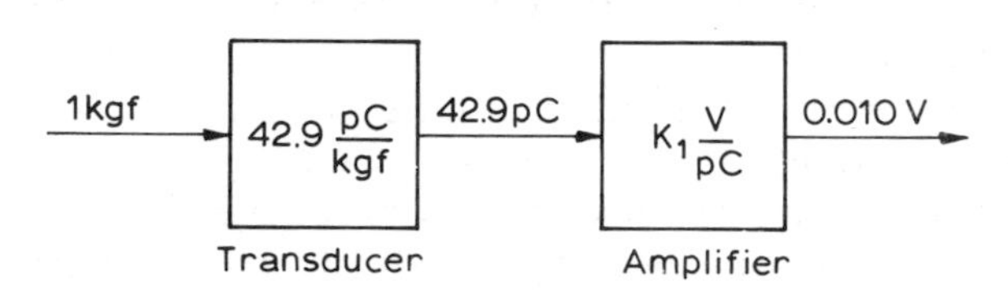

Fig. 2.49

Figure 2.49 shows the block diagrams for the original setting, with 1 kgf input. It is seen that

$$K_1 = \frac{0{\cdot}010}{42{\cdot}9}\ \text{V/pC} = 0{\cdot}237\ \text{mV/pC}$$

To convert the transducer sensitivity, which is unchanged, into the required units

$$42{\cdot}9\ \text{pC/kgf} = 42{\cdot}9\ \text{pC/kgf} \times \left[\frac{\text{kgf}}{9{\cdot}81\ \text{N}}\right] \quad \text{(see appendix A)}$$

$$= 4{\cdot}37\ \text{pC/N}$$

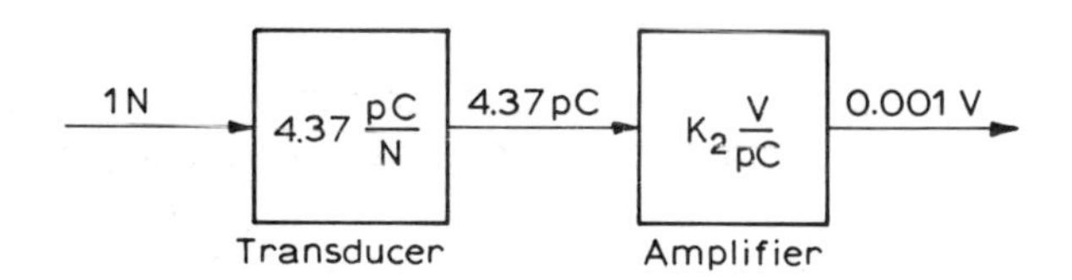

Fig. 2.50

The block diagram for the new setting is shown in fig. 2.50. It is seen that

$$K_2 = \frac{0{\cdot}001}{4{\cdot}37}\ \text{V/pC} = 0{\cdot}229\ \text{mV/pC}$$

The amplifier setting is changed from 42·9 to 43·7, giving a change of gain from 0·237 to 0·229 mV/pC.

2.9 Tutorial and practical work

2.9.1 On placing a thermometer into a hot liquid, the meniscus is sometimes seen to fall momentarily before it rises. Discuss the reason for this, and other causes for delay in indicating the temperature of the liquid.

2.9.2 Will the sensitivity of a mercury-in-glass thermometer be increased or reduced by
a) increasing the internal volume of the glass bulb,
b) increasing the internal diameter of the capillary tube,
c) making the instrument of a glass having a lower expansion rate,
d) reducing the glass thickness (i) of the bulb, (ii) of the stem?
Give reasons in each case.

2.9.3 For the Fortin barometer illustrated in fig. 2.5(a), show, working from first principles, that the atmospheric pressure is given by $P_a = \rho g h$, where ρ is the mercury density and g the local gravitational acceleration. State why it is essential that no gas or liquid be permitted to reach the top of the tube.

2.9.4 For the two accelerometers shown in fig. 2.7, state why (b) cannot be used in free space, and discuss the disadvantage that has to be overcome in using (a) in a gravity field. For (b), derive an expression relating θ to acceleration.

2.9.5 Explain why there is usually a temperature difference between the wet- and dry-bulb thermometers of fig. 2.13. What is implied if the two thermometers read the same? What is the purpose of the screen.
Sketch the outline of a system using this effect to automatically control the humidity of a room.

2.9.6 What is the Doppler effect? Explain how this can give a change of frequency between the transmitted and reflected waves as in fig. 2.14(b).

2.9.7 For the angular-speed sensing mechanism of fig. 2.7(c), show by differentiating the basic relationship that, if the shaft has angular acceleration α, then this is a function of both F and $\mathrm{d}F/\mathrm{d}t$.

2.9.8 Sketch a simple spring–mass arrangement to sense the angular acceleration of a shaft, showing a way of transmitting and indicating the signal.

2.9.9 Determine the fluid-amplification ratio of a Hounsfield Tensometer. Would this value or the overall sensitivity be affected if the capillary tube were positioned vertically instead of horizontally?

2.9.10 Determine the amplification ratio of a pressure-gauge mechanism.

2.9.11 Sketch a system of mirrors to give an amplification of eight times the angular change $\delta\theta$ of the input ray.

2.9.12 Examine the electronic amplifiers used in your laboratory (or works) instrument systems. Note from their literature the gain, frequency range, effect of load resistance, etc., and so compare their performance.

2.9.13 Compare the characteristics of continuous-recording devices which are available to you.

2.9.14 Discuss the difficulties in transmitting a signal from a rotating or oscillating component to a fixed point (e.g. from a strain-gauge on a piston to the stationary frame or casing). Suggest methods of achieving this.

2.9.15 The quantity of rainfall on a remote hilltop is to be measured automatically, and the signal is to be transmitted to the water-control centre several kilometres away. Two schemes are being considered:
a) a pivoted arm having two small containers, one of which is positioned under the rain-collecting funnel, until the weight of rain-water tips the arm, empties the water, and brings the other container into the collecting position, at the same time operating a switch; and
b) the collecting funnel delivers water to a cylinder containing a float which operates a variable resistance.

Sketch the devices, and discuss the transmission of the signal in each case. Outline the relative advantages and disadvantages of the systems.

2.9.16 State, with reasons, whether the following carry out operations of summing or integrating or differentiating with respect to time: (a) a motor-vehicle speedometer, (b) a household electricity meter, (c) a digital voltmeter (i.e. a voltmeter indicating voltage as a displayed number, rather than a pointer against a scale), (d) a tank collecting and measuring a fluid flowing at a variable rate, (e) a mechanical counter being tripped once for every rotation of a shaft.

2.9.17 For a simple float-type liquid-level control (such as the domestic ball-cock), draw the block diagram of the system, labelling each block with its function. State the input and output quantity of each block.

2.9.18 Draw the block diagram for the amplifier shown in fig. 2.31, stating the input and output variable from each element.

2.9.19 Discuss the possible advantages and disadvantages with might arise in using magnetic tape to record a measured quantity.

2.10 Exercises

2.10.1 For the mechanical amplification device shown in fig. 1.7(b), determine the radius of the scale from the centre of the roller R if the latter is 10 mm in diameter and a scale movement of 1 mm from the central position is required for 0·10 mm movement (x) of the rod.

Will the scale divisions be equal for equal movements of the rod? How could the device be calibrated? Give reasons. [95 mm]

2.10.2 A wire used as a torsion member in a miniature galvanometer is 0·10 mm in diameter, arranged as shown in fig. 2.51. If it is made of phosphor bronze for which the modulus of rigidity (G) is 40 GN/m^2, calculate the angular deflection per unit torque, and the maximum torque if the stress is limited to 100 MN/m^2.

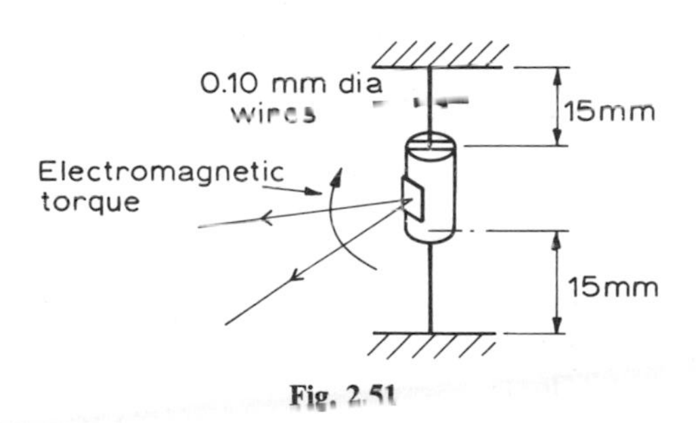

Fig. 2.51

A collimated light ray is reflected by the galvanometer mirror as in fig. 2.37(d), the radius from the mirror to the circular scale being 200 mm. Draw the system block diagram, and calculate the overall sensitivity of the instrument in m/Nm and the maximum possible scale deflection in each direction in mm from zero.

[$6{\cdot}38 \times 10^{-3}$ rad/Nm, $39{\cdot}3 \times 10^{-6}$ Nm, 3·82 m/Nm, 150·1 mm]

2.10.3 a) An accelerometer is to be constructed as shown in fig. 2.52, using the deflection of a steel strip 100 mm long by 10 mm × 1 mm cross-section due to a mass of 0·2 kg at the end. Calculate (i) the force (F) on the end of the strip per m/s^2 acceleration (a) of the 'fixed' end,

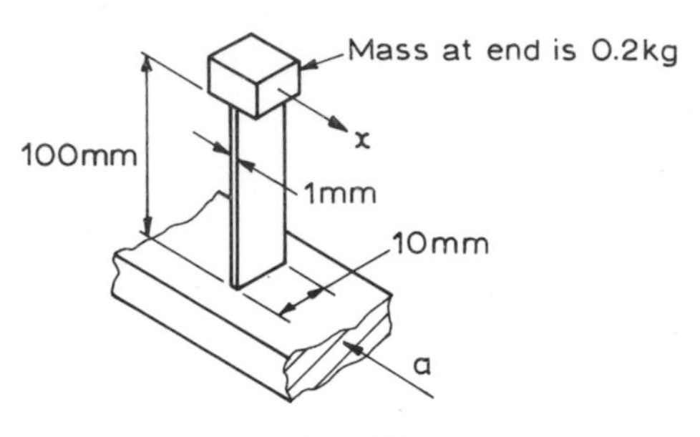

Fig. 2.52

(ii) the deflection (x) of the 'free' end per unit force, if the modulus of elasticity of the steel is 200 GN/m^2. (Refer to appendix D for the deflection equation.)

b) Suggest a method of measuring the deflection (x) of the end without applying appreciable force on the cantilever.

c) If the displacement-measuring unit is to give a current to a galvanometer instrument having a sensitivity of 10°/mA, and 150° deflection is to indicate $2g$ acceleration, draw the block diagram for the system. Calculate the required sensitivity of the displacement-measuring unit in A/m, and state all quantities, sensitivities, and units on the diagram. [0·2 N s/m^2, 2 mm/N, 1·92 A/m]

2.10.4 The mean velocity of fluid across a 100 mm diameter pipe is measured by a system which gives 0·5 V/(m/s). The output is connected to the y-axis of a strip-chart recorder whose sensitivity is 10 mm/V on this axis. The x-axis has a movement of 120 mm/hour. If a trace obtained has an overall length of 180 mm and an area of 6320 mm^2, calculate the mean velocity, the mean flow rate in m^3/s and litre/s, and the total volume of fluid passing in this period.

[7·02 m/s, 0·055 m^3/s, 55 l/s, 297 m^3]

2.10.5 The graph fig. 2.53 shows the water-thermometer readings θ_1 at inlet to and θ_2 at outlet from a heat exchanger on test, together with the rate of water flow through the exchanger as measured by a flow meter, both against a base of time. Calculate the rate of heat dissipation in watts at 500 s intervals, and determine the maximum heat-transfer rate.

Outline a measurement system which would indicate and record this heat-transfer rate instantaneously for the production quality-testing of exchangers. [750 W]

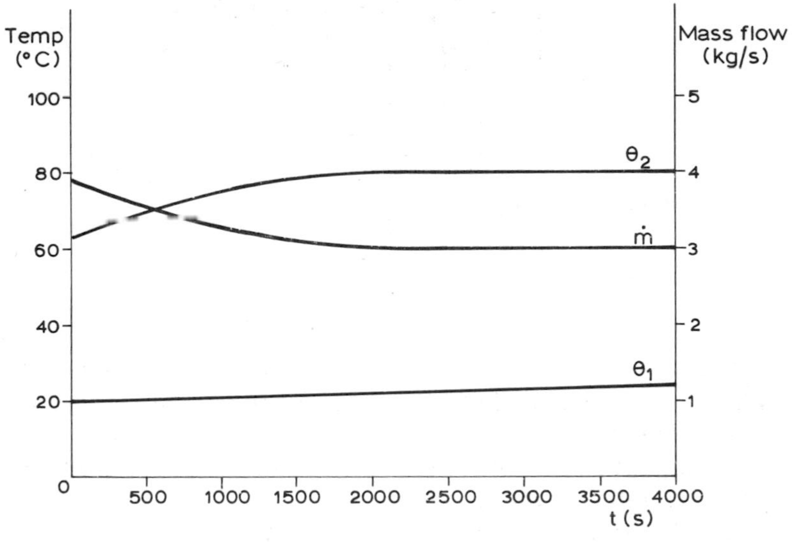
Temp
(°C)
100
80
60
40
20
0
Mass flow
(kg/s)
5
4
3
2
1
θ2
ṁ
θ1
500
1000
1500
2000
2500
3000
3500
4000
t (s)

Fig. 2.53

3

Dynamic Performance of Instrument Systems

3.1 Dynamic performance

It is relatively easy to measure with good accuracy many steady quantities, such as a constant voltage or the length of a component. It is less easy to measure accurately the instantaneous value of a rapidly varying voltage, or the instantaneous position of a moving machine-tool carriage. In this chapter, the form of the output signal is discussed when the input signal to an instrument changes as a step, as a ramp, or sinusoidally. These and other types of input are illustrated in fig. 1.1. The equations are developed using mechanical-system elements, but it is shown that similar relationships occur in electrical, pneumatic, hydraulic, and mixed systems, and generalised relationships may be derived.

3.2 Spring, mass, and damper systems

Many instruments use a spring element to balance a force or torque. The required sensitivity of the instrument will determine the *rate* or *stiffness* of the spring element, as a force per unit linear deflection or as a torque per unit angular deflection, etc. The moving parts will have *mass*, which may vary in magnitude according to the design. The combination of spring and mass gives a system which will oscillate naturally at a given frequency.

A factor which determines the *amplitude* of any oscillation, and also other important characteristics of the system, is the *damping*. Damping is a means of dissipating energy in a system. It may occur naturally, as hysteresis in materials or as viscous friction in bearings etc. (see Chapter 1). It may also be intentionally introduced, as in a moving-coil electric meter [fig. 2.24(b)] or as a dashpot device similar to the automobile damper. A damping force or torque opposes motion, and is taken to be proportional to linear or angular velocity.

The interaction between spring, mass, and damper determines the *dynamic response* of a mechanical system, i.e. the form of the output signal in response to inputs that may be varying such as indicated in fig. 1.1.

*3.2.1 Spring and damper**

In this idealised system, the mass is *assumed* to be zero. When the piston P of fig. 3.1 moves to the right, fluid flows past the clearance, and the drag exerts a damping force (F_d) proportional to the relative velocity (V) and in a direction opposite to the motion.

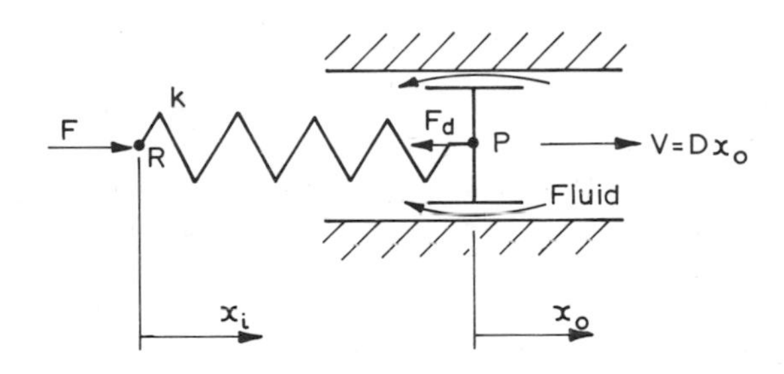

Fig. 3.1

i.e. $F_d = fV$, where f is the *damping constant*

$= f\,dx_0/dt$

$= f\,Dx_0$, where $D \equiv d/dt$ is the differential operator.

Also, $F = k(x_i - x_0)$.

Since there is no mass, $F = F_d$ for equilibrium, hence

$$f\,Dx_0 = k(x_i - x_0)$$

and

$$(fD + k)x_0 = kx_i$$

or

$$\frac{x_0}{x_i} = \frac{k/f}{D + (k/f)} \qquad 3.1$$

This is a transfer operator (see Chapter 2) if x_i is an input displacement and x_0 the corresponding output displacement. (Actual systems must have mass but, if the mass is small relative to the damping constant (f), the above is nearly true.)

It should be noted that when $D = 0$, then $x_0 = x_i$.

* The device consisting of a piston in an oil-filled cylinder is known as a 'dashpot'.

3.2.2 Spring, mass, and damper

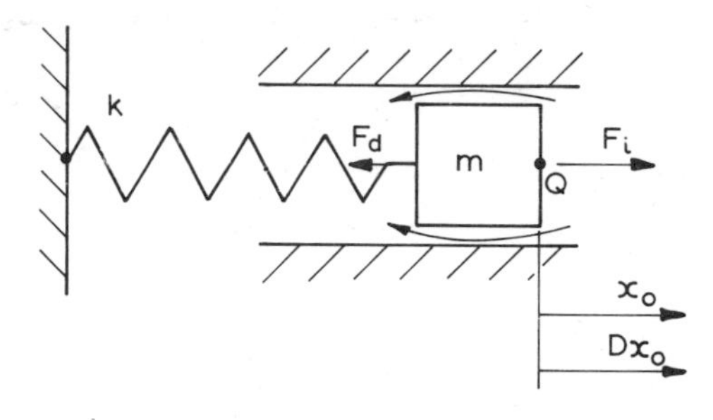

Fig. 3.2

Figure 3.2 represents a more practical system, having mass; in rotational form, this is exemplified by the d'Arsonval galvanometer movements of fig. 2.37. The equation of motion for the linear system shown in fig. 3.2 is

$$m\,\mathrm{d}^2x_0/\mathrm{d}t^2 = -kx_0 - f\,\mathrm{d}x_0/\mathrm{d}t + F_i$$

$$\therefore \quad (m\,\mathrm{D}^2 + f\,\mathrm{D} + k)x_0 = F_i$$

and

$$\frac{x_0}{F_i} = \frac{1/m}{\mathrm{D}^2 + (f/m)\,\mathrm{D} + (k/m)} \qquad 3.2$$

This is the transfer operator for the system if the input is a force (F_i) and the output a displacement (x_0). When $\mathrm{D} = 0$, $x_0 = F_i/k$, the static deflection under a steady force. However, F_i may be a suddenly applied force, a steadily increasing force, or an oscillating force, i.e. step, ramp, or sinusoidal inputs, or, in practical cases, of a more complicated nature.

If the damping constant (f) is zero, then a step input of force or displacement will cause a simple harmonic oscillation of circular frequency (ω_n) equal to $\sqrt{(k/m)}$ and periodic time $T_p = 2\pi/\omega_n = 2\pi\sqrt{(m/k)}$, having a constant amplitude A from the equilibrium position. When damping is present, these values are altered slightly, and are then referred to as the damped circular frequency (ω_d) and the damped periodic time (T_d). With damping, the oscillations reduce in amplitude, or decay (see fig. 3.3). Experiment shows that all spring–mass oscillations decay, and hence some form of damping is always present.

If the damping constant (f) is increased sufficiently, oscillation is completely prevented after a step input, and the output will move to the new value without overshoot (see fig. 3.5). Mathematical analysis

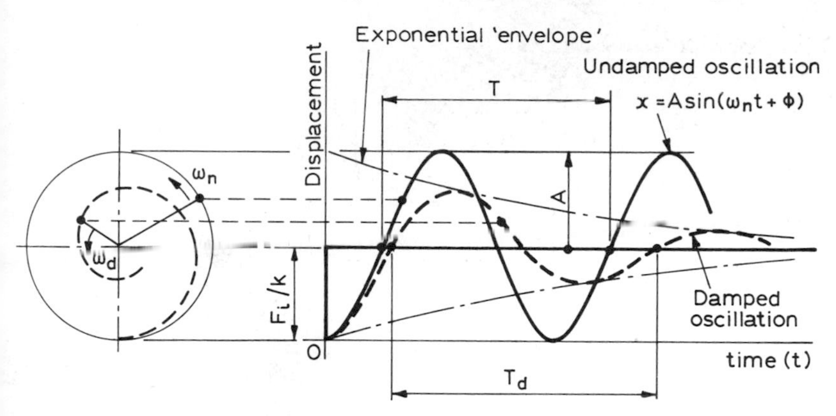

Fig. 3.3 Time response of a spring–mass system to a step input of force

(ref. 6) shows that the mass (m) and the stiffness (k) of the system also affect this, and in fact, when the damping constant is equal to $2\sqrt{(km)}$, the system just fails to overshoot,

i.e. the critical damping constant $(f_c) = 2\sqrt{(km)}$

Hence
$$\frac{f}{f_c} = \frac{f}{2\sqrt{(km)}} = \frac{f\sqrt{m}}{2m\sqrt{k}} = \frac{f}{2m\omega_n}$$

and
$$\frac{f}{m} = 2\left(\frac{f}{f_c}\right)\omega_n$$
$$= 2\zeta\omega_n$$

where $\zeta = \dfrac{f}{f_c} = \dfrac{\text{actual damping constant}}{\text{damping constant for just no overshoot}}$

ζ is the *damping ratio*, and is of importance in the design and selection of instrument systems. When ζ is less than unity, the system is *underdamped;* when $\zeta = 1$, *critically damped;* and when ζ is more than unity, *overdamped*.

Equation 3.2 may now be written in a standardised form, thus:

$$\frac{x_0}{F_i} = \frac{1/m}{D^2 + 2\zeta\omega_n D + \omega_n{}^2} \tag{3.3}$$

This may be compared with the numerical equation of a particular system, to determine the actual values of ζ and ω_n etc. (see example 3.5.1).

3.2.3 Time-constant

The time-constant (τ) of a system is a useful concept which, with the damping ratio (ζ), will indicate the nature of the dynamic response. The equations 3.1 and 3.3 (or 3.2) may be written in the following form.

Equation 3.1 becomes

$$\frac{x_0}{x_i} = \frac{1}{1 + (f/k)\,\mathrm{D}}$$

or

$$\frac{\theta_0}{\theta_i} = \frac{1}{1 + \tau\,\mathrm{D}} \qquad 3.4$$

Equation 3.3 becomes

$$\frac{x_0}{F_i} = \frac{1/m\omega_n^{\,2}}{1 + (2\zeta/\omega_n)\,\mathrm{D} + (1/\omega_n^{\,2})\,\mathrm{D}^2}$$

or

$$\frac{x_0}{F_i/k} = \frac{\theta_0}{\theta_i} = \frac{1}{1 + 2\zeta\tau\,\mathrm{D} + \tau^2\,\mathrm{D}^2} \qquad 3.5$$

The symbols θ_i and θ_0 are used to indicate input and output signals generally. They have the same units, giving dimensionless relationships. In eqns 3.4 and 3.5 they are displacements when applied to the systems of fig. 3.1 and 3.2; though for the latter, $\theta_i(= F_i/k)$ is the *static* displacement which would be produced by F_i (see example 3.5.6).

It is seen that, if equations are written in these standard forms, then τ is the coefficient of D in the first-order case, and τ^2 is the coefficient of D^2 in the second-order case. These forms of the equations are frequently used in control systems. In the case of the spring–damper system, $\tau = f/k$; and in the case of the spring–mass–damper system, τ is the reciprocal of the undamped circular frequency (ω_n), i.e. $\tau = 1/\omega_n$.

Equation 3.4 is a first-order differential equation, referred to as a *simple lag*, and eqn 3.5 is a second-order differential equation, referred to as a *complex lag*.

3.2.4 Response to step inputs

The solutions to equations such as 3.4 and 3.5 give the response of the system. Each solution consists of the *complementary function* (c.f.) part, which corresponds to the short-time or *transient* value of the output, and the *particular integral* (p.i.), corresponding to the long-time value. (It must be appreciated that short- and long-time are relative terms, and the transient may last only a part of a second in some cases.)

Figure 3.4 shows the response of a first-order system to a step change of input. The input and output signals are given the general symbols θ_i and θ_0. The solution of eqn 3.4 gives

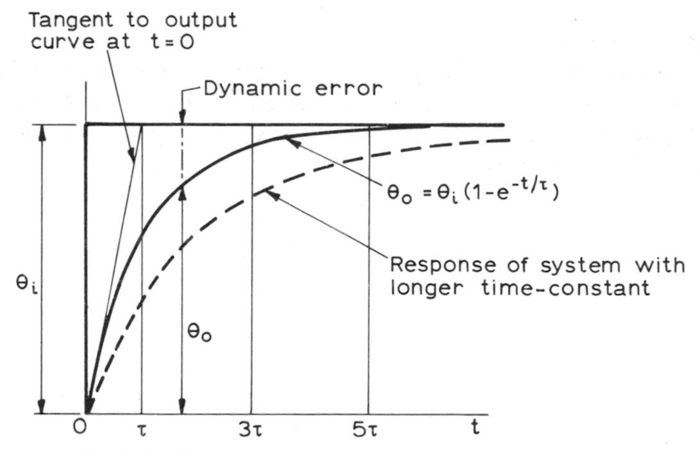

Fig. 3.4 Time response of a first-order system to a step input
When $t = 3\tau$, θ_0 is 95% of step. When $t = 5\tau$, θ_0 is 99% of step.

$$\theta_0 = \theta_i(1 - e^{-t/\tau}) = \theta_i - \theta_i e^{-t/\tau} \qquad 3.6$$

The second term on the right-hand side is time-dependent and decreases in value as t increases. When t is very large, its value is approximately zero; it is the c.f. part of the solution, giving the transient part of the response. The first term, the p.i. part, is the steady-state response, hence $\theta_0 \approx \theta_i$ when t is large. The vertical difference between the input and response curves is the *dynamic error*.

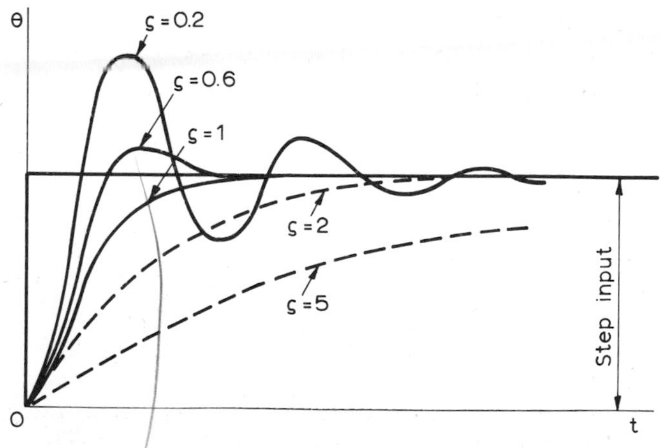

Fig. 3.5 Response of a second-order system to a step input, for different values of ζ

Figure 3.5 shows the response of a second-order system to a step input change. If the damping ratio (ζ) is less than unity, the output overshoots the step value, and the number of oscillations increases as ζ reduces. In instrument systems, ζ is usually less than unity, and for this condition the response equation is

$$\theta_0 = \theta_i - \theta_i \frac{e^{-\zeta\omega_n t}}{\sqrt{(1 - \zeta^2)}} \cos\{\sqrt{(1 - \zeta^2)}\omega_n t - \phi\} \qquad 3.7$$

Also, the damped circular frequency $(\omega_d) = \sqrt{(1 - \zeta^2)}\omega_n$; hence the equation may be written

$$\theta_0 = \theta_i - \frac{\theta_i\, e^{-\zeta\omega_n t}}{\sqrt{(1 - \zeta^2)}} \cos(\omega_d t - \phi) \qquad 3.8$$

where $\phi = \arcsin \zeta$.

The second term on the right-hand side is the transient-response (c.f.) term, giving an oscillatory output decaying within an exponential 'envelope' (fig. 3.9), and the first term is the long-time-value (p.i.) term, which is seen to be the same as the input. From eqns 3.6 and 3.8 it is seen that, after sufficient time has elapsed after a step input to a first- or second-order instrument, the output is virtually the same as the input; i.e. there is no steady-state error due to dynamic effects.

3.2.5 *Response to ramp inputs*

If the input signal varies linearly with time, e.g. having an equation $\theta_i = \Omega t$, where Ω is a constant, the output of a first-order system is as shown in fig. 3.6, and the solution of the differential equation gives

$$\theta_0 = \Omega t - \Omega\tau + \Omega\tau\, e^{-t/\tau} \qquad 3.9$$

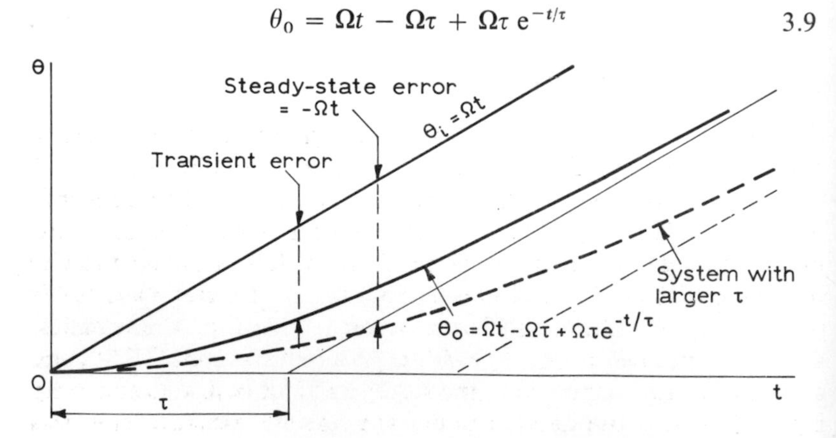

Fig. 3.6 Response of a first-order system to a ramp input

Here the third term on the right-hand side is the transient, and it is seen that the long-time output value (the p.i.) is $\Omega t - \Omega\tau$, so that the output has an error of $-\Omega\tau$. A longer time-constant increases the time taken to reach a steady state, and also gives a larger steady-state error of output.

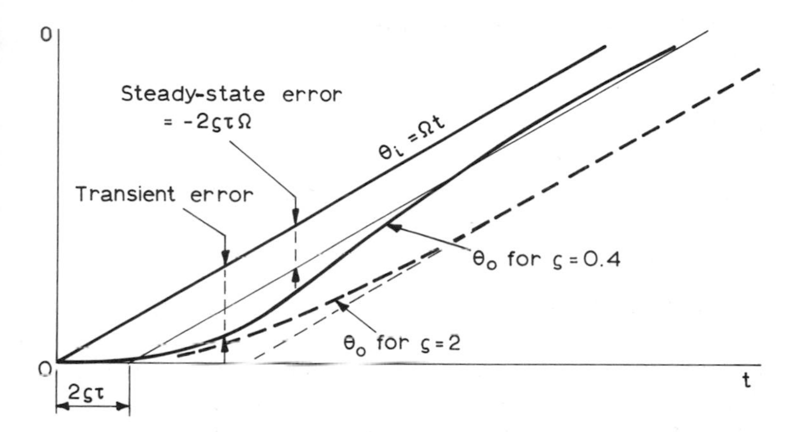

Fig. 3.7 Response of a second-order system to a ramp input

The response of a second-order system to this kind of input is shown in fig. 3.7, solution of the differential equation giving, for $\zeta < 1$,

$$\theta_0 = \Omega t - 2\zeta\Omega\tau + \Omega \frac{e^{-\zeta\omega_n\tau}}{\sqrt{(1-\zeta^2)}\omega_n} \cos\{\sqrt{(1-\zeta^2)}\omega_n t - \phi\}$$

$$= \Omega t - 2\zeta\Omega\tau + \frac{\Omega}{\omega_d} e^{-\zeta\omega_n t} \cos(\omega_d t - \phi) \qquad 3.10$$

where $\phi = \arcsin(2\zeta^2 - 1)$.

(Equations 3.8 and 3.10 for second-order systems may be written in several different forms which are equivalent to those given.)

Equation 3.10 shows that the steady-state error is $-2\zeta\Omega\tau$; it depends on the damping ratio (ζ) and the time-constant (τ), as well as on the rate of input (Ω). For a given value of Ω, the values of ζ and τ must be minimum for least error. However, from fig. 3.5 it is seen that, as ζ is reduced, the amount of oscillation becomes excessive. Mathematical analysis shows that a value $\zeta = 0{\cdot}7$ gives a first overshoot of 5% and a relatively rapid approach to the steady state for both step and ramp inputs. For this and another important reason, discussed later, this value of ζ is often adopted in instruments.

3.2.6 Determination of damping ratio and time-constant

The values of ζ and τ for an instrument may be found by observing its output signal corresponding to a step or ramp input of known magnitude. The accuracy of these values found will depend on how accurately the output can be observed, and will be better if a continuous output trace is available. For first-order systems the time taken to reach 95 or 99% of the step input value will be 3 and 5 time-constants respectively, and the rate of change of the output signal initially is θ_i/τ (see fig. 3.4).

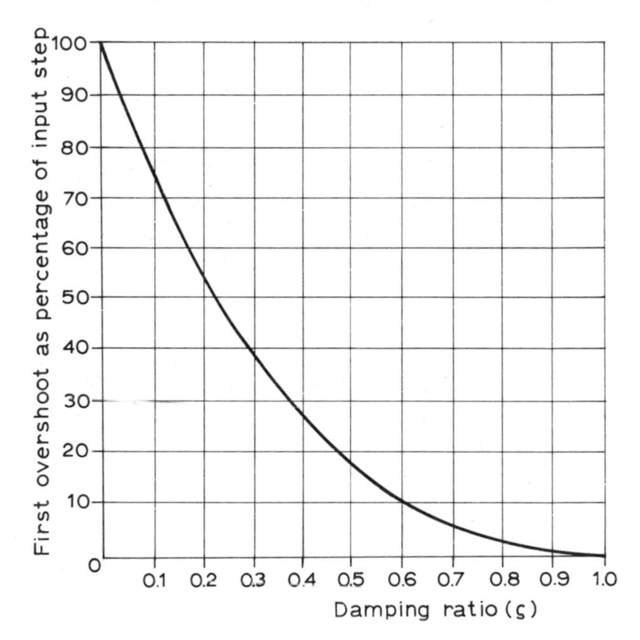

Fig. 3.8

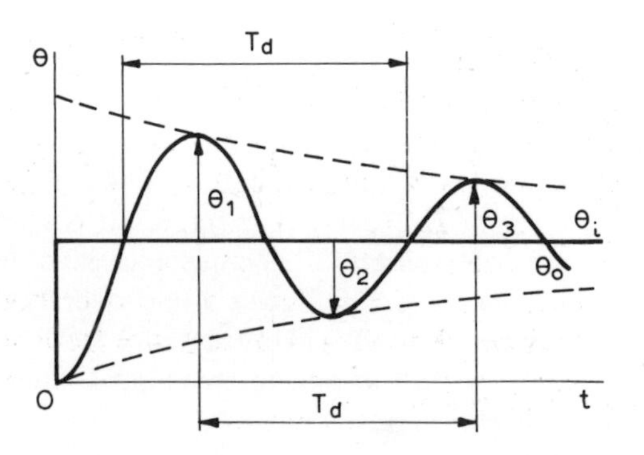

Fig. 3.9

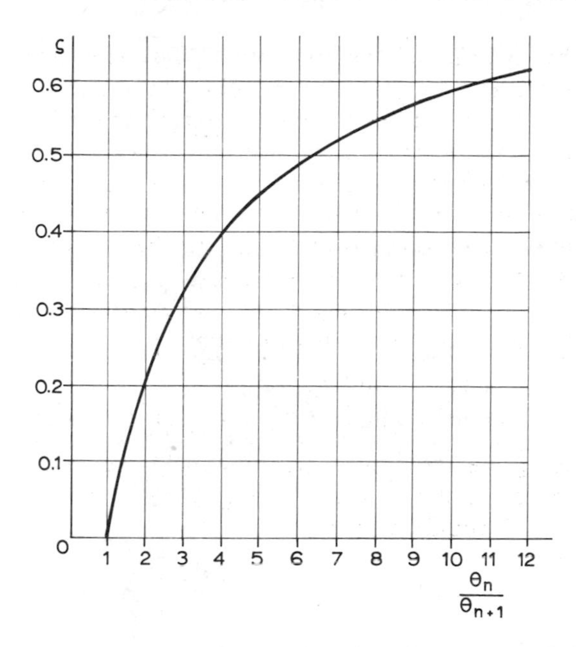

Fig. 3.10

In the case of an underdamped second-order system, if the first overshoot value is observed, the value of ζ may be found from a curve such as fig. 3.8. If subsequent peak values θ_1, θ_2, etc. from the step value can be measured, as in fig. 3.9, then the ratio of two successive values θ_n/θ_{n+1} may be used to find the value of ζ from a curve such as that in fig. 3.10. If the damped periodic time (T_d) can be measured, then τ may be found. Since $\omega_d = \omega_n\sqrt{(1 - \zeta^2)}$, $T_d = 2\pi/\omega_d$, and $\tau = 1/\omega_n$, then $\tau = T_d\sqrt{(1 - \zeta^2)}/2\pi$.

3.2.7 *Frequency response*

For instruments where the input may be oscillatory, it is necessary to know the effect of this on the accuracy of the output. The standard input is taken to be sinusoidal. (In fact any oscillatory signal of repetitive form may be expressed as the sum of sinusoidal signals, each of different amplitude and frequency.)

Considering the damped spring of fig. 3.1, if the point R were made to oscillate such that $x_i = A_i \cos pt$, where A_i is a constant amplitude and p a constant circular frequency, then eqn 3.1 would become

$$\frac{x_0}{A_i \cos pt} = \frac{k/f}{\mathrm{D} + k/f} = \frac{1}{1 + \tau \mathrm{D}}$$

and $(\tau \mathrm{D} + 1)x_0 = A_i \cos pt$.

The solution of this differential equation is

$$x_0 = C\,e^{-\zeta\omega_n t} + \frac{A_i}{\sqrt{\{1 + (p\tau)^2\}}}\cos(pt - \alpha) \qquad 3.11$$

where $\alpha = \arctan(p\tau)$.

The first term (the c.f.) is transient, and has a constant C which depends on the starting conditions of the oscillation. The second term (the p.i.) shows that the steady-state output oscillates at the same frequency as the input, but lags behind it by the phase-angle α. The main interest is in the amplitude ratio, thus

$$\frac{\text{output amplitude}}{\text{input amplitude}} = \frac{A_0}{A_i} = \frac{1}{\sqrt{\{1 + (p\tau)^2\}}} \qquad 3.12$$

Considering the second-order spring–mass–damper system of fig. 3.2, if a sinusoidal input force $F_i = F_{max}\cos pt$ were applied at point Q, then eqn 3.2 would become

$$\frac{x_0}{F_{max}\cos pt} = \frac{1/m}{D^2 + (f/m)\,D + k/m}$$

F_{max} may be converted to an equivalent static displacement A_i, since $F_{max} = kA_i$, and the equation may be written

$$\frac{x_0}{kA_i\cos pt} = \frac{1/m}{D^2 + (f/m)\,D + k/m}$$

or $$(D^2 + 2\zeta\omega_n D + \omega_n^2)x_0 = \omega_n^2 A_i \cos pt$$

The solution of this equation, for $\zeta < 1$, is

$$x_0 = Ce^{-\zeta\omega_n t}\cos(\omega_d t - \phi) + \frac{A_i}{\sqrt{[\{1 - (p/\omega_n)^2\}^2 + (2\zeta p/\omega_n)^2]}}\cos(pt - \alpha) \qquad 3.13$$

The first term, the transient, is not of interest, and again the second term gives a steady-state output of the same frequency as the input, but lagging by the phase-angle α. Writing, for convenience, $p/\omega_n = \mu$, the amplitude ratio is given by

$$\frac{A_0}{A_i} = \frac{1}{\sqrt{\{(1 - \mu^2)^2 + (2\zeta\mu)^2\}}} \qquad 3.14$$

and the phase-angle $\alpha = \arctan\{2\zeta\mu/(1 - \mu^2)\}$.

Figure 3.11(a) shows the input and output displacements plotted against time, and fig. 3.11(b) shows the amplitude ratio against the

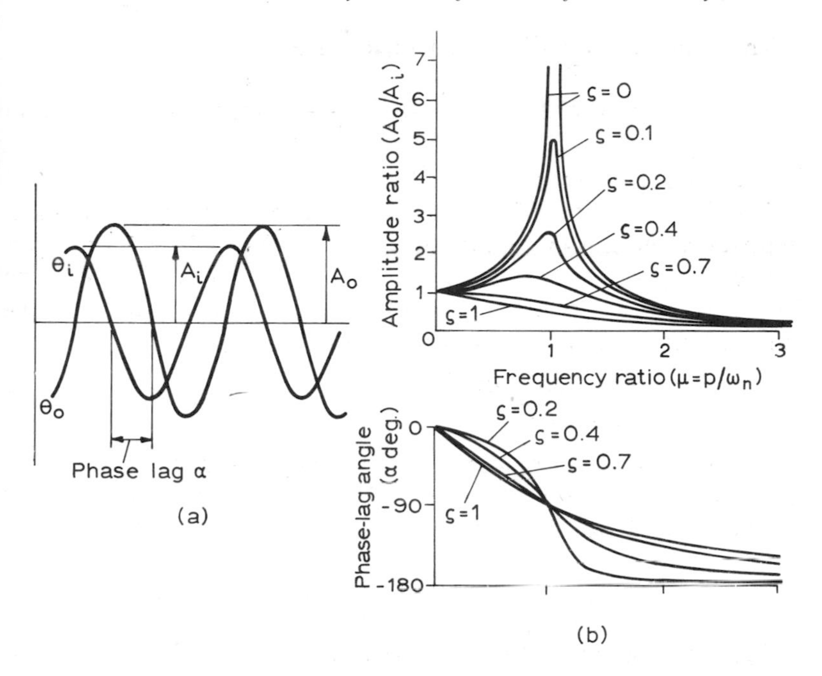

Fig. 3.11 Frequency response characteristics – second-order system: (a) Response to sinusoidal input (b) Variation of amplitude ratio and phase-angle for various μ and ζ values

frequency ratio for different values of the damping ratio (ζ), both for the second-order system. For values of $\zeta > 0{\cdot}707$, it may be shown that the amplitude ratio never exceeds unity. For values of ζ less than 0·707, the amplitude increases as $\mu \approx 1$; and for $\zeta = 0$ and $\mu = 1$, the amplitude would be infinite, except that, in spring systems, elastic ranges would be exceeded and energy dissipated in plastic deformations.

3.3 Dynamic performance of other systems

The mechanical system has been used to illustrate the dynamic response characteristics, but the equations derived are equally applicable to any other physical system which results in a first- or second-order differential equation of the type shown. Thus eqns 3.4 and 3.5, which are written in dimensionless form, are equally applicable to electrical, pneumatic, and heat systems, etc., and also to mixtures of these.

The resistance–capacitance network of fig. 3.12 may have an input signal $\theta_i = v_1$; and v_2 across the capacitance (C) may be the output θ_o. The current (i) through the resistance (R) is $(\theta_i - \theta_o)/R$. The rate of change $d\theta_o/dt = i/C = (\theta_i - \theta_o)/RC$.

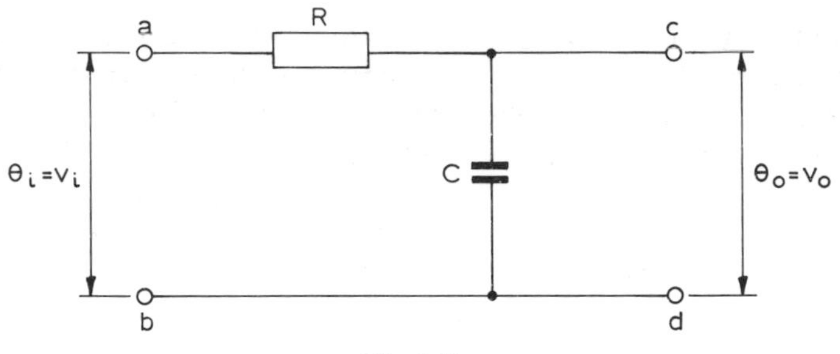

Fig. 3.12

Hence $$RC\,\mathrm{D}\theta_0 = \theta_i - \theta_0$$

and $$\frac{\theta_0}{\theta_i} = \frac{1}{1 + RC\,\mathrm{D}} = \frac{1}{1 + \tau\,\mathrm{D}}$$

This corresponds to eqn 3.4, and the time-constant (τ) of the network is seen to be *RC*.

It may be shown that the addition of an inductance (*L*) to the resistance–capacitance system can give a second-order equation corresponding to eqn 3.5.

Figure 3.13 shows a manometer tube with uniform bore of cross-sectional area *a* except at Q, where a restriction exerts a drag force on the liquid mass, proportional to its velocity, the drag elsewhere being

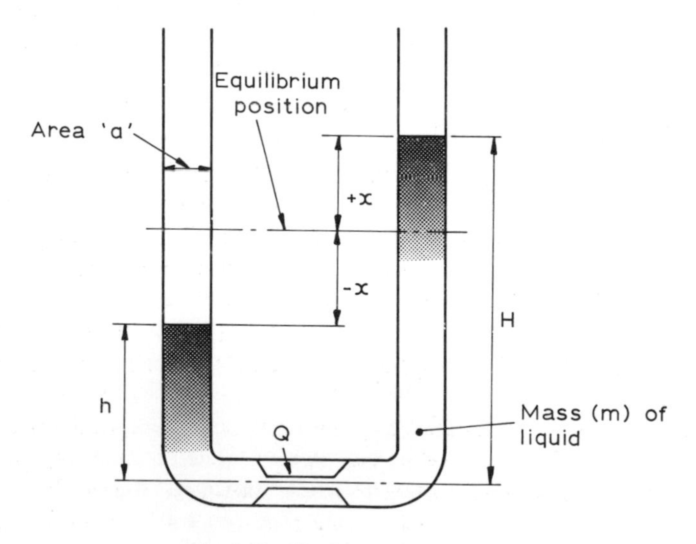

Fig. 3.13 U-tube manometer

negligible. If the total volume of the liquid is V and its density ρ, then the force exerted on it due to the difference in head is $\rho ag(H - h) = 2\rho agx$. The mass is ρV, and the drag force at Q is $f\,\mathrm{D}x$.

Net force applied = mass × acceleration

$$\therefore \quad f\,\mathrm{D}x + 2\rho agx = -\rho V\,\mathrm{D}^2x$$

and
$$(\rho V\,\mathrm{D}^2 + f\,\mathrm{D} + 2\rho ag)x = 0$$

or
$$\{\mathrm{D}^2 + (f/\rho V)\,\mathrm{D} + (2ag/V)\}x = 0$$

Comparing this with eqn 3.3,

$$\omega_n = \sqrt{(2ag/V)} \qquad \text{and} \qquad 2\zeta\omega_n = f/\rho V.$$

Hence a simple RC network has been shown to give a first-order system, and a simple hydraulic device a second-order system, both equations being of the same form as those of spring, mass, and damper combinations. Many other instrument elements give similar equations, and the responses to the standard inputs are given by eqns 3.6, 3.9, and 3.11 for first-order systems, and by eqns 3.8, 3.10, and 3.13 for second-order systems, for step, ramp, and sinusoidal inputs respectively.

3.4 Relationship between sensitivity, time-constant, and damping ratio

The requirements of an instrument to have a high sensitivity, quick response, small lag, and good frequency response are not always compatible. The sensitivity may be a fixed requirement, and then the linear or angular damping constant (f) and the mass (m) or rotational inertia (I) may be varied to suit the remaining requirements, by a compromise if necessary.

3.4.1 First-order systems

For the spring–damper system of fig. 3.1, eqn 3.4 shows that, if the stiffness k is fixed by the sensitivity requirements of the system, then only f can be varied to give a suitable value of τ. Referring to eqn 3.6 and fig. 3.4, it is seen that a low τ value is required for a rapid response to a step input. Equation 3.9 and fig. 3.6 show that the time-lag and the steady-state error increase with τ. Equation 3.12 shows that, for forced oscillation, the amplitude ratio decreases as τ increases. Hence τ should be low.

In resistance–capacitance systems, R may correspond to stiffness, and C may be altered to obtain a suitable τ value. In heat systems, τ may depend on such quantities as mass, specific heat capacity, thermal conductivity, area, etc., and it may not be easy to obtain a desired value.

3.4.2 *Second-order systems*

For a spring–mass–damper system as in fig. 3.2, if k is fixed to give a required sensitivity, since $\omega_n = \sqrt{(k/m)}$ then ω_n $(= 1/\tau)$ can be varied only by altering the mass of the system, which may be difficult. With k, m, and hence ω_n, fixed, the damping constant (f) may be varied to give a suitable value of the damping ratio (ζ), since $\zeta = f/2\sqrt{(km)}$. As mentioned in section 3.2.7, resonance is avoided if $\zeta > 0{\cdot}707$. Since this value also gives a fairly rapid response to a step input, it is often adopted in instruments where the inputs can be oscillatory.

To reduce steady-state errors arising from ramp inputs, ζ and τ should be small; though if ζ is too small, the settling time will be long.

Similar adjustments may be made to the variables in electrical, pneumatic, and heat systems, etc., to give suitable values of ζ and τ; but in many instruments the optimum of one value will not allow the optimum of the other, and a compromise may be necessary, or the range of use of the instrument may be restricted.

3.5 Worked examples

3.5.1 A linear instrument system as in fig. 3.2 has a mass of 6·0 grams and a stiffness of 1·0N/mm. Calculate the natural circular frequency (ω_n), and find the damping constant (f) necessary to just prevent overshoot when F_i is a step input of force.

$$\text{Natural circular frequency } (\omega_n) = \sqrt{(k/m)}$$
$$= \sqrt{\{(1{\cdot}0 \times 10^3)/(6{\cdot}0 \times 10^{-3})\}}$$
$$= 410 \text{ rad/s}$$

For just no overshoot, $\zeta = 1$. Therefore, comparing eqns 3.2 and 3.3,

$$f/m = 2\zeta\omega_n$$

hence

$$f = 2m\zeta\omega_n$$
$$f_c = 2 \times 6{\cdot}0 \times 10^{-3} \times 1 \times 410$$
$$= 4{\cdot}9 \text{ N s/m}$$

Alternatively,

$$f_c = 2\sqrt{(km)}$$
$$= 2\sqrt{(1{\cdot}0 \times 10^3 \times 6{\cdot}0 \times 10^{-3})}$$
$$= 4{\cdot}9 \text{ N s/m}$$

3.5.2 A d'Arsonval galvanometer movement such as illustrated in fig. 2.37(a) has a stiffness (k) such that 40·0 μNm torque gives a full-scale deflection of 90°. If the rotational inertia (I) is $0{\cdot}50 \times 10^{-6}$ kg m^2 and the damping constant (f) is 5·00 μN ms/rad, determine the damping ratio (ζ), the natural circular frequency (ω_n), the damped natural circular frequency (ω_d), and the time-constant (τ) of the movement.

$$\zeta = f/2\sqrt{(kI)}$$
$$= 5{\cdot}00 \times 10^{-6}/2\sqrt{\{40{\cdot}0 \times 10^{-6} \times (2/\pi) \times 0{\cdot}50 \times 10^{-6}\}}$$
$$= 0{\cdot}70$$

$$\omega_n = \sqrt{(k/I)}$$
$$= \sqrt{\{40{\cdot}0 \times 10^{-6} \times 2/(\pi \times 0{\cdot}50 \times 10^{-6})\}}$$
$$= 7{\cdot}15 \text{ rad/s}$$

$$\omega_d = \omega_n\sqrt{(1 - \zeta^2)}$$
$$= 7{\cdot}15\sqrt{(1 - 0{\cdot}70^2)}$$
$$= 5{\cdot}0 \text{ rad/s}$$

$$\tau = 1/\omega_n$$
$$= 1/7{\cdot}15$$
$$= 0{\cdot}14 \text{ s}$$

3.5.3 For the galvanometer movement of example 3.5.2, calculate the steady-state dynamic error and the time-lag when the input voltage is increasing steadily at the rate of 5 V/s.

Let the θ values be in volts, then $\theta_i = \Omega t$, where $\Omega = 5$ V/s.

From eqn 3.10, the steady-state error when the transient oscillation has died away is

$$\text{error} = -2\zeta\tau\Omega$$
$$= -2 \times 0{\cdot}70 \times 0{\cdot}14 \times 5$$
$$= -1{\cdot}0 \text{ V}$$

The time-lag is $2\zeta\tau = 2 \times 0{\cdot}70 \times 0{\cdot}14$ (see fig. 3.7)

$$= 0{\cdot}2 \text{ s}$$

3.5.4 When a step input was applied to a pen recorder similar to that shown in fig. 2.37(c), the values of the first two peaks θ_1 and θ_2 (see fig. 3.9) were 10·5 units and 1·0 unit respectively from the final value. The damped periodic time T_d was found to be 0·2 seconds. Determine (a) the damping ratio of the instrument, (b) the time-constant, and (c) the value of the steady-state lag which may be expected when a steadily increasing input of 2 units per second is applied.

a) $$\frac{\theta_n}{\theta_{n+1}} = \frac{10{\cdot}5}{1{\cdot}0} = 10{\cdot}5$$

From fig. 3·10, the value of the damping ratio (ζ) is 0·6

b) From section 3.2.6, the time-constanι is given by

$$\begin{aligned}\tau &= T_d\sqrt{(1 - \zeta^2)}/2\pi \\ &= 0{\cdot}2\sqrt{(1 - 0{\cdot}6^2)}/2\pi \\ &= 0{\cdot}025 \text{ seconds}\end{aligned}$$

c) From eqn 3.10, the steady-state error is $-2\zeta\Omega\tau$, and Ω is 2 units per second

$$\begin{aligned}\therefore \quad \text{steady-state error} &= 2 \times 0{\cdot}6 \times 0{\cdot}025 \\ &= -0{\cdot}06 \text{ units}\end{aligned}$$

3.5.5 In a test, the response of a thermocouple system was found to be a simple lag with a time-constant of 10 seconds. If the system is used to measure the temperature of a furnace which is fluctuating approximately sinusoidally between the temperatures 540 °C and 500 °C, with a periodic time of 80 seconds, determine the approximate maximum and minimum values the thermocouple will indicate, and find the phase-angle and corresponding time-lag between the input and output signals.

$$\begin{aligned}\text{The forcing frequency } (p) &= 2\pi/80 \\ &= 0{\cdot}078 \text{ rad/s}\end{aligned}$$

From eqn 3.12, when transient effects have died away, the amplitude ratio is

$$\begin{aligned}\frac{A_0}{A_i} &= \frac{1}{\sqrt{\{1 + (p\tau)^2\}}} = \frac{1}{\sqrt{\{1 + (0{\cdot}078 \times 10)^2\}}} \\ &= 0{\cdot}79 \\ A_0 &= 40 \times 0{\cdot}79 \\ &\approx 32\,^\circ\text{C}\end{aligned}$$

Hence $\theta_{max} = 536\,^\circ\text{C}$ and $\theta_{min} = 504\,^\circ\text{C}$.

From eqn 3.11, the phase-angle (α) = arctan($p\tau$),

$$\begin{aligned}\therefore \quad \alpha &= \arctan(0{\cdot}078 \times 10) \\ &\approx 38^\circ \\ &\approx 0{\cdot}66 \text{ rad}\end{aligned}$$

The corresponding time-lag is equal to

$$\alpha/p = 0{\cdot}66/0{\cdot}078$$
$$\approx 8{\cdot}5 \text{ seconds}$$

The input and output signals against time are shown in fig. 3.14. It is seen that at two points in every cycle there is no dynamic error.

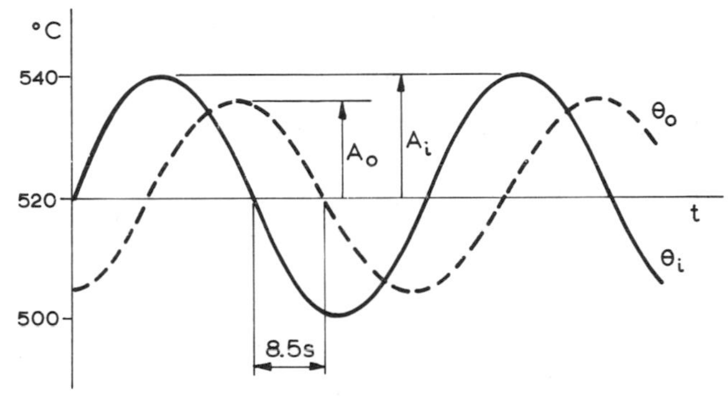

Fig. 3.14

3.5.6 The core of a solenoid is an iron mass of 0·5 kg, supported on a coil spring of stiffness 18 kN/m and immersed in oil such that the damping ratio for linear vibrations is 0·707. Determine the percentage overshoot, the actual overshoot, and the final deflection of the mass when a sudden force of 50 N is applied electrically.

Static (final) deflection $= F_i/k$
$= 50/18$
$= 2{\cdot}78$ mm

By eqn. 3.5, the equation of motion of the system is

$$\frac{x_0}{F_i/k} = \frac{x_0}{2{\cdot}78} = \frac{1}{1 + 2\zeta\tau\,\mathrm{D} + \tau^2\,\mathrm{D}^2}$$

The solution of this is given by eqn 3.8, but the information required may be found from fig. 3.8, from which it is seen that the first overshoot percentage, for $\zeta = 0{\cdot}7$, is 5%.

The actual overshoot is 5% of $2{\cdot}78 = 0{\cdot}14$ mm.

3.5.7 In the design of a d'Arsonval galvanometer movement, it is found that the damping ratio is 0·65 and the natural frequency of undamped oscillation is 4 Hz.

a) If the sensitivity of the movement is now doubled by fitting lower-rate restoring springs, calculate the new damping ratio and the new natural frequency.

b) If the damping ratio is now restored to its original value by alteration of the moment of inertia of the system, determine the final natural frequency of the system.

a) Since $\zeta_1 = f/2\sqrt{(k_1 I_1)}$, and $k_2 = k_1/2$,

$$\zeta_2 = f/2\sqrt{(k_1 I_1/2)}$$

Hence
$$\frac{\zeta_2}{\zeta_1} = \sqrt{\left(\frac{2k_1 I_1}{k_1 I_1}\right)}$$

and
$$\zeta_2 = \zeta_1 \times \sqrt{2}$$
$$= 0{\cdot}65 \times 1{\cdot}414$$
$$= 0{\cdot}92$$

Let F be the natural frequency of undamped oscillation of the system, in hertz,

then
$$F_1 = \frac{\omega_n}{2\pi} = \frac{\sqrt{(k_1/I_1)}}{2\pi}$$

and
$$F_2 = \frac{\sqrt{(k_2 I_2)}}{2\pi}$$

Hence
$$F_2 = F_1 \times \sqrt{\left(\frac{k_1/2I_1}{k_1/I_1}\right)}$$
$$= F_1 \times 1/\sqrt{2}$$
$$= 4 \times 0{\cdot}707$$
$$= 2{\cdot}83 \text{ Hz}$$

b)
$$\zeta_2 = f/2\sqrt{(k_2 I_2)}$$
$$\zeta_1 = \zeta_3 = f/2\sqrt{(k_2 I_3)}$$
$$\therefore \quad \frac{\zeta_2}{\zeta_1} = \sqrt{\left(\frac{I_3}{I_2}\right)}$$

and
$$I_3 = I_2 \times \left(\frac{\zeta_2}{\zeta_1}\right)^2 = I_2 \times \left(\frac{0{\cdot}92}{0{\cdot}65}\right)^2$$
$$\therefore \quad I_3 = 2I_2 \; (= 2I_1)$$

$$F_3 = \sqrt{(k_2/I_3)}/2\pi$$
$$F_1 = \sqrt{(k_1/I_1)}/2\pi$$
$$\therefore \quad F_3 = F_1\sqrt{\left(\frac{k_2 I_1}{k_1 I_3}\right)} = F_1\sqrt{\left(\frac{k_1 I_1}{2k_1 2I_1}\right)}$$
$$= F_1/2 = 4/2$$
$$= 2 \text{ Hz}$$

3.6 Tutorial and practical work

3.6.1 Show that, in torsional spring–mass–damper systems, if k is the stiffness in N/rad, f the damping constant in Nms/rad, and I the rotational inertia in kgm^2, then equations 3.2 and 3.3 apply if m is replaced by I.

3.6.2 Distinguish between damping constant and damping ratio. Discuss the reasons for preferring to use the latter. Why do many instruments have a damping ratio of about 0·6 to 0·7?

3.6.3 Distinguish between (a) frequency and circular frequency; (b) undamped natural frequency, damped natural frequency, and resonant frequency; (c) applied frequency and frequency ratio.

3.6.4 If the product km (or kI) is maintained constant for an instrument movement, and the ratio k/m (or k/I) is varied, discuss what effect this has on the time-response of the system.

3.6.5 Discuss the difficulties that might arise in measuring the instantaneous values of a rapidly fluctuating temperature such as that of the face of a piston in an internal-combustion engine.

3.6.6 From the literature relating to measuring systems available to you, determine their natural frequency, range, and damping ratio, and compare their dynamic performance.

3.6.7 Show that, for a simple mercury-in-glass thermometer, if m is the mass of mercury in the bulb, c its specific heat capacity, t the thickness of the glass bulb, A its surface area, and λ its thermal conductivity, then, if the heat capacity of the glass is neglected, the equation relating input to output is a simple lag with a time-constant (τ) = $mc/\lambda At$.

Discuss what steps could be taken in the design of a thermometer or pyrometer so that the time-constant is a minimum. State if any of these steps would lead to loss of sensitivity, accuracy, or reliability.

3.6.8 Sketch the values of the amplitude ratio (A_0/A_i) and the phase-angle (α) against values of $p\tau$ from 0 to 8 when a sinusoidal input of circular frequency p is applied to a first-order system.

3.6.9 Determine the damping ratio and the damped periodic time of a U-tube manometer by applying a step input of pressure and obtaining successive overshoot values and times. Determine the effects of (a) increasing the quantity of fluid, (b) using different fluids such as mercury, water, paraffin, etc.

Discuss the results and state your conclusions.

3.6.10 Determine the response characteristics of (a) a variety of temperature-measuring instruments, (b) a variety of d'Arsonval galvanometers.

3.6.11 Connect a resistance and a capacitance as shown in fig. 3.12, and apply a step input of voltage. Measure the output voltage with a high-impedance voltmeter, and compare the observed time-constant with that calculated from $\tau = RC$.

3.6.12 Why is it usually desirable to have a high natural frequency in an instrument system? If a system having $\zeta = 0{\cdot}6$ is to have maximum sensitivity in response to a sinusoidal input, at about what value of the frequency ratio (μ) would this occur?

3.7 Exercises

3.7.1 When a step-voltage of 20 V is applied at the terminals AB in the resistance–capacitance network of fig. 3.12, the voltage over *cd* is found to rise from 0 to 8 V in 6 seconds. Determine the time-constant of the system and the value of the capacitance (C) if the resistance (R) has a value 1 kΩ. [$\tau = 4{\cdot}33$ s, $C = 4{\cdot}33$ mF]

3.7.2 In a response test on a thermometer, it was thrust into a temperature-controlled bath of water maintained at 100 °C, and the time was observed as the indicated temperature reached preselected values, giving the following readings:

time (seconds)	0·0	1·2	3·0	5·6	8·0	11·0	15·0	18·0
temperature (°C)	20	40	60	80	90	95	98	99

Draw the response curve on graph paper, and show that it follows closely the form of a simple lag with a time-constant of 4 seconds.

Determine the steady-state error when the same thermometer is measuring the temperature of a liquid which is cooling at a constant rate of one degree every five seconds. [+0·8°C]

3.7.3 In a production-line weighing device illustrated in fig. 3.15, masses having a minimum value of 0·9 kg and a maximum of 1·1 kg are to be weighed on a platform having a mass of 0·2 gram.

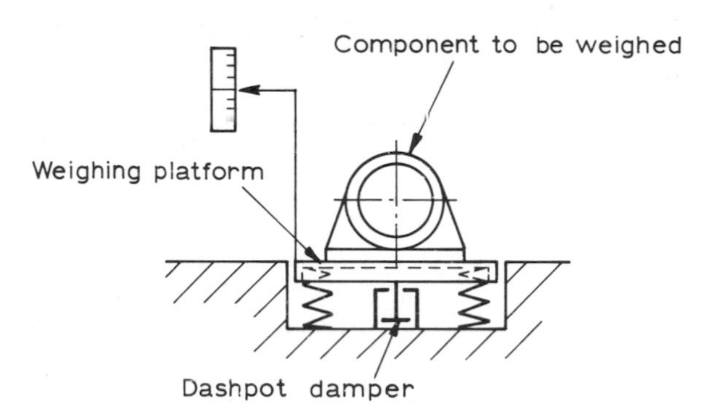

Fig. 3.15

a) If the scale deflection is to be ± 10 mm over the range, calculate the necessary spring stiffness (k).
b) If the damping ratio is to be 0·7 when the nominal (1 kg) mass is on the table, calculate the damping constant (f) required in the dashpot.
c) Calculate the damping ratio with (i) the minimum mass on the table, (ii) the maximum mass on the table.
d) Determine the percentage overshoot with (i) the minimum, and (ii) the maximum, masses when they are placed suddenly (not dropped) on the table.
e) Determine the changed values in (a), (b), (c), and (d) if the springs are changed to give a deflection of ± 6 mm over the same range of mass.

[0·0981 N/mm, 15·2 N s/m^2, 0·73, 0·67, 6%, 4%,
0·1635 N/mm, 19·6 N s/m^2, 0·74, 0·68, 6%, 4%]

3.7.4 Observation of a U-tube manometer shows that, when a step input (x in fig. 3.13) of 40 mm is applied, the first, second, and third overshoot values are 20 mm, 10 mm, and 5 mm respectively, and the observed periodic time of the oscillations is 1·2 s.
a) Determine the approximate values of the damping ratio and the time-constant, and estimate the dynamic error of pressure head and time when the pressure difference over the manometer is decreasing at the rate of 5 mm per second.
b) Determine the approximate values of ζ, τ, and the dynamic error and lag if the volume of fluid in the U-tube is halved.

[$\zeta_1 \approx 0{\cdot}25$, $\tau_1 \approx 0{\cdot}19$ s, error$_1 \approx +0{\cdot}5$ mm, time-lag $\approx 0{\cdot}1$ s;
$\zeta_2 \approx 0{\cdot}35$, $\tau_2 \approx 0{\cdot}13$ s, error and time-lag unchanged]

3.7.5 In a pressure-measuring transducer consisting of a number of strain-gauges bonded to a diaphragm (see Chapter 5), the undamped natural frequency of the system was found to be 60 kHz, and the damping ratio 0·6. Calculate the amplitude of the output signal in terms of pressure, if the input pressure to the transducer is fluctuating sinusoidally with an amplitude of 800 kN/m^2 at a frequency of 30 kHz. What is the percentage error in the output amplitude?

[832 kN/m^2, +4%]

3.7.6 Plot the amplitude ratio for a second-order system for values of the frequency ratio (μ) from 0 to 1, for a damping ratio (ζ) equal to 0·6, taking values at suitable intervals. From the graph, determine the range of μ for which the output amplitude is within 3% of the input amplitude. If the system has an undamped natural frequency of 3 kHz, determine the frequency range over which this applies.

[3% accuracy is maintained up to 2·5 kHz input frequency.]

3.7.7 Show that, when a step input is applied to a first-order system, the output reaches 0·63 of the step value after time τ.

4

Potentiometer and Bridge Circuits

4.1 Basic laws for networks

A series/parallel arrangement of resistances, capacitances, and inductances is referred to as a 'network'. If a source of electromotive force (e.m.f.) is contained in a network, the network is referred to as 'active'; if not, it is 'passive'.

Ohm's law applies to a complete circuit thus:

$$E = I_T R_T$$

where E is the e.m.f. (volts), I_T is the total current (amperes), and R_T is the total resistance of the circuit (ohms); or to an element of a circuit:

$$V = IR$$

where V is the potential difference (p.d.) over the element (volts), I is the current through the element (amperes), and R is the resistance of the element (ohms).

To determine the current, p.d., or resistance values in a network, the following laws are used.

4.1.1 Kirchhoff's laws

a) The algebraic sum of the currents flowing at a junction is zero [fig. 4.1(a)].

b) In a closed loop of a circuit, the algebraic sum of the products of the current and resistance of each part of the loop equals the resultant e.m.f. in that loop. Symbolically, $\sum E = \sum IR$ *in a loop of a circuit*.

Thus in fig. 4.1(b), taking positive values clockwise, the equation for the loop ABR_2R_1A is

$$E_1 = I_1R_1 + (I_1 - I_3)R_2 \qquad 4.1$$

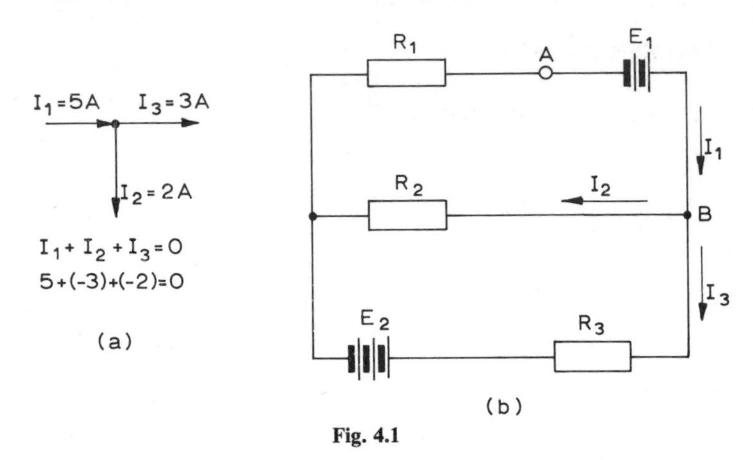

Fig. 4.1

For the circuit ABR_3R_1A, the equation is

$$E_1 - E_2 = I_1R_1 + I_3R_3 \qquad 4.2$$

Equations 4.1 and 4.2 may be solved simultaneously for two unknown values appearing in both (see example 4.5.1).

4.1.2 Thévenin's theorem

An active network having two terminals A and B behaves, as far as any load connected across those terminals is concerned, as if the network contained a single source of e.m.f. (E) with an associated resistance (R), where E is equal to the terminal voltage of the network with the load disconnected, and R is the resistance of the network.

Hence the active network of fig. 4.2(a) will behave as the simplified network of fig. 4.2(b), where R_L is the load resistance in each case. E will be the terminal voltage V_{AB} and R will be the resistance across AB, in both cases for the original network, without the load (see example 4.5.2).

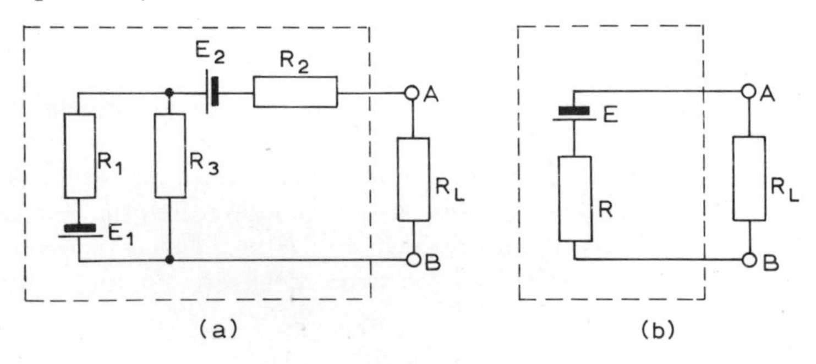

Fig. 4.2

4.2 Potentiometer circuits

4.2.1 Voltage-dividing potentiometer

The p.d. across the length of wire AB in fig. 4.3(a) may be divided in the ratio $V_{AC}:V_{AB}$ by a length division $x:l$ in the same ratio, if the wire is straight, of uniform diameter, uniform resistivity, and is at a uniform temperature; i.e. $V_{AC}/V_{AB} = x/l$. A similar relationship will hold for the wire wound into a coil, as in fig. 4.3(b), if the length of each turn is identical and the axial spacing of the turns is uniform.

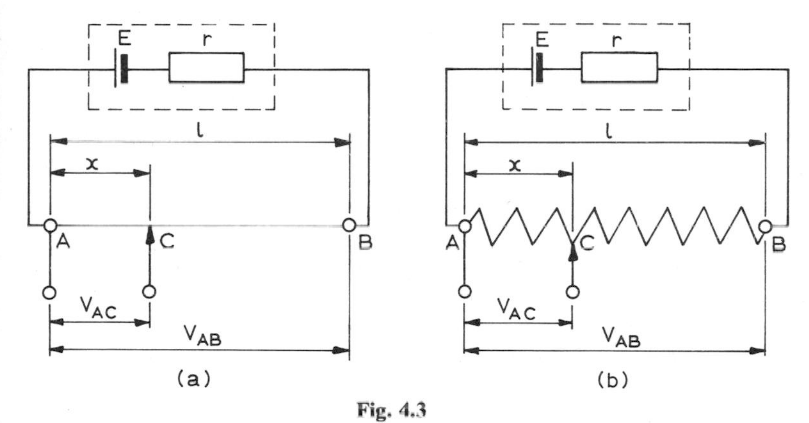

Fig. 4.3

The output p.d. (V_{AC}) may be used to measure axial displacement x (see Chapters 2 and 10), or an accurately determined fraction of V_{AB} may be obtained by accurate division of length. The method may be adapted to the measurement of angular displacement, or to voltage-division by accurately measured angular displacement, as indicated in fig. 4.4(a) and (b), using a circular wire and a toroidal wire respectively. No current should be taken from point C in these arrangements, and an instrument of very high input resistance should be used to measure V_{AC}, otherwise excessive errors will occur (see example 4.5.3).

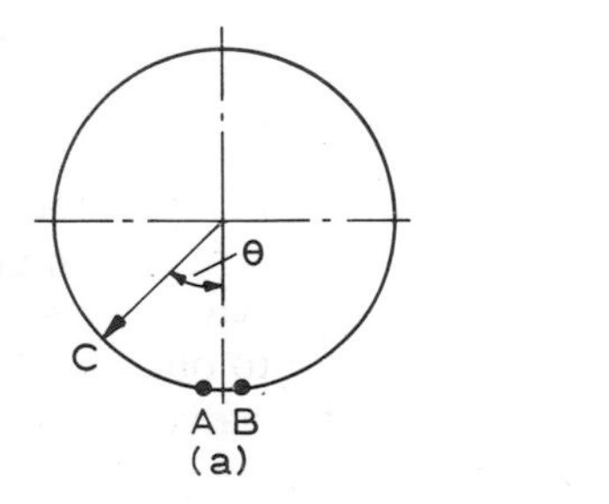

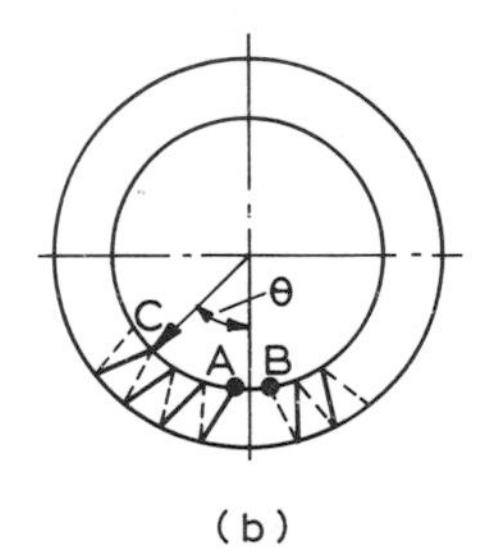

Fig. 4.4

4.2.2 *Voltage-balancing potentiometer*

This circuit is used to measure a p.d. by balancing it against a p.d. obtained from a part of the resistance wire (AC). Balance is indicated by a null (zero) reading on the galvanometer G when the p.d. to be measured (V) is equal and opposite to the divided p.d. (V_{AC}). When balance is obtained, then $V = (x/l) \times V_{AB}$ [fig. 4.5(a)].

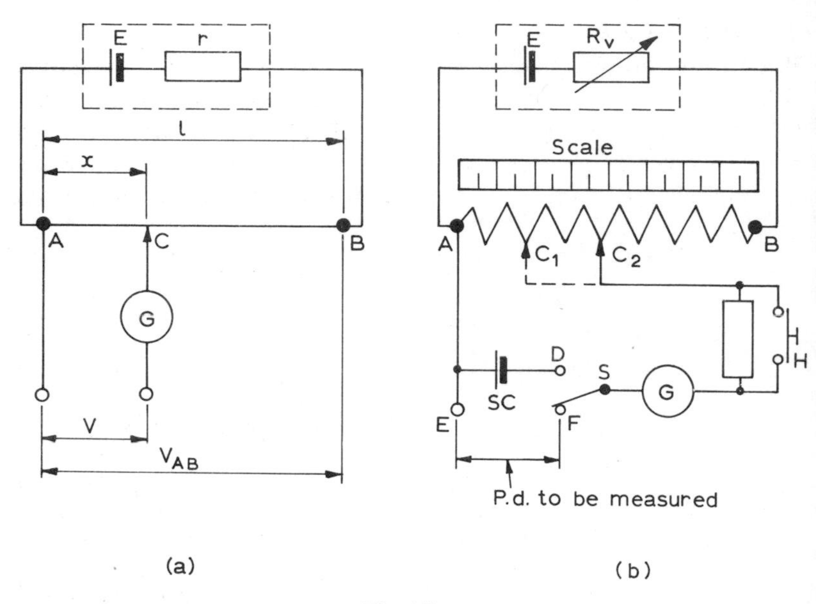

Fig. 4.5

The difficulty is in knowing the value of V_{AB} accurately. For very accurate measurements, the circuit of fig. 4.5(b) is used. Here a standard cadmium cell (SC) of e.m.f. 1·0186 V at 20°C is used to standardise the slide-wire AB. The scale may be graduated in volts, and, with the two-way switch S connected to the standard cell at D, the slider is set so that C_1 reads 1·0186 V (or some suitable multiple of it). The variable resistance R_v is now adjusted so that the galvanometer reads zero when the short-circuiting button H is depressed. Then, the p.d. from the slide-wire is 1·0186 V also, and, if the scale and slide-wire are accurately made, each division represents 1 volt (or a multiple of 1 volt). The p.d. to be measured is now connected to EF, and the switch S is connected to F. The slider is moved to obtain null balance again, and the new position C_2 gives the p.d. value on the scale in volts. The standardisation should be carried out just prior to measuring the p.d., since, if E is provided by a battery, its terminal voltage may change in use. In commercial instruments, the resistances are connected to a circle of contacts, and separate resistances are arranged for subdivision of the scales. The potentiometer is frequently used in the calibration of other electrical instruments (see example 1.5.1) and in the measurement of p.d.'s or e.m.f.'s, such as from thermocouples (see example 13.10.5).

4.3 D.C. bridges

Bridges are a particular kind of network. The d.c. ones, usually termed 'Wheatstone' bridges [fig. 4.6(a)], are used to measure resistance and the very small changes of resistance such as occur in strain- or temperature-measuring systems. The network is active, energised by a d.c. e.m.f. (E), the p.d. across AC being adjusted to a suitable value by the variable resistor R_v.

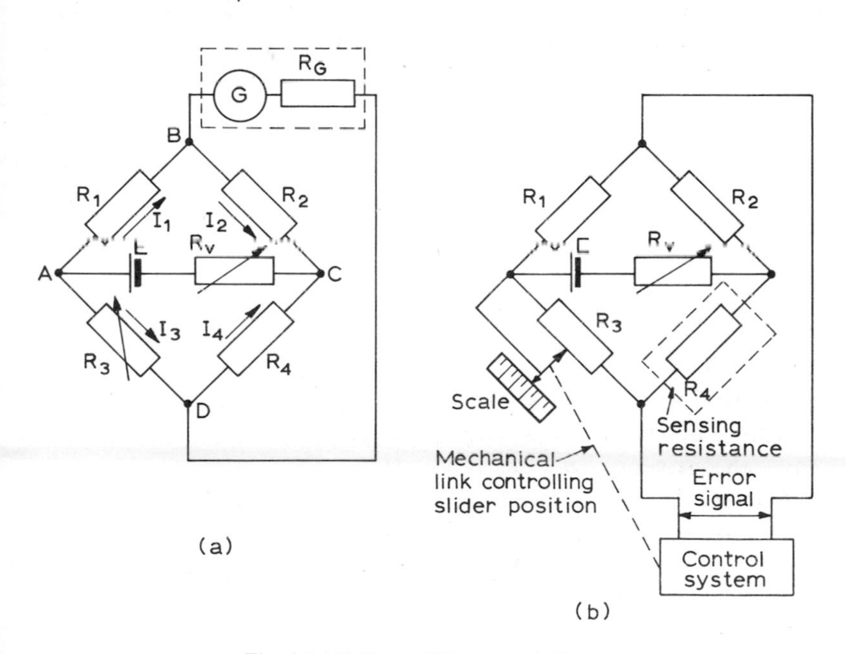

Fig. 4.6 Null-type Wheatstone bridges

4.3.1 Null-type bridge

The bridge is balanced by varying the resistance value R_3 until no current is indicated on the galvanometer, when $V_{BD} = 0$. Then

$$V_{AB} = V_{AD} \quad \text{and} \quad V_{BC} = V_{DC}$$

or

$$I_1R_1 = I_3R_3 \quad \text{and} \quad I_2R_2 = I_4R_4$$

but $I_1 = I_2$ and $I_3 = I_4$ at balance,

$$\therefore \quad I_1R_1 = I_3R_3 \qquad 4.3$$

and

$$I_1R_2 = I_3R_4 \qquad 4.4$$

Dividing eqn 4.3 by 4.4,

$$\frac{R_1}{R_2} = \frac{R_3}{R_4} \quad \text{and} \quad R_4 = \frac{R_2}{R_1} \times R_3 \qquad 4.5$$

If the resistance value of R_4 is to be measured, then, if R_3 is an accurately calibrated variable resistor, the value of R_4 may be calculated from eqn 4.5, using the value of R_3 at balance.

Frequently it is the change δR_4 in the resistance R_4 which is to be found. The change will unbalance the bridge, giving a galvanometer deflection, and R_3 has to be adjusted by an amount δR_3 to restore balance. The rebalanced condition will give

$$R_4 + \delta R_4 = (R_2/R_1)(R_3 + \delta R_3)$$

$$\delta R_4 = \frac{R_2R_3}{R_1} + \frac{R_2}{R_1}\delta R_3 - R_4$$

$$\delta R_4 = \left(\frac{R_2}{R_1}\right)\delta R_3 \qquad 4.6$$

A commercial bridge may have R_3 variable up to say 10000 Ω in 0·1 Ω steps, and also (R_2/R_1) variable in decade steps from say 10000:1 to 1:10000, as shown in fig. 4.7. This will enable measurements of resistance from 1 Ω to 100 MΩ to be made to five significant figures. Assuming the R_1/R_2 ratios are accurate, the *percentage* accuracy of

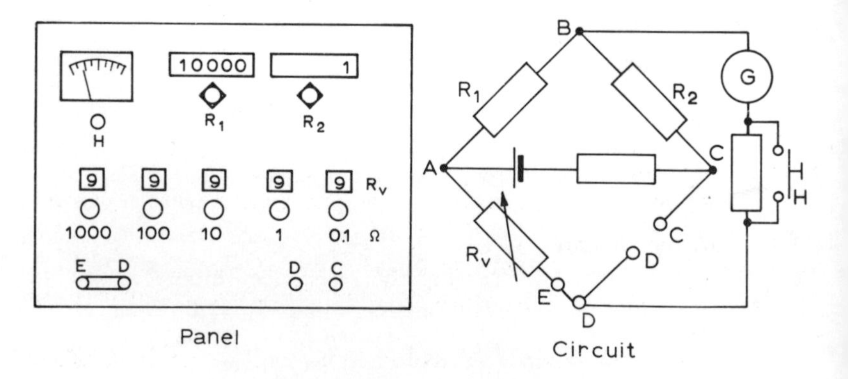

Fig. 4.7 Commercial Wheatstone bridge (null type)

the measurements will be of the same order as the percentage accuracy of the set values of the variable resistance (see example 4.5.5).

In some applications, such as in temperature-measuring systems, an automatic bridge-balancing control system may be used, as indicated in fig. 4.6(b). The voltage or current output over BD is used as an error-signal input to the control system, which then varies R_3 through the mechanical link to the slide-wire until the error signal disappears, i.e. when the bridge is balanced. The read-out scale may be calibrated in resistance-change values or in temperature units, etc.

The null-type bridge can be extremely accurate, but balancing, even if done automatically, is not instantaneous. For the measurement of rapidly changing signals, such as occur in dynamic strain measurement, a small output voltage or current from a slightly unbalanced bridge is used as a measure of the change of resistance. The derivation of the output voltage or current is complex, and, to facilitate analysis, the values of the bridge and indicating-instrument resistances may be expressed in terms of the resistance R of the sensing resistance, as indicated in fig. 4.8.

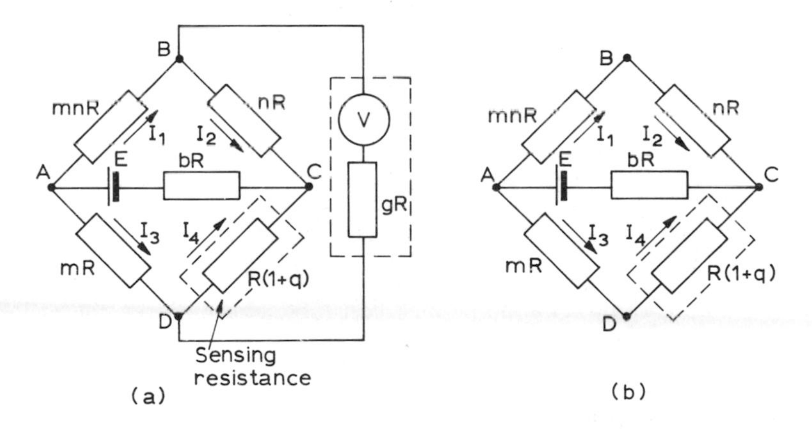

Fig. 4.8 Voltage-sensitive Wheatstone bridge

4.3.2 Voltage-sensitive bridge

The bridge is similar to the null-type bridge of fig. 4.6, except that the galvanometer is replaced by a voltmeter or oscilloscope of very high input resistance. The current (I) through this instrument will be very small indeed, and may be neglected, and the effective circuit is as shown in fig. 4.8(b).

The bridge is balanced initially when $q = 0$, giving $mnR/nR = mR/R = m$. The sensing resistance in arm CD then has a fractional change of resistance $q\ (=\delta R/R)$ which causes a difference of the currents flowing in the arms ABC and ADC of the bridge. A p.d. will appear over the points BD, equal to the difference of the p.d.'s AB and AD.

i.e. $$V_{BD} = I_1 mnR - I_3 mR \qquad 4.7$$

but $$I_1 = I_2 = V_{AC}/(mnR + nR) \qquad 4.8$$

and $$I_3 = I_4 = V_{AC}/\{mR + R(1 + q)\} \qquad 4.9$$

Substituting from eqns 4.8 and 4.9 in eqn 4.7,

$$V_{BD} = V_{AC}\left(\frac{mnR}{mnR + nR} - \frac{mR}{mR + R(1 + q)}\right)$$

$$= V_{AC}\left(\frac{m}{m + 1} - \frac{m}{m + 1 + q}\right)$$

$$= V_{AC}\frac{mq}{(m + 1)(m + 1 + q)} \qquad 4.10$$

$$\approx \frac{mV_{AC}}{(m + 1)^2} \times \left(\frac{\delta R}{R}\right) \qquad 4.11$$

The fractional change $(q) = \delta R/R$ of the sensing resistance is a convenient quantity, particularly in the case of strain measurement. Equation 4.10 shows that the output voltage V_{BD} does not change exactly linearly with the resistance change, but, since q is very small compared with unity, the equation is approximately linear, especially if m is greater than unity. However, the bridge sensitivity $V_{BD}/(\delta R/R)$ is seen to reduce with increasing m. For accurate measurements, the p.d. across AC must be maintained constant; alternatively, the ratio V_{BD}/V_{AC} may be used as a measure of q.

4.3.3 *Current-sensitive bridge*

If the output from an unbalanced bridge is to be indicated on a galvanometer-type instrument, the resistance of the indicating circuit will be relatively low, and the current flowing in it cannot be neglected. The circuit is shown in fig. 4.9(a).

The bridge is initially balanced with $q = 0$, and the fractional change $q\ (= \delta R/R)$ then occurs in the sensing resistance. The equivalent circuit D–AC–B–e–gR–D may then be drawn as in fig. 4.9(b) if the supply resistance bR is neglected, e being the equivalent e.m.f. due to the p.d. across BD caused by unbalance. The total resistance of the parallel/series resistances in BD is given by

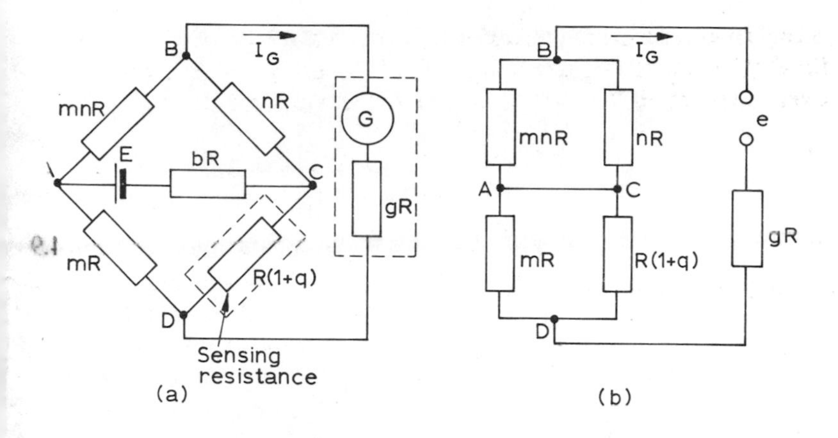

Fig. 4.9 Current-sensitive Wheatstone bridge

$$R_T = \frac{mR \times R(1+q)}{mR + R(1+q)} + \frac{mnR \times nR}{mnR + nR}$$

$$= \frac{mR(1+q)}{m+1+q} + \frac{mnR}{m+1}$$

$$= \frac{mR(1+q)(m+1) + mnR(m+q+1)}{(m+1+q)(m+1)}$$

$$= \frac{mR\{(m+1)(n+1)+q\}}{(m+1+q)(m+1)} \qquad 4.12$$

$$I_G = e/(R_T + gR)$$

$$= V_{BD}/(R_T + gR) \qquad 4.13$$

V_{BD} is given very nearly by eqn 4.10, as for the voltage-sensitive bridge, and substitution of this and eqn 4.12 in eqn 4.13 gives

$$I_G = \frac{V_{AC}mq}{(m+1)(m+1+q)}$$

$$\times \frac{(m+1+q)(m+1)}{mR\{(m+1)(n+1)+q\} + gR(m+1+q)(m+1)}$$

$$= \frac{V_{AC}mq/R}{m\{(m+1)(n+1)+q\} + g(m+1)(m+1+q)} \qquad 4.14$$

$$\approx \frac{V_{AC}mq/R}{g(m+1)^2 + m(m+1)(n+1)} \quad \text{since } q \text{ is small} \qquad 4.15$$

Eqn 4.15 may be more conveniently expressed as

$$I_G = \frac{V_{AC}}{KR} \times \left(\frac{\delta R}{R}\right) \qquad 4.16$$

where $K = g(m + 1)^2/m + (m + 1)(n + 1)$.

If R is high, the bridge sensitivity $I_G/(\delta R/R)$ is low. If V_{AC} is high, sensitivity is increased; but if m and n are made high to improve linearity (see eqn 4.14), sensitivity is reduced. Low sensitivity may be satisfactory, since only a small current is required to operate a galvanometer-type indicator or recording instrument, or a d.c. amplifier. The accuracy depends on the accuracy and stability of the values of m and n (i.e. of the values of the resistances of the bridge), and also on the maintainance of V_{AC} to a very constant value. This latter may be achieved by using a mains transformer and rectifier, or a semiconductor device known as a zener diode.

4.3.4 *Errors*

In addition to errors in the ratios m, n, and g, other errors may arise due to variable resistance of contacts and to thermal e.m.f.'s arising due to difference of the temperatures of the junctions of dissimilar metals, and also to indicating-instrument errors. However, the largest effect which may arise is that due to change of temperature of the resistances and their connecting leads. This may arise due to the heating effect of the currents in unbalanced bridges and in null-type bridges during balancing, or to external effects.

4.3.5 *Temperature errors and compensation*

The change of resistance of a conductor due to change of temperature is approximately linear, and is expressed by the simplified equation

$$\frac{\delta R_0}{R_0} = \alpha \theta_0$$

where R_0 is the resistance at 0°C, $\delta\theta_0$ is the temperature change from 0°C, and α is the temperature coefficient of resistivity of the material. From any temperature θ, at which the resistance is R,

$$\frac{\delta R}{R} = \alpha' \, \delta\theta$$

where $\alpha' = \alpha/(1 + \alpha\theta)$, the temperature coefficient of resistivity referred to temperature θ. Hence the fractional change of resistance is

$$q = \delta R/R = \alpha' \, \delta\theta$$

If the resistance and leads of each arm of the bridge of fig. 4.6(a) are made of the same material, and all have the same temperature change $\delta\theta$ from the nominal value θ, then

$$R_1 \text{ changes to } R_1(1 + \alpha' \,\delta\theta) = R_1(1 + q)$$

$$R_2 \text{ changes to } R_2(1 + q), \text{ etc.}$$

and the bridge is still balanced, since

$$\frac{R_1(1 + q)}{R_2(1 + q)} = \frac{R_3(1 + q)}{R_4(1 + q)}$$

In some bridge designs, care is taken to keep all arms of the bridge in close proximity, so that the likelihood of temperature differences is small. An example is the pressure transducer shown in fig. 5.28, where all the bridge arms are in the transducer head. If the temperature changes are unequal, then unbalance of the bridge will result, leading to inaccurate measurements (unless, of course, the only temperature change is in a sensing resistance which is measuring temperature).

Figure 4.10 shows a temperature-sensing element of resistance R_s, which has a change δR_s due to a change of temperature δT from the ambient temperature T to the temperature to be measured, T_s. However, the leads to the sensing resistance, having a resistance R_L, may be exposed to a temperature T_L due to being in close proximity to the heat source, and will have a resistance change δR_L. T_L will not usually be constant along the length of the wire. To compensate for the resistance

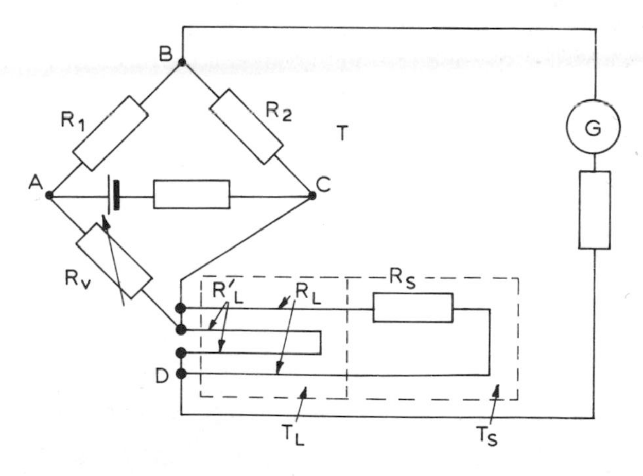

Fig. 4.10 Temperature-compensated bridge

change δR_L, which will give a spurious signal of change of temperature T_s, dummy leads having resistance R_L' at T may be placed alongside the active leads, and connected in the adjacent arm of the bridge. If δR_v is the change in R_v necessary to rebalance the bridge after the temperature change δT, then

$$\frac{R_1}{R_2} = \frac{R_v + \delta R_v + R_L' + \delta R_L'}{R_s + \delta R_s + R_L + \delta R_L}$$

and

$$R_s + \delta R_s + R_L + \delta R_L = \left(\frac{R_2}{R_1}\right)R_v + \left(\frac{R_2}{R_1}\right)\delta R_v + \left(\frac{R_2}{R_1}\right)R_L' + \left(\frac{R_2}{R_1}\right)\delta R_L'$$

But, at balance, $\dfrac{R_1}{R_2} = \dfrac{R_v + R_L'}{R_s + R_L}$, giving $R_s + R_L = \left(\dfrac{R_2}{R_1}\right)R_v + \left(\dfrac{R_2}{R_1}\right)R_L'$

$$\therefore \quad \delta R_s + \delta R_L = \left(\frac{R_2}{R_1}\right)\delta R_v + \left(\frac{R_2}{R_1}\right)\delta R_L'$$

Hence for temperature compensation of the active leads it is necessary for the resistance change $\delta R_L'$ per degree of the dummy leads to be R_1/R_2 times that of the active leads. It is usually found convenient to use an equi-armed bridge, so that $R_1/R_2 = 1$, and to make $R_L = R_L'$, with dummy and active leads of the same material.

A further complication arises in the measurement of strain by change of resistance, when strain occurs unintentionally due to different expansion rates of the sensing resistance and the material to which it is bonded. Compensation for this effect is discussed in Chapter 12.

The fixed and variable resistances in bridges are frequently made of manganin, an alloy of copper, nickel, and manganese which has a low negative temperature coefficient of resistivity (see fig. 12.8) and a low thermal e.m.f. when joined to copper.

4.4 A.C. bridges

Electronic amplifiers working on an a.c. input are more stable and are cheaper to construct than those working on a d.c. input. Also, changes of reluctance, inductance, and capacitance may be detected, in addition to changes of resistance, if an a.c. source of e.m.f. is used to energise a bridge circuit. For these reasons, a large number of a.c. bridges of differing types are used in modern instrument systems. The analysis of such bridges is complex, and the following outline of some basic types will give only a brief introduction to the subject and some typical examples in mechanical-engineering measurement.

Resistance (R), capacitive reactance (X_C), and inductive reactance (X_L) are combined to give impedance (Z).

$$X_C = 1/2\pi f C$$
$$X_L = 2\pi f L$$

where f is the frequency (hertz), C is the capacitance (farads), and L is the inductance (henrys).

For resistance, capacitance, and inductance connected *in series*, the total impedance (Z), in j-operator notation, is given by

$$Z = R + \mathrm{j}(X_L - X_C), \quad \text{where} \quad \mathrm{j} = \sqrt{(-1)}.$$

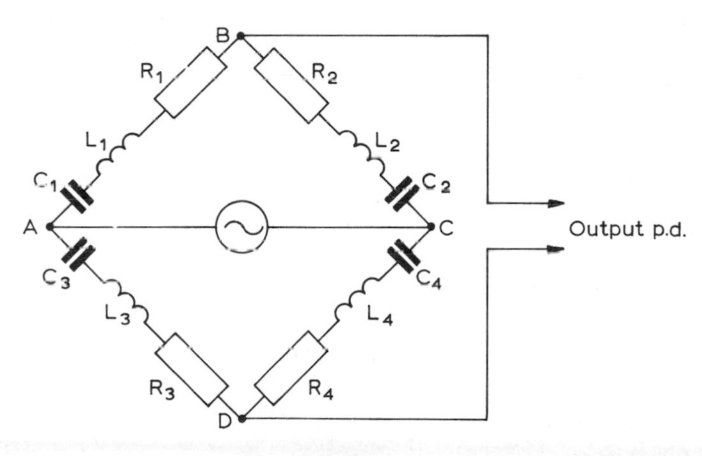

Fig. 4.11

For a bridge containing elements in series (fig. 4.11), for balance, i.e. $V_{AB} = V_{AD}$ *and* V_{AB} *in phase* with V_{AD},

$$\frac{Z_1}{Z_2} = \frac{Z_3}{Z_4}$$

or

$$\frac{R_1 + \mathrm{j}(X_L - X_C)_1}{R_2 + \mathrm{j}(X_L - X_C)_2} = \frac{R_3 + \mathrm{j}(X_L - X_C)_3}{R_4 + \mathrm{j}(X_L - X_C)_4}$$

The real and imaginary (j) parts are equated separately, to give

$$\frac{R_1}{R_2} = \frac{R_3}{R_4} \qquad \text{as in the purely resistive bridge,}$$

and
$$\frac{(X_L - X_C)_1}{(X_L - X_C)_2} = \frac{(X_L - X_C)_3}{(X_L - X_C)_4}$$

or simply
$$\frac{X_1}{X_2} = \frac{X_3}{X_4}$$

In most a.c. bridges, initial balance is obtained by varying R and L or C, and the measurement is obtained by detecting the unbalance of the bridge due to the change of resistance, capacitance, or inductance of the sensing element.

4.4.1 A.C. resistance bridge

If the voltage-sensitive bridge of fig. 4.8 is energised by an a.c. e.m.f. across AC, of the form $E \cos \omega t$, a p.d. of the same frequency will appear at the output terminals BD when the bridge is unbalanced. Although the bridge is resistive, small unintentional capacitances and inductances are likely to exist due to adjacent wires and components. These make the bridge impossible to balance by means of the variable resistance only, and a variable capacitance (C) and two-way switch (S) are introduced as shown in fig. 4.12(a). Adjustment of C with S in the appropriate position will enable the bridge to be balanced, i.e. output voltage $V_{BD} = 0$ when resistive balance ($R_1/R_2 = R_3/R_4$) *and* reactive balance ($X_1/X_2 = X_3/X_4$) are obtained.

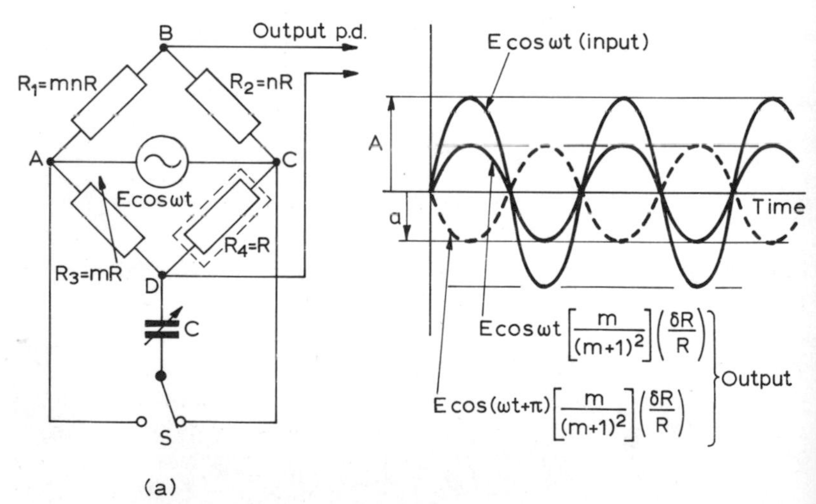

Fig. 4.12

If a change of R now occurs, equal to δR, the output voltage at BD is given by substituting $E \cos \omega t$ for V_{AC} in eqn 4.11, giving

$$V_{BD} = E \cos \omega t \times \left(\frac{\delta R}{R}\right)\frac{m}{(m+1)^2}$$

or

$$V_{BD} = EK \cos \omega t \times (\delta R/R) \qquad 4.17$$

where $K = m/(m+1)^2$. The frequency (ω) of the output wave is the same as the input. It may be in phase with the input, or of opposite phase, as shown in fig. 4.12(b), according to whether R has increased or decreased. Hence, from eqn 4.17, the amplitude of the output is proportional to the resistance change δR.

If the value of δR is varying, this gives a variation in the amplitude of the output wave as shown in fig. 4.13. Where changes from positive to negative values of δR occur, as at P, or the reverse at Q, it is necessary for the indicating system to show this if the sign of the resistance change is required. However, in many cases δR will have only positive values, and the output signal may be measured on a normal voltmeter after amplification.

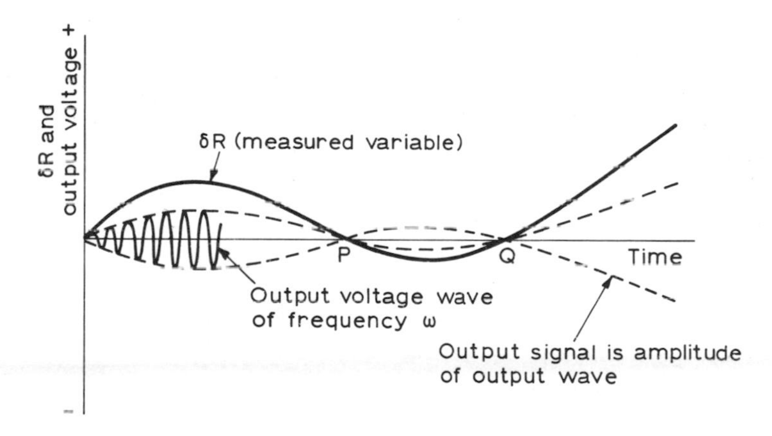

Fig. 4.13

4.4.2 *Capacitance bridge*

Change of capacitance is frequently used to sense displacement, liquid level, etc., and fig. 4.14 shows a typical capacitance bridge. The input e.m.f.'s across AB and BC are identical, equal to $E \cos \omega t$, and the variable capacitance C_3 is adjusted to give impedance balance at some datum condition of the sensing capacitance C_4. Variation (δC_4) of the sensing capacitance C_4 will unbalance the bridge, giving very nearly

$$v_0 = EK \cos \omega t \times \delta C_4$$

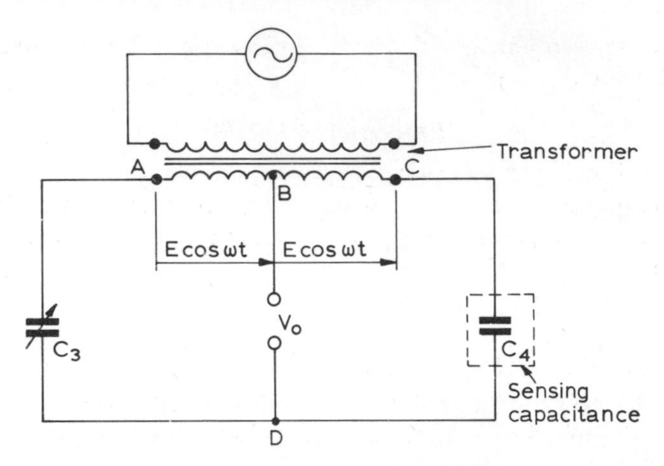

Fig. 4.14 **Capacitance bridge**

The constant K is found by calibration. The frequency $f(= \omega/2\pi)$ is usually high in capacitive bridges, because at low frequencies the impedance is high, giving low sensitivity.

4.4.3 Inductance bridge

Change of inductance may be used in various ways for sensing. A typical example is shown in fig. 4.15, where the displacement x of a

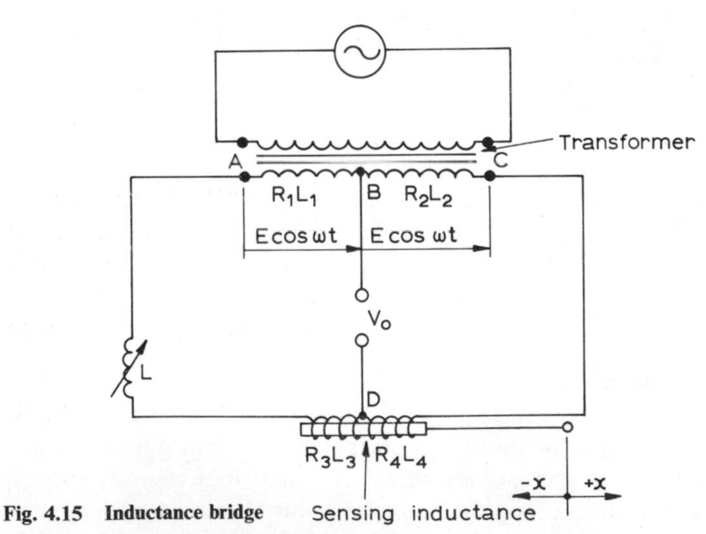

Fig. 4.15 Inductance bridge

ferrous rod in a coil causes a change of inductances L_3 and L_4. Impedance balance is first obtained for a datum position of the rod by making $R_1/R_2 = R_3/R_4$ and $L_1/L_2 = (L + L_3)/L_4$. Inductances L_3

and L_4 are then changed due to a displacement (x) of the rod. The output voltage v_0 is given by

$$v_0 = EK \cos \omega t \times x$$

and K is found by calibration.

Displacement in one direction will give an output in phase with the input, and an output of opposite phase is given by displacement in the opposite direction. Hence the direction of the displacement may be found by using a phase-detecting device.

4.5 Worked examples

4.5.1 Determine the magnitude and direction of the currents flowing in the network shown in fig. 4.16(a), and determine the p.d. across points BD, AC, and BC.

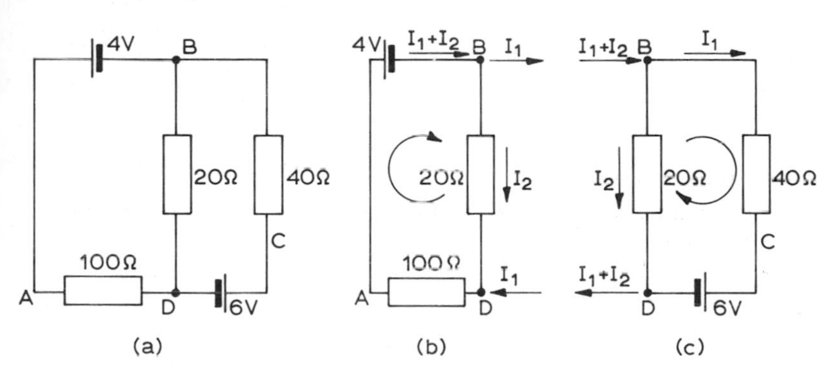

Fig. 4.16

Let the current through BC be I_1 and through BD I_2; the current through AB is then $I_1 + I_2$, since the algebraic sum of the currents at the junction B and the junction D is zero in each case. The circuit loops ABDA and BCDB as shown in (b) and (c) may be considered separately, using $\sum E = \sum IR$ and adopting a clockwise–positive sign convention.

Loop ABDA

$$-4 = 20I_2 + 100(I_1 + I_2)$$

or

$$100I_1 + 120I_2 = -4 \qquad \text{(i)}$$

Loop BCDB

$$-6 = -20I_2 + 40I_1$$

or

$$40I_1 - 20I_2 = -6 \qquad \text{(ii)}$$

Multiplying (ii) by 6 and adding to (i),

$$\begin{array}{l} 100I_1 + 120I_2 = -4 \\ 240I_1 - 120I_2 = -36 \\ \hline 340I_1 \qquad\qquad = -40 \end{array}$$

$$I_1 = -(2/17)\ \text{A} = -0{\cdot}118\ \text{A}$$
$$= -118\ \text{mA}$$

Substitution of I_1 in (i) gives $I_2 = (11/170)$ A
$= 65$ mA

These values may be checked by considering the loop ABCDA, for which

$$\sum IR = 40I_1 + 100(I_1 + I_2)$$
$$= 40\left(\frac{-2}{17}\right) + 100\left(\frac{-2}{17} + \frac{11}{170}\right)$$
$$= -10$$
$$\sum E = -4 - 6$$
$$= -10$$
$$V_{BD} = 20 \times I_2$$
$$= 20 \times 11/170$$
$$= 1{\cdot}30\ \text{V with B at the higher potential}$$

$$\text{p.d. across DA} = 100(I_1 + I_2)$$
$$= 100 \times \left(\frac{-2}{17} + \frac{11}{170}\right)$$
$$= -5{\cdot}3\ \text{V}$$

i.e. A is at a higher potential than D.

Also C is at 6 V higher potential than D, hence

$$V_{CA} = V_{CD} + V_{DA} \qquad \text{(clockwise round the diagram)}$$
$$= 6 - 5{\cdot}3$$
$$= 0{\cdot}7\ \text{V}$$
$$V_{CB} = V_{BD} + V_{DC}$$
$$= 1{\cdot}30 + (-6) = -4{\cdot}7\ \text{V}$$

4.5.2 For the active network shown in fig. 4.16(a), calculate the equivalent e.m.f. (E) and resistance (R) of the equivalent circuit shown in (b).

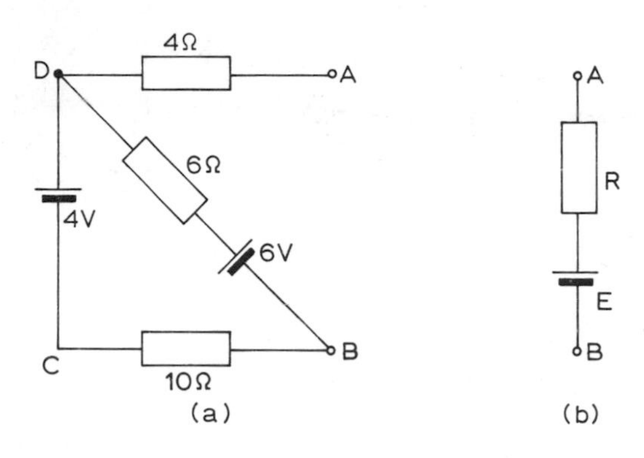

Fig. 4.17

Considering the loop BCDB, and letting the clockwise current be I,

$$4 - 6 = 10I + 6I$$

$$I = -0{\cdot}125 \text{ A}$$

The p.d. across BD is given by

$$\begin{aligned} V_{BD} &= V_{BC} + V_{CD} \\ &= -10 \times \tfrac{1}{8} - 4 \\ &= -5{\cdot}25 \text{ V} \end{aligned}$$

Hence the open circuit p.d. across AB is 5·25 V.

The total resistance of the circuit at AB is

$$\begin{aligned} R_T &= 4 + 1/(\tfrac{1}{6} + \tfrac{1}{10}) \\ &= 4 + 60/16 \\ &= 7{\cdot}75\ \Omega \end{aligned}$$

Hence, in the equivalent circuit, $E = 5{\cdot}25$ V and $R = 7{\cdot}75\ \Omega$, and A is the positive connection.

4.5.3 A voltmeter is used to measure the p.d. across the points AC in a voltage-dividing potentiometer, as shown in fig. 4.18(a). If the system is used to measure the displacement x of the slider C from the zero position at A, derive an expression for the percentage error in the

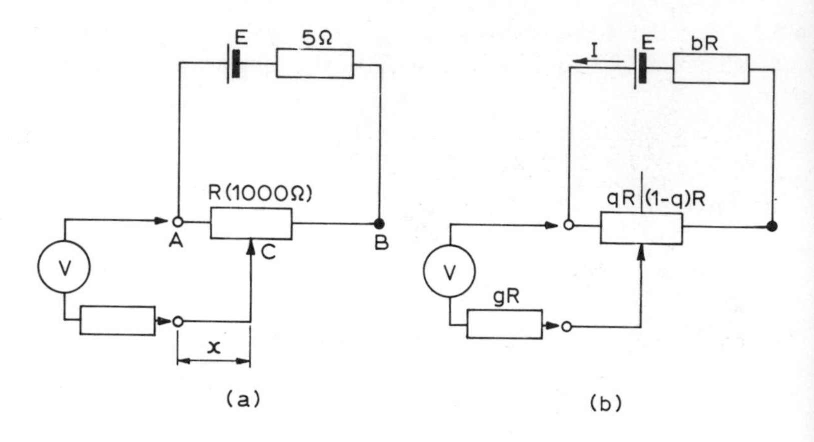

Fig. 4.18

measurement of this distance due to the application of a voltmeter of resistance gR. Calculate the percentage error when gR is (a) 100 Ω and (b) 10 000 Ω when x is 0·25 of the full-scale displacement AB.

It is convenient to express the resistances of the circuit as fractions of the overall slide-wire resistance R, as shown in fig. 4.18(b).

Without the voltmeter connected

$$I = E/(bR + R) = E/R(b + 1)$$

$$V_{AC} = I \times qR = Eq/(b + 1) \qquad \text{(i)}$$

With the voltmeter connected

$$I' = E\bigg/\left\{bR + (1 - q)R + \frac{1}{(1/gR + 1/qR)}\right\}$$

$$= E\bigg/\left\{(b + 1 - q)R + \frac{gqR}{g + q}\right\} = E\bigg/R\left\{b + 1 - q + \frac{gq}{g + q}\right\}$$

$$V_{AC}' = I' \times \text{resistance across AC}$$

$$= I'\bigg/\left(\frac{1}{qR} + \frac{1}{gR}\right)$$

$$= I' \times gqR/(g + q)$$

$$= \frac{E}{R} \times \frac{1}{b + 1 - q + gq/(g + q)} \times \frac{gqR}{g + q}$$

$$= E \times \frac{g + q}{(b + 1 - q)(g + q) + gq} \times \frac{gq}{g + q}$$

$$= Eq\left\{\frac{1}{q + (b + 1 - q)(g + q)/g}\right\} \qquad \text{(ii)}$$

The voltage error is eqn (ii) − eqn (i).

$$\text{Error} = Eq\left\{\frac{1}{q + (b + 1 - q)(g + q)/g}\right\} - Eq\left\{\frac{1}{b + 1}\right\} \qquad \text{(iii)}$$

Percentage error = eqn (iii) divided by eqn (i) × 100%

$$= \left\{\frac{b + 1}{Eq}\right\}$$

$$\times \left\{\frac{1}{q + (b + 1 - q)(g + q)/g} - \frac{1}{b + 1}\right\}$$

$$\times\, Eq \times 100\%$$

$$= \left\{\frac{b + 1}{q + (b + 1 - q)(g + q)/g} - 1\right\} \times 100\% \qquad \text{(iv)}$$

This expression shows that, if the voltmeter resistance is large compared with the resistance of AC, i.e. $g \gg q$, then $(g + q)g \rightarrow 1$, and the error approaches zero.

If $b = 5/1000 = 0{\cdot}005$, $g = 100/1000 = 0{\cdot}100$, $q = 0{\cdot}25$, then the percentage error in the measurement of x is

$$\text{error} = \left\{\frac{0{\cdot}005 + 1}{0{\cdot}25 + (0{\cdot}005 + 1 - 0{\cdot}25)(0{\cdot}100 + 0{\cdot}25)/0{\cdot}100} - 1\right\} \times 100\%$$

$$= (0{\cdot}339 - 1) \times 100\%$$

$$\approx -66\%$$

For $b = 0{\cdot}005$, $g = 10\,000/1000 = 10$, $q = 0{\cdot}25$,

$$\text{error} = \left\{\frac{1{\cdot}005}{0{\cdot}25 + (0{\cdot}005 + 1 - 0{\cdot}25)(10 + 0{\cdot}25)/10} - 1\right\} \times 100\%$$

$$= (0{\cdot}980 - 1) \times 100\%$$

$$\approx -2\%$$

4.5.4 A voltage-balancing potentiometer similar to those of fig. 4.5 has a single straight slide-wire AB. When standardised using a Weston standard cell, balance was obtained when distance AC_1 was measured as 121·4 mm. When the p.d. to be measured (V) was connected in place of the standard cell, the distance AC_2 was measured as 582·6 mm at balance. If all the equipment used had been in a temperature of 22 °C for 24 hours prior to the measurements, and the length measurements were made to an accuracy of $\pm 0{\cdot}1$ mm, calculate the value of V and the accuracy of the measurement. The temperature/e.m.f. values of the standard cell are as follows:

Temp. (°C)	*e.m.f.* (V)	*Temp.* (°C)	*e.m.f.* (V)
0	1·018 93	25	1·018 37
5	1·018 94	30	1·018 11
10	1·018 88	35	1·017 81
15	1·018 76	40	1·017 46
20	1·018 58		

From a plot of the above values, fig. 4.19, the value of the cell e.m.f. at 22 °C is 1·018 50 volts.

$$V = 1{\cdot}018\,50 \times 582{\cdot}6/121{\cdot}4$$

$$= 4{\cdot}887 \text{ volts}$$

The value of V cannot be quoted to this accuracy because of the inaccurate length measurement.

$$V = V_s \times l_2/l_1$$

where l_1 = length AC_1,

l_2 = length AC_2,

or

$$\ln V = \ln V_s + \ln l_2 - \ln l_1$$

Differentiating,

$$\frac{dV}{V} = \frac{dV_s}{V_s} + \frac{dl_2}{l_2} - \frac{dl_1}{l_1}$$

This may be expressed in differential form

$$\frac{\delta V}{V} = \frac{\delta V_s}{V_s} + \frac{\delta l_2}{l_2} - \frac{\delta l_1}{l_1}$$

where δV_s, δl_2, and δl_1 are the errors in the measurement of V_s, l_2, and l_1.

$$\therefore \quad \frac{\delta V_s}{V_s} = \pm \frac{5 \times 10^{-6}}{1{\cdot}018\,500} = 0{\cdot}000\,0049$$

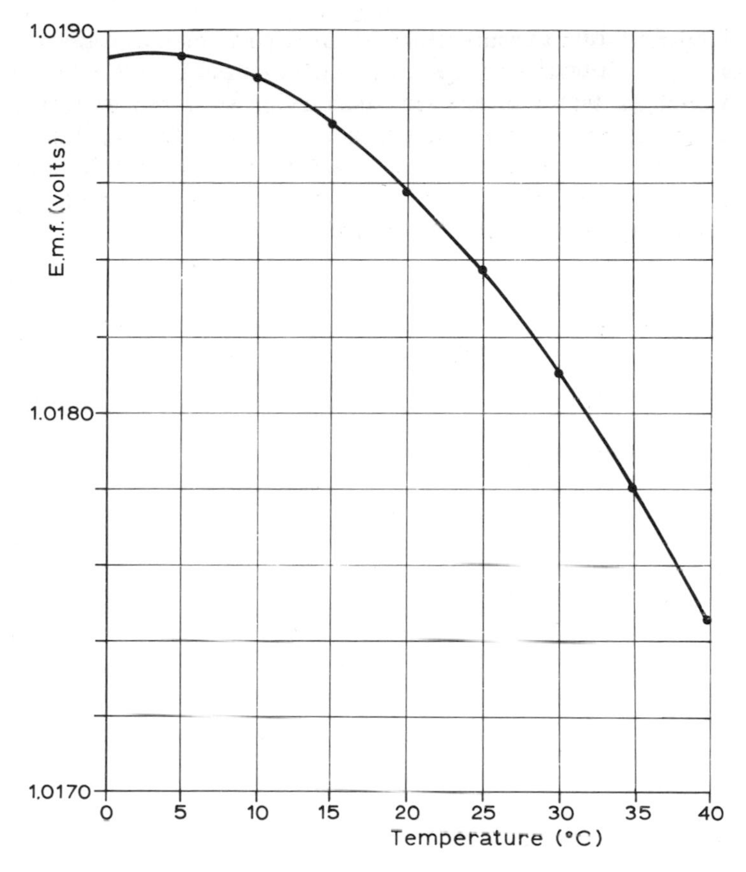

Fig. 4.19 Temperature/e.m.f. characteristics – Weston cell

$$\frac{\delta l_2}{l_2} = \pm \frac{0{\cdot}1}{582{\cdot}6} = 0{\cdot}000\,171\,6$$

$$\frac{\delta l_1}{l_1} = \pm \frac{0{\cdot}1}{121{\cdot}4} = 0{\cdot}000\,823\,7$$

$$\therefore \quad \frac{\delta V}{V} = \pm 0{\cdot}001\,000\,2$$

$$\delta V = \pm 0{\cdot}001\,000\,2 \times 4{\cdot}887$$

$$= \pm 0{\cdot}005 \text{ volts}$$

Hence $\quad V = 4{\cdot}887 \pm 0{\cdot}005$ volts.

The inaccurate measurement of length, particularly on the shorter length, has reduced the accuracy of the measured value. If *any* length of wire could be measured to an accuracy of $\pm 0{\cdot}1$ mm (assuming the inaccuracy to be in the reading of the scale and not, for example, due to temperature change), then a more accurate result could be obtained by using longer portions of the wire by decreasing the overall voltage V_{AB}.

4.5.5 In a commercial null-type Wheatstone bridge, such as shown in fig. 4.7, the decade resistance values R_v are accurate to within $\pm 0{\cdot}02\%$, and the ratios R_1/R_2 are accurate also to $\pm 0{\cdot}02\%$. Calculate the value and the error in the measurement of a resistance R_4 connected over the points CD, if null balance is obtained when $R_1/R_2 = 1000$, and the decade values R_v are set at $2534{\cdot}2\ \Omega$.

$$\text{At balance, } R_4 = \frac{R_2}{R_1} \times R_v$$

$$= \frac{2534{\cdot}2}{1000}$$

$$= 2{\cdot}5342\ \Omega$$

$$R_4 = K \times R_v$$

$$\therefore \quad \ln R_4 = \ln K + \ln R_v$$

and

$$\frac{dR_4}{R_4} = \frac{dK}{K} + \frac{dR_v}{R_v}$$

or, in differential form,

$$\frac{\delta R_4}{R_4} = \frac{\delta K}{K} + \frac{\delta R_v}{R_v}$$

where δR_4, δK, and δR_v are the errors of the resistances

$$\therefore \quad \delta R_4/R_4 = \pm 0{\cdot}02 \pm 0{\cdot}02\%$$

$$= \pm 0{\cdot}04\%$$

$$\therefore \quad \delta R_4 = 2{\cdot}5342 \times (\pm 0{\cdot}04/100)$$

$$= 0{\cdot}001014\ \Omega$$

Hence

$$R_4 = 2{\cdot}534 \pm 0{\cdot}001\ \Omega$$

4.5.6 Determine the output p.d. and current respectively of the voltage- and current-sensitive bridges of figs. 4.8 and 4.9 when measuring a 1% change in the value of a 100 Ω resistance (R) in the circuit. The bridges both have V_{BC} equal to 6 volts, and all resistances are equal at balance. The galvanometer resistance is 200 Ω.

For equal resistances, $m = n = 1$.

$$gR = 200$$

$$\therefore \quad g = 200/100 = 2$$

Voltage-sensitive bridge (eqn 4.11 of section 4.3)

$$V_{BD} = V_{AC} \times \frac{m}{(m+1)^2} \times \frac{\delta R}{R}$$

$$= 6 \times (1/4) \times 0{\cdot}01$$

$$= 0{\cdot}015 \text{ volts}$$

$$= 15 \text{ mV}$$

Current-sensitive bridge (eqn 4.16 of section 4.3)

$$K = g(m+1)^2/m + (m+1)(n+1)$$

$$= 2 \times 2^2 + 2^2$$

$$= 12$$

$$I_G = \frac{V_{AC}}{KR} \times \frac{\delta R}{R}$$

$$= \frac{6}{12 \times 100} \times 0{\cdot}01$$

$$= 50\ \mu\text{A}$$

4.6 Tutorial and practical work

4.6.1 Examine the manufacturers' literature relating to the current- and voltage-measuring instruments used in your laboratory work. Compare the values of their resistance, range, accuracy, discrimination, etc., and note the temperature at which each was calibrated, if this is stated.

4.6.2 Examine the potentiometers available to you, and their manufacturers' related literature. Compare their range, accuracy, and discrimination.

4.6.3 Compare the discrimination of a potentiometer slide-wire consisting of a coil against that of a single wire.

4.6.4 Obtain the circuit diagrams of d.c. bridges used for measurement in your laboratory work, and determine whether they are null-type or current- or voltage-sensitive. Determine the bridge sensitivity in each case.

4.6.5 For the bridges in the previous question, determine and discuss the methods used to balance the bridges and to standardise or calibrate them where applicable.

4.6.6 Show that, if $R_\theta = R_0 (1 + \alpha\theta_0)$, then

$$\delta R_\theta / R_\theta = \delta\theta \times \alpha/(1 + \alpha\theta)$$

where R_θ = resistance at temperature θ, R_0 = resistance at temperature $\theta_0 = 0\,°\text{C}$, α = temperature coefficient of resistance from 0 °C.

4.6.7 Measure the change of resistance of (a) the compensating leads of a laboratory platinum-resistance thermometer, and (b) the active leads and sensing resistance, over the range of temperature from freezing- to boiling-point. From the readings, discuss the significance of the change of lead resistance.

4.6.8 Discuss the effect of the heating of the resistance wires of a null-type bridge whilst the bridge is being balanced.

4.6.9 Wheatstone bridges are not usually used to measure resistances of less than about 1 Ω. Discuss the reasons for this.

4.6.10 Examine the circuit diagrams of a.c. bridges in the laboratories, e.g. those in electrical comparators, surface-texture and roundness measuring devices, etc. Describe the way in which each circuit operates.

4.7 Exercises

4.7.1 Determine the currents in the network shown in fig. 4.20 and the p.d.'s across AC and BD. [I_{AB} = 3·8 mA, I_{BD} = 86·5 mA, I_{CB} = 82·7 mA, V_{AC} = −8·1 V, V_{BD} = −4·3 V]

4.7.2 Calculate the equivalent e.m.f. and the resistance of the network of fig. 4.21 related to the points AB. [7·4 V, 3·9 Ω]

4.7.3 If in the voltage-dividing potentiometer of fig. 4.3(a) the p.d. across AB is maintained constant, show that the percentage error in the measurement of V_{AC} due to the application of the voltmeter is $\{1/(q + (1 - q)(g + q)/g) - 1\} \times 100\,\%$.

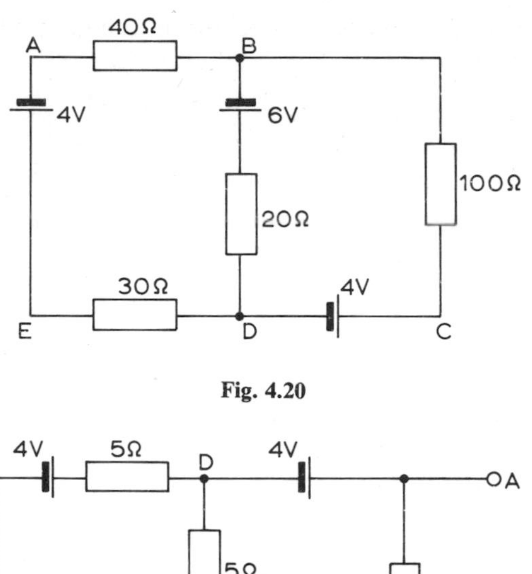

Fig. 4.20

Fig. 4.21

Calculate the necessary voltmeter resistance if the error in measuring V_{AC} when the slider is in mid-position is not to exceed 0·1%. Using this voltmeter-resistance value, calculate the percentage error when q is 0·10, 0·30, 0·70, 0·90, and 1·00, and sketch the application-error curve. [250 Ω]

4.7.4 A voltage-dividing potentiometer is used to measure an angular displacement θ as shown in fig. 4.4(b). The reading on a voltmeter connected across AC is used as a measure of θ. If the angle AOC (θ) is exactly 60°, and AB subtends 355° at the centre, calculate the p.d. across AC without a voltmeter applied, and the percentage error in reading this with a voltmeter of 1 MΩ resistance. The p.d. across AB is constant at 60 V, and the resistance between AB is 1 kΩ uniformly distributed. [10·14 V, −0·014%]

4.7.5 The current flowing through a standard 1 Ω resistor during the calibration of an instrument is determined by measuring the p.d. across the resistor with a voltage-balancing potentiometer consisting of a single straight slide-wire and a Weston standard cell. Sketch the circuit arrangement.

With a constant p.d. across the slide-wire, the standard cell, of e.m.f. 1·0186 V, was balanced against the p.d. across 253·5 mm of wire, and immediately afterwards the p.d. across 147·2 mm of wire balanced the p.d. across the standard 1 Ω resistor. Determine the current through the standard resistor and the limits of accuracy of this value, if the standard resistor is accurate to ±0·05%, the standard cell e.m.f. is known to a certainty of ±0·0001 V, and the lengths are measured to an accuracy of ±0·15 mm. [0·5914 ± 0·0013 A]

4.7.6 Calculate the possible error in the measurement of a resistance of 1 Ω using a null-type Wheatstone bridge as in fig. 4.7, if the accuracy of the bridge ratio R_1/R_2 is ±0·05% and of the R_V values ±0·03%. Discuss other errors which may arise in making the measurement.

[±0·0008 Ω]

4.7.7 In a current-sensitive bridge, the galvanometer of the ultra-violet recorder used to record the output has a resistance of 100 Ω. The bridge values are $m = 10$ and $n = 1$, using the notation of fig. 4.9. Determine the bridge sensitivity in amperes per percentage resistance change when the resistance changes of a 500 Ω sensing resistance are being measured and the bridge supply is 4 volts.

[5·96 μA/% change of *R*]

4.7.8 The output from a d.c. Wheatstone bridge is to be recorded on an instrument having 1 MΩ input resistance and a maximum sensitivity of 1 mm/20 μV. Calculate the value of the bridge ratio *m* if a change of 0·1% in the sensing resistance is to give an instrument deflection of 50 mm when the bridge supply is 6 volts.

[3·73 or 0·27]

5

Measurement of Force, Torque, and Pressure

5.1 Measurement of force, torque, and pressure

The SI force unit, the newton (N), is the force which when applied to a mass of one kilogram (kg) causes an acceleration of one metre per second squared (m/s^2). The unit of torque is the newton metre (N m), the equivalent of the torque produced by a force of 1 N acting at a radius of 1 m. Force intensity or pressure is the force per unit area, the unit being the newton per metre squared (N/m^2), otherwise known as the pascal (Pa). The bar (b), equal to 10^5 Pa, is also used. Pressure- and force-measuring devices may be similar, the connection being the area over which the force is distributed.

5.2 Force measurement

5.2.1 Gravity-balance methods

The gravitational force on a given mass varies due to geographical position, since g varies by about $\pm 0{\cdot}3\%$ from the standard value of $9{\cdot}806\,65\ m/s^2$. For any given position it is virtually constant, and hence may be used as a reference force.

Fig. 5.1 shows a typical arrangement of a beam and mass (m) to apply a force to a test specimen. The force (F_t) is applied by a screw or lever at Z to the beam through a knife edge p until the pointer indicates that the beam is horizontal, pivoted on the knife edge q. For balance of moments,

$$F_t \times a = F_g \times b$$

$$\therefore \quad F_t = F_g \times b/a = m \times g \times b/a$$

$$= \text{constant} \times b \text{ (if } g \text{ is constant)}$$

Hence the force (F_t) is proportional to distance (b) of the mass from the pivot, and the right-hand side of the beam may be calibrated in force

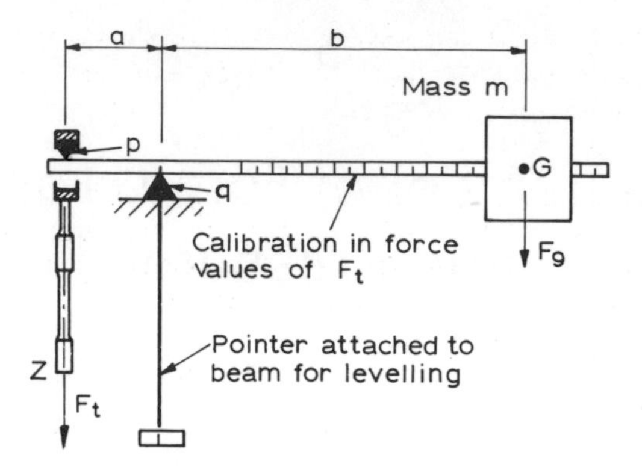

Fig. 5.1

values of F_t. For use in a gravitational field of different magnitude, a slight correction could be made.

The arrangement is used extensively in materials-testing machines, where, with suitable modifications, compression, shearing, and bending forces are easily applied. Very accurate force measurement is obtained with well-designed machines, but more compact machines are obtained by other methods.

It should be noted that the method is equally suited to measurement of mass, the measured mass (m') applying a force $F_t = m'g$ newtons. It is the basis of countless weighing (i.e. mass-measuring) machines.

5.2.2 *Fluid-pressure methods*

Fluid (usually liquid) pressure acting on a given area may be used to measure the force applied. Figure 5.2 shows an example. The oil from

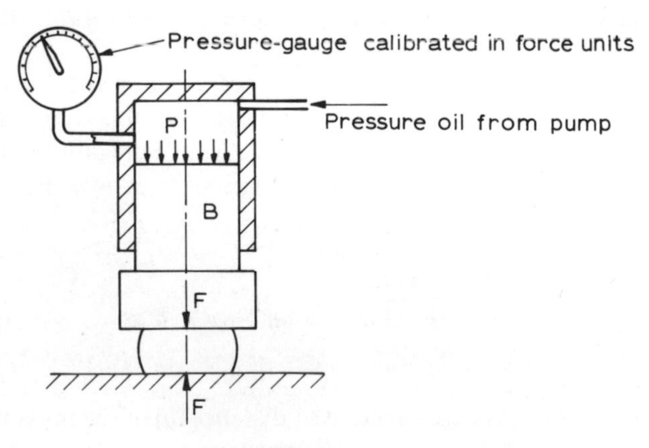

Fig. 5.2

the hand or motorised pump exerts pressure on the face of piston B. If the piston area is A and the fluid pressure P, then the force exerted on the compression specimen is given by

$$F = P \times A - \text{friction-drag force}$$

The friction drag is small compared with F, and the pressure-gauge may be calibrated in force units, since A is constant.

Tension or compression force-measuring links may be used, again the gauges being calibrated in force units. Such links may for example be used to measure the tractive effort of a vehicle, or to indicate load on a crane or lift, see fig. 5.3.

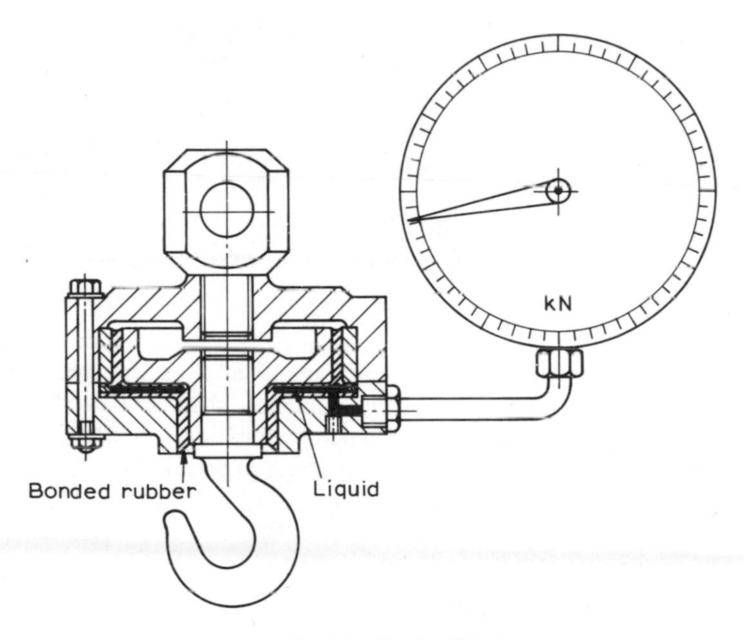

Fig. 5.3 Tension link

5.2.3 *Deflection or strain of elastic elements*

Many different shaped elements may be deflected by a force. If the resulting stress does not exceed the elastic limit of the material, then the deflection or the strain of the element may be used to measure the force. If the limit of proportionality is not exceeded, then the deflection is proportional to the applied force. Metals are usually used, although other materials, e.g. quartz, are possible. The desirable material properties are a large and proportional elastic range and freedom from hysteresis (see Chapter 1).

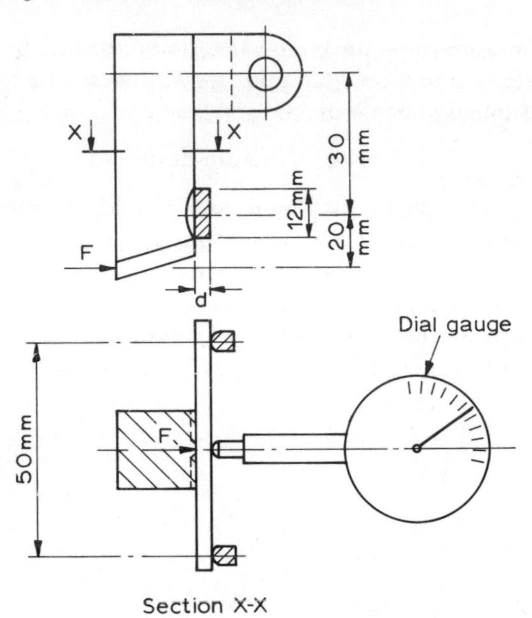

Fig. 5.4 Simply-supported beam as elastic-deflection element

Figure 5.4 shows a method of measuring the force (F) on a shaping-machine cutter, using the deflection of a simply-supported beam, measured by a dial indicator. Many different shapes of element are used, some of the more common being illustrated in fig. 5.5. The elastic-force/deflection formulae are given in appendix D.

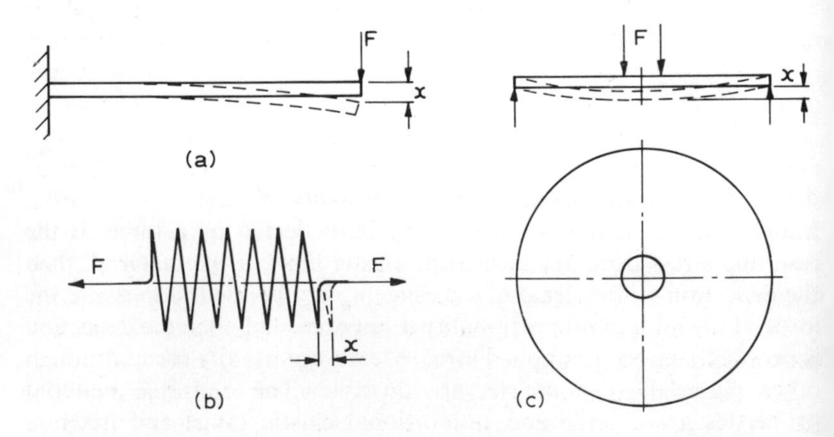

Fig. 5.5 (a) Cantilever (b) Helical spring (c) Simply-supported disc

Many different methods are used to measure the deflection of the element, such as air-jet sensing, optical methods, and electrical methods based on the change of resistance, capacitance, and inductance (see Chapter 10).

Instead of measuring the deflection, the strain of an elastic member may be measured. The sensing element is usually of the electrical-resistance type, although other methods such as those described in Chapter 12 are possible.

Figure 5.6 shows a strain-gauge bonded to the upper surface of a cantilevered cutting tool. The signal from the bridge circuit could be used to measure the magnitude of F, or even to control F by varying the feed rate or the depth of cut.

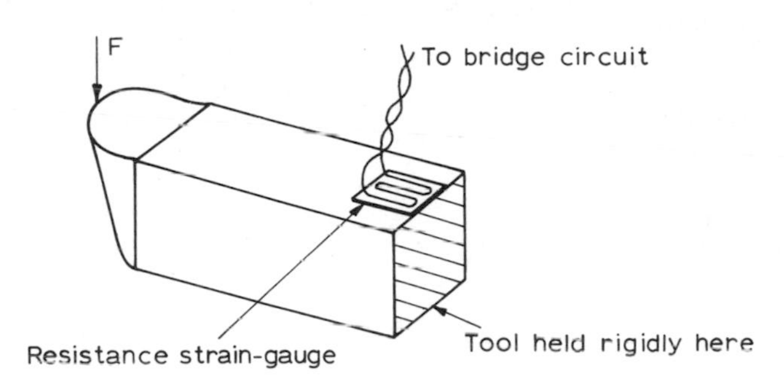

Fig. 5.6

5.2.4 Piezoelectric elements

Force applied to the faces of specially prepared crystals causes a separation of charge on the opposite faces. The crystal acts as a source of a small e.m.f., and also as a capacitor, and, with the input resistance of the amplifier, the equivalent input circuit is as shown in fig. 5.7(b).

The circuit has resistance and capacitance, and hence a time-constant $\tau = RC$, which may be varied by varying R. A large value of R gives a long time-constant, but decay of the output signal always occurs (see fig. 5.8 and section 3.3); hence the method is suitable for dynamic force measurement or short-term static measurements. Care has to be taken to use correctly screened connecting leads, and a high standard of cleanliness must be observed or unexpected leakage of charge will occur, leading to rapid loss of signal.

The crystals are usually enclosed in a stainless-steel casing having flat and parallel faces, and care must be taken to distribute the force evenly over the faces, to prevent damage to the crystal. Figure 5.9

shows a typical piezoelectric force washer, and fig. 5.10 shows its application in measuring the initial force in a cylinder-head bolt and the additional force due to combustion.

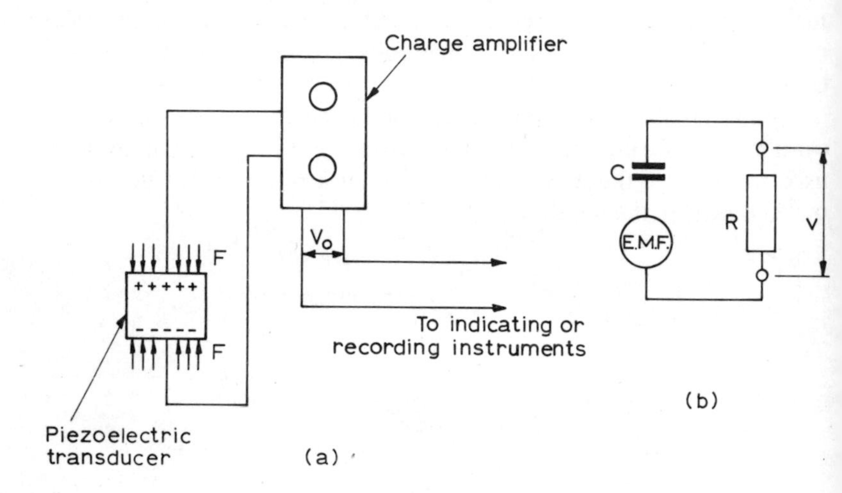

Fig. 5.7

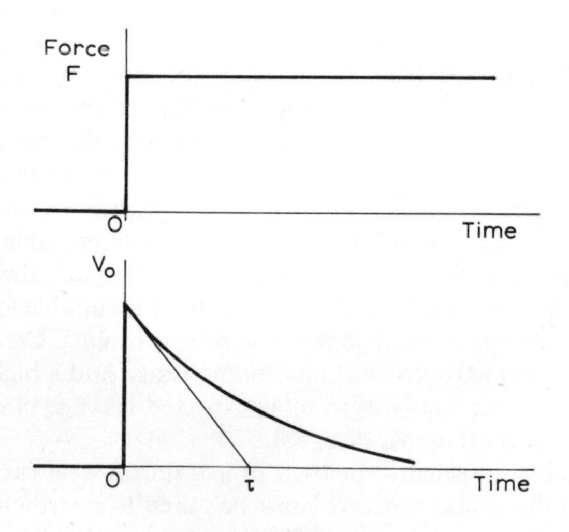

Fig. 5.8

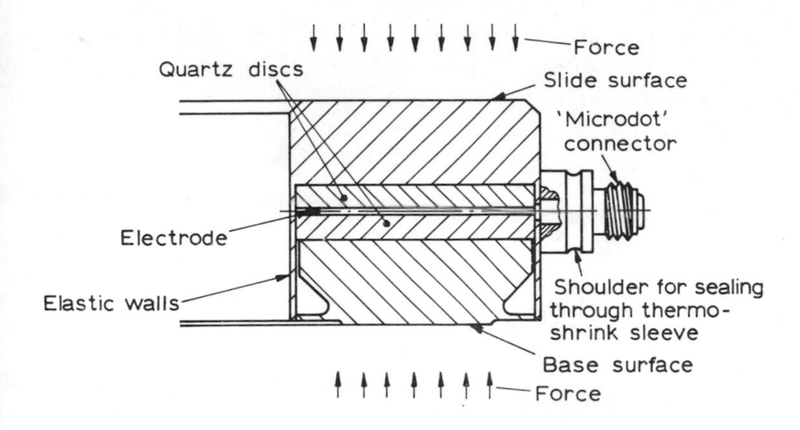

Fig. 5.9 Diagrammatic section of a quartz load washer

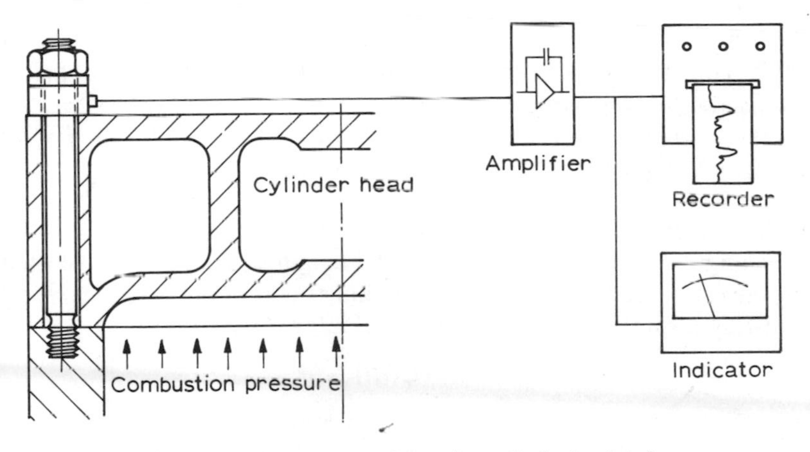

Fig. 5.10 Measurement of force in a cylinder-head stud

5.3 Torque measurement

5.3.1 Gravity-balance methods

A mass (m) may be moved along an arm as in fig. 5.11(a) until the value of the torque (Fr) equals the torque (T) which is to be measured. Alternatively, the magnitude of the mass may be varied at a constant radius. In either case, the arm must be horizontal, so that the moment-arm distance is perpendicular to the line of action of the force. The shaft is usually supported by a bearing at B, and the force (F) on the

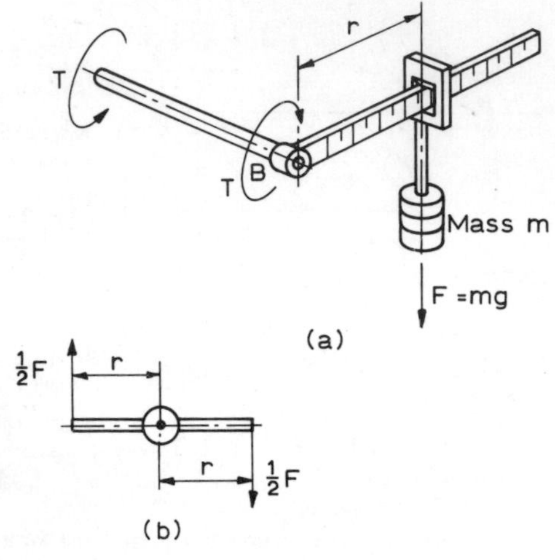

Fig. 5.11

bearing causes a friction torque, leading to error in the measurement of T. This may be obviated by arranging to apply equal and opposite forces to give the same torque T, as shown in fig. 5.11(b), when the bearing friction-drag due to F is eliminated. Small errors due to variation of g may occur, as discussed in section 5.2.1.

5.3.2 Deflection of elastic elements

The angular deflection (θ) of a parallel length of shaft (l) may be used to measure the torque in a shaft, as illustrated in fig. 5.12. The shaft twisting formula $T/I_p = G\theta/l$ gives

$$T = (I_p G/l) \times \theta$$

i.e. $T = \text{constant} \times \theta$ if I_p, G, and l are constant.

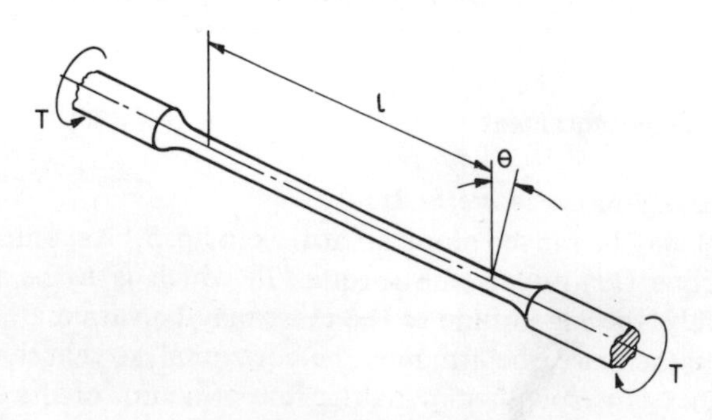

Fig. 5.12

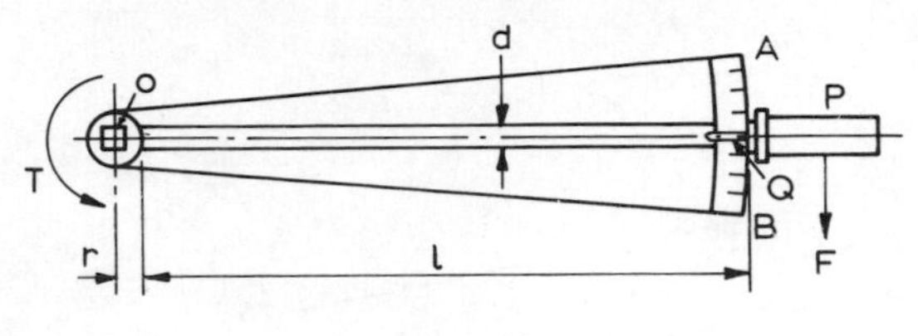

Fig. 5.13

Figure 5.13 shows a torque wrench where the bending of the bar OP as a cantilever gives a deflection of the mark Q on OP relative to the scale AB, which is rigidly attached to the boss at O. For balance,

$$T = F(l + r)$$
$$= \text{constant} \times \text{scale deflection}$$

The torque in a rotating shaft has often to be measured to determine the power transmitted. The angular deflection of A relative to B in fig. 5.14 may be found by using an angular-movement transducer C to measure the angular displacement of the outer sleeve D relative to the shaft at A. Sleeve D is rigid, attached firmly without backlash to the shaft at B, and carries no torque except that which may arise due to inertia or vibration of the sleeve. Slip rings, sometimes mercury-filled as illustrated at E in fig. 5.14, transmit the signal to a stationary member where it may be amplified and displayed or recorded.

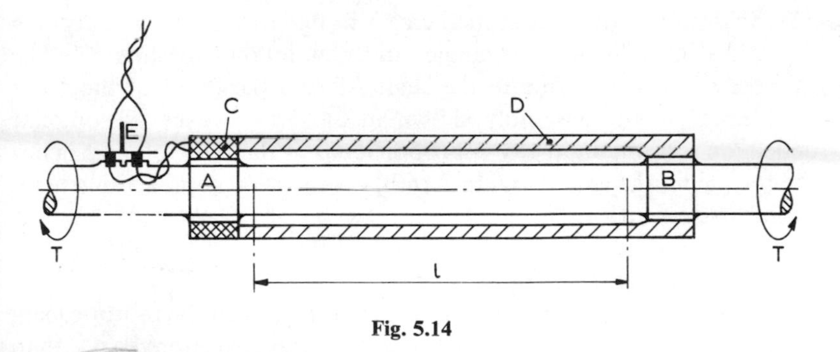

Fig. 5.14

Several electrical methods are employed, using two magnetic or photoelectric transducers as shown in fig. 5.15. The impulses from either slotted wheel may be counted electronically to give frequency or shaft speed (see Chapter 6). The time between pulses from the two wheels may be measured to give a signal proportional to twist θ. The two signals T and ω may be combined to measure power. These

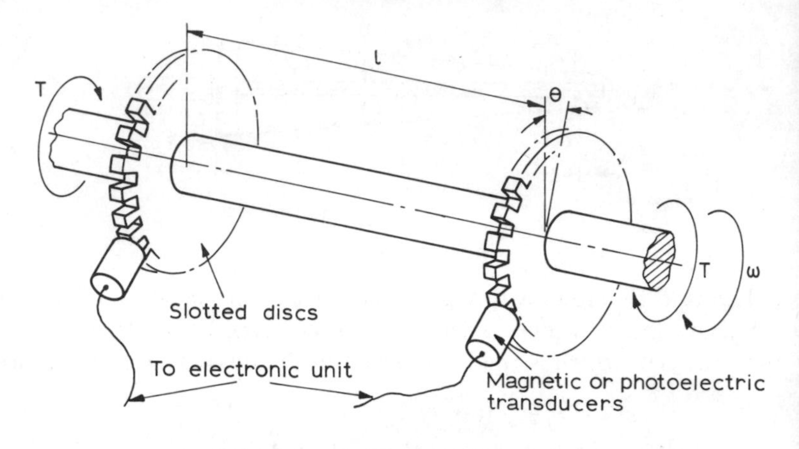

Fig. 5.15

methods eliminatc the errors arising from the use of slip rings, where leakage of the signal and 'noise' may occur. Stroboscopic methods (see Chapter 6) may also be used to observe twist, if a suitable scale-and-pointer arrangement is devised.

An ingenious torquemeter uses a light ray to transmit the twist signal to a stationary scale. The basic principle is that, if two mirrors A and B are parallel, then the outgoing ray *rs* is parallel to the incoming ray *pq even though the mirrors are rotated, still parallel*. However, if a mirror is moved out of parallel, e.g. B in fig. 5.16, then the outgoing ray is deflected through an angle, and now follows the line *r's'*. The mirrors are attached one to the shaft AB and the other to the outer member D of the assembly shown in fig. 5.14. Hence the angular deflection θ of the light ray is proportional to the twist of, and hence the torque in, the shaft. Figure 5.16(b) shows the general arrangement of the instrument.

5.3.3 *Torque-reaction methods*

If a motor M is delivering an output torque (T_o), then the motor casing has a reaction torque (T_o') which may be measured more easily than the torque T_o in the shaft. If the motor *casing* is mounted on bearings at A and B, as shown in fig. 5.17, then the force (F) at radius r may be measured by one of the force-measuring methods discussed in section 5.2. The method shown in fig. 5.18 eliminates the need to support the casing on bearings, the motor platform being mounted on force transducers, and

$$T_o' = (F_1 + F_2)r$$

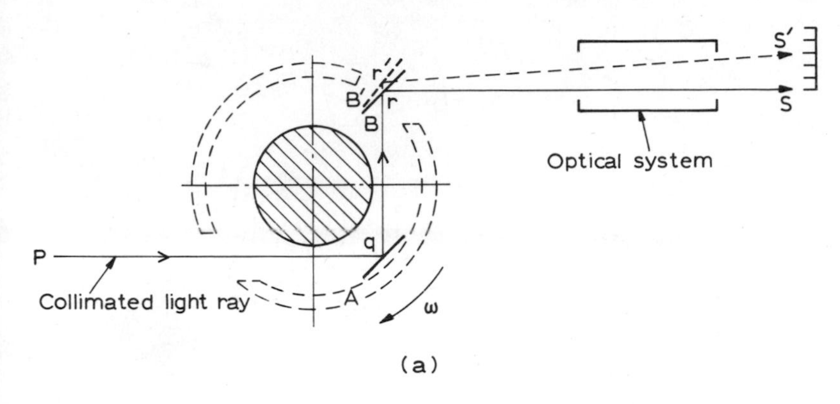

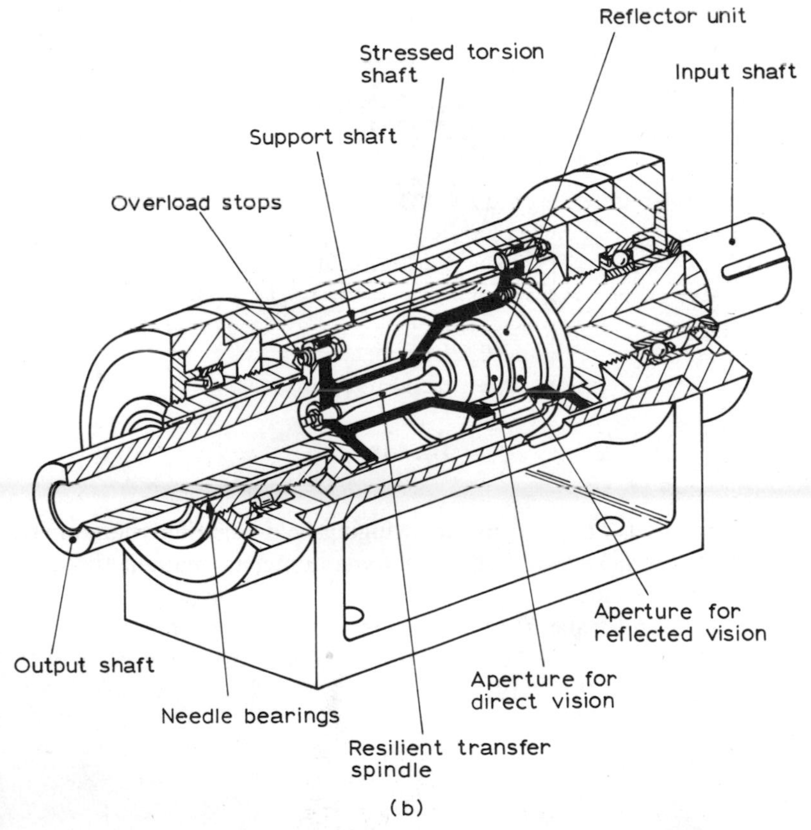

Fig. 5.16 (a) Torquemeter principle (b) Section through an optical torquemeter

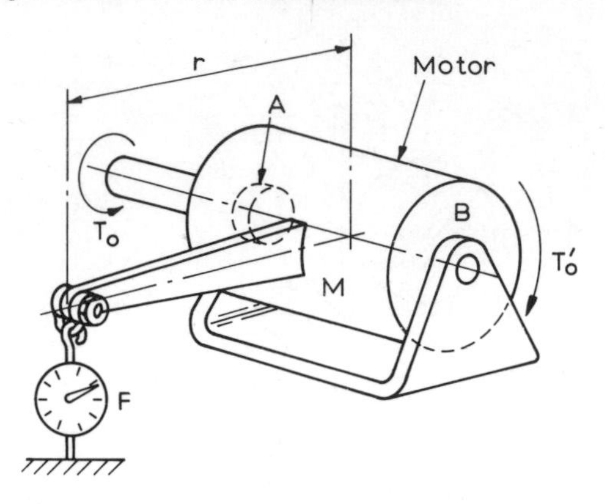

Fig. 5.17

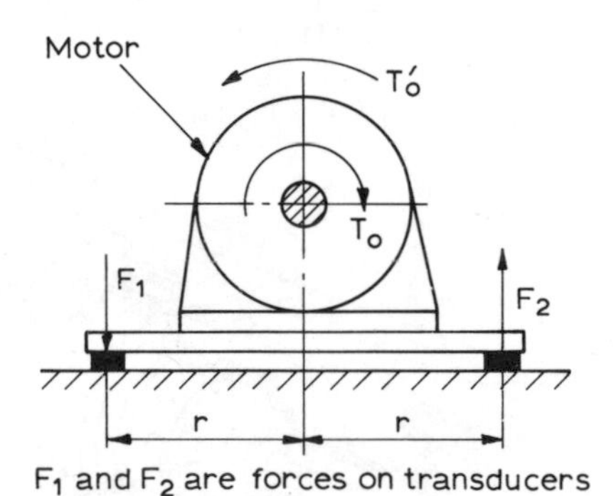

Fig. 5.18

Similar methods are used in the swinging-field electric dynamometer and in the well-known hydraulic dynamometers manufactured by Heenan and Froude.

The reaction torque T_r on a gearbox may be measured, and, if the gear ratio (n) and the efficiency (η) are known, the value of T_i or T_o may be calculated (fig. 5.19).

$$\omega_o = n\omega_i$$

$$\therefore \quad T_o = +\left(\frac{T_i}{n} \times \eta\right) = -T_o'$$

where T_i is the input torque, T_r is the torque on the casing, and T_o' is the reaction torque from the output.

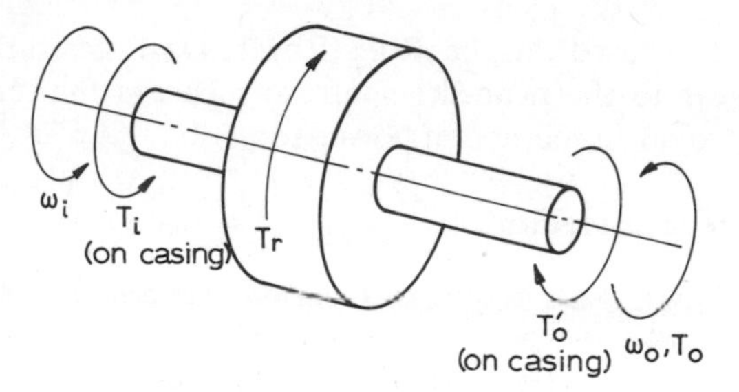

Fig. 5.19 Torque-reaction gearbox

For balance of the torques,

$T_i - T_r - T_o' = 0$ (positive sign indicates same direction as input)

or $$T_i - (n/\eta)T_i = T_r$$

Hence $$T_i = T_r/(1 - n/\eta) \qquad 5.1$$
$$= \text{constant} \times T_r$$

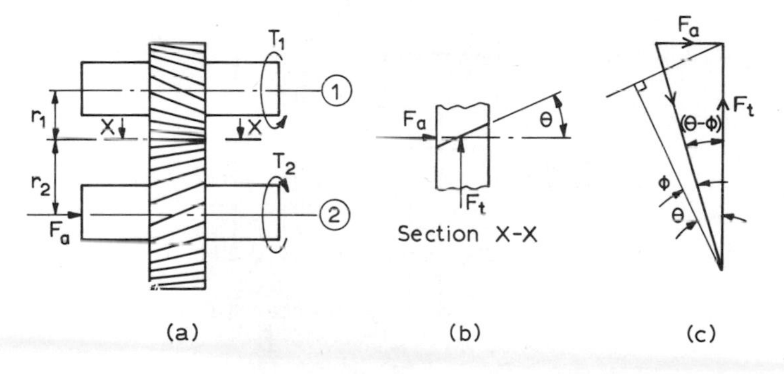

Fig. 5.20

Another method uses the axial thrust on the inclined tooth of a helical spur gear to measure torque. At a constant speed, the wheel (2) in fig. 5.20(a) is in equilibrium under the action of the tangential tooth force F_t and the axial thrust F_a. The relationship is shown in fig. 5.20(c), giving

$$F_a = F_t \tan(\theta - \phi)$$

where $\phi = \arctan \mu$, and μ is the coefficient of friction. Hence

$$T_2 = F_t \times r_2$$
$$= F_a \times r_2/\tan(\theta - \phi) \qquad 5.2$$
$$= \text{constant} \times F_a, \text{ if } \mu \text{ is constant.}$$

The method is used in the Rolls-Royce Dart gas-turbine engine reduction gear to the propeller shaft, to measure the torque output and hence provide a method of power control.

5.4 Pressure measurement

5.4.1 Piezometer and manometer

The fluid pressure in a pipe may be measured by a simple piezometer tube as shown in fig. 5.21(a). The absolute pressure in the fluid at section XX is $P = \rho g h + P_a$. The gauge pressure is

$$P_g = \rho g h$$

$$\therefore \quad P_g = \text{constant} \times h$$

if the fluid has constant and uniform density in a constant gravitational field. It is frequently convenient to express pressure as a head of fluid, often either water or mercury. (A table of other liquids used is given in ref. 9.)

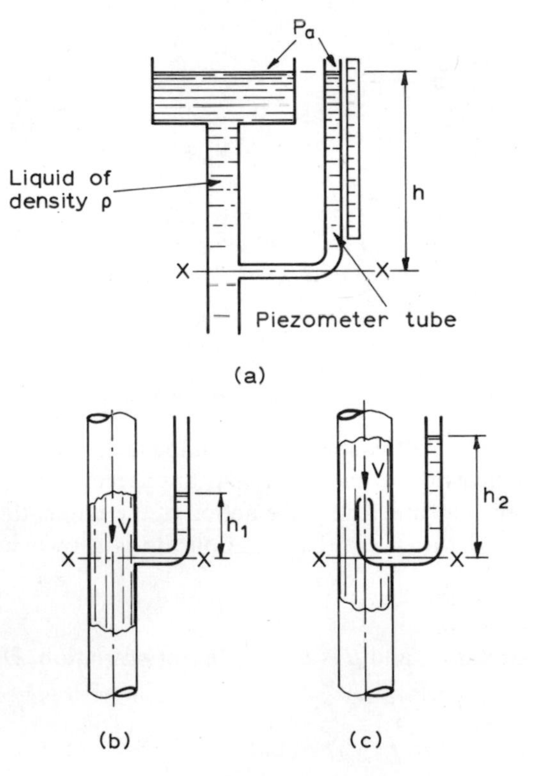

Fig. 5.21

In the case of a fluid in motion, different readings are obtained for the pressure head h for the arrangements shown in fig. 5.21(b) and (c). In (b), the pressure indicated is the static pressure. In (c), the inlet faces the fluid stream, and velocity energy is transformed to pressure energy, giving an increased head h_2. Arrangement (c) is known as a 'Pitot tube' (see Chapter 7).

The piezometer tube utilises the working fluid, and hence is not suitable for gases. The U-tube manometer (fig. 5.22), may be used for liquid- or gas-pressure measurement, provided the working and manometric fluids do not mix, or react chemically. Pressure at XX in both arms is the same,

hence $$P_1 = \rho gh + P_2$$

and $$P_1 - P_2 = \rho gh$$

If P_1 is the pressure being measured, then the U-tube may be used to measure (a) absolute pressure, if $P_2 = 0$ (i.e. vacuum); (b) gauge pressure, if P_2 = atmospheric pressure; (c) differential pressure, if P_2 is a second pressure to be compared with P_1.

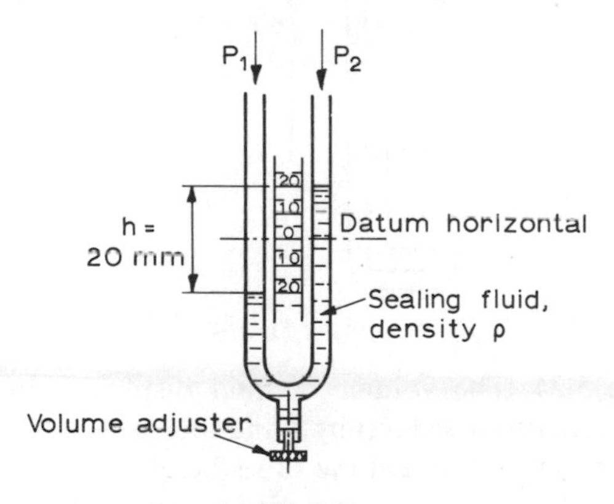

Fig. 5.22

A fixed single scale may be used for the piezometer tube as in fig. 5.21(a), but the U-tube manometer of equal cross-sectional area in both legs must have a double scale as in fig. 5.22, with provision to bring the scale zero to the level of the liquid surfaces when $P_1 = P_2$. This may be done either by a volume-adjusting device similar to that shown, or by a movable scale. If the U-tube scale is to be read directly, then it will be graduated in half-size increments, as indicated in fig. 5.23.

The level in one leg of the U-tube may be made nearly constant by making its cross-sectional area large compared with the other leg, as

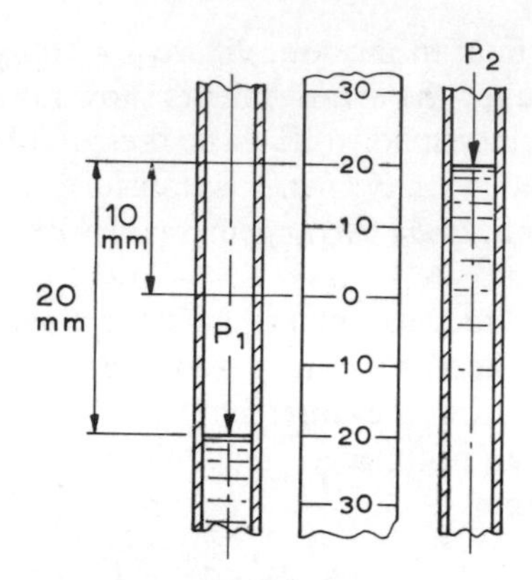

Fig. 5.23

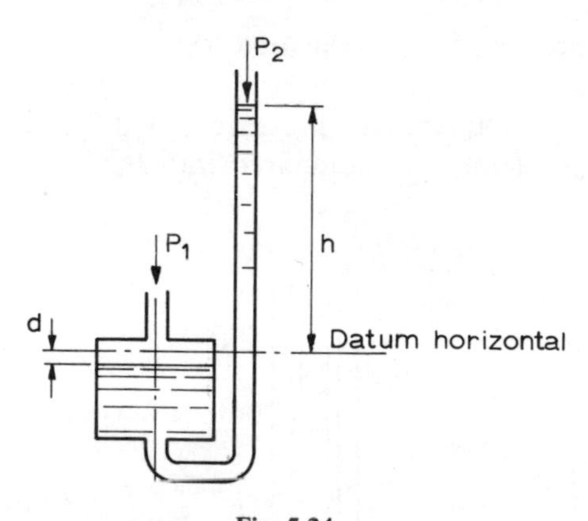

Fig. 5.24

shown in fig. 5.24. The volume of fluid displaced due to pressure difference reduces one arm level by d and increases the other by h, and the pressure head is $h + d$. Let the tubes be of uniform cross-sections, the areas being A for the large and a for the small tube.

Then $$A \times d = a \times h$$

and $$d = ah/A \qquad 5.3$$

Equating pressures at the lower surface level,

$$P_1 = P_2 + \rho g(h + d) \qquad 5.4$$

Combining eqns 5.3 and 5.4,

$$P_1 - P_2 = \rho g(h + ah/A)$$
$$= \rho gh(1 + a/A) \qquad 5.5$$

According to the degree of accuracy required, the a/A term may be neglected, or corrected for by calculation, or the scale divided so that true distance along the scale $\times$ $(1 + a/A)$ gives marked value.

The sensitivity of the upright piezometer and of the U-tube manometers depends on the fluid used, greater sensitivity (i.e. scale movement per unit pressure change) being obtained with less dense fluids. The sensitivity may be increased by inclining the piezometer tube and the measuring tube of the U-tube manometer, as indicated in fig. 5.25.

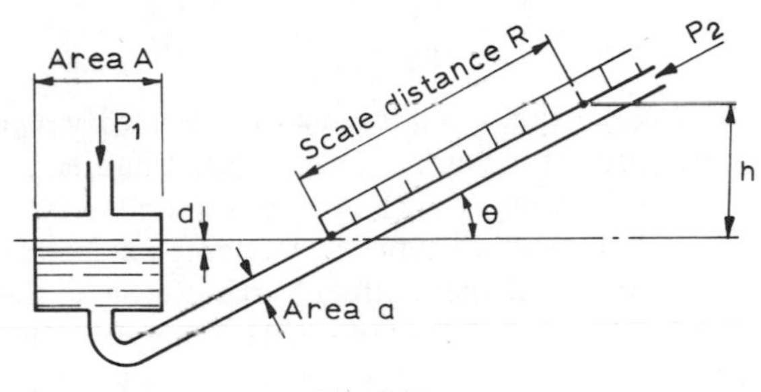

Fig. 5.25

Equating volumes displaced,

$$A \times d = R \times a$$
$$d = Ra/A \qquad 5.6$$

Equating pressures at the lower surface,

$$P_1 = P_2 + \rho g(h + d)$$
$$= P_2 + \rho g(R \sin \theta + d) \qquad 5.7$$

Combining eqns 5.6 and 5.7,

$$P_1 - P_2 = \rho g(R \sin \theta + Ra/A)$$
$$= \rho g R(\sin \theta + a/A) \qquad 5.8$$

Hence the scale should be divided so that true distance along the scale $\times$ $(\sin \theta + a/A)$ gives marked value.

Many refinements of the basic types shown here are made for special applications. For example, the rise and fall d of the well level may be used to operate a float, the movement of the float being amplified mechanically, electrically, or optically to give an enlarged scale movement. In the case of fluctuating pressure, it may be necessary to damp oscillations of the manometer fluid by means of an orifice through which the fluid has to pass (fig. 5.26). The orifice may be in the form

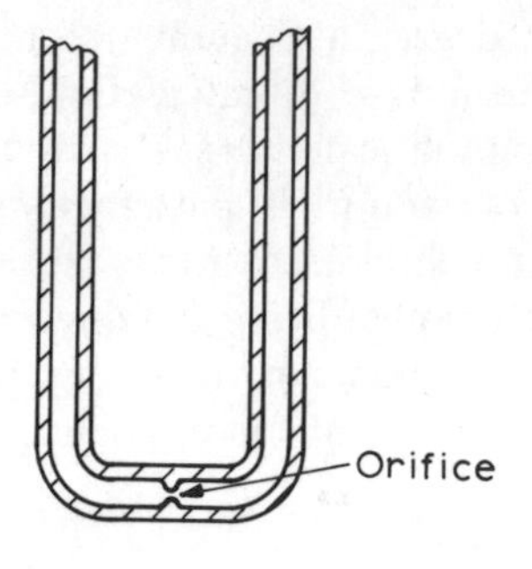

Fig. 5.26

of a rotatable plug, so that damping may be selected when required, and possibly the amount of damping varied (see Chapter 3).

Reference 9 describes many variations of the above basic types. It should be noted that, with all types of manometric instruments, the tubes must be correctly aligned, either vertically or at the angle (θ) according to type. Also, the indicated pressure values are dependent on g, and variations from the standard value will necessitate corrections. Also, where one of the pressures is atmospheric, the indicated value of the measured pressure will vary with atmospheric-pressure changes, and the barometer reading must be taken to calculate the absolute pressure = gauge pressure + atmospheric pressure *at the time*.

5.4.2 *Deflection and strain of elastic elements*

The deflection of helical springs, diaphragms, bellows, etc., such as are shown in section 5.2.3 for force measurement, may be adapted to pressure measurement as indicated in fig. 5.27.

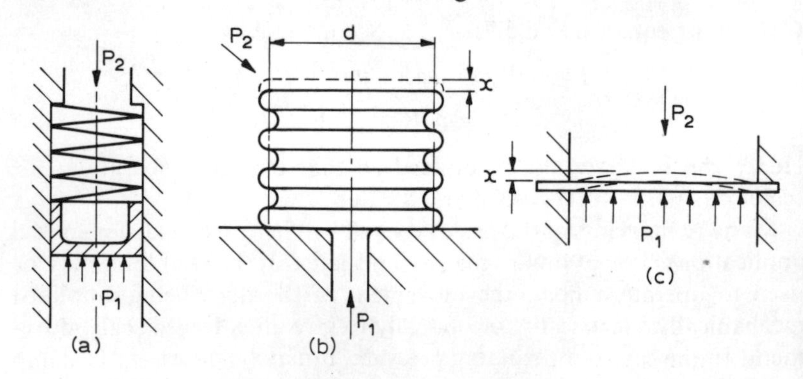

Fig. 5.27

A variety of methods are used to sense and transmit the deflection, a typical industrial instrument being shown in fig. 5.28. In this transducer, the deflection of the diaphragm due to pressure causes a change of reluctance of a miniature bridge circuit in the housing. Thus the effect of loss of signal or of interference in the connecting leads is eliminated, and the bridge is temperature-compensated.

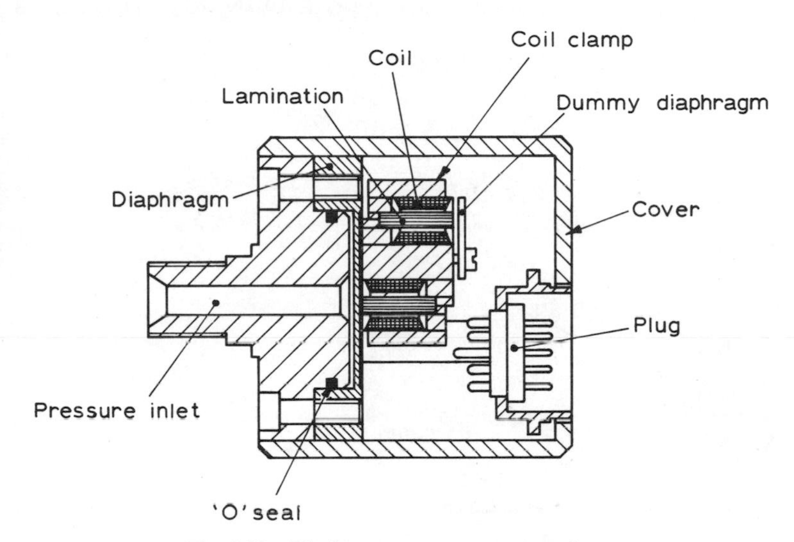

Fig. 5.28 Diaphragm-type pressure transducer

A further type, which occurs more frequently than any other, is the Bourdon tube, the simple C-type being illustrated in fig. 5.29. The applied pressure P_1 acts inside the tube, and P_2 acts at the outside. The result is a bending moment applied to the tube which tends to straighten it out. The deflection x of the end C is a function of $P_1 - P_2$, and is amplified by the lever, quadrant, pinion, and pointer arrangement. A flat spiral spring is commonly used to take up the backlash between the quadrant gear and the pinion (fig. 1.8).

Increased sensitivity may be obtained by extending the length of the Bourdon tube in the form of a flat spiral or a helix, as indicated in fig. 5.30(a) and (b). The movement of the end C of these elements may be amplified, transmitted, displayed, or recorded in a variety of ways, as discussed in Chapter 2. The resistance of these auxiliary elements should be as small as possible compared with the force exerted on the elastic element, to minimise error.

The strain of an elastic element may be used to measure pressure, as in the case of force, discussed in section 5.2.3. The electrical-resistance

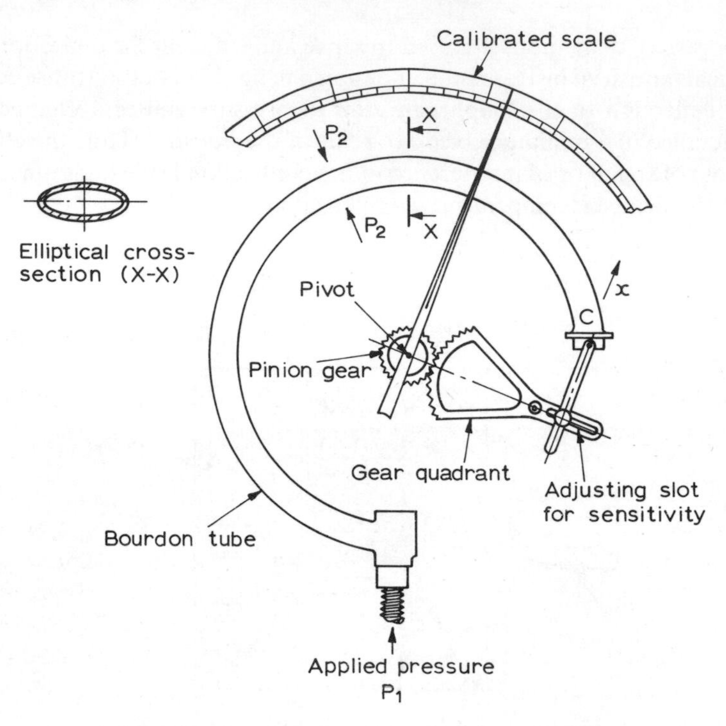

Fig. 5.29 Bourdon tube with mechanical amplification

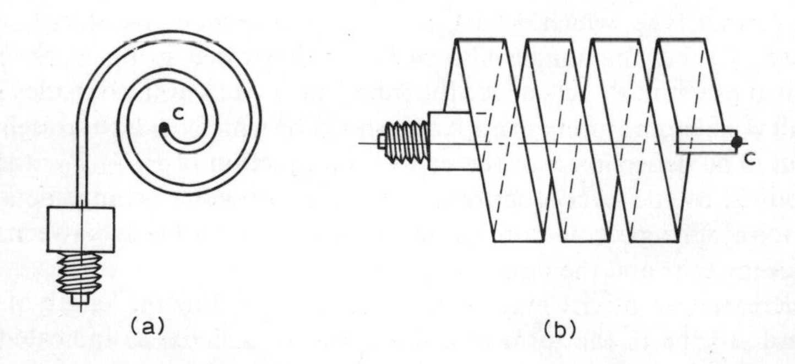

Fig. 5.30

strain-gauge is commonly used, as shown in figs 5.31 and 12.22(b) and (c). The gauge integrates the strain over the area that it covers, and hence the unit must be calibrated.

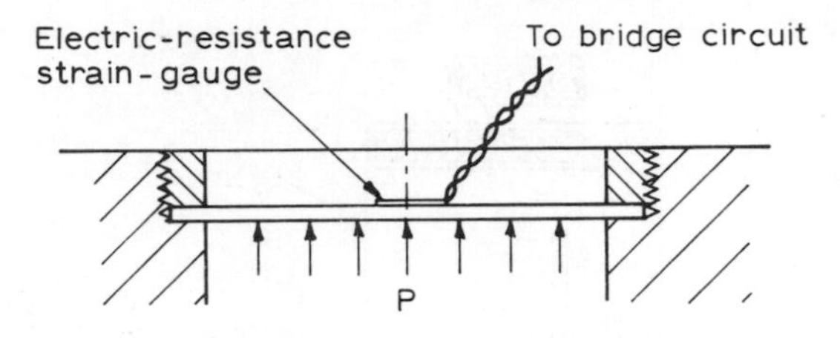

Fig. 5.31 Pressure transducer

5.4.3 Piezoelectric elements

If the force on the face of a piezoelectric transducer is due to uniform fluid pressure, then the system may be calibrated in pressure units.

5.4.4 Dynamic pressure measurement

Measurement of the continuously varying pressure in fluid systems, such as in reciprocating pumps and compressors and in hydraulic and other fluid systems, presents problems due to the high rate of change of the signal with time. Difficulties may also arise due to high temperatures, corrosive media, and vibration. Pressure in the cylinders of steam engines, gas engines, and the earlier slow-speed oil engines was measured by 'engine indicators' such as the Dobbie-McInnes instrument illustrated in fig. 5.32. The cylinder pressure connects to the underside of the piston when a connecting cock is opened. Deflection of the spring depends on the force applied to it and its stiffness. The linkage magnifies the piston displacement at the stylus, which has straight-line vertical motion. A cord connects the drum to a reciprocating part of the engine, hence the angle rotated through by the drum should be proportional to the stroke of the piston. The stylus traces a line which should be proportional to the cylinder pressure on the vertical axis and to the piston stroke (and therefore volume) on the horizontal axis. This line is marked on the 'indicator card' fixed round the drum. A typical trace for a four-stroke internal-combustion cycle is shown in fig. 5.33.

The gas pressure compressing the spring has also to accelerate the piston during the early part of its stroke, and the inertia of the piston has to be resisted by the spring on the later part of the stroke. The links in the amplifying elements are slender, but nevertheless have mass and rotational inertia; hence a torque is required to accelerate and decelerate them. Some bending occurs in the links due to this and to stylus drag, and with possible play in the pivots, some whip in the cord, friction, and the spring and cylinder inertias, considerable error

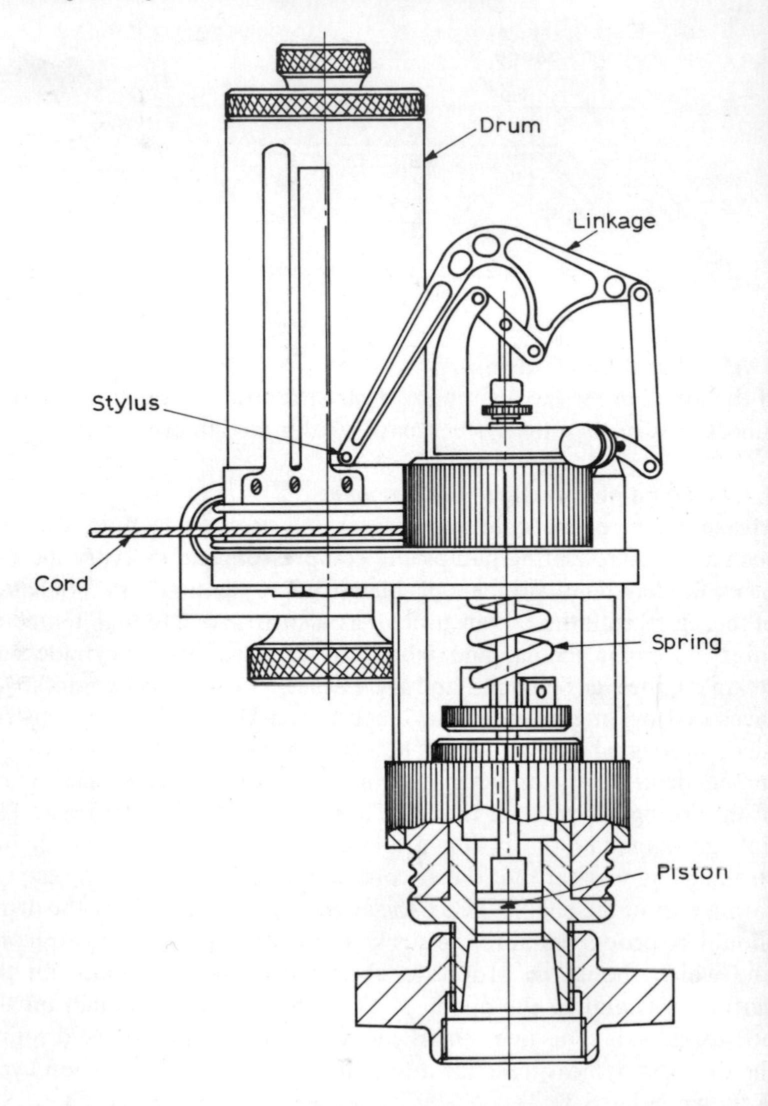

Fig. 5.32 Spring-and-piston-type engine indicator

can result. This is increased as the engine speed increases, and the piston acceleration and the rate of pressure rise increases. The instrument is thus unsuitable for use in higher-speed engines.

Piezoelectric transducers are becoming more widely used for engine-cylinder and other rapidly fluctuating pressures. Apart from very slight compression and expansion of the quartz and metal parts of the

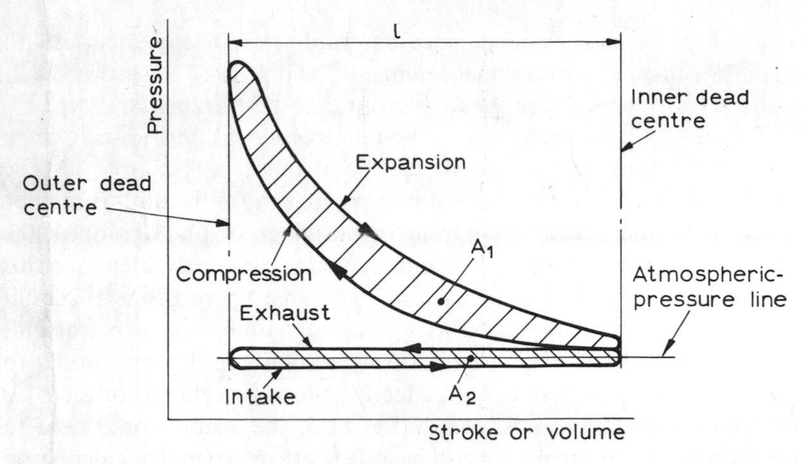

Fig. 5.33

transducer, the only moving parts of the system are electrons, and the system may be designed to follow a fluctuating pressure very accurately. The transducers vary in size, a miniature one mounted on a special sparking plug being illustrated in fig. 5.34. The system is as shown in fig. 5.7, the output usually being displayed on a CRO. Alternative

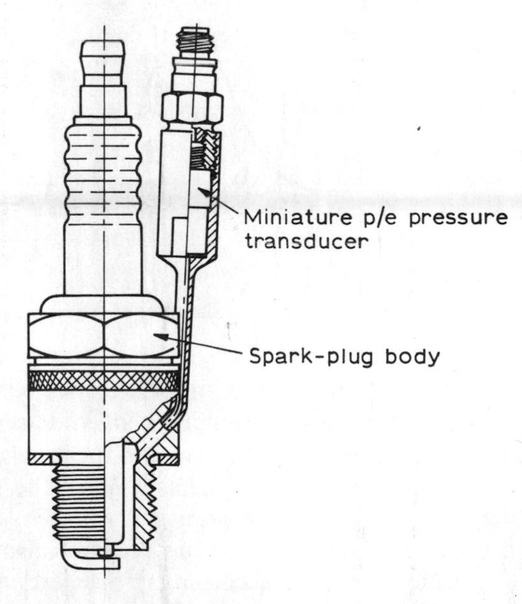

Fig. 5.34

systems use elastic elements such as diaphrams in conjunction with strain- or displacement-sensing elements, as discussed in section 5.4.2, but the transducers have greater inertia than the piezoelectric type.

The output from such systems would be observed on an oscilloscope on which timing marks from an electromagnetic pick-up may also be displayed (fig. 5.35). To obtain a permanent graph, the outline may be traced from the screen, using transparent paper, or photographed; the Polaroid camera, giving immediate prints, is extremely useful for this purpose. Better still, a trace may be made on a UV recorder. It should be noted that the horizontal axis is usually time, not displacement, and, in the normal engine mechanism, stroke is not proportional to time. Hence, to obtain a mean effective pressure value from the CRO or UV-recorder trace as shown in fig. 5.35, the graph would need to be redrawn with stroke as the base, and then a similar calculation carried out as for the engine indicator. More frequently, the trace is observed for variations in the cycle with varying ignition advance and other parameters.

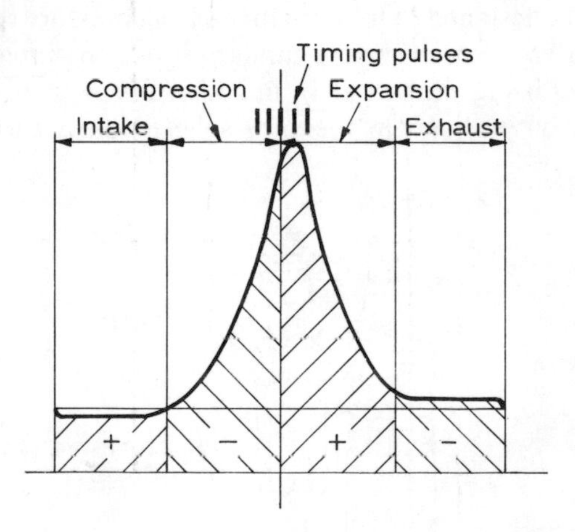

Fig. 5.35 CRO or UV-recorder trace of engine pressure cycle

The 'Farnboro'' indicator traces a line electrically on paper mounted on a rotating drum fixed to the engine shaft. A diaphragm is subjected to a controlled pressure on one side and to the pressure from an engine cylinder on the other, as in fig. 5.36(a). When the two pressures are equal, the sensor operates the primary of an h.t. coil, and a spark marks the paper. The controlled air pressure also operates a lever system to position the spark according to pressure. As the air pressure varies from above the maximum cylinder pressure to atmospheric

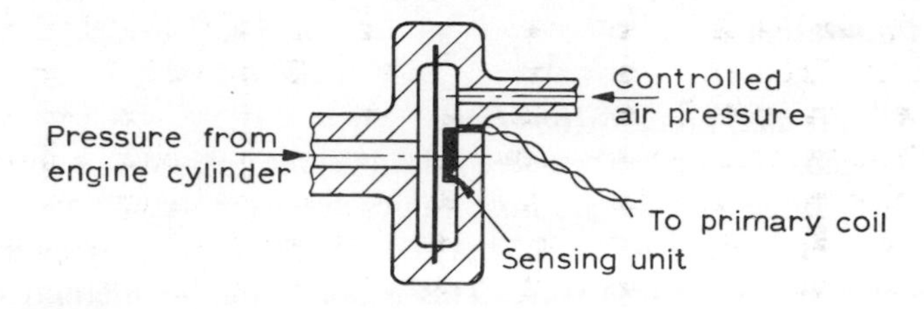

Fig. 5.36(a)

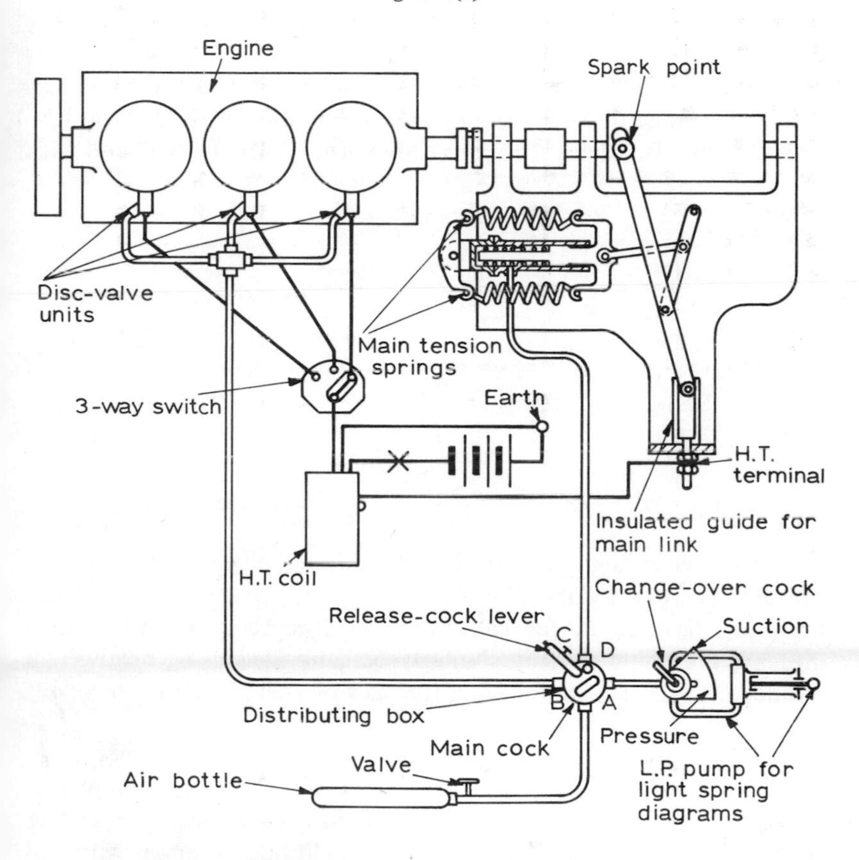

Fig. 5.36(b) Farnboro' indicator

pressure, a line is traced by sparks, the axes being pressure and crank angle. The trace is produced over several engine cycles, thus averaging out the values. The diaphragm unit may easily be subjected to pressure from different cylinders at the control of a cock. The general arrangement is shown in fig. 5.36(b).

5.5 Calibration

5.5.1 Force and torque calibration

Small devices may be calibrated by deadweight methods, i.e. by using the gravity force on masses whose values are accurately known. Thus the tool dynamometer of fig. 5.6 may be calibrated by applying masses to the end, in a suitable fixture. This is not a suitable method where larger forces are involved.

Manufacturers of instruments for force measurement work with the National Physical Laboratory to provide load cells and proving rings which have been very accurately calibrated against primary standards. These may be used to calibrate apparatus initially, or to check calibration at intervals. Examples of these are illustrated in figs 5.37 and 5.38.

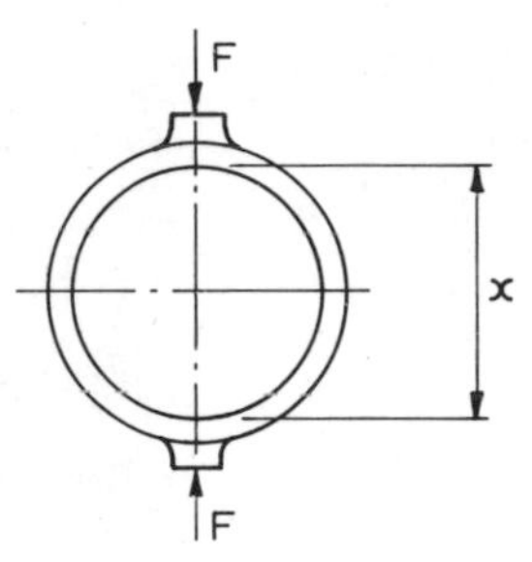

Fig. 5.37 Proving ring

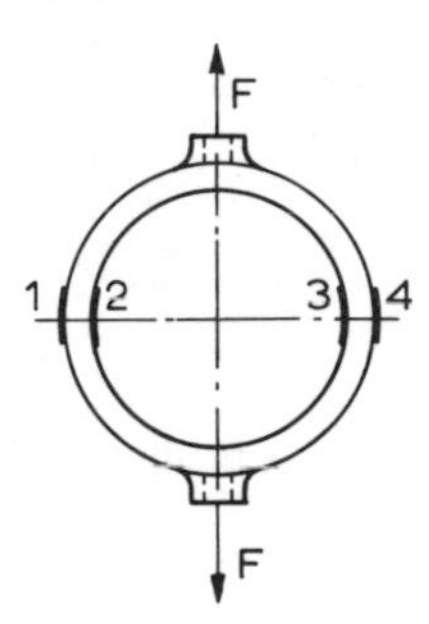

Fig. 5.38 Load cell

Other devices, e.g. piezoelectric transducers, may be obtained from the manufacturers with calibration charts and guaranteed accuracy, and may be used in the calibration of other equipment. Torque-measuring equipment may similarly be calibrated using deadweight methods, load cells, and proving rings, or piezoelectric transducers.

5.5.2 Pressure calibration

A simple piezometer tube or U-tube needs no calibration except that of length and the specific weight of the fluid, plus zeroing, at a known pressure. These devices may be used to calibrate other instruments. However, the pressure range is very low, and for higher values the deadweight tester shown in fig. 5.39 is used. In this precision instrument, the piston T is wound in by the handle H to pressurise the hydraulic fluid. When the spindle S is lifted by the fluid pressure, the pressure force is balanced against the gravity force on the mass M of the calibrated masses, plus the spindle and table, and a friction force. If the cross-sectional area of the spindle is A, then

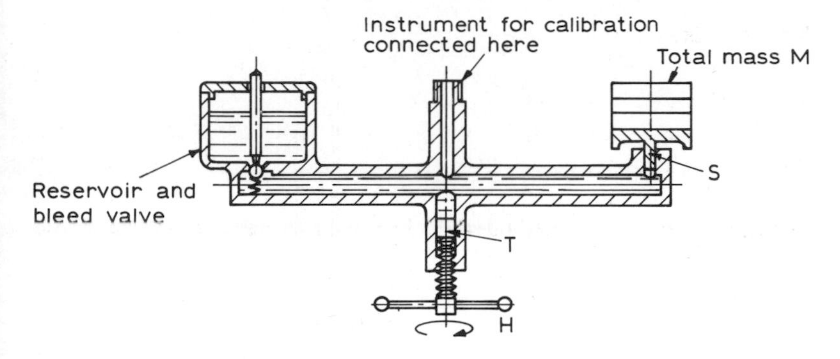

Fig. 5.39 Deadweight pressure-calibration unit

$$PA = Mg + \text{friction drag}$$
$$P = (Mg + \text{friction drag})/A$$

The friction drag is minimised by good surface finish and fit between the spindle S and its bore, and by rotating the table so that kinetic friction applies rather than static friction, with the probability of stick-slip conditions. Many kinds of pressure-measuring devices may be calibrated, such as industrial pressure-gauges, piezoelectric pressure transducers, engine indicators, etc. The tester is calibrated initially by the manufacturer for a given g value, usually the standard 9·806 65 m/s^2, and local variations in g may be corrected for if significant. Variations of the type illustrated are used to calibrate vacuum gauges and for oxygen gauges, where special precautions must be taken to prevent oil from entering the gauge, to prevent explosion risk.

Well-type manometers and similar fixed-scale devices depend on accurate well and tube sizes, and should be calibrated against simple U-type manometers or other accurate standards. A recent introduction is a helical Bourdon tube made of fused quartz, which has lower hysteresis, creep, and fatigue than any other tube material, and which is chemically inert to most gases. The deflection of the end may be measured optically, so that no force is applied to the end of the element. The instrument has a possible accuracy of 0·015% of reading, and is suitable for calibrating other instruments to a high degree of accuracy.

Surface-tension and adhesion effects between the liquid and the tube of a manometer cause a difference of level (h) as shown in fig. 5.40. The value of h increases as the size of bore (d) decreases, but it also depends on the type of surface, the liquid used, and the degree of contamination (cleanliness). Mercury barometers are corrected for this in the initial calibration, but obviously contamination can affect the accuracy. In

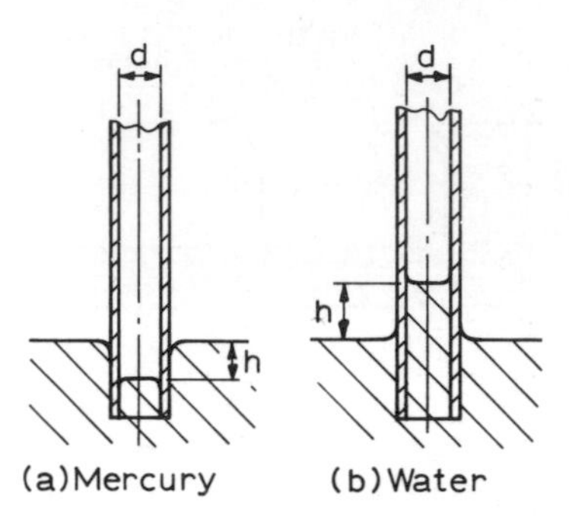

Fig. 5.40

U-tube manometers, h may not be the same in both legs. In well-tube manometers, the effect will be minute in the well, but may be considerable in the tube if d is small (ref. 27 gives data for mercury). However, if the scale zero is adjusted before taking readings, error due to this should be negligible. Other errors arise due to change of the gravitational acceleration (not more than 1 in 1000 in the UK), and temperature change causing expansion and density change of the fluid (ref. 27 gives data for mercury).

5.6 Worked examples

5.6.1 For the shaping-tool dynamometer shown in fig. 5.4, calculate the thickness of the measuring beam if it is made of steel for which the modulus of elasticity (E) = 200 GN/m². The maximum cutting-tool force is 1 kN, and the maximum beam deflection is 0·1 mm. Check whether the stress is in the elastic range. Calculate the overall sensitivity if the indicator pointer is 20 mm long and one revolution represents 1 mm travel of the plunger.

$$\text{Force on beam, } F_b = 1000 \times 50/30$$
$$= 1667 \text{ N}$$

For a simply-supported beam of span l, central deflection $y = Fl^3/48EI$. Also, $I = bd^3/12$ and $E = 20000$ MN/m² or N/mm².

$$\text{Hence } y = \frac{F_b l^3 \times 12}{48Ebd^3}$$

$$d = \sqrt[3]{\left(\frac{F_b l^3 \times 12}{48Eby}\right)} = \sqrt[3]{\left(\frac{1667 \times 50^3 \times 12}{48 \times 200 \times 10^3 \times 12 \times 0{\cdot}1}\right)}$$
$$= 6{\cdot}0 \text{ mm}$$

$$\sigma = \frac{My}{I} = \frac{1667}{2} \times 25 \times \frac{6}{2} \times \frac{12}{12 \times 6^3}$$

$$= 288 \text{ N/mm}^2 = 288 \text{ MN/m}^2$$

This is below the elastic limit for steels.

$$\text{Sensitivity} = \frac{\text{scale deflection}}{\text{force applied}}$$

$$= \frac{2\pi \times 20 \times 0{\cdot}1}{1000}$$

$$= 4\pi \text{ mm/kN}$$

5.6.2 The electromagnetic torque of a moving-coil instrument is 30×10^{-6} Nm for a current of 15 mA. This is to be balanced by the torque from two flat spiral springs made from copper strip 1 mm wide and 0·05 mm thick. If the maximum reading on a 90° scale with a pointer 90 mm long is to be 15 mA, calculate the length of strip required for each spring. Modulus of elasticity of copper (E) = 100 GN/m^2. Draw a block diagram for the system, giving the transfer function for each part, and state the overall transfer function or sensitivity of the system.

From appendix D, for a flat spiral spring,

$$\text{stiffness } k = \frac{T}{\theta} = \frac{EI}{l}$$

$$I \text{ for strip} = \frac{bd^3}{12} = \frac{1 \times 0{\cdot}05^3}{12} \text{ mm}^4$$

$$E = 100 \times 10^3 \text{ N/mm}^2, \quad T = 30 \times 10^{-3}/2 \text{ mNm per spring}$$

$$l = EI\theta/T$$

$$= 100 \times 10^3 \frac{\text{N}}{\text{mm}^2} \times \frac{1 \times 0{\cdot}05^3}{12} \text{mm}^4 \times \frac{\pi}{2}$$

$$\times \frac{2}{30 \times 10^{-3}} \frac{1}{\text{mNm}}$$

$$= \frac{1000\pi}{28{\cdot}8} \text{ mm}$$

$$= 109 \text{ mm for each spring}$$

$$\text{Transfer function of electromagnetic system} = \frac{\delta T}{\delta I}$$

$$= \frac{30 \times 10^{-6}}{15} \frac{\text{Nm}}{\text{mA}}$$

$$= 2 \times 10^{-6}\ \text{Nm/mA}$$

$$\text{Transfer function for springs} = \frac{\delta \theta}{\delta T} = \frac{\pi}{2} \times \frac{1}{30 \times 10^{-6}} \frac{\text{rad}}{\text{Nm}}$$

$$= \frac{\pi}{60} \times 10^{6} \frac{\text{rad}}{\text{Nm}}$$

$$\text{Transfer function for pointer} = \frac{\delta x}{\delta \theta}$$

$$= \frac{\theta r}{\theta} = r$$

$$= 90\ \text{mm/rad}$$

$$\text{Overall transfer function} = 2 \times 10^{-6} \frac{\text{Nm}}{\text{mA}} \times \frac{\pi}{60} \times 10^{6} \frac{\text{rad}}{\text{Nm}} \times 90 \frac{\text{mm}}{\text{rad}}$$

$$= 3\pi \frac{\text{mm}}{\text{mA}}$$

$$= 9{\cdot}43\ \text{m/A}$$

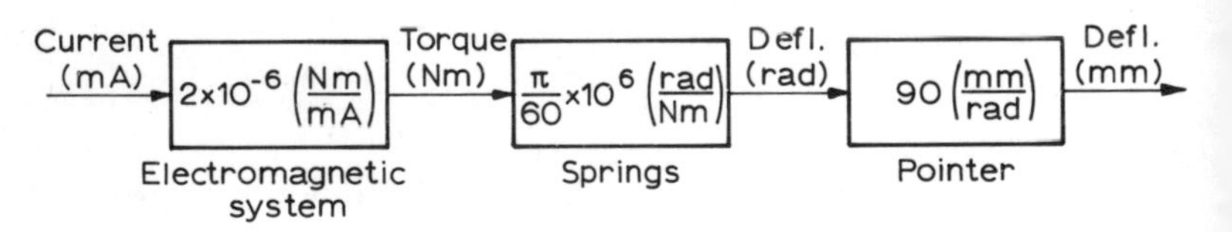

Fig. 5.41

5.6.3 An epicyclic gearbox gives a speed reduction of 8:1, and the output shaft rotates in the opposite direction to the input shaft. If the casing is mounted on bearings and a pressure capsule at 0·3 m radius indicates a force of 120 N, calculate the input torque, assuming a gearbox efficiency of 90%.

$$\text{Input torque} + \left\{\begin{matrix}\text{reaction torque} \\ \text{from outputshaft}\end{matrix}\right\} + \left\{\begin{matrix}\text{reaction torque} \\ \text{on casing}\end{matrix}\right\} = 0$$

or

$$T_i + T_o' + T_r = 0$$

$$T_i + T_i \times 8 \times 0{\cdot}9 + 0{\cdot}3 \times 120 = 0$$

$$T_i(1 + 0{\cdot}72) + 36 = 0$$

$$T_i = \frac{-36}{1{\cdot}72}$$

$$= -20{\cdot}9 \text{ Nm}$$

The input torque and the output reaction torque are in the same direction. The negative sign indicates that the input torque and the casing reaction torque arc in opposite directions.

5.6.4 A well-type U-tube manometer using mercury as the measuring fluid has a well 30 mm in diameter and a tube of 1 mm bore. If a scale graduated correctly in millimetres is used, and the datum level is at 0 mm, calculate the reading on the scale when a pressure difference of 1·0 metre of mercury is applied across the manometer. State the percentage error in the reading, and calculate the actual error in pascal and bar units. What disadvantages arise due to tubing as small as 1 mm bore?

$$\frac{P_1 - P_2}{w} = h + d = h\left(1 + \frac{a}{A}\right)$$

$$\therefore \quad 1000 = h\left(1 + \frac{\pi d^2}{\pi D^2}\right)$$

$$= h\left(1 + \frac{1}{900}\right)$$

$$h = 998{\cdot}9 \text{ mm}$$

$$\text{Percentage error} = \frac{1{\cdot}1}{1000} \times 100$$

$$= 0{\cdot}11\,\% \text{ below the true value}$$

Actual error is 1·1 mm of mercury.

$$\text{Error in } P_1 - P_2 = \rho g \times 1{\cdot}1/1000$$

$$= 13{\cdot}56 \times 1000 \times 9{\cdot}81 \times \frac{1{\cdot}1}{1000}$$

$$= 146 \text{ Pa} = 1{\cdot}46 \text{ mb below the true value}$$

Capillary depression can be large in the small-bore tube, and variable due to the state of cleanliness of the tube and mercury. The actual head $h + d$ for a given pressure might vary by several millimetres due to this, and frequent zero setting would be necessary.

5.6.5 A manometer is to use transformer oil of relative density 0·864 as the measuring liquid, but the scale is to be graduated in millimetres of water. If one leg is a 2 mm bore tube and the other a 20 mm diameter well, calculate the angle to the horizontal at which the tube and scale must be inclined to give 4 mm scale deflection for a pressure difference $(P_1 - P_2)$ the equivalent of 1 mm head of water.

Let R be the scale deflection.
Let suffix w refer to water.
Let suffix o refer to oil.

$$P_1 - P_2 = w_w R_w \left(\sin\theta + \frac{a}{A}\right) \text{ for the water manometer}$$

$$= w_o R_o \left(\sin\theta + \frac{a}{A}\right) \text{ for the oil manometer}$$

If $\dfrac{P_1 - P_2}{w_w} = \dfrac{1}{1000}$ m, then

$$\frac{1}{1000} = \frac{w_o R_o}{w_w}\left(\sin\theta + \frac{a}{A}\right)$$

$$\frac{1}{1000} = \frac{0{\cdot}864}{1} \times \frac{4}{1000}\left\{\sin\theta + \left(\frac{2}{20}\right)^2\right\}$$

$$\sin\theta = 0{\cdot}2892 - 0{\cdot}01$$

$$= 0{\cdot}2792$$

$$\theta = 16^\circ\ 13'$$

5.6.6 From analysis of the pressure trace from an 'engine indicator' attached to a slow-speed oil engine, it was estimated that the maximum acceleration of the piston of the indicator was $2{\cdot}5g$ near outer dead centre. The mass of the indicator piston plus spindle, with a proportion of the indicator-spring mass, amounted to 30 grams. Calculate the maximum error in the indicated pressure due to the inertia effect of this mass, and the vertical distance this is represented by on the trace, if the piston bore is 12 mm and the indicator spring constant is 120 kPa/mm. What would be the corresponding values at three times this speed?

Force required to accelerate piston assembly $= M \times a$

i.e.

$$F = M \times a$$

$$= \frac{30}{1000} \times 2{\cdot}5 \times 9{\cdot}81$$

$$= 0{\cdot}735 \text{ N}$$

Pressure represented by this force is $P = F/A$

$$= \frac{0{\cdot}735}{\pi \times 6^2}$$

$$= 6{\cdot}5 \times 10^{-3} \text{ N/mm}^2$$

$$= 6{\cdot}5 \text{ kPa}$$

Distance on vertical axis of graph $= 6{\cdot}5 \text{ kPa} \times \dfrac{1}{120}\dfrac{\text{mm}}{\text{kPa}}$

$$= 0{\cdot}05 \text{ mm}$$

In the engine mechanism, the rate of change of compression and expansion depends on the engine piston acceleration, which is proportional to the square of the shaft speed. Hence the indicator piston acceleration $\approx k\omega^2$, where ω is the shaft angular velocity. If $\omega_2 = 3\omega_1$, then the new inertia force $\approx 3^2$ times the original value. Hence

$$\text{pressure error} \approx 6{\cdot}5 \times 9 = 58{\cdot}5 \text{ kPa}$$

and

$$\text{vertical-trace error} \approx 0{\cdot}05 \times 9 = 0{\cdot}45 \text{ mm.}$$

5.6.7 A diagram from a spring-and-piston-type indicator had a maximum height of 31·5 mm above the atmospheric line. If the spring rate was 120 kPa/mm and the barometer at the time read 76·2 cm of mercury, calculate the maximum pressure in the cylinder in pascal and bar units.

Gauge pressure $= 120 \dfrac{\text{kPa}}{\text{mm}} \times 31{\cdot}5 \text{ mm}$

$$= 3780 \text{ kPa} = 37{\cdot}8 \text{ bar}$$

Atmospheric pressure $= \rho g h$

$$= 13\,546 \frac{\text{kg}}{\text{m}^3} \times 9{\cdot}81 \frac{\text{m}}{\text{s}^2} \times \frac{76{\cdot}2}{1000} \text{ m}$$

$$= 101{\cdot}2 \text{ kN/m}^2 = 1{\cdot}012 \text{ bar}$$

Maximum cylinder pressure $= 3780 + 101$

$= 3881$ kPa $= 38{\cdot}8$ bar

5.7 Tutorial and practical work

5.7.1 Choose a dial indicator suitable for use with the tool dynamometer of example 5.1 (fig. 5.4) and devise a gravity-balance method of measuring the force to depress the spindle. (Vibration will probably be necessary to overcome static friction in the mechanism.) Plot the force–deflection graph, and derive an equation for the force applied by the indicator for a given spindle deflection. Refer to BS 907, and check whether or not the indicator complies with its recommendations.

5.7.2 Explain the difference between the spring-and-pointer system of the moving-coil instrument of example 5.2 and of the dial indicator in fig. 5.3.

5.7.3 Measure the relevant dimensions of a spring element such as shown in appendix D, and, using laboratory standard masses, check the force–displacement linearity and the value against that obtained from the relevant equation. Discuss reasons for any non-linearity and difference of stiffness.

5.7.4 Sketch the application of a piezoelectric force washer (figs 5.9 and 5.10) to measure the tensile and compressive dynamic forces in a link. Discuss the procedure to be carried out in calibration and measurement.

5.7.5. Sketch a simple rig to calibrate a torque wrench, and give a good estimation of the accuracy of the readings.

5.7.6 Sketch an arrangement using a vacuum pump and a mercury-filled U-tube manometer to check the calibration of a Bourdon-tube suction gauge measuring up to 8 metres of water negative head.

5.7.7 Discuss the reasons for using oils or other fluids rather than water in some manometers, although the scale is graduated in units of water head.

5.7.8 Insert small-bore glass tubes into mercury, water, paraffin, and light transformer oil, and observe the extent of capillary elevation or depression.

5.7.9 Vary the angle to the horizontal of tubes containing various manometer fluids. Note the angle at which each fluid meniscus is at

90° to the tube wall. Is there any advantage in arranging the manometer tube at this angle?

5.7.10 Discuss the relative accuracy and usefulness of the following in measuring the pressure in the cylinders of a high-speed reciprocating IC engine: (a) the Farnboro' indicator, (b) the piezoelectric pressure transducer, amplifier, CRO, and UV-recorder system.

5.7.11 Sketch a torque-measuring device using a stroboscope (see Chapter 6) to measure the torque transmitted by a shaft rotating at speeds between 500 and 5000 rev/min.

5.8 Exercises

5.8.1 A hydraulic testing machine as shown in fig. 5.2 is to apply a maximum force of 300 kN, and the ram diameter is 130 mm. If the force is to be indicated on a pressure-gauge having a scale over an arc of 270°, calculate the required range and sensitivity of the gauge.

[22·6 MPa, 11·9°/MPa]

5.8.2 For the torque wrench shown in fig. 5.13, the maximum torque is to be 100 N m, the length (l) 400 mm, and the boss radius (r) 15 mm. Calculate the minimum round-bar diameter required if the maximum allowable stress due to bending is 600 MN/m^2. Calculate the maximum scale deflection using bar of this size if the modulus of elasticity (E) of the steel is 200 GN/m^2. Sketch the scale arrangement in detail.

[11·8 mm diameter, 27·2 mm]

5.8.3 The axial force on a drill is to be measured by the deflection of a simply-supported disc of 40 mm effective diameter and 1·60 mm thick. Assuming that the force is applied round the perimeter of a circle 4 mm radius from the centre of the disc, calculate the proportional force range, the corresponding maximum deflection, and the transfer function of the element. Suitable formulae are listed in appendix D, and the following constants apply for the disc material: proportional limit stress = 400 MN/m^2, modulus of elasticity (E) = 200 GN/m^2, Poissons ratio (ν) = 0·3.

Sketch a suitable arrangement for the dynamometer.

[882 N, 0·22 mm, 0·25 μm/N]

5.8.4 An epicyclic gearbox gives a reduction ratio of 12:1 with a constant input speed of 3000 rev/min. The output shaft rotates in the same direction as the input shaft, and the output power is to be measured by measuring the force F at radius r on an arm attached to the casing, which is supported on bearings. Calculate the radius at which the force-measuring transducer must be positioned so that the output power from the gearbox is given by the equation $P = 5F$ watts, where F is in newtons. Assume a gearing efficiency of 0·9. Find the required force-transducer/amplifier transfer function if the output is to be on a digital voltmeter reading directly in watts. [173 mm, 5 V/N]

5.8.5 A pair of helical spur gears as shown in fig. 5.20 has a helix angle $\theta = 15°$, and the friction coefficient μ between the gear-teeth is 0·05. If the pitch-circle diameter of gear 2 is 125 mm, calculate the value of the axial force (F_a) per kilowatt power transmitted at 3000 rev/min. Sketch a way of using fluid pressure on the end of the shaft to balance and measure F_a, and determine the pressure per kilowatt power transmitted if the shaft diameter is 30 mm. [12·8 N, 18·1 kN/m^2]

5.8.6 A manometer has a well 20 mm in diameter, and the tube of the other leg is 4 mm diameter bore. It is proposed to use a scale graduated accurately in millimetres to measure the pressure head directly, i.e. a 1 mm scale division indicates a 1 mm pressure-head change. Calculate the angle at which the tube must be inclined to the vertical to do this. [15° 56′]

5.8.7 A mercury manometer is to use a well of 30 mm diameter bore and a vertical tube of 5 mm diameter bore. Calculate the distance between the zero and 1 bar scale marks. Is the instrument more or less sensitive than a simple U-tube manometer?
[732·5 mm, less sensitive than a U-tube using the full head (h), in the ratio 1:1·028]

5.8.8 A mercury U-tube manometer used to measure a low steam pressure shows a positive head of 35·0 mm of mercury above atmospheric pressure, but 48·0 mm of water has condensed above the lower level of mercury in the tube. If the barometer reading is 76·21 cm of mercury, calculate the absolute pressure of the steam in pascal units. What percentage error is incurred if the head of condensed water is not taken into account? [105 kPa, +0·45 %]

5.8.9 A miniature piezoelectric pressure transducer was connected to a deadweight tester, and the electrical output was connected through a charge amplifier to a valve voltmeter. For various deadweight pressures on the tester, the voltage and time values were recorded thus:

Deadweight pressure (lbf/in^2)	200	400	600	800	1000	Time elapsed (min)
Valve voltmeter readings (V)	1·20	2·20	3·60	4·90	6·00	0·00
	0·60	1·20	2·40	2·40	3·20	0·25
	0·32	0·70	1·10	1·41	1·85	0·50
	0·18	0·40	0·65	0·75	1·10	0·75
	0·09	0·22	0·40	0·50	0·65	1·00
	0·03	0·13	0·22	0·30	0·39	1·25
	0·02	0·10	0·15	0·18	0·20	1·50

a) Plot the initial voltage values (at $t = 0$) against deadweight pressure, and obtain the sensitivity of the system in volts/pascal.
b) Plot voltage–time curves for each pressure value on the same axes, and determine the time-constant of the system (τ). (Refer to Chapter 3.)

[0·93 V/MPa, $\tau = 24$ s]

6

Timing, Counting, Frequency, and Speed Measurement

6.1 Time standards and measurement

6.1.1 Standards

Time was first measured more or less crudely by ancient man by observation of the motions of the sun and stars relative to the earth. More recently, accurate observations have led to the adoption of a definition of the second in terms of the mean solar day, which in turn was defined in terms of the tropical year 1900. Following the advances in atomic physics, it is now possible to measure the frequency of the radiation from certain elements with great precision. Hence the base unit the *second* in the SI is now measured with reference to the frequency of the radiation from the element caesium 133 under very carefully controlled conditions (see appendix A).

6.1.2 Measuring methods

Having obtained a standard of time, the measurement may be carried out by a variety of means. History records examples of time measurement using the rate of flow of liquid through an orifice under a constant head, and the use of sundials and of sand-glasses is well known. More modern devices use the characteristic of many mechanical, electrical, hydraulic, etc. systems of oscillating at a particular frequency, or of charging or discharging in a known way. The summation of the successive periodic times or charge times may be indicated by a digital or analogue device, to give a measure of time elapsed. Thus it is seen that time measurement and frequency measurement are closely allied.

6.1.3 Mechanical oscillators

Mechanical systems rely on either pendulum or spring-and-mass devices to give a constant circular frequency (ω_n) and periodic time

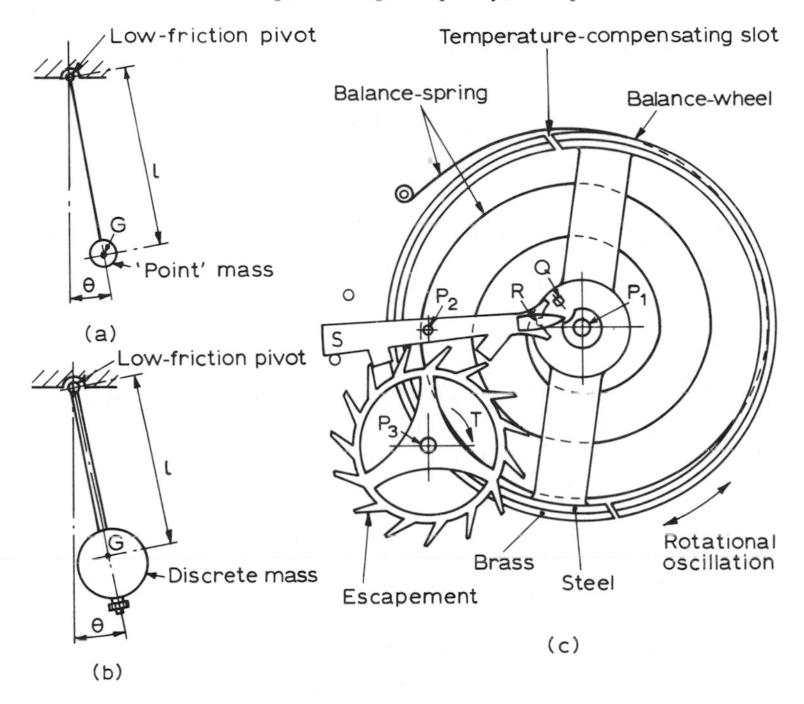

Fig. 6.1 Mechanical clocks: (a) Simple pendulum (b) Compound pendulum (c) Balance-wheel and lever escapement

(T_p). For systems oscillating with simple harmonic motion, the relationship is (see fig. 3.3)

$$T_p = 2\pi/\omega_n$$
$$= 2\pi\sqrt{(\text{displacement/acceleration})} \qquad 6.1$$

For a *simple pendulum* [fig. 6.1(a)] this gives, for small oscillations from the vertical,

$$T_p = 2\pi\sqrt{(l/g)} \qquad 6.2$$

Actual pendulums cannot have a point mass and massless arm, as assumed in a simple pendulum, and the relationship becomes, for a *compound pendulum*,

$$T_p = 2\pi\sqrt{\{(k_g{}^2 + l^2)/g\}} \qquad 6.3$$

where k_g is the *radius of gyration* of the total mass about its centre of gravity [fig. 6.1(b)].

For spring–mass systems, eqn 6.1 becomes

$$T_p = 2\pi\sqrt{(\text{inertia/stiffness})}$$

For linear spring–mass systems, this is

$$T_p = 2\pi\sqrt{(m/k)} \tag{6.4}$$

and for systems having rotational oscillation,

$$T_p = 2\pi\sqrt{(I/k)} \tag{6.5}$$

where m is the mass, I is the moment of inertia about the pivot, and k is the spring stiffness, linear or torsional.

A typical timekeeping spring–mass balance movement is shown in fig. 6.1(c). The balance-wheel oscillates about the pivot (P_1), against the restoring torque of the flat spiral balance-spring, and has a particular periodic time (T_p). As the wheel passes through the equilibrium position (P_1P_2), the peg Q on the wheel engages in the slot R in lever S, rotates the lever, and allows the escapement-wheel to rotate by one tooth. For the remainder of the oscillation, the lever is not in engagement with the balance-shaft. The externally applied torque (T) rotates the clock indicating mechanism, and also gives the balance-wheel a slight impulse, to sustain its oscillation. T may be applied by a hanging weight, by a coil spring, by an electric motor, etc. The counting of the elapsed time may be indicated on the familiar twelve-hour clock dial, on a 24 hour or other type of dial, or displayed in a digital manner using a mechanical or electrical counter.

The discrimination of the timing depends on the time of one oscillation, and is usually not less than 0·5 s. This is not important in time measurement over long periods, but it is not good enough for short-time measurements. Where timepieces are well designed, so that compensation for temperature changes etc. is made, and manufactured to high standards of accuracy, errors may be as small as 0·1 s per day.

6.1.4 Electrical oscillators

A circuit containing resistance (R) and inductance (L) or capacitance (C) has a natural frequency of oscillation (f) which depends on the values of R and L or C. However, these values may vary with changes of temperature and humidity etc., resulting in changes of f. The frequency may be stabilised by introducing into the circuit some device whose natural frequency is very stable under varying ambient conditions.

It is possible to use a tuning fork with a capacitive or inductive pick-off in the circuit to give stability, but a better device is a quartz crystal. This has a natural frequency of mechanical vibration which depends on its size and shape, and which is very stable. The forces in the crystal during the oscillation give piezoelectric charges across the faces of the crystal (see section 5.4.3). If these faces are connected into an *RC* or an *RL* oscillator circuit having a natural frequency close to that of the crystal, the charges will stabilise the oscillations in the circuit to the same frequency as the crystal. Together with an amplifier and a synchronous motor driving a clock, this gives a very accurate time-keeping device. The changes in the natural frequency of the crystal are small and predictable, and hence can be allowed for. At best, the accuracy may be within 0·1 seconds in a year. The device is shown schematically in fig. 6.2.

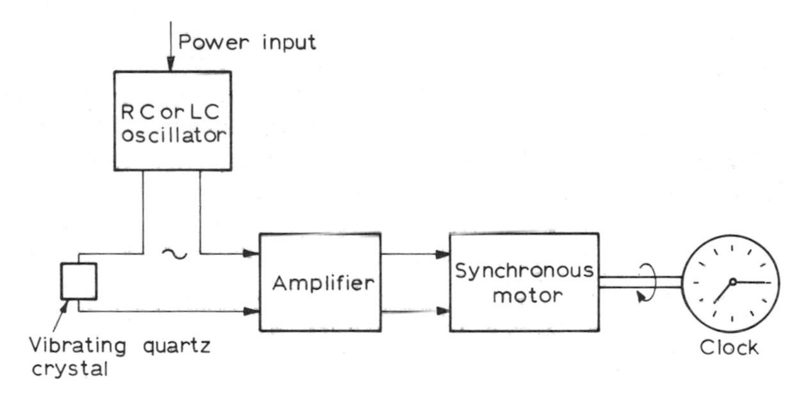

Fig. 6.2 Quartz clock

6.1.5 Atomic clocks

The frequency of oscillation of atomic radiation from some elements is even more stable than that of a quartz crystal. Atomic clocks compare the frequency of quartz oscillators with the frequency of the atomic radiation, and use the difference to correct the oscillation. By this method, the accuracy can be as good as 0·01 s per year; however, accuracies of this order are seldom required in industrial applications, and the expense of obtaining them could not be justified.

6.1.6 Industrial timing methods

Simple timing may be carried out by using stop-watches and stop-clocks, started and stopped by hand. However, errors arise due to poor discrimination of the timepiece, and more so in operator error in carrying out the starting and stopping operations at the correct instant.

Synchronous electric clocks operating from the mains supply are very useful. They can be started and stopped very conveniently by making or breaking an electric circuit. Their accuracy is no better than that of the mains supply frequency, which by law must be within $\pm 1\%$

Fig. 6.3 Laboratory timer

of 50 Hz, and which seldom reaches these limits. A typical electric clock used in laboratory work is shown in fig. 6.3.

In industrial applications, timing may be carried out by clockwork mechanisms (for example in some chart recorders), by synchronous-motor devices, resistance–capacitance circuits (see Chapter 3), by pneumatic devices, etc.

6.2 Counting devices

Many counting operations at slow and medium speeds may be carried out by mechanical or electromechanical counting devices. However, as with many other operations, the inertia of the moving parts sets a limit to the speed of operation. The force required to operate mechanical counters, and the corresponding reaction force, may also be undesirable in some applications. To overcome these difficulties, electronic counters are used, though they are much more expensive than mechanical ones.

6.2.1 Mechanical counters

These typically consist of a number of small drums, each numbered from 0 to 9 round the periphery, as shown in fig. 6.4(a). The first drum may be rotated continuously, or rotated in increments of 36° by a ratchet. As each rotation of drum one is completed, a transfer-tooth segment engages with a transfer pinion, to rotate drum 2 by 36°. A complete rotation of drum 2 rotates drum three by 36°, and so on.

The device is frequently encountered, in a variety of forms, e.g. in the vehicle odometer, in component counters, shaft-revolution counters, etc. It is cheap, compact, and reliable, and it may be situated at some distance from the measuring point, connected by a rod or a flexible shaft. Alternatively, operation may be by electric solenoid,

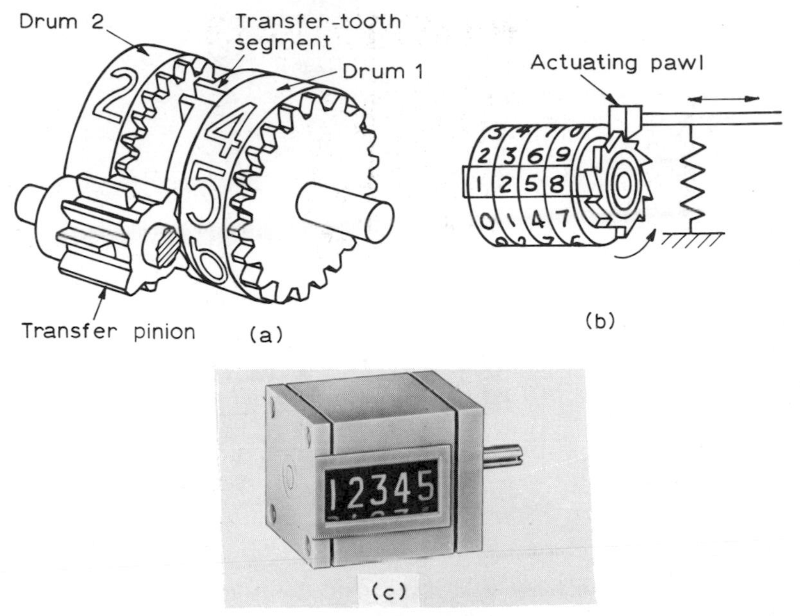

Fig. 6.4 Mechanical counter: (a) Basic mechanism (b) Operating ratchet (c) Typical casing

actuated by a pulse from a switch or transducer. This may give counting rates of up to 1000 per minute. Variations of the basic counters may be used to subtract digits as well as add, or to operate a switch after a preset number of pulses (mechanical or electrical) or rotations have been counted.

6.2.2 *Electronic counters*

The electronic counter is capable of extremely high rates of counting. Its input is a series of electric pulses (see fig. 6.8) which are shaped and amplified to a suitable form in the input circuits. Electronic counting circuits operate in a manner analagous to the rotating drums of the mechanical counter, and the numbers are displayed by one of the devices shown in fig. 6.5. At (a) is shown the cold-cathode display, where a glow at one of ten points around a tube indicates a number. At (b) is illustrated a stack of panels each having a number engraved on a clear plastics disc. The number required to be displayed is illuminated at the edge, and thus becomes visible from the front, whilst the remaining numbers cannot be seen. The type at (c) consists of columns of numbers from 0 to 9, one letter in each column being illuminated from the rear. At (d) is shown the type where a number is displayed by illuminating a number of squares to form the required figure. Types

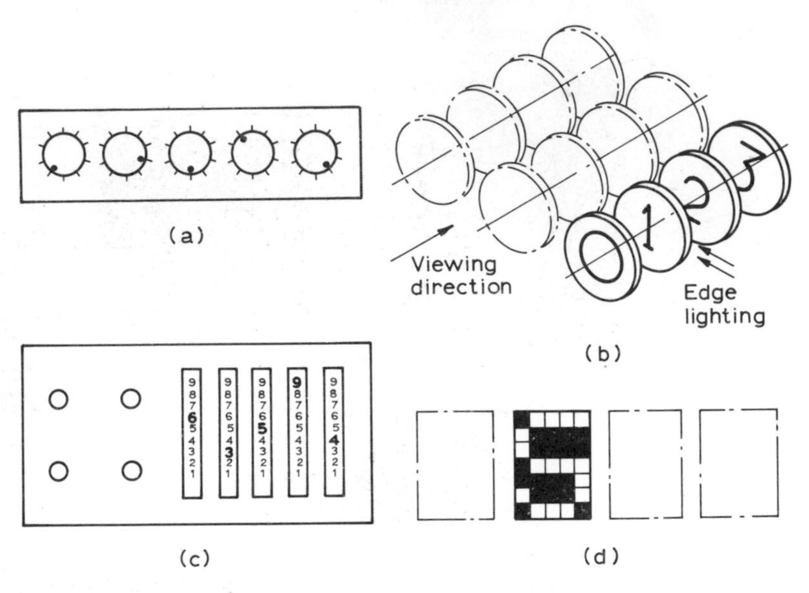

Fig. 6.5 Timer/counter digital display units: (a) Cold-cathode tube display (b) Edge-lit panels, exploded view (c) Column display (d) Built-up figures

(b) and (d) are more easily read than types (a) and (c), but are more costly.

6.3 Frequency and time measurement

If the incoming pulses occur at equal time-intervals, then, if a suitable timing method is available, the number of cycles arriving *in one second* may be counted to give the frequency in hertz. Industrial instruments use the mains frequency to measure the time, either for one second, for frequency measurement, or for time-interval over a longer period. The typical unit (see fig. 6.6) is a timer/counter, having inputs for transducers for counting and frequency measurement and connections for starting and stopping counting and timing, by making or breaking circuits, applying and removing voltages, etc.

6.4 Measurement of linear and angular speed

The mean linear velocity of an object, or of a point in a mechanism, may be found by measuring the time (t) taken for it to travel a known distance (s), then $V_{mean} = s/t$. For a rotating shaft, the time to rotate R times gives a mean rotational speed $N = R/t$. However, the *instantaneous* linear or angular velocity is usually of greater interest, and this may be measured in several ways.

Fig. 6.6 Electronic timer/counter

6.4.1 Mechanical tachometer

The simple mechanical hand tachometer may be held against a rotating shaft to give instantaneous speed, but its accuracy is poor. It suffers from slip and wear of the driving wheels, which are usually of rubber.

A similar device may be belt- or gear-driven from a shaft, for permanent indication of shaft speed.

6.4.2 Tachogenerator

This is an accurately made a.c. or d.c. generator giving an output voltage or voltage amplitude directly proportional to the angular velocity of the armature [see fig. 2.21(b)]. It may be directly coupled to a shaft, or it may be connected by a rack-and-pinion or cable etc. to a point having linear motion, as shown in fig. 6.7.

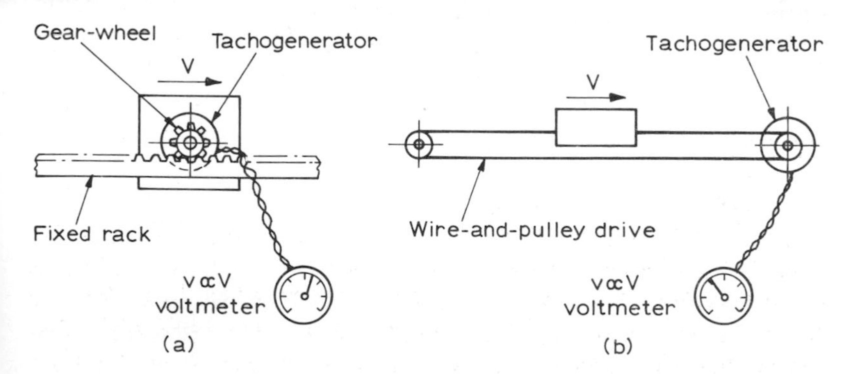

Fig. 6.7 Linear-velocity measurement

6.4.3 Digital pick-ups

A common type of transducer, known as a velocity pick-up (or pick-off or probe) has a coil wound round a permanent magnet. If a ferrous material moves near the end of the magnet, the magnetic field is changed, thus inducing a voltage in the coil proportional to the rate of change of the magnetic flux.

If a toothed wheel is rotated past the probe, a number of voltage pulses will result. These are amplified and squared, and may then be fed into a counter or frequency-measuring unit, as shown schematically in fig. 6.8. The method may also be applied to linear velocity, using a fine-toothed rack, a wire drive, or a screw. If a 60 tooth gear is used on a shaft, the frequency meter will give speed directly in revs/min; however, the speed measurement is not instantaneous, as the frequency meter will take a short time, usually one second, to count the pulses.

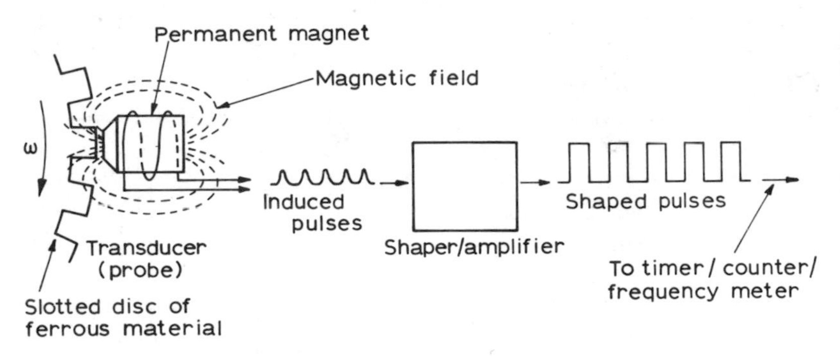

Fig. 6.8 Shaft-speed measurement

Another transducer uses the phototransistor principle [see fig. 2.28(b)]. Light falling directly or by reflection on a transistor causes a current to flow. Intermittent light falling on the transducer gives pulses which are amplified, squared, and counted as for the magnetic transducer. A typical portable device has its own light source whose ray is directed onto a small reflective surface on the rotating object. The intermittent light-reflections give pulses which, after amplifying and shaping, are converted to an analogue signal of revs/min.

6.4.4 Stroboscopic measurements

If a rotating object is illuminated once every revolution, or once every n or $1/n$ revolutions, where n is a whole number, when the object is in the same position each time, then, due to the persistence of human vision, the lighting will appear to be continuous except at very slow speeds. Hence the object will appear to be stationary, and this may be used to measure its speed, if the flashes occur once every revolution and the rate of flashing is known. A *stroboscope* powered from the mains electric supply, and using the mains frequency as a standard, has the flashing frequency (f) set by the operator, and its light beam is directed onto the rotating object, on which a reflective mark has been made. To ensure that the light flashes once every revolution, i.e. $n = 1$, the frequency is first adjusted so that *two* images of the mark appear, as in fig. 6.9(a). The frequency setting is then halved and adjusted until the mark is stationary, then the frequency reading is taken [fig. 6.9(b)]. In addition to this, the stroboscope can also be used to measure the oscillation frequency of vibrating machinery, and, if set very near to this frequency, a slow-motion image of the vibration can be observed. This is useful in observing the operation of rapidly moving devices such as the valve motion and contact-breaker mechanism on internal-combustion engines; approximate timing and other measurements may be carried out from the slow-motion observations. A typical stroboscope is illustrated in fig. 6.9(c).

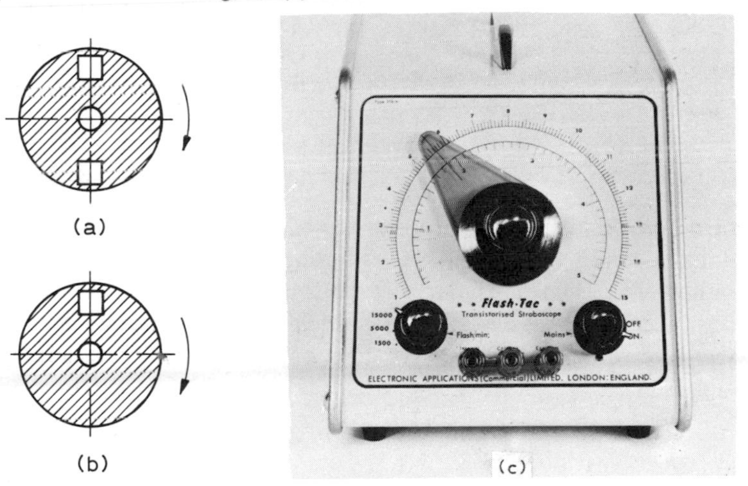

Fig. 6.9 (c) shows a Xenon stroboscope

6.4.5 *Radar methods*

The Doppler effect, whereby, for example, the whistle of an approaching train has a different frequency and hence pitch from when it is receding, has been experienced by most people. In a similar way, the frequency sent out from a stationary source is altered if it is received by a moving object. Similar effects occur in optics, for example the 'red shift' of the light received from receding stars, and in electromagnetic waves. In radar devices, a very high-frequency electromagnetic wave is used, and the difference in frequency (Doppler frequency) between a transmitted wave and its reflected wave is proportional to the relative velocity V between the transmitter and the reflector. The use of radar in road-vehicle speed measurement is well known, but a recent introduction is a small portable radar set which may be used for linear- or rotational-speed measurement. For linear-speed measurement, as illustrated in fig. 6.10(a), the Doppler frequency $(f_d) = 2V\cos\theta/\lambda = kV$, where $V\cos\theta$ is the component of velocity of the 'reflector' object in the direction of the radar beam from the measuring instrument and λ is the wavelength of the beam, usually 0·0225 m. In rotational motion, $V = \omega R$ [fig. 6.10(b)], giving $f_d = 2\omega R\cos\theta/\lambda$; hence $f_d = k\omega$. The

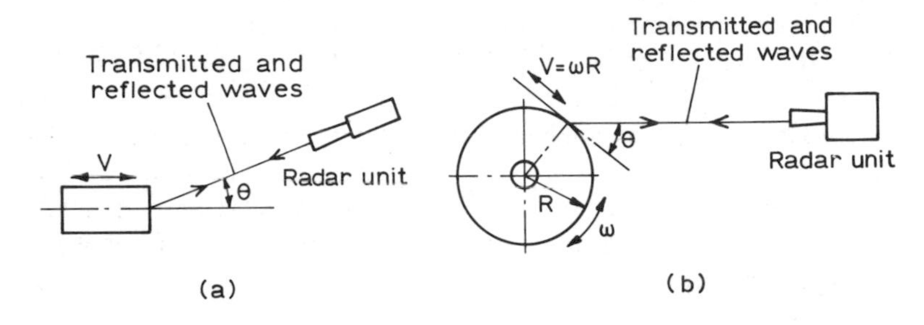

Fig. 6.10 (a) Radar measurement of velocity (b) Radar measurement of shaft speed

instrument is scaled so that a variety of ranges of linear and angular velocities may be measured. In addition, the device may be used with toothed wheels in a similar way to the magnetic and photoelectric transducers already described, in which case the output is connected to an electronic counter. A typical example of such a device is shown in fig. 6.11.

Fig. 6.11 Miniature radar speedmeter

6.5 Tutorial and practical work

6.5.1 Sketch a compound-pendulum clock using a battery-operated solenoid as the power source.

6.5.2 Detail the effects that a rise in temperature would have on the periodic time of the clock in question 1.

6.5.3 How could the effects of temperature change be minimised in the design of a mechanical clock? Discuss the effect of temperature change on the moment of inertia of the balance-wheel shown in fig. 6.1(c), and explain the function of the bimetal strip which forms the rim of the wheel.

6.5.4 Determine the discrimination of a mechanical or electrical stop-watch or stop-clock.

6.5.5 Discuss the reasons why periodic times of less than about 0·5 second are not practicable in mechanical movements.

6.5.6 What is the relationship between rotational speed and the flashing-rate of a stroboscope when a) a single small reflecting surface shows (i) three static images, (ii) four static images; b) two small reflecting surfaces opposite each other show (i) two static images, (ii) six static images.

6.5.7 Discuss possible application and operator errors (see Chapter 1) which may arise in shaft-speed measurement by the methods described in this chapter.

6.5.8 When a vibrating component was illuminated with a stroboscope flashing at 1600 Hz, it appeared to be oscillating 20 times every minute. What are the possible vibration frequencies of the component?

6.5.9 The response curve of displacement against time is required for a pneumatic ram whose air supply is controlled by the on/off operation of a 12 V electric-solenoid-operated piston valve. The timing is required from the instant of connecting the 12 V supply to the solenoid. Sketch out the pneumatic system and a timing system using an electric or electronic timer, showing the start/stop arrangements for the timer. The maximum time likely to be measured is 0·5 s. Explain how the required curve may be built up.

6.5.10 Discuss the possible physical effects which may affect the short- and long-term natural frequency of a tuning fork.

6.5.11 Discuss methods of measuring the accuracy of a camera-shutter exposure time of 0·001 s, (a) as a laboratory test, (b) as a production test.

7

Flow and Viscosity Measurement

7.1 Flow measurements

The measurement of the velocity, or of the volume or mass flow rate, of solids (in the form of particles) or of liquids or gases, is very widely necessary in the development and production processes in industry and in everyday commercial transactions. It may be necessary to know the rate of cooling-water flow to an engine, the mass flow rate of gravel on a conveyor, or the ventilation air flow in a mine. Alternatively, or sometimes in addition to this, the total quantity of the flow may be required, such as in the case of a gas meter or a petrol pump.

A large variety of velocity meters and flow meters use a number of physical principles to give a signal such as force, pressure, speed of rotation, etc. which is a function of velocity or of volume or mass flow rate. The signal may be integrated by mechanical or electrical devices, to give the total quantity in a given time. Thus, if $\mathrm{d}Q/\mathrm{d}t$ is the volume flow rate, then

$$\text{total volume } (Q) = \int \frac{\mathrm{d}Q}{\mathrm{d}t}\,\mathrm{d}t$$

or,

$$Q = \sum \frac{\mathrm{d}Q}{\mathrm{d}t}\,\delta t$$

Or, if $\mathrm{d}m/\mathrm{d}t$ is the mass flow rate, then

$$\text{total mass } (m) = \int \frac{\mathrm{d}m}{\mathrm{d}t}\,\mathrm{d}t$$

or

$$m = \sum \frac{\mathrm{d}m}{\mathrm{d}t}\,\delta t$$

Other devices may measure the total flow, such as the volume or mass of fluid in a collecting tank or the quantity of gas delivered from a gas-holder (see fig. 7.3), or may measure an *increment* of mass or volume, such as the device described in section 2.9.15 and those shown in figs 7.1 and 7.2. These latter may give a mechanical or electrical pulse for each increment, and the pulses may be summed to give the total volume or mass which has passed. These devices may give an indication of the flow *rate* from the frequency of the pulses.

7.2 Positive-displacement devices measuring quantity

Figure 7.1 illustrates a reciprocating-piston type of meter used for clean liquids. The incoming liquid is pushing the piston to the right. The liquid from the right-hand side of the piston is being pushed through the outlet via the slide-valve D. At the end of the stroke, the piston rod operates a mechanism which moves the slide-valve to the left position (shown broken), and incoming fluid fills the right side of the piston, whilst that at the left is exhausted. The motion is continuous, each oscillation moving a digit on a counter. The pressure loss across this type of meter is not very large, and very good accuracy is obtainable.

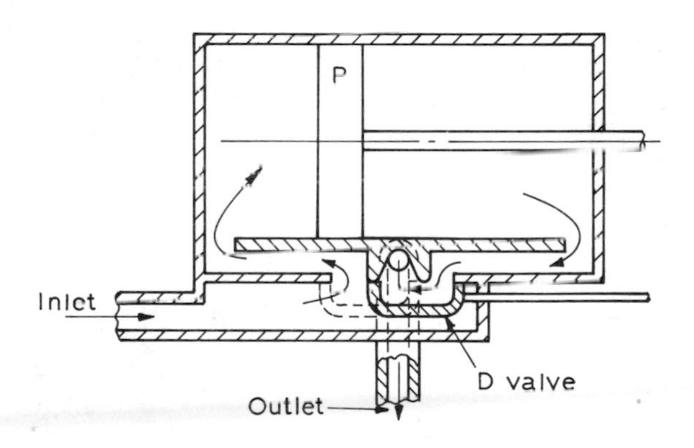

Fig. 7.1 Reciprocating-piston meter

Figure 7.2 illustrates the principle of a positive-displacement meter used for gas. In (a), gas is admitted through port E, moving disc 2 to the right and delivering gas outside diaphragm 2 through port F. At the end of the stroke, disc 2 operates a mechanism which moves the slide-valves to the positions shown in (b). Gas entering through port D causes delivery through port C. Diagrams (c) and (d) show the further motions of the four-cycle operation, which delivers the same volume of gas to the outlet at each stroke. A counter is incorporated to indicate total flow. Fluctuations of pressure and temperature may affect the accuracy of the meter, but devices to compensate for these may be

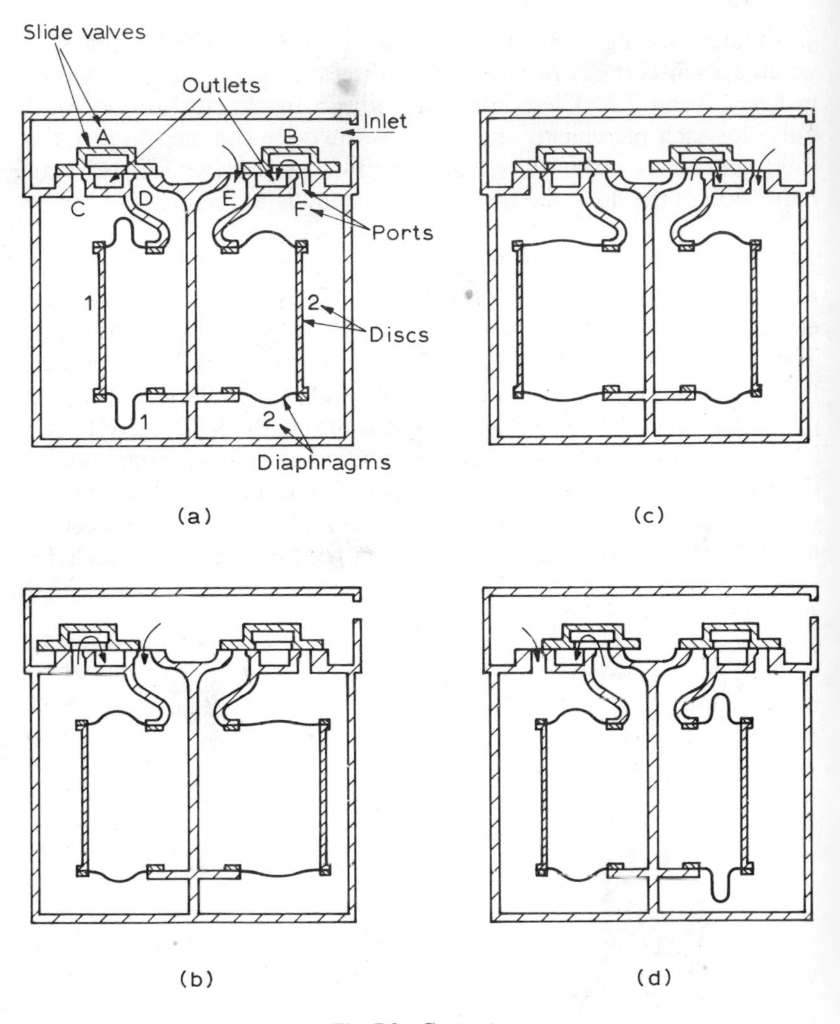

Fig. 7.2 Gas meter

incorporated. When used to measure town gas, the accuracy of the meter, and the fluctuation in delivery pressure due to its operation, must be within limits specified in government regulations.

The air supply to, for example, an engine or compressor under test may be measured by using a gas-holder type of device illustrated in fig. 7.3. After pumping air in, stopcock A is closed, B is opened, and the quantity used is read from the volume scale. The delivered air is at slightly above atmospheric pressure, due to the weight (F_g) of the air-holder.

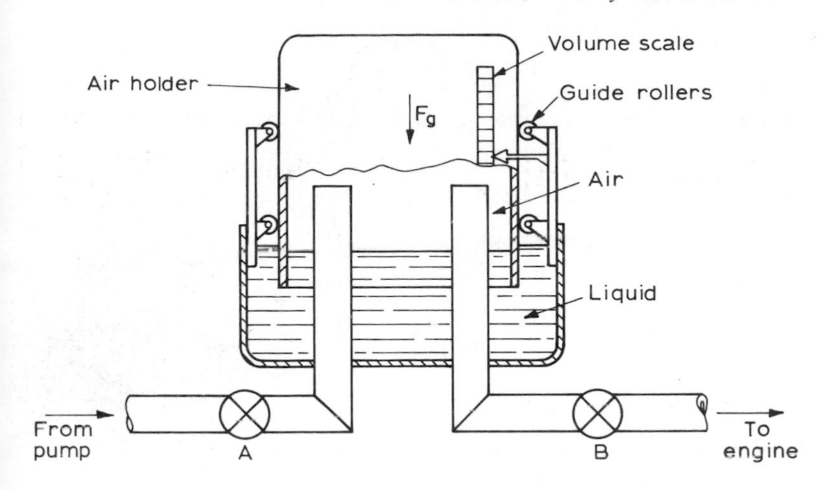

Fig. 7.3 Gas-volume measurement

7.3 Rate-of-flow meters for fluids

Instantaneous rates of flow may be measured by positive-displacement devices of rotating type, consisting of geared or lobed members engaging together. Other devices sense the force on stationary or pivoted members, or sense the pressure changes in the fluid due to momentum changes in the fluid.

Whichever method is used, the signal is a function of the fluid velocity, this usually being taken as the mean velocity ($\overline{V}$) given by

$$\overline{V} = \dot{Q}/A$$

where $\dot{Q} = \mathrm{d}Q/\mathrm{d}t$ is the volume of fluid passing per second. Hence $\dot{Q} = \overline{V}A$, and the rate of flow may be found from the velocity signal.

7.3.1 Positive-displacement meters

Figure 7.4(a) illustrates a lobed type of meter used for measuring gas flows. Each rotation transfers a definite volume of gas from the inlet to the outlet. The speed of rotation of the rotors indicates the rate of flow ($\dot{Q}$), and the total number of revolutions indicates the total volume (Q) passed through. The gas drives the rotors in a meter, but the same device may be used to pump gas and meter it at the same time. Similar devices for liquids use a variety of geared rotors, oscillating pistons, nutating discs, etc. (see refs 9 and 11). One type is illustrated in fig. 7.4(b).

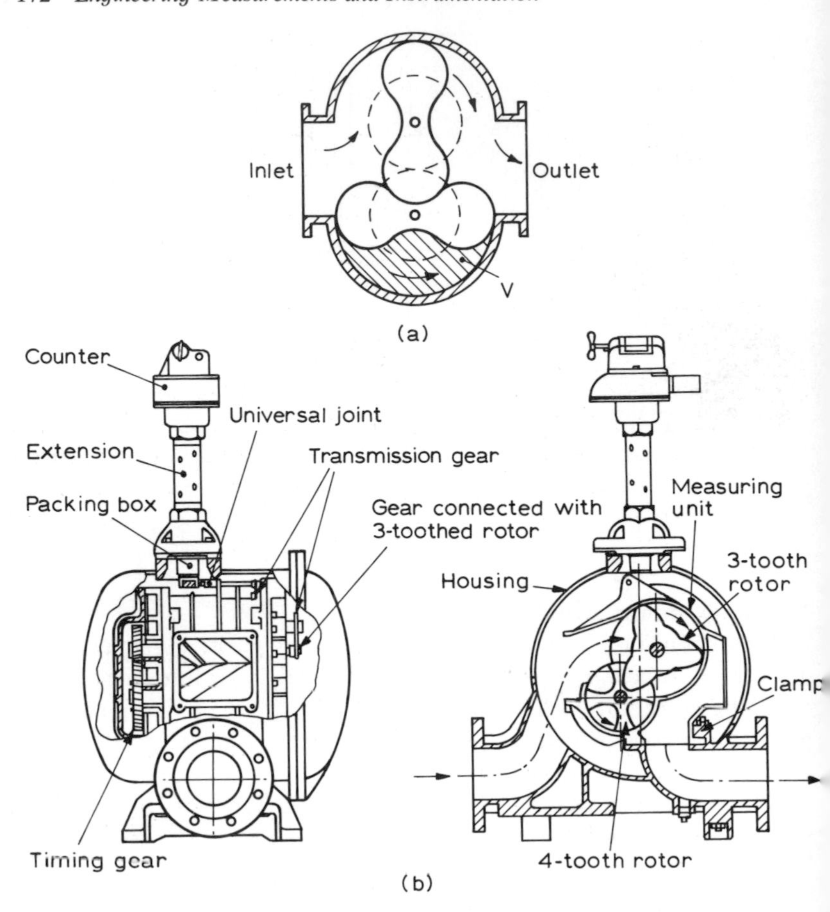

Fig. 7.4 (a) Lobed-rotor meter (b) Fluted-spiral rotor-type meter

7.3.2 *Vane- and turbine-type meters*

A simple type of static-vane flow meter for liquids is illustrated in fig. 7.5. The liquid impinging on the vane is diverted and suffers a change of momentum which exerts a force on the vane. When the vane is deflected through an angle θ, the gravitational force tends to return it to the vertical position. A position of balance is reached for each velocity value for a particular fluid, and the scale is calibrated in flow units.

Other single-vane devices have very small vane deflection, and the force on the vane is measured by the deflection of a spring or by some other force-measuring device. The force is a function of velocity for a particular liquid.

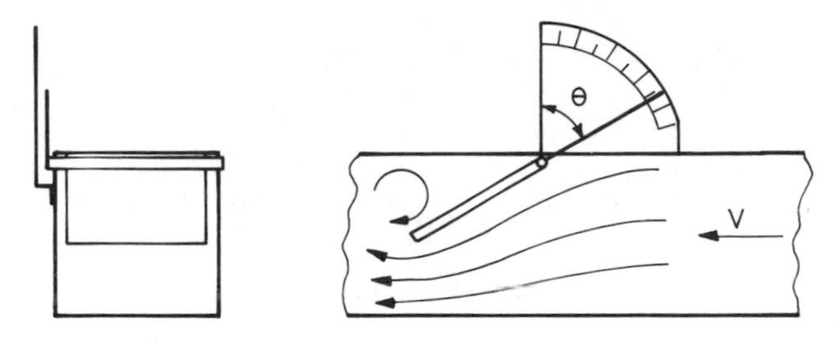

Fig. 7.5 Vane flow meter

The interruption of flow in these devices is rather severe, and an improvement is effected by using rotatable vanes, as shown in fig. 7.6. The speed of rotation depends on the flow velocity and on the reaction torque due to friction or to the counting mechanism etc. The total rotation is an indication of the total flow.

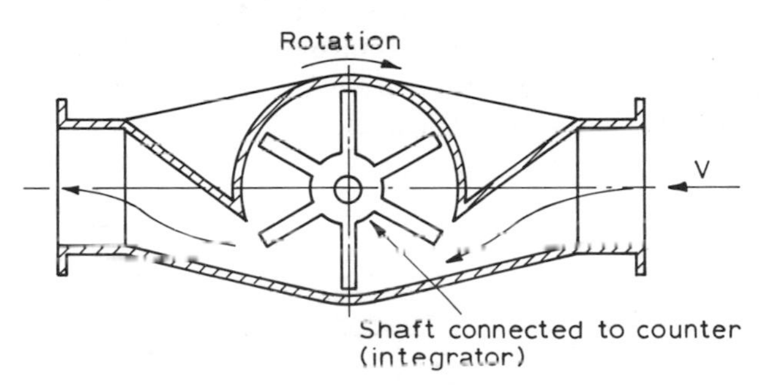

Fig. 7.6 Rotating-vane flow meter

Further reduction of losses in the flow meter may be obtained by using rotors with helical vanes, or by using aerofoil-shaped rotor vanes. In the latter case, the signal is usually taken off by an electrical pick-off, and care is taken to minimise, by design and manufacture, the friction losses in the rotor bearings. This results in a meter with an extremely small pressure loss, giving a flow rate to an accuracy within $\pm 1\%$ in some cases, with good linearity and repeatability. This type is usually referred to as a turbine flow meter, and modern types have a digital read-out instrument giving almost instantaneous flow values available in a variety of units. Figure 7.7 illustrates the principle of this instrument.

In vane- and turbine-type meters, the viscosity of the fluid affects the calibration, and increased viscosity gives a greater pressure loss for the same velocity. Turbine-type instruments may be used on gases in

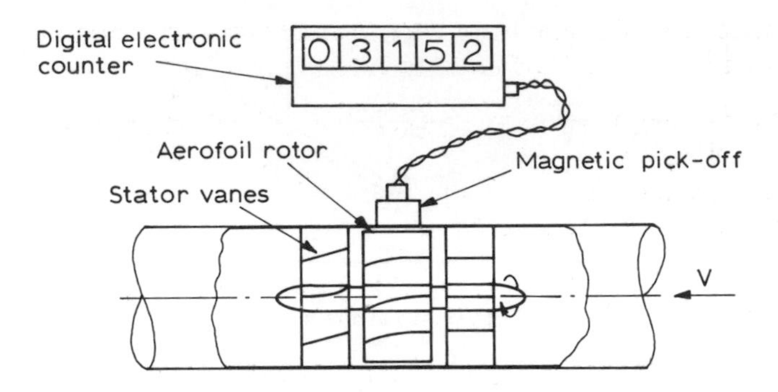

Fig. 7.7 Turbine flow meter

addition to liquids, when a magnetic pick-off is recommended rather than a mechanical counter.

For measuring flow in larger pipes, a part of the flow may be made to deviate by causing a restriction in the flow. The flow rate through the by-pass is measured, using a turbine or vane flow meter or a differential-pressure meter (see section 7.3.3) etc., and this is proportional to the main flow rate. The principle is illustrated in fig. 7.8.

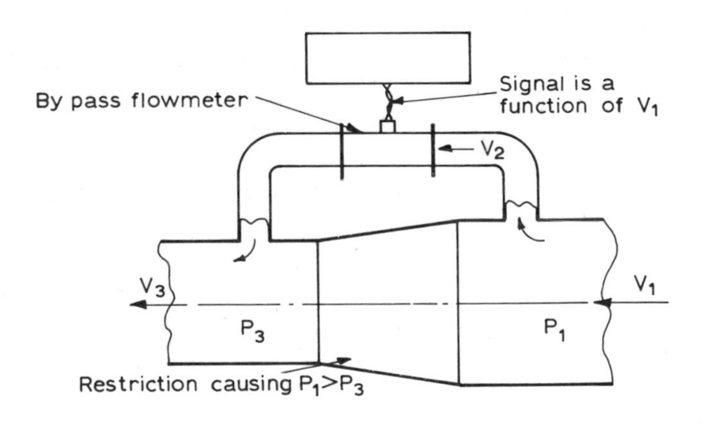

Fig. 7.8 By-pass flow meter

7.3.3 *Pressure-differential flow measurements*

The flow of liquids or gases may be described as streamline, when there is no velocity component normal to the flow direction (i.e. there is no cross flow), or turbulent, where a continual motion of fluid particles exists in all directions, although the mean velocity may be constant in magnitude and direction. An intermediate state can exist, described as transitional. Which of these states actually exists in a flow depends, in a circular pipe, on the following variables:

$\overline{V}$, the mean flow velocity, $= \dot{Q}/A$;
d, the pipe bore;
η, the dynamic viscosity of the fluid; and
ρ, the density of the fluid.

These are combined to give a dimensionless parameter, the Reynolds number, (Re), which describes the flow.

$$(Re) = \frac{\overline{V}d\rho}{\eta} = \frac{\overline{V}d}{\nu} \qquad 7.1$$

where $\nu = \eta/\rho$ is the kinematic viscosity.

Usually, streamline or laminar flow exists for $(Re) < 2000$, and turbulent flow for $(Re) > 3000$. These values are approximate and depend on other factors, such as a disturbance in the flow due to a bend or orifice etc.

The energy of a fluid may be in the form of velocity (kinetic) energy, pressure energy, or potential energy. These are interchangeable, but give a constant total energy if no losses occur. This is expressed by Bernoulli's equation thus:

velocity head + pressure head + potential head = constant

or

$$\frac{V^2}{2g} + \frac{P}{\rho g} + z = \text{constant} \qquad 7.2$$

where P is the pressure, and z is the height above some datum.

If losses occur, as they do in practice, then

$$\frac{V_1^{\,2}}{2g} + \frac{P_1}{\rho g} + z_1 = \frac{V_2^{\,2}}{2g} + \frac{P_2}{\rho g} + z_2 + \text{losses} \qquad 7.2(a)$$

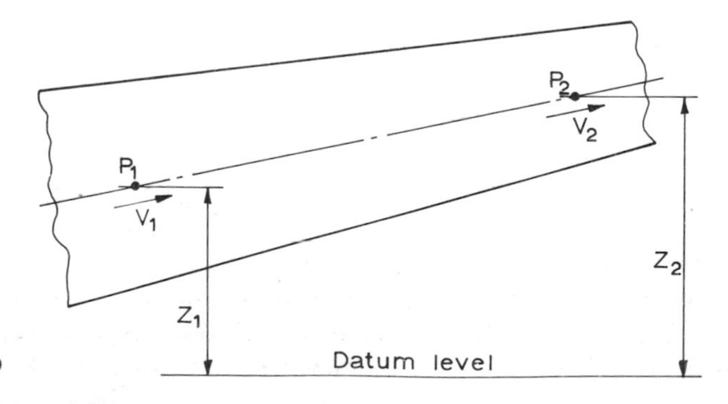

Fig. 7.9

where P_1 and V_1 are the pressure and velocity respectively at a point 1, and P_2 and V_2 are the values at a point 2, as indicated in fig. 7.9.

If the flow of a liquid is at a constant rate through a converging pipe, as shown, then the velocity must gradually increase as the area of cross-section reduces. Hence the pressure will reduce if the axis of the pipe is horizontal, and *may* do so if it is sloping. This loss of pressure with increasing velocity at a reduced area section is used in many flow-measuring devices. One very common device is the carburettor, where the pressure at the throat is a function of the velocity of air into the engine. The measurement is not indicated, but is used directly to control the quantity of petrol entering the airstream. Although Bernoulli's equation is for incompressible fluids, it may be applied to air and other gases at low velocities without appreciable error. In applying the equation to gases, the potential head (z) is always omitted, as its effect is very small.

7.3.3.1 The Pitot tube and the Pitot-static tube. Fig. 7.10(a) [and also fig. 5.21(b) and (c)] illustrates the difference in head obtained when a pressure tapping is pointed upstream, compared with one flush with the tube wall. A properly designed static tapping will measure the static pressure of the flow, i.e. P in eqn 7.2. In the upstream-pointing tapping, or Pitot tube, some of the fluid at an initial velocity V_1 is brought to rest at point 2 in the tube. Applying eqn 7.1,

$$\frac{V_1^2}{2g} + \frac{P_1}{w} + 0 = 0 + \frac{P_2}{w} + 0$$

where $w = \rho g$ is the specific weight of the fluid.

Hence
$$\frac{P_2}{w} = \frac{V_1^2}{2g} + \frac{P_1}{w}$$

where P_2/w is the *total head*, or *stagnation head*, or *impact pressure;* $V_1^2/2g$ is the *velocity head;* P_1/w is the *static head;* and P_2 is the *stagnation pressure*. This gives

$$\frac{V_1^2}{2g} = \frac{P_2 - P_1}{w} = h$$

or
$$V_1 = \sqrt{(2gh)}$$

where h is the head of the *measured* fluid of density ρ.

If h_m is a differential pressure head of a manometer fluid of density ρ_m, then, since $P = \rho gh$ (see section 5.4.1),

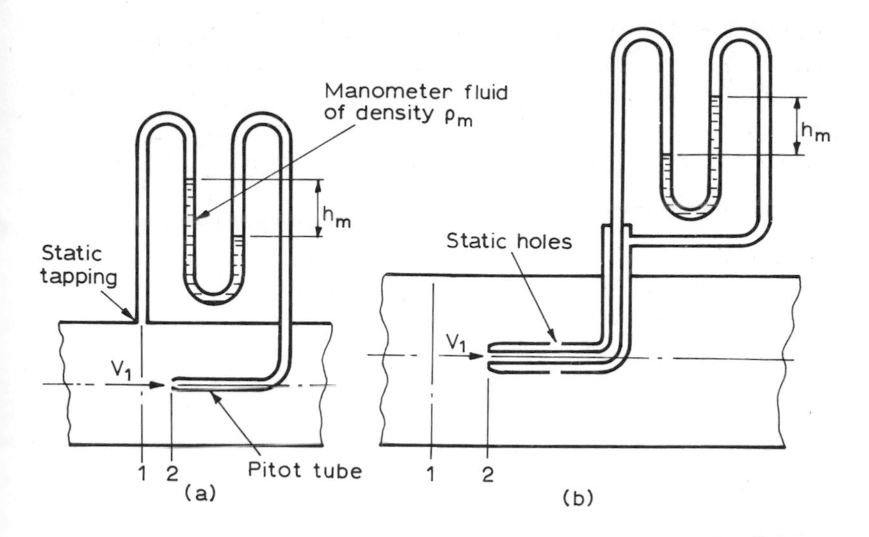

Fig. 7.10 Velocity measurement: (a) Separate static and Pitot tappings (b) Pitot-static tube

$$\delta P = h\rho g = h_m \rho_m g \quad \text{and} \quad h = h_m \rho_m / \rho$$

$$\therefore \quad V_1 = \sqrt{2gh_m\rho_m/\rho}$$

A correction coefficient (C) has to be applied, to take energy losses into account, and hence

$$V_1 = C\sqrt{2gh_m\rho_m/\rho} \qquad 7.3$$

By combining the Pitot and static tubes in a double-walled arrangement, as in fig. 7.10(b), a more convenient 'Pitot-static' tube is obtained. The static tappings are around the outer tube wall, and are nearer to the velocity-measuring point than would be a pipe-wall tapping. Pitot-static connections are an essential fitting in aircraft, to measure air velocity relative to the aircraft. A further use is in the measurement of ventilating-duct air flows in mines, where an easily transportable tube and manometer are used.

When measuring fluid flow in pipes and ducts, there is a layer of still fluid at the walls, and the velocity is normally maximum at the centre. The velocity distribution varies according to the type of flow, as indicated by the Reynolds number, and is as shown in fig. 7.11. If the flow rate $\dot{Q} = \bar{V}A$ is required from the Pitot-static head h_m, it is seen that a different value will be obtained for different positions of the probe across the fluid stream. The flow pattern may be unsymmetrical if the section is near to (within six diameters of) a bend, valve, or other

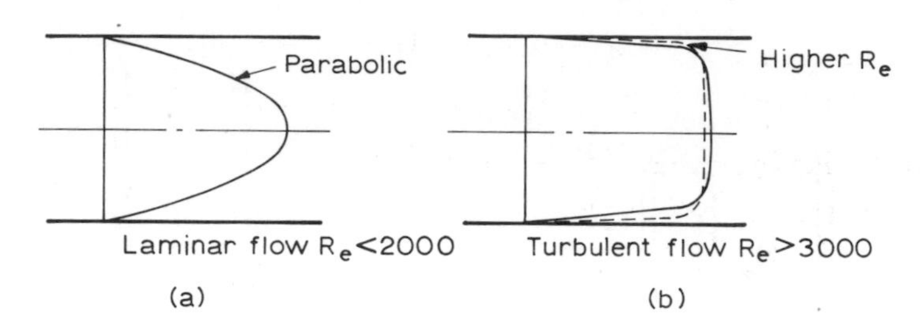

Fig. 7.11 Velocity distribution in pipes

obstruction. To obtain accurate flow rates, the probe should be used to measure velocity at a number of points across the section and an average value should be taken. BS 848: part 1: 1963 gives details of velocity-measuring points in circular and rectangular airways, and methods of computing flow rates from the measured values.

7.3.3.2 The venturimeter. Figure 7.12 shows the basic form of the venturimeter, which can measure to a high degree of accuracy the mean flow velocity ($\overline{V}$) or the flow rate ($\dot{Q}$) of liquids or gases. The proportions of the meter are given in BS 1042: part 1: 1964. The entrance cone has an included angle of 21° ± 2°, and the outlet or discharge cone is 5° to 15° included angle.

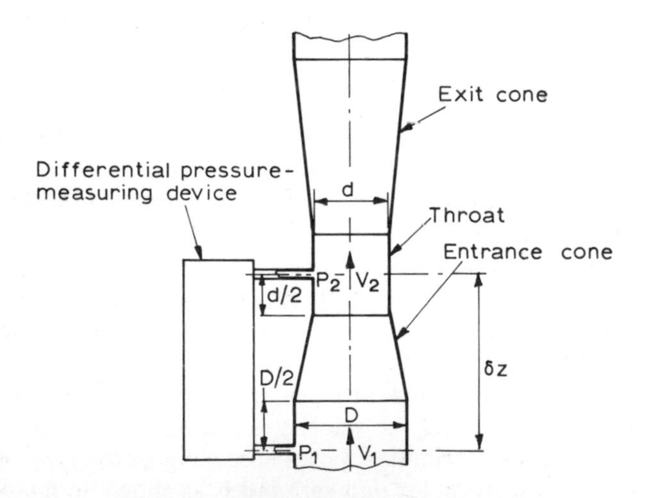

Fig. 7.12 Venturimeter

Assuming that liquid is flowing at a constant rate $\dot{m}$ or $\dot{Q}$ through the meter, the continuity equation may be written

$$\dot{Q} = A_1V_1 = A_2V_2$$
$$\therefore \quad V_1 = V_2A_2/A_1 = V_2(d_2/d_1)^2 \qquad 7.4$$

Substituting this value of V_2 in eqn 7.2 gives

$$\frac{V_2^2}{2g}\left(\frac{d_2}{d_1}\right)^4 + \frac{P_1}{\rho g} + z_1 = \frac{V_2^2}{2g} + \frac{P_2}{\rho g} + z_2$$

hence
$$\frac{V_2^2}{2g}\left\{1 - \left(\frac{d_2}{d_1}\right)^4\right\} = \left(\frac{P_1 - P_2}{\rho g}\right) + (z_1 - z_2)$$

and
$$V_2 = \left\{2g\,\frac{(P_1 - P_2)/\rho g + (z_1 - z_2)}{1 - (d_2/d_1)^4}\right\}^{1/2}$$

and
$$\dot{Q} = A_2V_2$$

Wherever possible, venturimeters are installed in a horizontal position, giving $z_1 - z_2 = 0$. The term $1/\{1 - (A_2/A_1)^2\}^{1/2}$ or $1/\{1 - (d_2/d_1)^4\}^{1/2}$ is termed the 'velocity of approach' factor (E). Losses always occur, and the actual discharge is always less than the theoretical value.

C_D = coefficient of discharge

= (actual discharge)/(theoretical discharge)

Hence the discharge rate ($\dot{Q}$) for a *horizontal* meter is given by

$$\dot{Q} = C_DEA_2\sqrt{2g\,\delta P/\rho g}$$
$$= C_DEA_2\sqrt{2\,\delta P/\rho} \qquad 7.5$$

The value of C_D is in the region 0·9 – 0·97, and varies with the Reynolds number for the flow. It is also affected by installation factors such as adjacent bends, valves, etc.

The advantages of the venturimeter are high accuracy and low pressure loss, this being about 10% of δP for the 5° outlet angle and 20% of δP for the 15° outlet. The disadvantages are high cost and a long installation length.

7.3.3.3 Nozzles and orifices. These devices are in general not as accurate as a venturimeter, and have a higher pressure loss; however, they are relatively cheap, and are small enough in many applications to fit between an existing pair of pipe flanges. Figure 7.13 shows a flow tube which has a shaped inlet but no outlet cone, and is suitable for use in high-velocity flows.

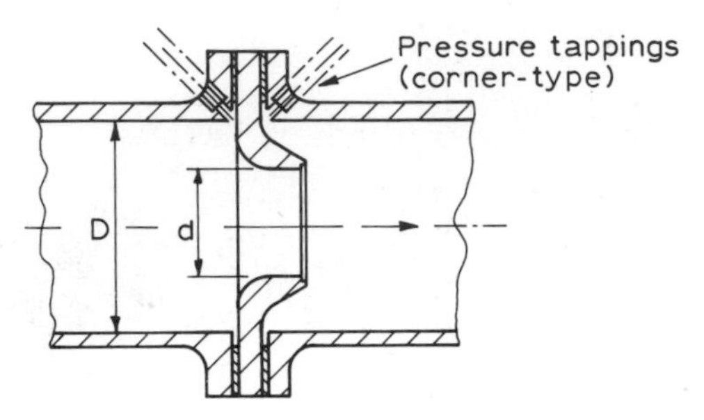

Fig. 7.13 Flow nozzle

The simple sharp-edged orifice plate shown in fig. 7.14 is the cheapest device. A *vena-contracta* effect occurs, giving minimum stream area and maximum velocity a little downstream of the orifice, and these effects are compensated for in the coefficient of discharge (C_D).

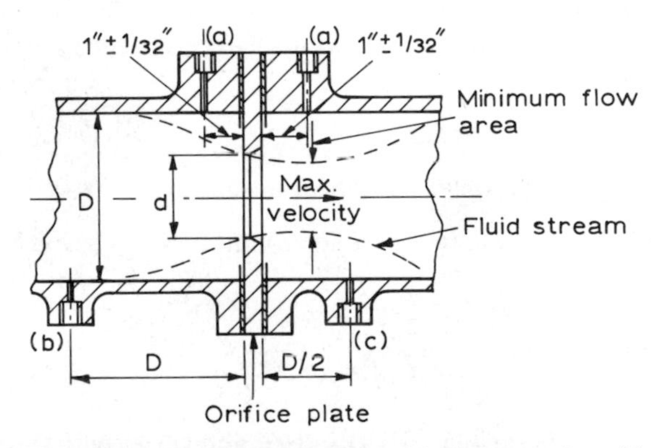

Fig. 7.14 Orifice plate with pressure tappings: (a) Flange tappings (b) '*D*' tappings (c) '*D*/2' tappings

Pressure tappings may be 'corner tappings' as in fig. 7.13 or 'flange tappings' or '*D* and *D*/2 tappings' as in fig. 7.14, any of these types being applicable to the various orifices. Details of the tappings and details of the orifices, discharge coefficients etc. are given in BS 1042:part 1:1964.

The flow rate $\dot{Q}$ for nozzle and orifice meters is given by eqn 7.5, as for the venturimeter, but the coefficients of discharge are much lower, in the region of 0·6.

Orifices are also used in more specialised flow-measuring devices such as the niveau, illustrated in fig. 7.15. For a constant flow rate

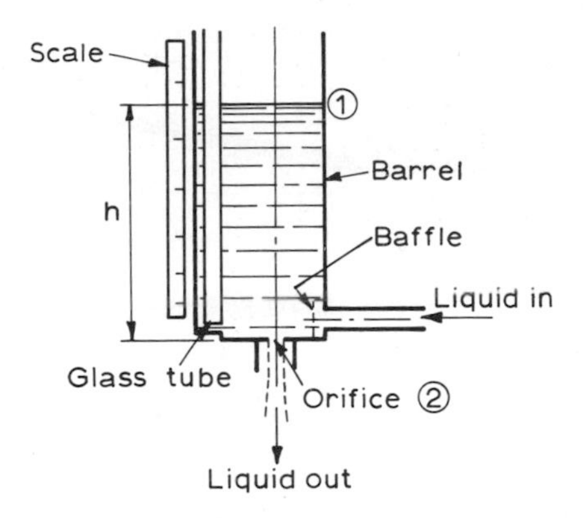

Fig. 7.15 The niveau

into the barrel, the outlet flow rate through the orifice will be the same, if the head (h) is constant. Applying eqn 7.2, using point 2 at the orifice, point 1 at the liquid surface,

$$0 + \frac{P_1}{\rho g} + z_1 = \frac{V_2^{\,2}}{2g} + \frac{P_2}{\rho g} + z_2$$

But $$P_1 = P_2 = \text{atmospheric pressure,}$$

and $$z_1 \quad z_2 - h$$

$$\therefore \quad V_2 = \sqrt{(2gh)}, \text{ neglecting losses}$$

or $$\dot{Q} = C_D A_0 \sqrt{(2gh)} \qquad 7.6$$

where C_D is the coefficient of discharge, and A_0 is the area of the outlet orifice.

A typical use of the device is in measuring the cooling-water flow through an engine during laboratory tests.

Venturi, nozzle, and orifice meters may be used to measure gas flows, but a 'compressibility factor' (ε) is introduced which depends on the ratio of the areas ($m = A_1/A_2$), the ratio of the specific heats of the gases ($\gamma = c_p/c_v$), and the pressure ratio ($r = P_1/P_2$). Where $r > 0{\cdot}97$ the flow may be considered incompressible. An example of the orifice used for air-flow measurement may be found in many laboratories, measuring the air flow into internal-combustion engines; this is the airbox, illustrated in fig. 7.16. Reduced pressure in the drum is applied to an inclined-tube manometer which gives the pressure difference $(P_1 - P_2)/\rho g = h$. The ratio P_1/P_2 is in the region of $0{\cdot}99$, and so the flow may be taken as incompressible. Applying eqn 7.2 and assuming condition 1 is atmospheric air at rest, and 2 is just inside the box, then

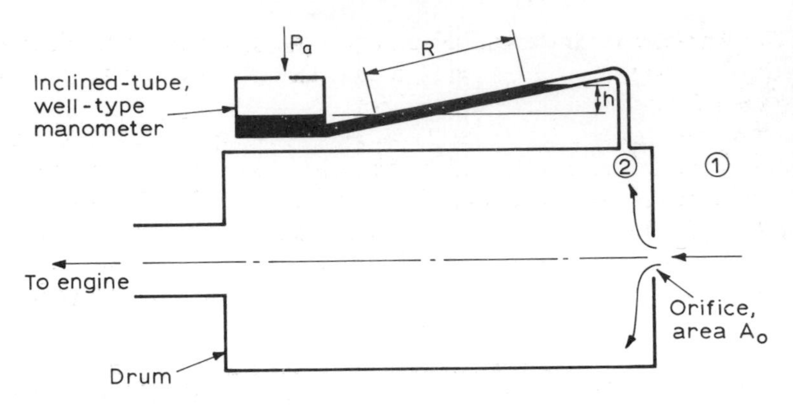

Fig. 7.16 Airbox

$$0 + \frac{P_1}{\rho g} + 0 = \frac{V_2^{\,2}}{2g} + \frac{P_2}{\rho g} + 0$$

hence $\quad V_2 = \sqrt{2(P_1 - P_2)/\rho_a}$, neglecting losses

and $\quad \dot{Q} = C_D A_0 \sqrt{2\,\delta P/\rho_a} \qquad$ 7.7

But $\quad \delta P = (P_1 - P_2) = \rho_a g h_a = \rho_m g h_m$

where suffix 'a' indicates air, and suffix 'm' indicates manometer fluid.

Hence $\quad \dot{Q} = C_D A_0 V_2$

$$\dot{Q} = C_D A_0 \sqrt{2 g h_m \rho_m / \rho_a} \qquad \text{7.7(a)}$$

The density of air varies slightly with temperature and pressure fluctuations, the relationship being given by the characteristic gas equation

$$PV = mRT$$

where P is the absolute pressure, V is the volume of a mass m of gas, R is the gas constant, and T is the temperature (kelvin).

$$\therefore \quad \rho_a = \frac{m}{V} = \frac{P}{RT}$$

Taking R for air as 287 J/kg K, normal atmospheric pressure as 101 kN/m^2, and the temperature as 15 °C = 288 K, then

$$\rho_a = \frac{101 \times 10^3}{287 \times 288} = 1{\cdot}23 \text{ kg/m}^3$$

The characteristic gas equation may also be used to convert the air flow rate $\dot{Q}$ at the particular test conditions to an equivalent air flow under standard conditions (see example 7.9.5).

The box is necessary when the flow is pulsating; its function is to damp out the pulsations which otherwise would cause 'root-mean-square' errors and oscillation of the manometer fluid. The larger the engine, the slower is its speed; and the smaller the number of cylinders, the larger is the necessary size of the airbox (see Kastner, 'The airbox method of measuring air consumption', *Proc. I. Mech. E.* 1947, vol. 157).

7.3.3.4 Variable-area meters. The principle of the float type of meter for liquids and gases is shown in fig. 7.17(a). The fluid flows upward past the 'float', which is denser than the fluid. The upward force due to the fluid on the float depends on the flow area past the float, the float volume, and the fluid flow rate. At any particular flow rate, the height (h) of the float is steady when a balance of the forces on the float is obtained.

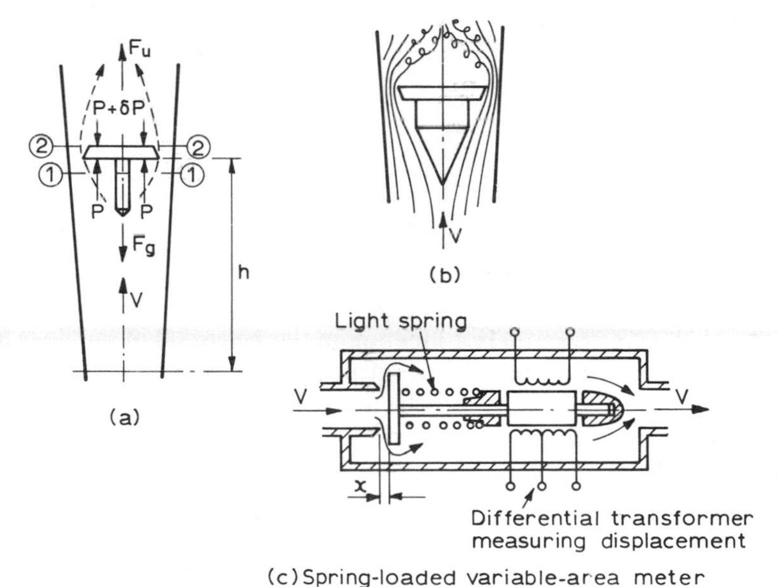

(c) Spring-loaded variable-area meter

Fig. 7.17 Variable-area meters

If P is the pressure on the underside of the float, δP is the change of pressure over the float, ρ is the density of the fluid, ρ_f is the density of the float material, A is the maximum cross-sectional area of the float,

a is the minimum cross-sectional area of flow, V_f is the volume of the float, then

$$\begin{Bmatrix}\text{gravity}\\ \text{force}\end{Bmatrix} + \begin{Bmatrix}\text{downward}\\ \text{pressure force}\end{Bmatrix} = \begin{Bmatrix}\text{upthrust due}\\ \text{to immersion}\end{Bmatrix} + \begin{Bmatrix}\text{upward}\\ \text{pressure force}\end{Bmatrix}$$

$$\therefore \quad \rho_f V_f g + (P + \delta P)A = V_f \rho g + PA$$

and

$$\rho_f V_f g + \delta P \times A = V_f \rho g$$

hence

$$\delta P = V_f g(\rho - \rho_f)/A \qquad 7.8$$

From eqn 7.2,

$$\frac{V_1^2}{2g} + \frac{P_1}{\rho g} + 0 = \frac{V_2^2}{2g} + \frac{P_2}{\rho g} + 0$$

and

$$\frac{V_2^2}{2g} = \left(\frac{V_1^2}{2g} + \frac{P_1}{\rho g}\right) - \frac{P_2}{\rho g} \qquad 7.9$$

Since the underside pressure is the 'impact' pressure (see section 7.3.3.1), then the right-hand side of this equation is the head difference $(-\delta P/\rho g)$; hence substituting from eqn 7.8 in eqn 7.9 gives

$$\frac{V_2^2}{2g} = \frac{V_f g(\rho_f - \rho)}{A\rho g}$$

and

$$V_2 = \sqrt{2gV_f(\rho_f - \rho)/A\rho}$$

Hence the volume flowing past the float is

$$\dot{Q} = C_D a\sqrt{2gV_f(\rho_f - \rho)/A\rho} \qquad 7.10$$

where C_D is a coefficient of discharge.

Since the flow area (a) is a function of the height (h) of the float in the taper tube, then the flow-rate scale may be marked on the tube. (Note that it will not necessarily be linear.)

The coefficient of discharge C_D will depend on the Reynolds number for the flow, and hence will vary with the area (a), the velocity (V_2), and the kinematic viscosity (ν) for any tube-and-float combination. Float shapes vary widely in different meters, and can be made to be more or less sensitive to viscosity changes in the measured fluid. Figure 7.17(b) shows a typical float shape used.

A variety of secondary transducers are used to give a signal which is a function of either flow rate or float height and which is capable of being transmitted to a remote point. The variable-area meter is also itself used as a secondary transducer, e.g. with an orifice-plate in a by-pass arrangement similar to fig. 7.8.

Figure 7.17(c) shows a slightly different type of variable-area meter, using a spring force instead of gravity. The float-type meter has a constant pressure difference δP across it, but in this type the spring introduces a variable force in place of the constant gravity force, resulting in a non-linear relationship between the flow rate ($\dot{Q}$) and the displacement (x). The displacement is measured electrically, using a linear variable-differential transformer (see Chapter 10).

7.3.4 *The viscous-flow meter*

The head loss (h_f) due to friction in a parallel pipe is a function of velocity; hence

$$h_f = kV^n$$

For laminar flow $n = 1$, and for turbulent flow n varies from about 1·7 to 2. If laminar flow is ensured by restricting the flow to a Reynolds number of less than 2000, then $h_f = kV$. The viscous-flow air meter shown in fig. 7.18 uses this principle by making the air passages very small in cross-section and relatively long. By using laminar (viscous) flow, the root-mean-square error present when measuring pulsating flow using a simple orifice is eliminated; hence $\dot{Q} = Kh$, where h is the manometer head.

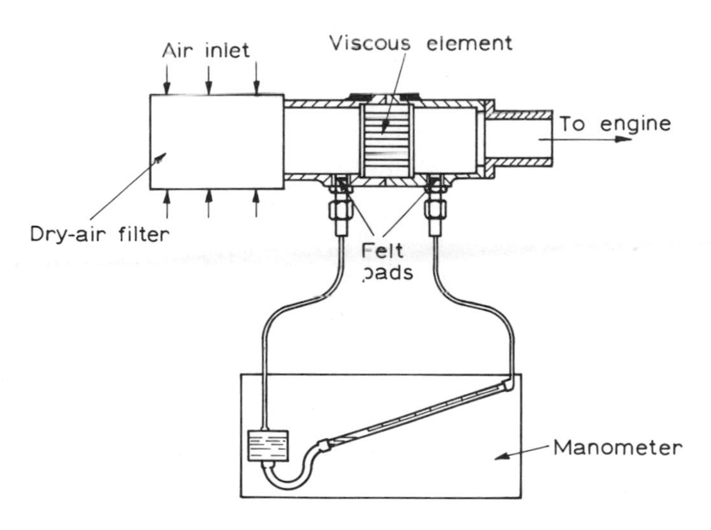

Fig. 7.18 Allcock viscous-flow air meter

7.3.5 *Weirs*

Rate of liquid flow may be measured by measuring the height (H) of the fluid as it passes over a weir. Figure 7.19 illustrates the principle,

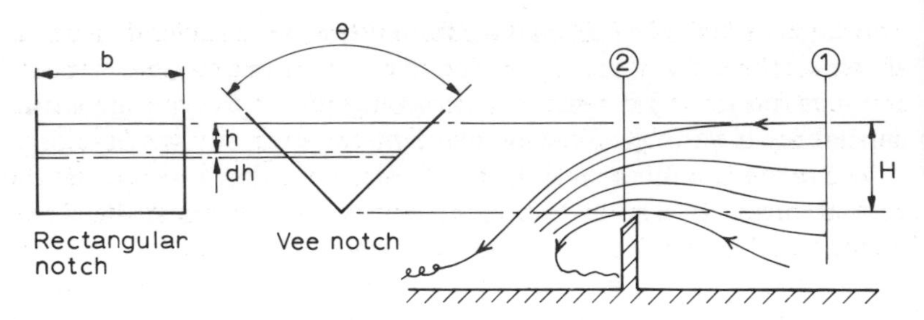

Fig. 7.19 **Sharp-crested weir**

which is used for the measurement of the flow rates in streams and rivers.

By assuming zero approach velocity, and that the head H applies to the fluid at plane 2, Bernoulli's equation (7.2) may be applied at points 1 and 2 to obtain the velocity at a level h in plane 2. Integration gives

$$\dot{Q} = C_D K H^n$$

where n is 3/2 for the square notch and 5/2 for the symmetrical vee-notch; hence flow scales are not linear. C_D varies with the head H and with the proportions of the notch.

7.4 Measurement of flow direction

We are familiar with the weather or wind vane, where unequal pressure is exerted on two sides of a flat plate when the wind direction has a component normal to it. A refinement of this basic principle is used in a more convenient instrument illustrated in fig. 7.20.

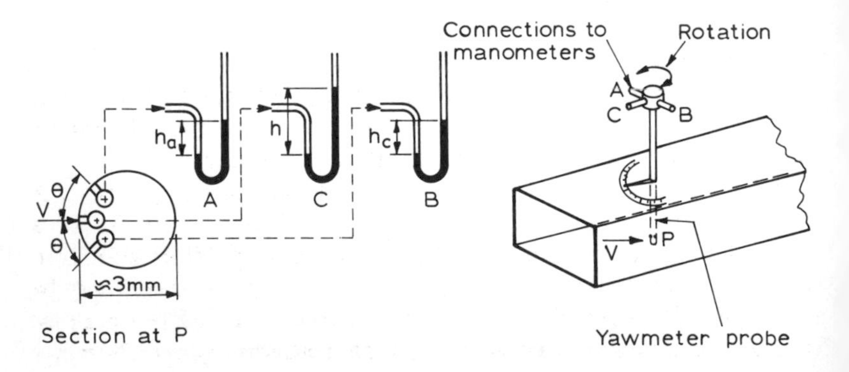

Fig. 7.20 **Yawmeter**

The yawmeter consists of three pressure tappings, as shown, in a thin rod having three longitudinal holes (produced by extruding or drawing a drilled billet). Radial drillings are made at angles θ apart, as shown, at the end P of the probe. The pressure reading on each manometer consists of static head plus a proportion of the velocity head depending on the angle of incidence of the gas stream on the hole.

The probe is rotated until the head readings A and B are equal; the central pressure tapping C then is pointing directly into the local direction of the fluid velocity. The angle may be read off a scale, as indicated, and the total head value may be read from the manometer C. The yawmeter may be traversed in X, Y, and Z directions to measure gas or liquid velocity directions and magnitudes around components such as aerofoil sections and pump and turbine blades.

Other methods of determining flow directions include the injection of lines of smoke into gases, and of jets of die and particles of light solids such as aluminium into liquids. The method is illustrated in fig. 7.21.

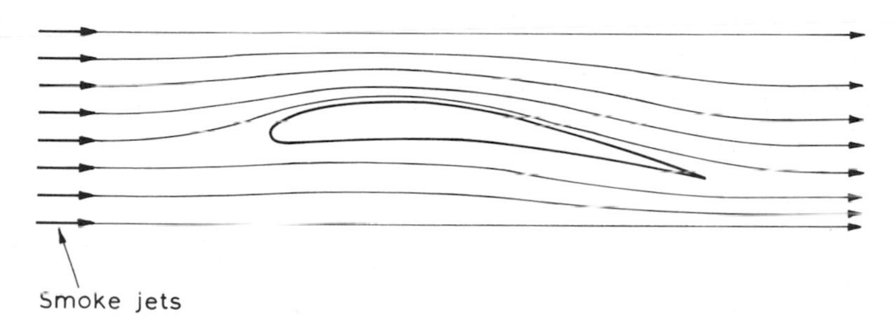

Fig. 7.21 Smoke jets used in flow visualisation

7.5 Rate-of-flow meters for solids

The measurement of the flow rate of particulate solids such as gravel is important in some industries, for example in the automatic plants coating gravel with bituminous materials for roadworks.

Figure 7.22 illustrates the principle of the weighing conveyor. A frame supporting a length (l) of conveyor belt is pivoted at O, and a load cell measures the force (F) at the other end. F is proportional to the mass (m) in length l. The output signal from the load cell may be a voltage (v_1), proportional to F. The conveyor speed (V) may be variable, in which case its value may be measured by a tachometer connected to a roller, giving a voltage (v_2) proportional to velocity V.

Hence
$$m = k_1 v_1$$
$$V = k_2 v_2$$

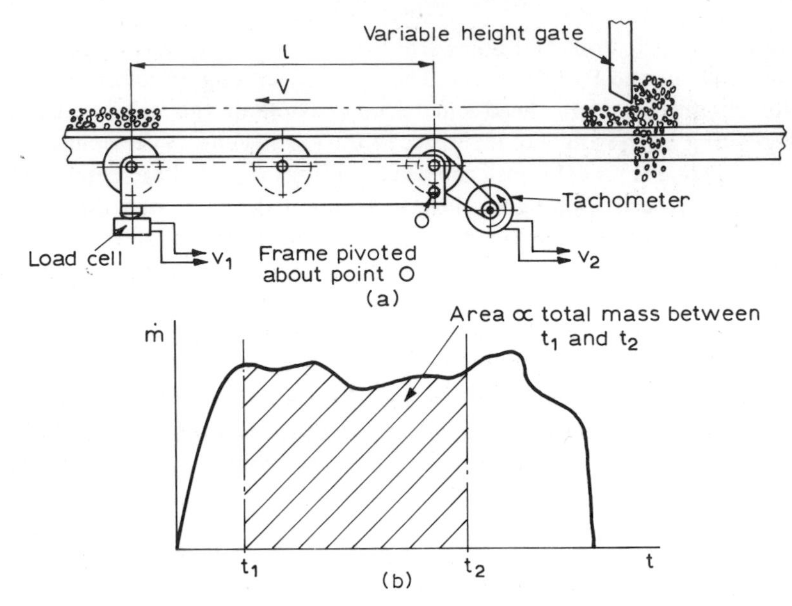

Fig. 7.22 (a) Weighing conveyor (b) Recorder trace

The mass flow rate is given by

$$\begin{Bmatrix}\text{mass flow}\\ \text{rate}\end{Bmatrix} = \begin{Bmatrix}\text{mass per}\\ \text{unit length}\end{Bmatrix} \times \begin{Bmatrix}\text{conveyor}\\ \text{velocity}\end{Bmatrix}$$

or

$$\dot{m} = \frac{m}{l} \times V$$

$$= \frac{k_1 v_1}{l} \times k_2 v_2$$

$$= K v_1 v_2$$

Hence the voltage signals must be multiplied to give a signal proportional to the mass flow rate.

The total mass may be found by integration of this expression:

$$m = \int K v_1 v_2 \, \mathrm{d}t$$

The integration may be carried out from the trace of $\dot{m}$ against time obtained from a recorder [see fig. 7.22(b)], or by an integrating device.

Another device for measuring the mass flow rate of a particulate solid is shown in fig. 7.23. An electric motor (M) drives a disc (D) onto

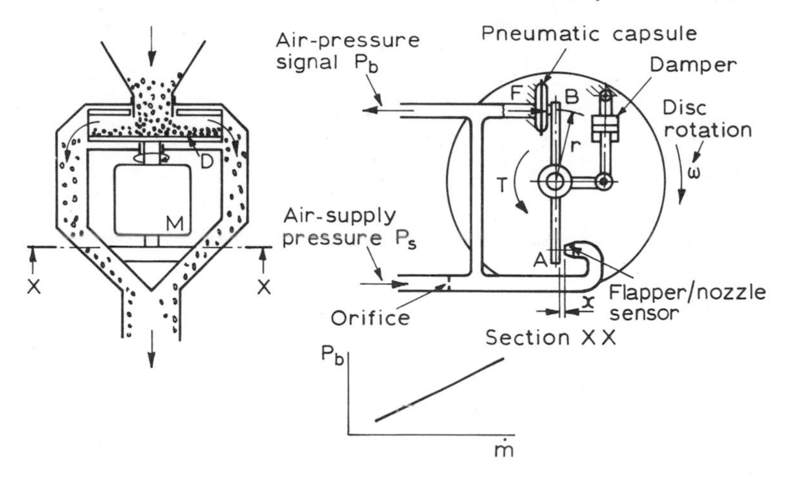

Fig. 7.23 Mass-flow-rate measurement

which the particles are delivered. The centrifugal effect flings the particles out tangentially so that they leave the disc at a mass flow rate $\dot{m}$. The reaction torque (T) is proportional to the mass flow rate, i.e. $T = k\dot{m}$. The reaction torque causes rotation of arm A, closing the gap x and increasing the pneumatic back-pressure P_b (see Chapter 10). This increases the force F exerted on the arm B by a pneumatic capsule, until a balance point is reached. For every value of torque T (and hence of force F) there is a corresponding value of x and P_b. The overall relationship is $\dot{m} = KP_b$. The signal pressure P_b may control an indicator or recorder, or may be used to initiate control action to regulate $\dot{m}$. The damper controls the oscillations of the arms AB (see Chapter 3).

7.6 Viscosity measurement

Viscosity is a measure of the flow characteristics of fluids. Figure 7.24 illustrates an infinitely thin film of fluid of thickness dy, one face of it having a velocity V and the other $V + \mathrm{d}V$. The change of velocity (dV) is due to the shear stress (τ) caused by forces at the faces exerted by adjacent solid surfaces or fluid. For an ideal or Newtonian fluid, the relationship is

$$\frac{\text{shear stress}}{\text{velocity gradient}} = \text{constant} = \textit{dynamic viscosity}\ (\eta)$$

i.e. $$\text{dynamic viscosity } (\eta) = \tau/(\mathrm{d}V/\mathrm{d}y)$$

$$\therefore \quad \eta = (F/A)/(\mathrm{d}V/\mathrm{d}y) \qquad 7.11$$

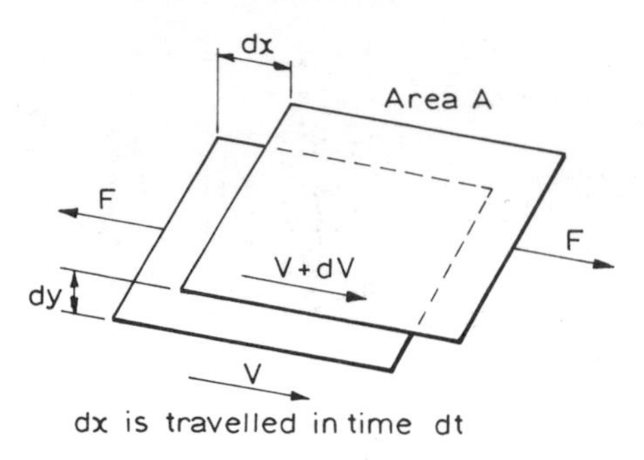

Fig. 7.24

where F is the shearing force on an area A. In the SI, η has units of $N\,s/m^2$.

At a constant temperature value, the dynamic viscosity is fairly constant for many fluids at low pressures and flow rates, but it varies at higher values. With increasing temperature, viscosities decrease for liquids and increase for gases.

The quantity (dynamic viscosity/density) occurs frequently in fluid calculations, and this is given the name *kinematic viscosity* (ν); i.e. kinematic viscosity $= \nu = \eta/\rho$, and the units in the SI are m^2/s. Instruments for viscosity measurement of fluids may be classified as follows, according to their principle of operation:

a) falling sphere,
b) falling piston or rotating cylinder,
c) flow through an orifice,
d) flow through a capillary tube,
e) flow through a variable area.

In the falling-sphere viscometer, fig. 7.25, the sphere is released into the liquid and accelerates under the gravity force until it reaches a maximum (terminal) velocity V, when the gravity force is just balanced by the buoyancy and the viscous drag force. BS 188:1957 gives details of viscometers of this type and procedures to be followed in testing. Timing of the fall through a distance of 150 or 200 mm enables the terminal velocity to be calculated. The kinematic viscosity (ν) of the liquid is given by

$$\nu = Fd^2g(\rho_s - \rho_l)/(\rho_l \times 0{\cdot}18V) \text{ centistokes} \qquad 7.12$$

where F is a constant correcting for the tube-wall effect, ρ_s is the density of the sphere (gm/cm^3), ρ_l is the density of the liquid (gm/cm^3), V is the terminal velocity (cm/s), d is the sphere diameter (cm), and g is the local gravitational acceleration (cm/s^2).

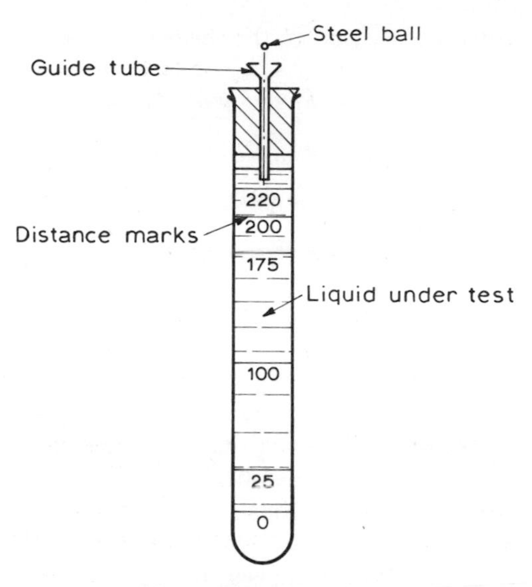

Fig. 7.25 Falling-sphere viscometer

The centistoke is 0·01 cm^2/s, and the above equation must be modified to give the kinematic viscosity in SI units, thus:

$$\nu' = 10^{-6}\,\nu\ (m^2/s)$$

If in the viscometers shown in fig. 7.26 the terminal velocity of the piston is V, the tangential velocity of the cylinder is $V\,(=\omega r)$, and y is very small. Then $V/y \approx dV/dy$, and eqn 7.11 becomes

$$\eta = (F/A)/(V/y)$$

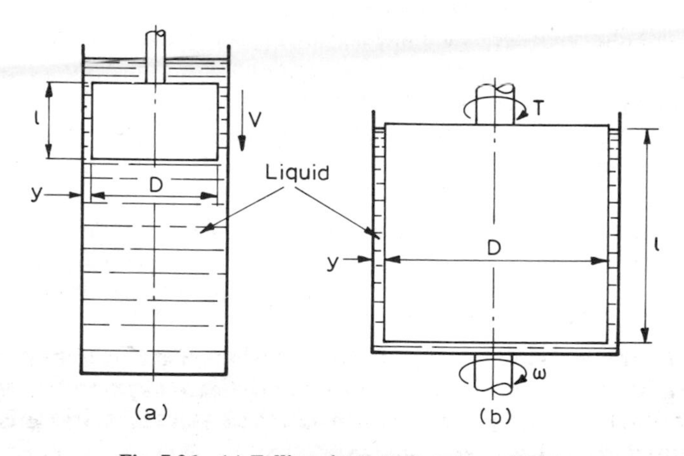

Fig. 7.26 (a) Falling piston (b) Rotating cylinder

For the falling-piston viscometer, the shear stress is

$$\tau = F/A = mg/\pi Dl$$

Hence
$$\eta = (mg/\pi Dl)/(V/y)$$
$$= mgy/\pi DlV$$

or
$$\eta = K/V$$

Hence η is inversely proportional to V, as in the falling-sphere viscometer.

In the rotating-cylinder viscometer [fig. 7.26(b)], one member is rotated at a constant angular velocity (ω). The other member is stationary, and the space between them is filled with the fluid. The torque (T) on the stationary member is measured in some suitable way. If y is small, then eqn 7.11 may be used to derive a relationship between the torque and the viscosity of the fluid. It is difficult, however, to eliminate the effect of oil on the end of the cylinder, and this causes an additional drag which should be taken into account.

Orifice-type viscometers are widely used for practical measurement of oil and fuel viscosity. The type illustrated in fig. 7.27 is the Redwood (Britain), but the Engler (Europe) and Saybolt (America) types are very similar. The essential items are a level-indicator (A), an outlet orifice (B), a valve (C), a temperature-control jacket (D), thermometers (E), and a collecting flask (F). When the appropriate quantity of oil at the correct temperature is allowed to flow through the orifice, the time (t) in seconds is measured for 50 ml to flow into the flask. The viscosity is then t Redwood seconds. The flow rate through the orifice is not given by any simple law, but the dynamic viscosity may be calculated from the Redwood viscosity by empirical formulae. However, this type of instrument is more suitable for comparative measurement than for the determination of absolute values.

For laminar-flow conditions in a constant-bore tube, Poiseuille's equation applies:

$$\eta = \pi w h_f d^4/(128\dot{Q}l) \qquad 7.13$$

where $w(= \rho g)$ is the specific weight of the fluid, h_f is the friction head loss of the flow, d is the pipe or tube diameter, $\dot{Q}$ is the rate of flow, and l is the length of pipe or tube.

This equation may be used to find η in the apparatus shown in fig. 7.28(a). However, errors will occur due to the variation of h_f as the flow proceeds, due to losses at the sudden changes of section at inlet and outlet, and due to the kinetic energy of the flow at the outlet. The

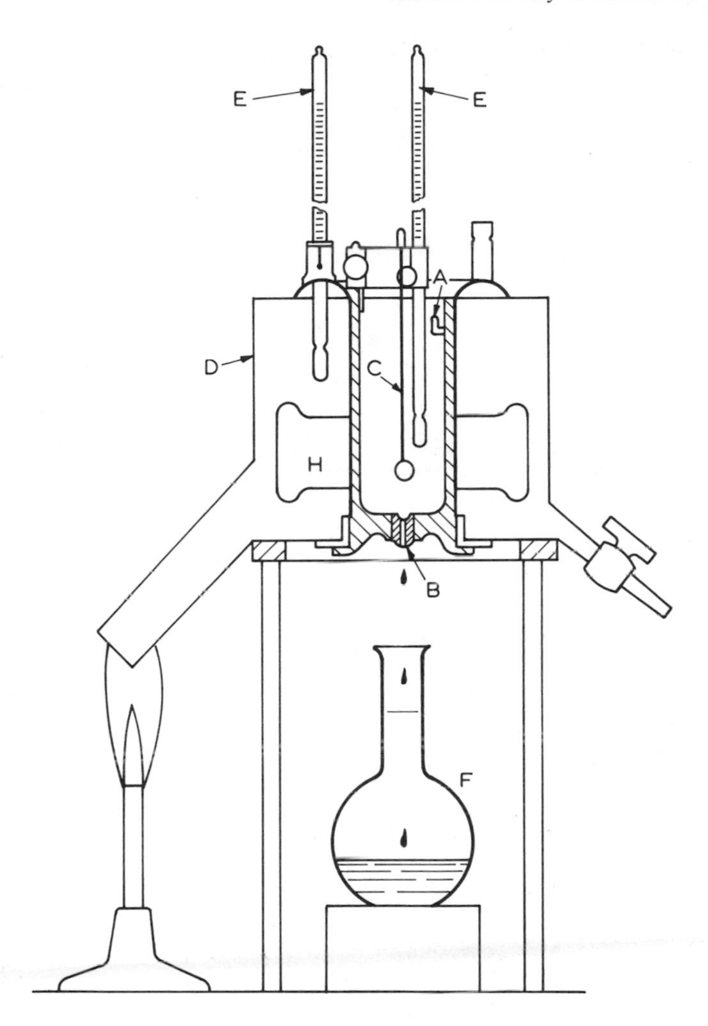

Fig. 7.27 Redwood viscometer

apparatus used in practice is the British Standard U-tube viscometer to BS 188 : 1957, shown in fig. 7.28(b). The time (t) for the given quantity of liquid filling the U-tube to flow so that the level drops from E to F is measured. Then

$$v = Ct$$

where t is the time in seconds and C is a constant determined by calibration. The constant contains terms from Poiseuille's equation, and also accounts for the changing head during the timing, and for variation of the capillary bore etc. It does not take into account the

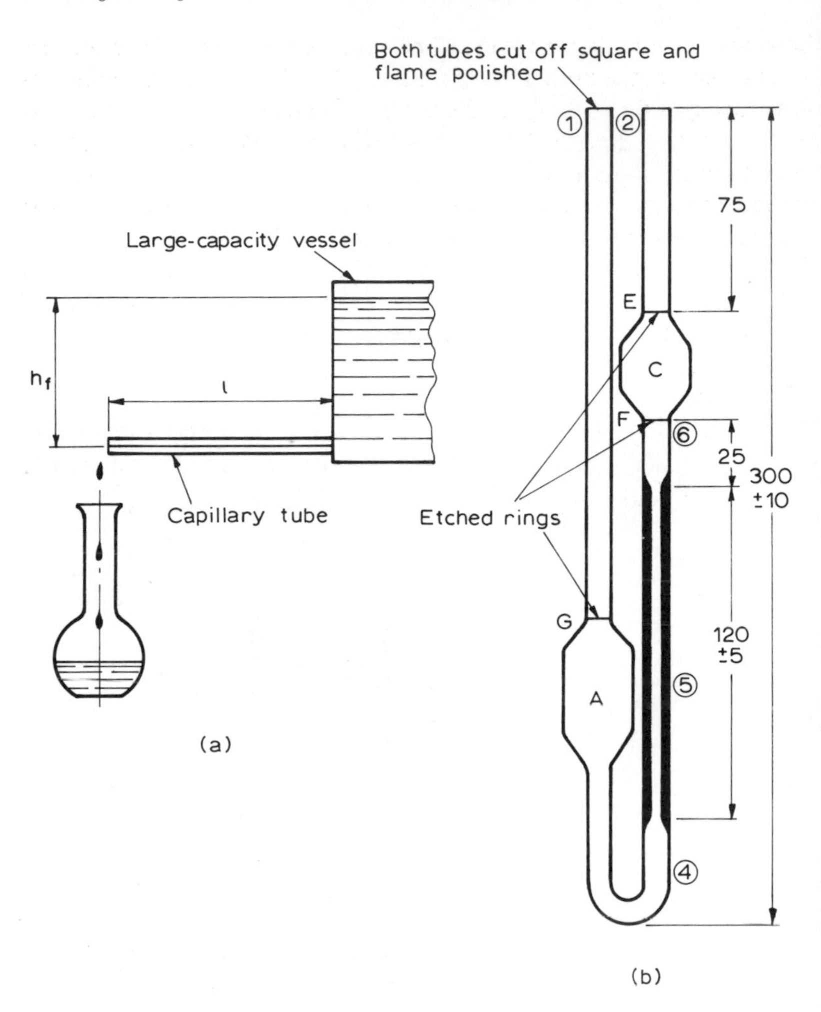

Fig. 7.28 (a) Flow through a capillary tube (b) U-tube viscometer

kinetic energy at outlet, and if this is significant, as in lower-viscosity fluids, a compensating term is added thus:

$$v = Ct - B/t \tag{7.14}$$

Calculated values of C and B are given in BS 188.

Viscometers of the types (a), (b), (c), and (d) are used with a sample of the fluid. In each of them, the temperature may be maintained at the correct value to within close limits by temperature-controlled water-baths or similar arrangements. The variable-area flow meter, however, is suitable for taking continuous measurement of viscosity, as all or part of the fluid flows through it. The float of the variable-area flow meter in fig. 7.17(a) is affected by viscous flow forces as well as pressure forces, but in flow meters the float is designed so that the viscous effect is minimal. For use as a viscosity meter, the float is designed so that the viscous effect is as large as possible. However, the pressure effects cannot be eliminated, and the height h of the float in the tube depends on both viscosity and flow rate. If the flow rate is maintained at a constant value, the change of h is due to change of viscosity, and the scale may be marked in viscosity units.

7.7 Examples

7.7.1 Air is flowing in a duct, and its static head, velocity head, and temperature are measured by pressure tappings and a temperature probe, as indicated in fig. 7.29.

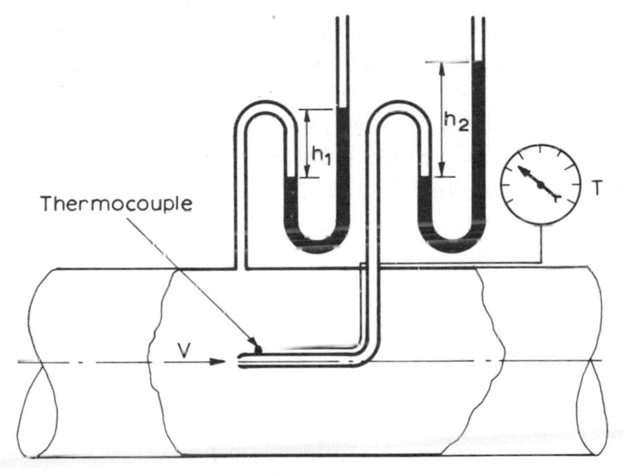

Fig. 7.29

If the readings are $\theta = 27\ °C$, $h_1 = 56$ mm, and $h_2 = 80$ mm of water, determine the local velocity of the fluid if the barometer reads 760 mm of mercury. Take the gas constant (R) for air as 287 J/kg K and the relative density of mercury as 13·6, and assume no losses occur in the system.

$$\text{The static pressure in the duct} = \rho_m g h_m + \rho_w g h_w$$
$$= 9{\cdot}81(13{\cdot}6 \times 10^3 \times 0{\cdot}760 + 10^3 \times 0{\cdot}056)$$
$$= 101{\cdot}9\ \text{kN/m}^2$$

From $PV = mRT$,

$$\rho_a = \frac{m}{V} = \frac{P}{RT}$$

$$= \frac{101\,900}{287 \times (273 + 27)}$$

$$\therefore \quad \rho_a = 1{\cdot}18 \text{ kg/m}^3$$

By eqn 7.2,

$$\text{velocity} = C\sqrt{2gh_w\rho_w/\rho_a}$$

$$= 1\sqrt{\{2 \times 9{\cdot}81 \times (80 - 56) \times 10^{-3} \times 10^3/1{\cdot}18\}}$$

$$\approx 20 \text{ m/s}$$

7.7.2 In a vertically mounted venturimeter measuring water flow, as shown in fig. 7.30, the vertical distance between the tapping points 1 and 2 is 112 mm, the diameter of the meter at 1 is 100 mm, and that at 2 is 80 mm. If the pressure difference $P_1 - P_2$ is 30·6 kN/m^2, calculate (a) the velocity in the throat if the meter coefficient is 0·97, (b) the Reynolds number for the flow in the throat if the kinematic viscosity is 1×10^{-6} m^2/s, (c) the height, h_m, of mercury in a U-tube manometer connected to the same levels in the meter.

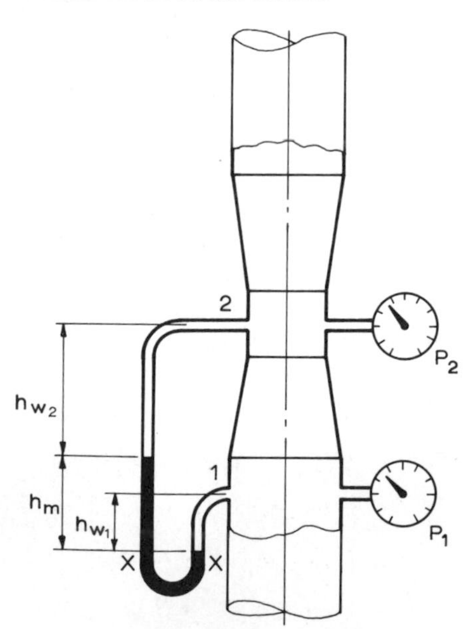

Fig. 7.30

a) Applying the continuity equation,

$$V_1 A_1 = V_2 A_2$$

$$\therefore \quad V_1 = V_2 A_2 / A_1 = V_2 (d_2/d_1)^2$$

Bernoulli's equation (7.2), neglecting losses, gives

$$\frac{V_1^2}{2g} + \frac{P_1}{w} + z_1 = \frac{V_2^2}{2g} + \frac{P_2}{w} + z_2$$

$$\therefore \quad \frac{V_2^2}{2g}\left(\frac{d_2}{d_1}\right)^4 + \frac{P_1}{w} + z_1 = \frac{V_2^2}{2g} + \frac{P_2}{w} + z_2$$

and

$$\left(\frac{P_2 - P_1}{w}\right) + (z_1 - z_2) = \frac{V_2^2}{2g}\left\{1 - \left(\frac{d_2}{d_1}\right)^4\right\}$$

Hence

$$V_2^2 = 2g \frac{(P_2 - P_1)/w + (z_1 - z_2)}{1 - (d_2/d_1)^4}$$

$$= 2 \times 9{\cdot}81 \frac{(30\,600/9810) - 0{\cdot}112}{1 - 0{\cdot}8^4}$$

$$= 100$$

$$\therefore \quad V_2 = 10 \text{ m/s}$$

This is the theoretical value; hence the actual mean velocity is

$$\bar{V} = 0{\cdot}97 \times 10$$

$$= 9{\cdot}7 \text{ m/s}$$

b) The Reynolds number at the throat is given by

$$(Re) = \bar{V} d_2 / \nu$$

$$= 9{\cdot}7 \times 0{\cdot}08/(1 \times 10^{-6})$$

$$= 7{\cdot}76 \times 10^5$$

c) Considering the pressure in the two legs of the manometer at level XX, then

$$\rho_w g h_{w_1} + P_1 = \rho_m g h_m + \rho_w g h_{w_2} + P_2$$

$$P_1 - P_2 = (\rho_m h_m + \rho_w h_{w_2} - \rho_w h_{w_1})g$$

$$= (13{\cdot}6\rho_w h_m + \rho_w h_{w_2} - \rho_w h_{w_1})g$$

$$= \{12{\cdot}6\rho_w h_m + \rho_w(h_m + h_{w_2} - h_{w_1})\}g$$

$$= \{12{\cdot}6\rho_w h_m + \rho_w(z_2 - z_1)\}g$$

$$\frac{P_1 - P_2}{\rho_w g} = 12{\cdot}6h_m + (z_2 - z_1)$$

$$\therefore \quad h_m = \frac{1}{12{\cdot}6}\left\{\left(\frac{P_1 - P_2}{\rho_w g}\right) - (z_2 - z_1)\right\}$$

$$= \frac{1}{12{\cdot}6}\left\{\frac{30\,600}{9810} - 0{\cdot}112\right\}$$

$$= 0{\cdot}239 \text{ m} = 239 \text{ mm}$$

7.7.3 An orifice-plate similar to that of fig. 7.14 is situated in a water pipe 50 mm diameter and has an orifice 30 mm diameter. The pressure tappings are corner type, as shown in fig. 7.13. For a Reynolds number of 100000, BS 1042 gives a discharge coefficient of 0·61. Calculate the nominal velocity $\overline{V}_2$ at the orifice to give this flow value, and the corresponding pressure difference at the tappings. Take the density (ρ) as 1000 kg/m^3 and the kinematic viscosity (ν) as 10^{-6} m^2/s.

The Reynolds number is given by

$$(Re) = \overline{V}_2 d_2/\nu$$

$$\therefore \quad \overline{V}_2 = (Re)\nu/d_2$$

$$= 100\,000 \times 10^{-6}/0{\cdot}030$$

$$= 3{\cdot}33 \text{ m/s}$$

From eqn 7.5,

$$\dot{Q} = \overline{V}_2 A_2 = C_D E A_2 \sqrt{(2\,\delta P/\rho)}$$

where $\overline{V}_2$ is the nominal velocity at the orifice.

$$\therefore \quad \overline{V}_2 = C_D E \sqrt{(2\,\delta P/\rho)}$$

and

$$\delta P = \left(\frac{\overline{V}_2}{C_D E}\right)^2 \frac{\rho}{2}$$

$$= \left(\frac{3{\cdot}33}{0{\cdot}61}\right)^2 \times \left(\frac{1 - 0{\cdot}6^4}{2}\right) \times \rho$$

$$= 29{\cdot}8 \times \frac{0{\cdot}87}{2} \times 1000$$

$$= 13{\cdot}0 \text{ kN/m}^2$$

7.7.4 The flow rate of water used in irrigation is to be measured using the fall of a horizontal jet issuing from an orifice as shown in fig. 7.31. Derive an expression for the flow rate in m^3/s and in litres/second when $l = 300$ mm, the orifice is 20 mm diameter, and its coefficient of discharge is 0·60. Hence calculate the flow rate when $h = 50$ mm.

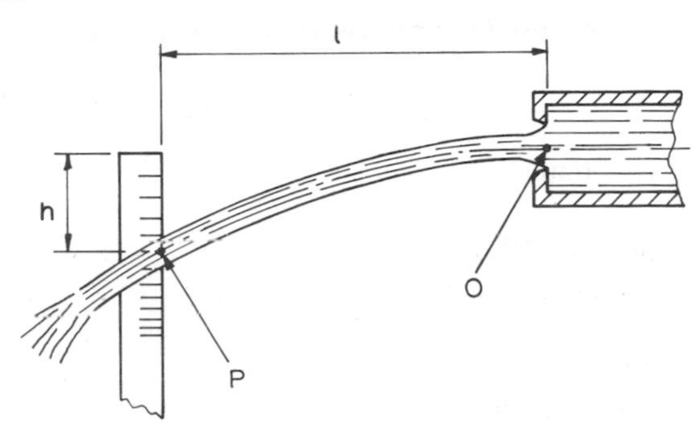

Fig. 7.31 Flow-rate gauge

Let t be the time for a particle of fluid to travel from O to P. The horizontal travel is due to the velocity at the orifice, and the vertical travel is due to gravitational acceleration.

Then $\quad h = 0 + \frac{1}{2}gt^2$

and $\quad t = \sqrt{(2h/g)}$

Also $\quad V = l/t$

$$= l\sqrt{(g/2h)}$$

Flow rate $\quad \dot{Q} = C_D \times VA$

$$= 0{\cdot}60 \times \frac{\pi}{4} \times 0{\cdot}02^2 \times 0{\cdot}3 \times \sqrt{(9{\cdot}81/2h)}$$

$$= 0{\cdot}125 \times 10^{-3}/\sqrt{h} \quad \text{m}^3/\text{s}$$

$$= 0{\cdot}125/\sqrt{h} \quad \text{litres/second}$$

When $\quad h = 0{\cdot}050$ m

then $\quad \dot{Q} = 0{\cdot}125/\sqrt{0{\cdot}05}$

$$= 0{\cdot}56 \text{ litres/second}$$

7.7.5 An airbox of the type shown in fig. 7.16 is used with a well-type inclined-tube water-sealed manometer, as shown. If the orifice is

35 mm diameter and has a coefficient of discharge of 0·60, calculate the volume flow rate through the box referred to standard conditions, i.e. 0 °C and 101 kN/m^2, when the reading (R) from the zero position is 200 mm, the barometer reads 760 mm of mercury, and the ambient temperature is 17 °C. The inside diameter of the manometer tube is 5 mm, the inside of the well is 100 mm square, and the tube is inclined at 15° to the horizontal. Take R for air as 287 J/kg K.

By eqn 5.8,

$$P_1 - P_2 = \rho_w g R(\sin\theta + a/A)$$
$$= 10^3 \times 9{\cdot}81 \times 0{\cdot}200[\sin 15^\circ + \{\pi \times (0{\cdot}005^2/4)/0{\cdot}100^2\}]$$
$$= 9{\cdot}81 \times 200(0{\cdot}2588 + 0{\cdot}0020)$$
$$= 508 \text{ N/m}^2$$

Atmospheric pressure $(P_a) = \rho_m g h_m$

$$= 13{\cdot}6 \times 10^3 \times 9{\cdot}81 \times 0{\cdot}760$$
$$= 101 \text{ kN/m}^2$$

From the characteristic gas equation, $P/\rho_a = RT$,

$$\therefore \quad \rho_a = P_1/RT_1 = 101 \times 10^3/(287 \times 300)$$
$$= 1{\cdot}17 \text{ kg/m}^3\text{, outside the airbox}$$

From eqn 7.7, assuming ρ_a is constant,

$$\dot{Q} = C_D A_0 \sqrt{(2\,\delta P/\rho_a)}$$
$$= 0{\cdot}60 \times (\pi \times 0{\cdot}035^2/4) \times \sqrt{(2 \times 508/1{\cdot}17)}$$
$$= 0{\cdot}017 \text{ m}^3\text{/s at 17 °C and 101 kN/m}^2$$

To refer this to standard conditions: from the characteristic gas equation, PV/T = constant,

$$\therefore \quad \frac{P_1 V_1}{T_1} = \frac{P_2 V_2}{T_2}$$

and $$V_2 = V_1 \times \frac{P_1}{P_2} \times \frac{T_2}{T_1}$$
$$= 0{\cdot}017 \times 1 \times 273/300$$
$$= 0{\cdot}015 \text{ m}^3$$

Hence $\dot{Q} = 0{\cdot}015$ m^3/s at standard conditions.

7.7.6 A variable-area meter of the type illustrated in fig. 7.17(a) and (b) has a tube with an included-angle taper of 5°. The float has a volume of 560 mm³, with an effective diameter of 12 mm and a vertical range of movement of 200 mm, and is made from aluminium of relative density 2·70. If the internal diameter of the measuring tube at the bottom of the range is 16 mm, estimate the range of flow which can be measured using paraffin of relative density 0·80. At what height (h) will the mean flow rate occur?

Let A be the effective area of the float, A_1 be the area of the tube bore at the lower scale position, A_2 be the area of the tube bore at the upper scale position, a_1 be the effective flow area at the lower scale position, a_2 be the effective flow area at the upper scale position, then

$$
\begin{aligned}
a_1 &= A_1 - A \\
&= \pi(8^2 - 6^2) \\
&= 28\pi \text{ mm}^2
\end{aligned}
$$

The radius of the tube bore at position 2 is

$$
\begin{aligned}
r_2 &= 8 + 200 \tan 2{\cdot}5° \\
&= 8 + 8{\cdot}74 \\
&= 16{\cdot}74 \text{ mm}
\end{aligned}
$$

$$
\begin{aligned}
\therefore \quad a_2 &= A_2 - A \\
&= \pi(16{\cdot}74^2 - 6) \\
&= 244\pi \text{ mm}^2
\end{aligned}
$$

For the lower float position, eqn 7.10 gives

$$
\begin{aligned}
\dot{Q} &= C_D a_1 \sqrt{\{2gV_f(\rho_f - \rho)/\rho A\}} \\
&= 1 \times 28\pi \times 10^{-6} \sqrt{\{2 \times 9{\cdot}81 \times 560} \\
&\quad \overline{\times 10^{-9}(2{\cdot}7 - 0{\cdot}8)/(0{\cdot}8 \times 36\pi \times 10^{-6})\}} \\
&= 42{\cdot}2 \times 10^{-6} \text{ m}^3/\text{s} \\
&= 42{\cdot}2 \times 10^{-3} \text{ l/s} \\
&= 2{\cdot}53 \text{ l/min}
\end{aligned}
$$

For the flow rate at position 2,

$$
\begin{aligned}
\dot{Q}_2 &= \dot{Q}_1 \times a_2/a_1 \\
&= 2{\cdot}53 \times 244\pi/28\pi
\end{aligned}
$$

$$= 22{\cdot}0\ \text{l/min}$$

Hence the flow range is 2·5 to 22 l/min approx.

The mean flow rate will be when the effective area of flow is

$$a = (a_1 - a_2)/2$$
$$= (28\pi + 244\pi)/2$$
$$= 136\pi\ \text{mm}^2$$

Hence $\quad \pi r^2 - \pi \times 6^2 = 136\pi$

and $\quad r = \sqrt{136 + 36}$

$$= 13{\cdot}1\ \text{mm (tube inside diameter)}$$

$$\frac{13{\cdot}1 - 8}{h} = \tan 2{\cdot}5^\circ = 0{\cdot}0437$$

and $\quad h = 5{\cdot}1/0{\cdot}0437$

$$= 117\ \text{mm}$$

Hence the scale is not linear. (Variation of the coefficient (C_D) due to changes of the Reynolds number (*Re*) for the flow tends to compensate for this variation due to the area/height relationship, and for many instruments the scales are linear.)

7.7.7 In a falling-sphere-viscometer test using the BS equipment (fig. 7.25), filtered oil of density 810 kg/m^3 was placed in a 40 mm inside-diameter tube. A 3·00 mm diameter steel ball of density 7700 kg/m^3 was found to travel between the 175 and 25 mm marks in 20·4 s. If the constants are $F = 0{\cdot}834$ and $g = 980{\cdot}7$ cm/s^2, calculate the kinematic viscosity of the oil.

Referring to eqn 7.12,

$$d = 0{\cdot}30\ \text{cm}$$
$$V = 15/20{\cdot}4\ \text{cm/s}$$
$$(\rho_s - \rho_l)/\rho_l = (7700 - 800)/800 = 6900/800$$

Hence $\quad \nu = 0{\cdot}834 \times 0{\cdot}30^2 \times 980{\cdot}7 \times 6900$

$$\times\ 20{\cdot}4/(0{\cdot}18 \times 800 \times 15)$$
$$= 4800\ \text{cS} = 4800 \times 0{\cdot}01\ \text{cm}^2/\text{s}$$
$$= 4{\cdot}8 \times 10^{-3}\ \text{m}^2/\text{s}$$

7.7.8 Acetic acid of density 1050 kg/m^3 was passed through an accurate tube 0·001 m bore and 0·300 m long under a head (h) of 0·200 m, as shown in fig. 7.28(a). A quantity of 50 cm^3 was found to flow in 520 seconds. Calculate the dynamic viscosity of the acid, and check that the flow is laminar.

Using Poiseuille's equation (7.13),

$$\eta = \pi w h_f d^4/(128\dot{Q}l)$$

$$= \pi \times 9{\cdot}81 \times 1050 \times 0{\cdot}200 \times \frac{0{\cdot}001^4 \times 10^6 \times 520}{128 \times 50{\cdot}0 \times 0{\cdot}300}$$

$$= 1{\cdot}75 \times 10^{-3}\ \mathrm{N\,s/m^2}$$

$$= 1{\cdot}75\ \mathrm{mN\,s/m^2}$$

The Reynolds number is given by

$$(Re) = \bar{V}d/v$$

$$= \frac{\dot{Q}}{A} \times d \times \frac{\rho}{\eta}$$

$$= \frac{50 \times 10^{-6} \times 0{\cdot}001 \times 1050}{(\pi/4) \times 0{\cdot}001^2 \times 520 \times 1{\cdot}75 \times 10^{-3}}$$

i.e. (Re) = 74, and the flow is laminar.

7.8 Tutorial and practical work

7.8.1 Sketch an arrangement showing a fixed vane positioned normal to a fluid flow, together with a suitable force-measuring and indicating device, to give a signal which is a function of velocity. Compare the functioning of this device with the variable-area devices of section 7.3.3.4.

7.8.2 Discuss the reasons for using, in the niveau shown in fig. 7.15, (a) a baffle at the fluid inlet, (b) a relatively large barrel.

7.8.3 Discuss the suitability of the various pressure-measuring methods shown in Chapter 5 in measuring the pressure difference in venturimeters, orifice meters, and similar devices.

7.8.4 Discuss the effect of using a venturimeter on a fluid (a) more dense than that for which it was calibrated, (b) more viscous than that for which it was calibrated.

7.8.5 Why is the turbine-type flow meter more accurate than the helical-vane type? What other advantages and disadvantages has it?

7.8.6 Examine eqn 7.10 and discuss the factors which may affect the accuracy of flow measurement when the temperature of the fluid increases by about 10 °C, the volume flow rate remaining the same.

7.8.7 Sketch a system for the remote indication of (a) a venturimeter signal, (b) a float-type flow-meter signal.

7.8.8 Show that for the variable-area meter of fig. 7.17(c), the flow rate is given by $\dot{Q} = C_D \times x^{3/2}\sqrt{(2k/\rho A)}$, where k is the spring rate. What advantage will be gained by making the spring rate very low, so that the spring force is virtually constant?

7.8.9 What is the purpose of the felt pads at the pressure tappings of the viscous air meter shown in fig. 7.18?

7.8.10 Poiseuille's equation for laminar flow gives the friction head loss (h_f) $= 32\eta\overline{V}l/\rho d^2$, where l is the pipe length and d is its internal diameter. Discuss the effects of temperature change on the accuracy of the viscous air meter of section 7.3.4.

7.8.11 Discuss the purpose of (a) the 'gate' on the weighing conveyor of fig. 7.22, (b) the dashpot on the mass-flow measuring device of fig. 7.23.

7.8.12 Sketch a viscosity-measurement system suitable for measuring the viscosity of liquids under very high pressures, using a falling sphere whose terminal velocity is measured electrically.

7.8.13 Would it be correct to classify the viscometer of fig. 7.26(a) as a constant-force, varying-velocity type, and that of fig. 7.26(b) as a constant-velocity, varying-force type?

7.8.14 Rewrite eqn 7.12 so that each term is in SI units, giving kinematic viscosity in m^2/s.

7.8.15 Discuss methods of calibrating the flow meters described in this chapter.

7.8.16 Discuss the effect of variation of liquid density and viscosity on the accuracy of the variable-area flow meters.

7.8.17 Investigate 'root-mean-square' errors. (Electrical textbooks may be helpful.)

7.9 Exercises

7.9.1 A Pitot-static probe with a thermocouple attached is used to measure the velocity of air in a duct (see fig. 7.29), and a differential head (h) of 120 mm of water is recorded. The thermocouple shows an air temperature of 320 K in the duct, and a separate static tapping shows that the air pressure in the duct is 9·5 kN/m^2 gauge. The local barometer reads 762 mm of mercury. Calculate the density and velocity of the air, if the coefficient C for the Pitot-static probe is 0·99. Take R for air as 287 J/kg K and the relative density of mercury as 13.6.

[1·21 kg/m^3, 43·7 m/s]

7.9.2 Calculate the stagnation pressure in an aircraft Pitot tube when the aircraft is travelling at 800 km/h relative to the air, and the ambient conditions are atmospheric pressure 92·5 kN/m^2 and temperature −5 °C. Take R = 287 J/kg K.

What is the reason for positioning the Pitot probe on a slender snout well forward of the fuselage on high-speed aircraft? [123 kN/m^2]

7.9.3 The flow of fuel of density 820 kg/m^3 and dynamic viscosity 2·4 × 10^{-3} N s/m^2 through a horizontally mounted venturimeter otherwise similar to that of fig. 7.30 gives a head (h_m) of mercury of 144 mm. If the pipe and throat diameters are 80 mm and 40 mm respectively,

a) calculate the flow rate in litres/min if the coefficient of discharge (C_D) is 0·97;

b) calculate the Reynolds number at the throat if the dynamic viscosity is 2·4 × 10^{-3} N s/m^2. [453 litres/min, 85 000]

7.9.4 The density of air in a duct is to be 1·2 kg/m^3, and its maximum velocity 20 m/s. The duct is 100 mm diameter, and the velocity is to be measured using an 80 mm diameter orifice and a differential manometer. If the coefficient of discharge is assumed to be 0·6, calculate the necessary range of the manometer in mm of water, and suggest a suitable type. [≈100 mm water]

7.9.5 During calibration of a niveau (fig. 7.15), water at 15 °C was passed through whilst the head (h) remained constant at 300 mm, and 80 litres were collected in 11 min 25 s. If the brass orifice has a diameter 10 mm, calculate the discharge coefficient (C_D) and the Reynolds number at the orifice if the dynamic viscosity is 1·15 × 10^{-3} N s/m^2. What is the maximum flow rate the meter will measure if the maximum head (h) is 600 mm? If the meter is used to measure the flow rate of water at 90 °C from an engine jacket, discuss the likely changes affecting the measurement accuracy. [0·62, ≈21 000, 9·9 litres/min]

7.9.6 For the rotating-cylinder viscometer of fig. 7.26(b), show that, by substitution of values in eqn 7.11, the dynamic viscosity is given by $\eta = 2Ty/\pi D^3 lw$, if y is small.

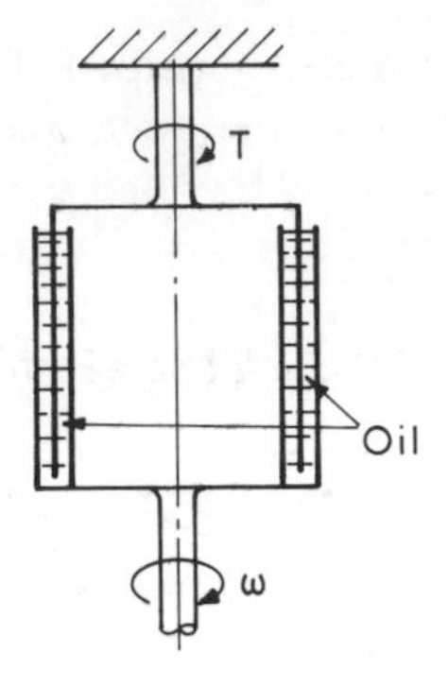

Fig. 7.32

In the meter shown in fig. 7.32, both sides of the top (static) cylinder are active. $r = 50$ mm, $y = 0{\cdot}8$ mm, $l = 100$ mm, and the lower cylinders rotate at 20 rad/s. If the torque (T) measured on the upper cylinder is 0·32 N m, calculate the dynamic viscosity of the oil which completely fills the space between the members. [81 mN s/m^2]

7.9.7 Three samples of oil having the same density are contained in three identical fall tubes having accurate and identical bores, and the terminal velocities of three identical spheres dropping in the oil were measured and found to be A – 0·82 cm/s, B – 0·69 cm/s, C – 0·22 cm/s. If the kinematic viscosity of oil B is known to be $5{\cdot}2 \times 10^{-3}$ m^2/s, calculate the kinematic viscosities of oils A and C.

[4·4 and 16·3 $\times 10^{-3}$ m^2/s]

7.9.8 Show by integration that, for a square-notch weir, the flow rate is given by $\dot{Q} = C_D \times \frac{2}{3} b \sqrt{(2gH^3)}$, and for a symmetrical vee-notch of semi-angle θ, $\dot{Q} = C_D \times (\frac{8}{15}) \tan(\theta/2) \sqrt{(2gH^5)}$.

8

Measurement of Power and Efficiency

8.1 Power and efficiency

Power is defined as the rate of doing work, or the rate of dissipation of energy, such as by heat, light, electricity, friction, etc. In the SI, power is measured in watts, so that the power available at an engine shaft, the heat dissipated in its cooler, the electrical power developed by its dynamo, the power used in circulating its cooling water, etc. are all measured in this same unit. The power absorbed in cutting operations such as turning, milling, or drilling may be measured to find the cutting characteristics of the material or to control the rate of metal removal. The power of a manufactured machine may be measured at the end of a production line, such as in the case of the acceleration and braking tests carried out on motor cars, using roller dynamometers.

Efficiency (η) may be defined as the ratio

$$\eta = \text{(useful power output)/(total power input)} \qquad 8.1$$

This equation may be applied to a component, a machine, or a system, and may be an instantaneous value or a mean. The measurement of efficiencies is a necessary part of the development work on a new machine, and may be made as part of the performance check on a production machine.

8.2 Measurement of mechanical power

Mechanical power may be transmitted, as in the case of a drive shaft, or dissipated as in the case of a metal-cutting operation. Power may be due to a force causing or opposing linear motion, or due to a torque causing or opposing rotation.

8.2.1 Measurement of linear power

If, for example, it is necessary to find the power required on a planing machine to cut a material at a given rate, then the cutting force and also the relative velocity of the component and tool must be measured. The power dissipated in cutting is then given by

$$P = FV \text{ watts} \tag{8.2}$$

where F is the instantaneous cutting force in newtons, and V is the instantaneous relative velocity in metres per second.

The same equation may be used to determine the power used in turning, if V is the tangential or cutting velocity and F is the tangential force. In measuring the tractive power of a locomotive, F would be measured using a draw-bar force dynamometer, V would be measured using one of the methods of section 6.4, and the power may then be calculated from eqn 8.2.

8.2.2 *Measurement of shaft power*

In development and production tests on prime movers such as internal-combustion engines, or on electric or hydraulic motors etc., the power developed has usually to be dissipated; hence an *absorption dynamometer* (see fig. 8.1) such as a rope brake, an electric generator and heater, an air blower, or a water brake, etc. may be used. At the same time, the reaction torque is measured as shown in section 5.3.3. If the rotational speed at which the torque is transmitted is also measured, then the power may be calculated thus:

$$\text{shaft power } (P) = T\omega \text{ watts} \tag{8.3}$$

$$= \frac{2\pi NT}{60} \text{ watts} \tag{8.4}$$

where T is the torque transmitted at ω rad/s, or N rev/min.

If the power output is not to be absorbed, the torque may be measured by a *transmission dynamometer* (see fig. 8.1), such as those using the elastic twist of a shaft, as described in section 5.3.2, or the torque-reaction ones such as the gearbox and gear-wheels shown in section 5.3.3.

Hence the dynamometer will measure the torque, and the rotational speed will have to be measured separately, such as by one of the methods described in section 6.4. The power may then be calculated using eqn 8.3. If the speed is constant within close limits, then power (P) $\propto$ torque (T) and an instrument indicating a signal proportional to torque may be used to indicate power, as in examples 5.8.4 and 5.8.5. Where the torque- and speed-measuring systems give electrical outputs, these may be multiplied together as shown in section 2.6.2 to give an electrical output signal proportional to instantaneous power.

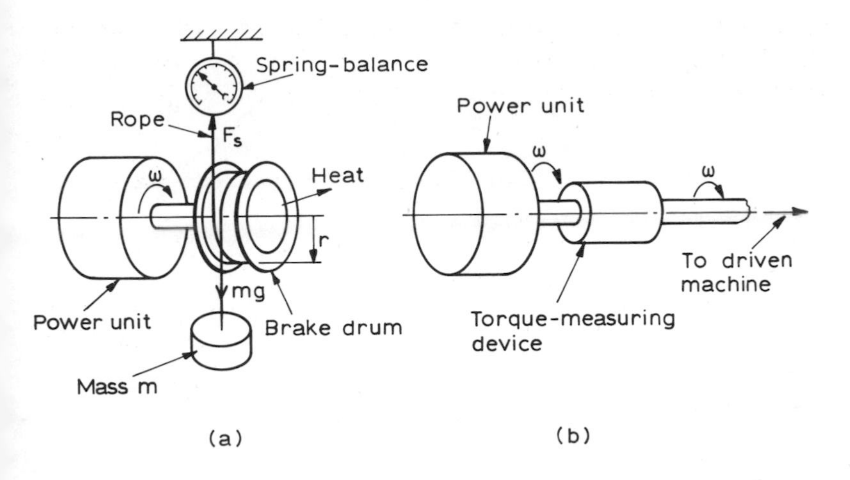

Fig. 8.1 Shaft-power measurement: (a) Absorption dynamometer (rope brake) (b) Transmission dynamometer

In many machines, particularly those with reciprocating motions, the torque and speed will not be uniform over each revolution, i.e. there will be *cyclic fluctuation*. To what degree the indicated values of torque, speed, and power follow the fluctuations depends on the dynamic-response characteristics of the measuring system (see Chapter 3); however, the mean value of these over the cycle may be quite adequate for most purposes.

8.2.3 Measurement of cylinder power and friction power

In research-and-development work on internal-combustion engines, it is necessary to know how the energy of the fuel used is accounted for – as power at the shaft and friction in the bearings, valve mechanism, cylinder bore, etc. The power developed at the piston of a reciprocating engine may be determined from the pressure–displacement measurements made continuously using the engine indicator, Farnboro' indicator, or other methods described in section 5.4.4, where typical traces of a low-speed engine (fig. 5.33) and a high-speed engine (fig. 5.35) are shown.

The work done on the piston is $F\,\mathrm{d}l$, where F is the instantaneous force on the piston and $\mathrm{d}l$ is an incremental movement of the piston, as indicated in fig. 8.2(a).

But $F = PA_\mathrm{p}$, where P is the instantaneous pressure, and A_p is the area of the piston face. Therefore the total work done during the stroke is

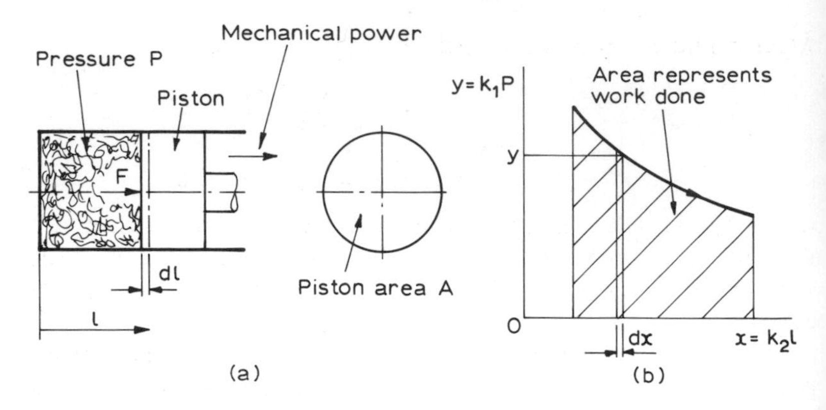

Fig. 8.2 Piston work (gas)

$$W = \int_{l_1}^{l_2} F\,\mathrm{d}l = A_p \int_{l_1}^{l_2} P\,\mathrm{d}l$$

and this is represented by the area under the pressure–stroke line on the indicator trace, as in fig. 8.2(b).

Hence work done $= A_p \times K \times$ area under pressure trace, where K is the work represented by unit area of the diagram and is positive when done *by the gas*. When the piston is doing work *on the gas*, as in compression and exhaust strokes, then the work may be regarded as negative, as indicated in fig. 5.35, showing an oscilloscope trace. The net work per cycle is the algebraic summation of the work of each stroke, and in terms of the diagram areas is given by

$$\text{work at piston} = A_p K \times \sum \text{(work areas)}$$

Where enclosed areas are formed, as on the engine-indicator trace in fig. 5.33, these may be positive such as area A_1, which represents the difference between the work done by the gas during expansion and the work done on the gas during compression, or negative such as area A_2, which represents the difference between the work done on the gas by the piston during exhaust and the work done by the air on the piston during induction. The negative work area A_2 is known as the 'pumping loop', and a light spring may be used to obtain this more accurately.

Using the enclosed areas,

$$\text{net work on piston per cycle} = A_p K(A_1 - A_2)$$

Mean cylinder power during cycle = i.p.

$$= \frac{\text{net work per cycle}}{\text{time taken per cycle}}$$

$$\therefore \quad \text{i.p.} = A_p K(A_1 - A_2)C/60 \qquad 8.5$$

where A_1 and A_2 are the positive and negative work areas respectively, C is the number of combustion cycles per minute, K is the work represented by unit area of the pressure–stroke diagram, and A_p is the engine piston area.

Often a *mean effective pressure* (m.e.p.) is used, this being the constant pressure, acting during the expansion stroke only, which would do the same work on the piston as the varying pressure during the complete combustion cycle. Hence

$$\text{m.e.p.} = \frac{\text{net area of indicator diagram}}{\text{length of indicator diagram}} \times \text{spring constant}$$

$$= (\textstyle\sum A/L) \times \theta \qquad 8.6$$

where L is the diagram length and θ is the spring constant.

The mean effective pressure for a single-stage reciprocating compressor is illustrated in fig. 8.3. In this case the whole diagram area represents pumping work, and may be regarded as negative.

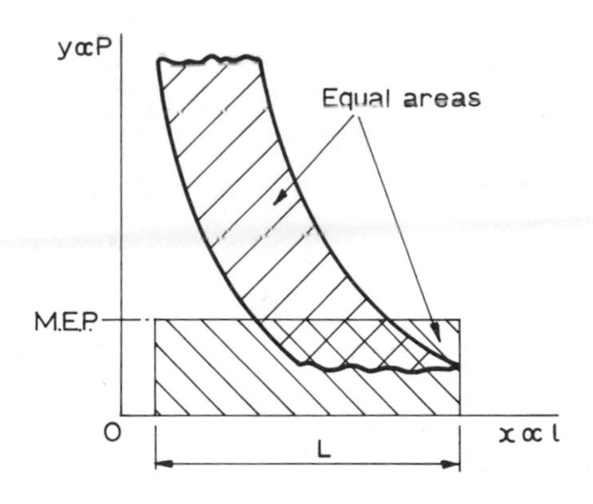

Fig. 8.3 Compressor indicator diagram

Using the mean effective pressure,

$$\text{i.p.} = \text{m.e.p.} \times l \times A \times C/60 \qquad 8.7$$

where l is the piston stroke and the other symbols are as in eqn 8.5.

Power-balance equations may be written down for engines, compressors, and other machines thus:

$$\text{for an engine, } I' = P_p + P_f + P \qquad 8.8$$

$$\text{for a compressor, } P = P_p + P_f \qquad 8.9$$

where I' is the *positive* indicated power, P_p is the pumping power, P_f is the total friction power (mechanical), and P is the shaft output or input power.

Some dynamometers, notably the electric swinging-field type, may be used to rotate (or 'motor') an engine, and the reverse torque on the dynamometer may be used to calculate the friction power and the pumping power combined. This should be carried out soon after a power run, so that temperatures have not changed excessively. The sum of the shaft output power and the 'motoring' power measured at the same speed gives the *positive* indicated power (I') of the engine at that speed.

Where other types of dynamometer are used, so that motoring is not possible, a Morse test may be carried out on multi-cylinder engines, whereby the power output is measured initially with all cylinders operative, and then with one cylinder made inoperative by removing a spark-plug lead or by blanking off a fuel line in the case of a compression-ignition engine. The engine speed is maintained constant during the power measurements, and each cylinder is made inoperative in turn.

Let P represent the shaft power with all cylinders operative,
P_1 represent the shaft power with no. 1 cylinder inoperative,
P_2 represent the shaft power with no. 2 cylinder inoperative,
etc.

then, for all cylinders operative,

$$P = \{I_1' - (P_p + P_f)_1\} + \{I_2' - (P_p + P_f)_2\} + \{I_3' - (P_p + P_f)_3\} + \{I_4' - (P_p + P_f)_4\} \qquad 8.10$$

With cylinder no. 1 inoperative,

$$P_1 = \{0 - (P_p + P_f)_1\} + \{I_2' - (P_p + P_f)_2\} + \{I_3' - (P_p + P_f)_3\} + \{I_4' - (P_p + P_f)_4\} \qquad 8.11$$

If it is assumed that the friction and pumping powers are the same for each cylinder and are unchanged with one cylinder inoperative, then subtracting eqn 8.11 from eqn 8.10 gives

$$\text{the positive indicated power of no. 1 cylinder} = I_1' = P - P_1 \qquad 8.12$$

This is repeated for each cylinder in turn, until all the I' values are known; these may then be used in eqn 8.10 to obtain the friction and pumping power loss of the engine (see example 8.5.4).

8.3 Measurement of power in hydraulic circuits

If a piston is being moved by a hydraulic pressure P exerting a total force F over an area A as shown in fig. 8.4, then the instantaneous power is

$$\text{power} = F\,\mathrm{d}l/\mathrm{d}t$$

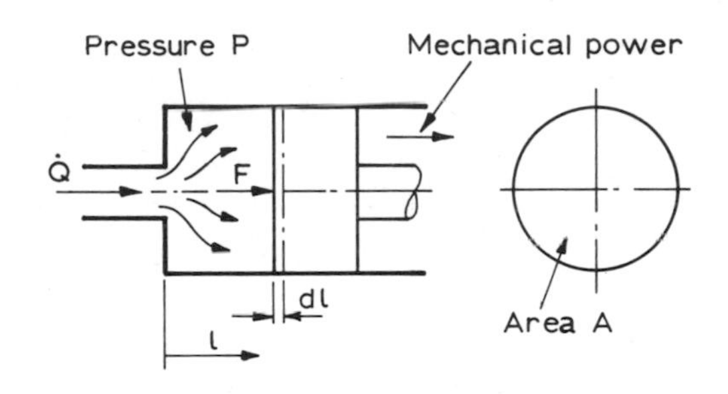

Fig. 8.4 Piston work (liquid)

But $F = PA$ and $A\,\mathrm{d}l = \mathrm{d}Q$, the volume displaced,

$$\therefore \quad \text{power} = PA\,\mathrm{d}l/\mathrm{d}t$$

$$= P\,\mathrm{d}Q/\mathrm{d}t$$

$$= P\dot{Q}\text{ watts} \qquad 8.13$$

where $\dot{Q}$ is the volume flow rate in m^3/s.

Equation 13 may also be expressed in terms of a head (h) of the liquid, giving

$$\text{power} = wh\dot{Q}\text{ watts} \qquad 8.14$$

where w is the specific weight of the fluid in N/m^3, and h is the head change, or the 'lift', in metres of liquid.

Equation 8.14 may be used for finding the fluid-power output of reciprocating or of rotating pumps (see example 8.5.5), or the power dissipated against friction in a hydraulic circuit, using the change of total head (δh) as shown in fig. 8.5;

i.e. $$\text{power dissipated against friction} = w\,\delta h\dot{Q}$$

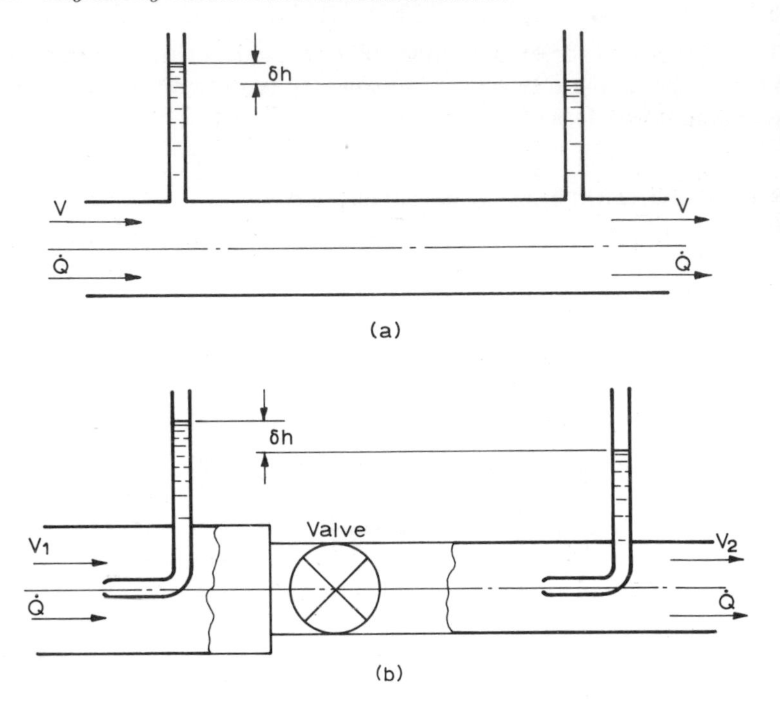

Fig. 8.5 Head loss due to friction

8.4 Electrical-power measurement

In a d.c. circuit, power is the product of current (I) and potential difference (V);

$$\begin{aligned} \text{power} &= IV \text{ watts} \\ &= I^2R \\ &= V^2/R \end{aligned} \Bigg\} \text{since } V = IR$$

The power may be measured by taking voltmeter and ammeter readings.

Referring to fig. 8.6(a), the voltmeter, having resistance r_1, may be connected either (i) across AB or (ii) across AC, and both the ammeter and the voltmeter will have different readings for connections made as (i) or (ii). The true power dissipated by resistor R is given by $I_R V_{AB}$. By connections (i), the power value obtained is proportional to $(I_{r_1} + I_R)V_{AB}$ [fig. 8.6(b)]; by connections (ii) it is $I_R V_{AC}$ [see fig. 8.6(c)]. In neither case is the true power given, but it can be shown that connections (i) give the least error (see examples 8.5.6 and 8.7.8) when the ammeter resistance is greater than the reciprocal of the voltmeter resistance.

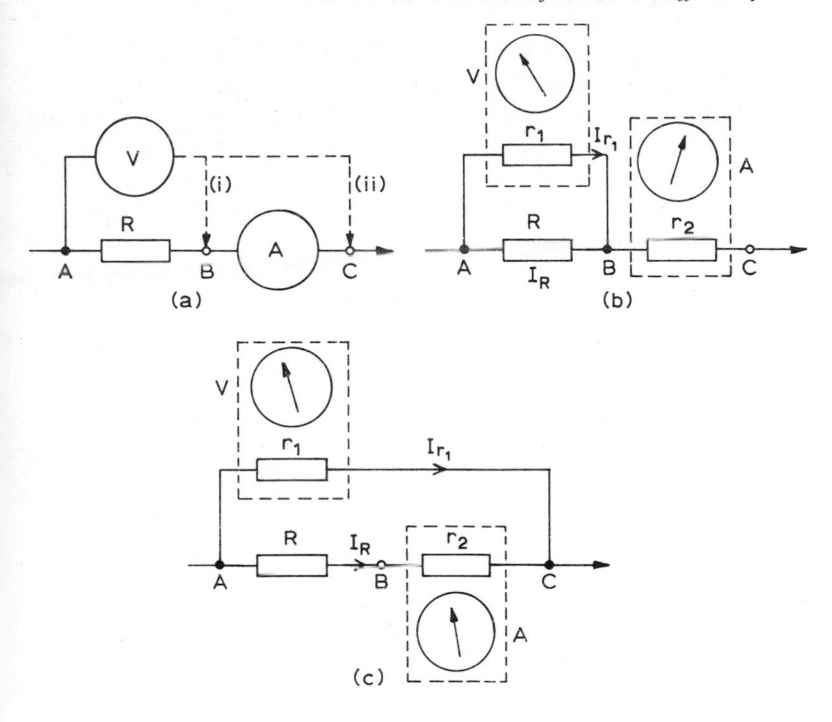

Fig. 8.6

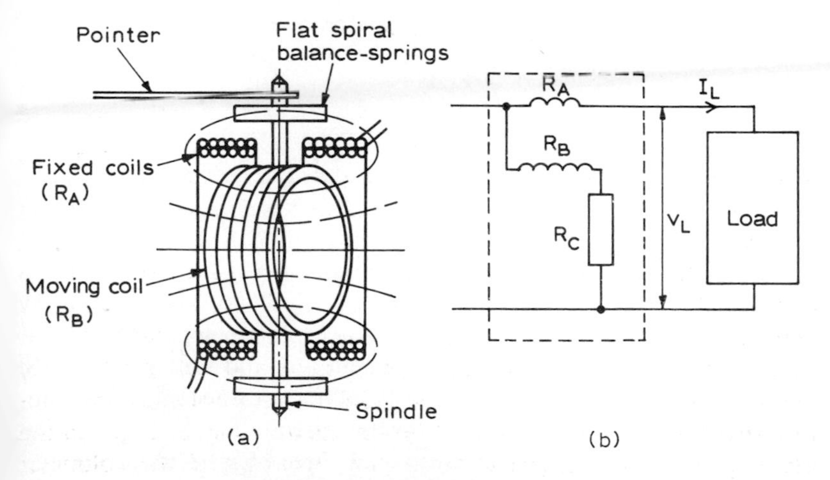

Fig. 8.7 Dynamometer wattmeter

By using a dynamometer wattmeter [fig. 8.7(a)] connected as shown in fig. 8.7(b), power may be measured by a single instrument.

If power in a purely resistive (i.e. non-reactive) circuit having alternating current is to be measured, the current (I_L) and voltage (V_L) *r.m.s. values* may be read from suitable a.c. instruments connected as in fig. 8.6(a), and again the power is the product of the two readings. Alternatively, power may be read directly from a dynamometer wattmeter connected as in fig. 8.7(b) or fig. 2.43 (see section 2.6.2).

However, if an a.c. circuit contains capacitive and/or inductive components or elements, then the rise and fall of the current will not be in phase with the rise and fall of voltage, and will either lead it or lag behind it by the phase-angle ϕ, as shown in fig. 8.8. It may be shown that

$$\text{average power value} = V_L I_L \cos \phi \qquad 8.15$$

and is not given by the product of ammeter and voltmeter readings.

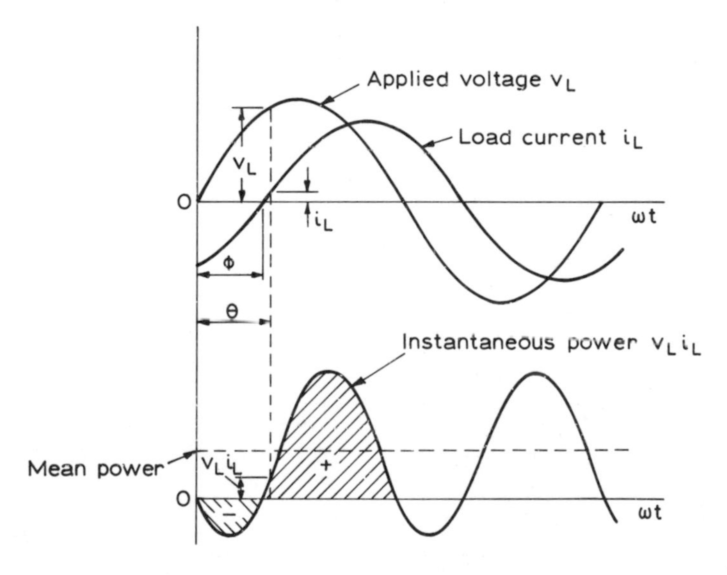

Fig. 8.8 A.C. voltage, current, and phase-angle

Since the torque on the spindle of the dynamometer wattmeter is due to the instantaneous values of current (i_L) and voltage (v_L), the phase lead or lag, ϕ, is automatically taken into account, and the instrument reads power directly in the a.c. circuit.

Many industrial machines operate on a three-phase electric supply, which may have either star or delta connections, and which may or may not be balanced (i.e. have the same impedance and power factor in each phase). The following methods may be used to measure the total power.

a) Star-connected balanced load, with neutral point accessible
The wattmeter is connected into the circuit as shown in fig. 8.9(a), and the total power is three times the wattmeter reading; i.e. total power = $3P$.

b) Star-connected load, balanced or unbalanced
Two wattmeters are connected, as shown in fig. 8.9(b).

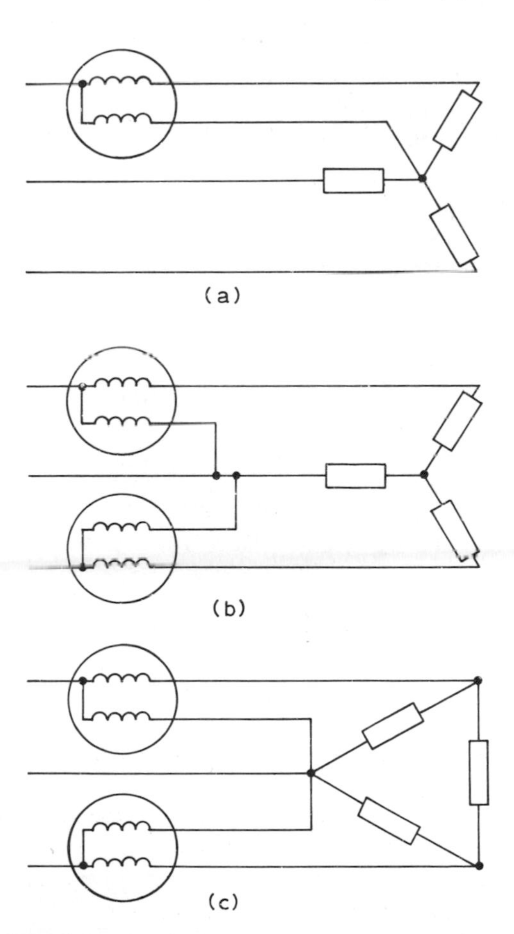

Fig. 8.9 Three-phase wattmeter connections: (a) Balanced star load (b) Balanced or unbalanced star load (c) Balanced or unbalanced delta load

c) Delta-connected load, balanced or unbalanced
Two wattmeters are connected, as shown in fig. 8.9(c).

It may be shown (ref. 39) that in cases (b) and (c) the total power is the sum of the wattmeter readings, i.e. total power $= P_1 + P_2$, irrespective of which two phases are connected to the wattmeters.

8.5 Examples

8.5.1 In a lathe cutting test, the following readings were made. From a cutting-force dynamometer,

tangential force, 795 N
axial (feed) force, 88 N

Spindle speed, by hand tachometer, 300 rev/min
Feed rate, 0·8 mm per rev
Mean diameter of cut, 0·100 m
Motor input power per phase (three-phase motor), 873 W [connections as fig. 8.9(a)].

Calculate the power absorbed in rotating the workpiece and in feeding the tool along the workpiece, and the overall efficiency of the lathe under these conditions.

Power used in turning the workpiece $= T\omega$

$$\therefore \quad P_1 = 780 \times 0{\cdot}050 \times \frac{300}{60} \times 2\pi$$

$$= 1250 \text{ W}$$

Power used in feeding the tool $= FV$

$$\therefore \quad P_2 = 88 \times 0{\cdot}8 \times 10^{-3} \times 300/60$$

$$= 0{\cdot}352 \text{ W}$$

Total power used in cutting $= P_1 + P_2$

$= 1250 + 0{\cdot}352$

$= 1250$ W, the feed power being negligible

The three-phase motor is a balanced load, hence

$$\text{power input to system} = 873 \times 3$$

$$= 2619 \text{ W}$$

$$\text{Overall efficiency of lathe} = \frac{\text{output power}}{\text{input power}}$$

$$= \frac{1250}{2619}$$

$$= 0{\cdot}48 \text{ or } 48\,\%$$

8.5.2 A three-phase caged-rotor induction motor rated at 3·75 kW output was tested by measuring the output torque and speed by a brake dynamometer and stroboscope respectively, and the input power by the two-wattmeter method illustrated in fig. 8.9(c). The spring-balance used measured the force tangentially at the end of an arm at 300 mm radius (r) from the shaft axis. The following readings were taken.

	Wattmeter readings		*Spring force*	*Speed*
Test	P_1 (watts)	P_2 (watts)	F (newtons)	N (rev/min)
1	+52·3 × 25	−20·2 × 25	0·0	1498
2	+87·0 × 25	−30·0 × 25	16·6	1493
3	+97·0 × 25	−18·0 × 25	25·6	1490
4	+112·0 × 25	0·0 × 25	42·4	1483
5	+120·0 × 25	+2·0 × 25	50·3	1482
6	+128·0 × 25	+12·0 × 25	60·3	1477
7	+140·0 × 25	+22·0 × 25	70·9	1473
8	+165·0 × 25	+47·0 × 25	90·3	1467
9	+184·0 × 25	+65·0 × 25	113·6	1457

Taking the calculations for test 2,

$$\text{input power } (P_i) = P_1 + P_2$$

$$= (87 - 30) \times 25$$

$$= 1425 \text{ W}$$

$$\text{output torque } (T_0) = F \times r$$

$$= 16{\cdot}6 \times 0{\cdot}3$$

$$= 4{\cdot}98 \text{ N m}$$

$$\text{output power } (P_0) = T_0\omega = T_0 \times 2\pi N/60$$

$$= 4{\cdot}98 \times 2\pi \times 1493/60$$

$$= 777 \text{ W}$$

Efficiency = (output power)/(input power)

$$= P_0/P_i$$

$$= 777/1425 = 0{\cdot}545 \text{ or } 54{\cdot}5\,\%$$

Completing the calculations for the series of tests, the following values are obtained.

Test	*Speed* (rev/min)	*Torque* (N m)	*Efficiency* (%)
1	1498	0·0	0·0
2	1493	4·98	54·5
3	1490	7·68	61·0
4	1483	12·72	70·5
5	1482	15·09	76·5
6	1477	18·09	79·6
7	1473	21·27	81·0
8	1467	27·09	82·0
9	1457	34·08	83·5

These values are shown plotted in fig. 8.10.

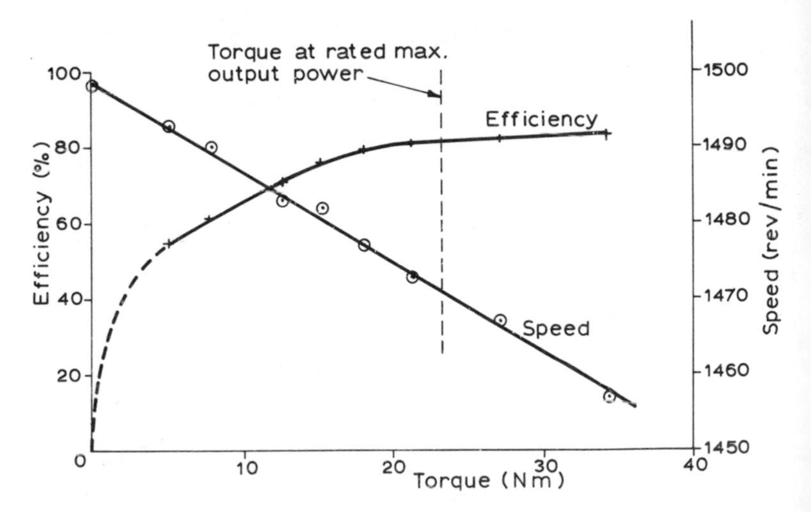

Fig. 8.10 A.C. motor characteristics

8.5.3 A single-cylinder compression-ignition engine was tested with a rope brake, and the following values were observed:

engine shaft speed (N), 450 rev/min
load on rope (m), 41 kg
spring-balance reading (F_s), 39·1 N
fuel used (r.d. = 0·88), 30 ml in 29 s
cooling-water flow rate, 4·5 kg/min
cooling-water inlet temperature (θ_1), 16 °C
cooling-water outlet temperature (θ_2), 84 °C

Engine indicator cards were taken using normal and light springs with a Dobbie-McInnes indicator, and are shown in fig. 8.11, the spring constants being 125 and 20 kN/m^2 per mm respectively.

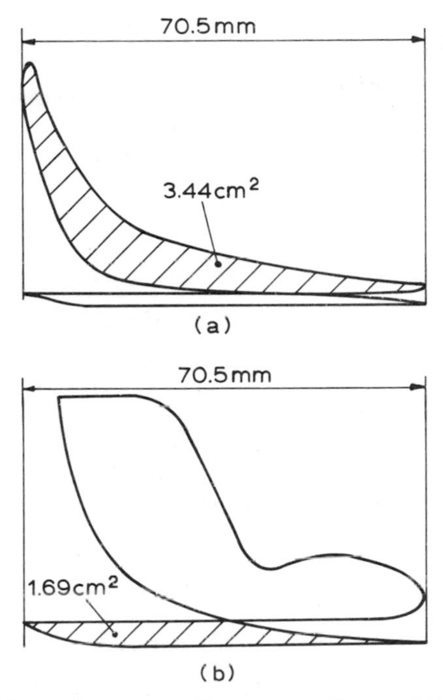

Fig. 8.11 Engine indicator diagrams: (a) Normal spring diagram (b) Light spring diagram

Area of positive loop by planimeter (A_1), 3·44 cm^2.
Area of pumping loop by planimeter (A_2), 1·69 cm^2.

The following constants apply:

cylinder bore (d), 300 mm
stroke (l), 300 mm
circumference of brake drum, 3·20 m
calorific value of fuel (c.v.), 44 MJ/kg

Calculate for the test (a) the brake power, (b) the cylinder power, (c) the friction power, (d) the power to the cooling water, (e) the thermal efficiencies, (f) the mechanical efficiency, and (g) the brake specific fuel consumption.

a) Brake power (b.p.) $= T\omega$

$$= (mg - F_s) \times r \times 2\pi N/60$$

But $2\pi r$ is the circumference of the brake drum = 3·20 m.

Hence $$\text{b.p.} = (41 \times 9{\cdot}81 - 39{\cdot}1) \times 3{\cdot}20 \times 450/60$$
$$= 8{\cdot}7\ \text{kW}$$

b) From eqn 8.6,

$$\text{m.e.p.} = (A_1 - A_2)\theta/L$$
$$= \left(\frac{3{\cdot}44 \times 10^2 - 1{\cdot}69 \times 10^2 \times 20/125}{70{\cdot}5}\right) \times 125 \times 10^3$$
$$= 560\ \text{kN/m}^2$$

From eqn 8.7,

$$\text{i.p.} = \text{m.e.p.} \times l \times A \times C/60$$
$$= 560 \times 10^3 \times 0{\cdot}300 \times \frac{\pi}{4} \times 0{\cdot}150^2 \times \frac{225}{60}$$
$$= 11{\cdot}1\ \text{kW}$$

c) Friction power (f.p.) = i.p. − b.p.

$$= 11{\cdot}1 - 8{\cdot}7 = 2{\cdot}4\ \text{kW}$$

d) Power to cooling water = mass flow rate × specific heat × temperature rise

$$= \frac{4{\cdot}5}{60} \times 4187 \times (84 - 16)$$
$$= 21{\cdot}3\ \text{kW}$$

e) Indicated thermal efficiency (η_I) $= \dfrac{\text{indicated power}}{\text{rate of energy from fuel}}$

$$= \frac{11\,100 \times 29}{0{\cdot}030 \times 0{\cdot}88 \times 44 \times 10^6}$$
$$= 0{\cdot}277 \text{ or } 28\,\%$$

Brake thermal efficiency (η_b) $= \dfrac{\text{brake power}}{\text{rate of energy from fuel}}$

$$= \frac{8700 \times 29}{0{\cdot}030 \times 0{\cdot}88 \times 44 \times 10^6}$$
$$= 0{\cdot}217 \text{ or } 22\,\%$$

f) Mechanical efficiency (η_m) = b.p./i.p.

$$= 8{\cdot}7/11{\cdot}1 = 0{\cdot}78 \text{ or } 78\,\%$$

g) Brake specific fuel consumption $= \dfrac{\text{mass of fuel used per hour}}{\text{brake power}}$

$$= \frac{0{\cdot}030 \times 0{\cdot}88 \times 3600}{29 \times 8{\cdot}7}$$

$$= 0{\cdot}38 \text{ kg/kWh}$$

8.5.4 During a Morse test on a four-cylinder petrol engine, the following readings were taken:

spring force on torque arm with all cylinders firing, 201·5 N
spring force on torque arm with cylinder 1 not firing, 143·5 N
spring force on torque arm with cylinder 2 not firing, 139·5 N
spring force on torque arm with cylinder 3 not firing, 141·5 N
spring force on torque arm with cylinder 4 not firing, 145·5 N
engine speed, 1 500 rev/min (constant)
radius of torque arm, 320 mm.

Calculate the brake power (P), the positive indicated power (I'), the friction and pumping power ($P_f + P_p$), and the mechanical efficiency (η_m) of the engine.

Brake power (b.p.) $= T\omega$

$$= Fr \times 2\pi N/60$$

$$= 201{\cdot}5 \times 0{\cdot}32 \times 2\pi \times 1500/60$$

$$= 10\,100 \text{ W}$$

Carrying out a similar calculation for the four readings with one cylinder inoperative, in turn, and using eqn 8.12, the following tabulation is obtained:

Cylinder not firing	*Brake power* (watts)	($P - P_n$) (watts)
none	$P = 10\,100$	
1	$P_1 = 7200$	2900
2	$P_2 = 7000$	3100
3	$P_3 = 7100$	3000
4	$P_4 = 7300$	2800
		$I' = 11\,800$ W

Friction power + pumping power = indicated power − brake power

$$\therefore \quad P_f + P_p = 11\,800 - 10\,100$$
$$= 1700 \text{ W}$$

Mechanical efficiency = (brake power)/(indicated power)

$$= 10\,100/11\,800$$
$$= 0{\cdot}85$$

8.5.5. A centrifugal pump was tested, using an electric motor mounted on a cradle as in fig. 5.17, to measure the shaft power. Delivery and suction gauges and a flow meter were mounted as shown in fig. 8.12.

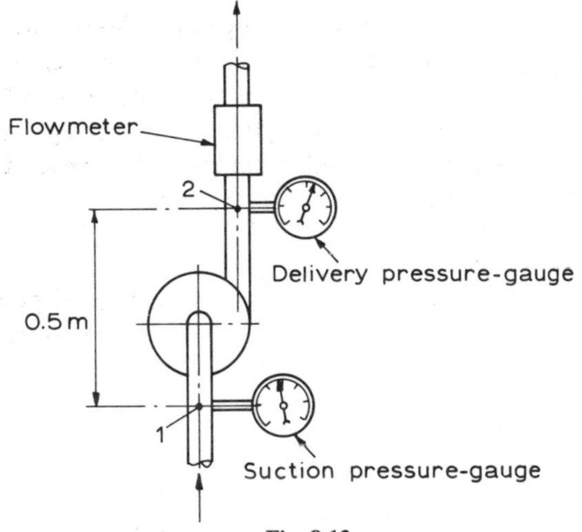

Fig. 8.12

The suction and delivery pipes were the same diameter, so that no change of velocity head occurred between the points 1 and 2. The following readings were taken:

torque-arm radius (r) = 150 mm
shaft speed (N) = 1400 rev/min
spring force (F_s) = 3·4 N
flow rate ($\dot{Q}$) = $0{\cdot}67 \times 10^{-3}$ m^3/s
suction head (h_s) = 1·56 metres of water (below atm. pressure)
delivery head (h_d) = 13·3 metres of water (above atm. pressure).

Calculate the water power, the shaft power, and the efficiency of the pump for these conditions.

$$\begin{aligned}\text{Shaft power} &= T\omega \\ &= 2\pi N F_s r/60 \\ &= 2\pi \times 1400 \times F_s \times 0{\cdot}15/60 \\ &= 22F_s \\ &= 22 \times 3{\cdot}4 \\ P_s &= 75{\cdot}5\ \text{W}\end{aligned}$$

Water power $= w\,\delta h \dot{Q}$ (eqn 8.14)

where

$$\delta h = \frac{V_2{}^2 - V_1{}^1}{2g} + \frac{P_2 - P_1}{w} + z_2 - z_1 = \text{the 'lift' of the pump}$$

Since $V_1 = V_2$, then

$$\begin{aligned}\delta h &= \frac{P_2 - P_1}{w} + z_2 - z_1 \\ &= (h_d + h_a) - (h_a - h_s) + z_2 - z_1,\end{aligned}$$

where h_a is the atm. pressure head

$$\begin{aligned}&= h_d + h_s + z_2 - z_1 \\ &= 13{\cdot}3 + 1{\cdot}56 + 0{\cdot}5 \\ &= 15{\cdot}4 \text{ metres of water}\end{aligned}$$

$$\therefore \quad \text{Water power} = 9810 \times 15{\cdot}4 \times 0{\cdot}67 \times 10^{-3}$$

$$= 101\ \text{W}$$

$$\begin{aligned}\text{Pump efficiency} &= 75{\cdot}5/101 \\ &= 0{\cdot}75 \text{ or } 75\%\end{aligned}$$

8.5.6 The power dissipated in a resistance of 10 Ω is to be measured using the two methods illustrated in fig. 8.6, the p.d. across AB being maintained at 10 V. The voltage and current values are measured by instruments of resistance 600 Ω and 1 Ω respectively.

Calculate (a) the true power dissipated, (b) the power measured by connections (i), and (c) the power measured by connections (ii).

a) True power dissipated $= V_{AB}{}^2/R$

$$= 10^2/10 = 10\ \text{W}$$

b) Resistance of parallel circuit AB is given by

$$R_p = R \times r_1/(R + r_1)$$
$$= 10 \times 600/(10 + 600) = 600/61$$
$$= 9{\cdot}84\ \Omega$$
$$I = V_{AB}/R_p$$
$$= 10/9{\cdot}84$$
$$= 1{\cdot}016\ \text{A}$$

$$\text{Power measured} = IV_{AB}$$
$$= 1{\cdot}016 \times 10$$
$$= 10{\cdot}16\ \text{W}$$

c)

$$I_R = V_{AB}/R = 10/10$$
$$= 1\ \text{A}$$
$$V_{AC} = IR_{AC} = 1 \times 11$$
$$= 11\ \text{V}$$

$$\text{Power measured} = I_R V_{AC}$$
$$= 1 \times 11$$
$$= 11\ \text{W}$$

Hence connections (i) givc an error (in this case) of $+1{\cdot}6\%$, and connections (ii) give an error of $+10\%$.

8.6 Tutorial and practical work

8.6.1 In ground testing of jet engines, the thrust is measured. Could a jet-power value be determined?

8.6.2 Sketch out a brake device to measure the torque available at the nose of a lathe spindle.

8.6.3 An epicyclic gearbox is to be used to measure torque as discussed in section 5.3.3. Sketch a system suitable for measuring the efficiency of the gearbox under different operating conditions.

8.6.4 Discuss the difficulties encountered in measuring the cylinder power in low- and high-speed reciprocating internal-combustion engines.

8.6.5 Discuss reasons for variation of the pumping power in the cut-off cylinder during a Morse test. Is the friction power likely to change?

8.6.6 Determine the resistance of voltmeter and ammeter instruments used in the laboratory, and hence determine whether any voltmeter + ammeter combination would give more accurate power values by connections as in fig. 8.6(c) rather than (b). (See problem 8.7.8.)

8.6.7 Referring to fig. 8.5, explain why Pitot tappings are necessary in case (b) and not in case (a). If ordinary Bourdon-tube gauges are used (static tappings) in (b), how is δh determined?

8.7 Exercises

8.7.1 A rope brake was rigged as shown in fig. 8.13 to test the power output from a machine. Torque balance was obtained when m_1 was 30 kg, m_2 was 10 kg, and the spring-balance read 8·5 N, when the shaft was rotating at 600 rev/min. Calculate the output power and the cooling water evaporated per minute if the drum temperature was 100 °C and the enthalpy of evaporation (h_{f_g}) is 2270 J/kg.

[212 W, 5·6 grams]

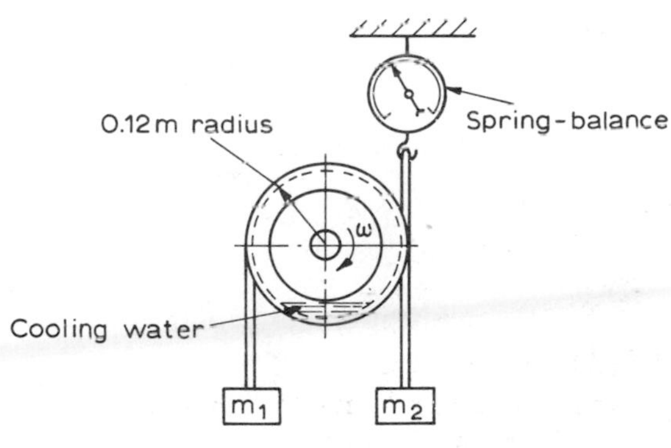

Fig. 8.13

8.7.2 As part of an investigation into a machine-tool drive, it is necessary to determine the efficiency of the mechanical part of the drive, in a particular gear ratio, at varying torques. The drive motor was tested separately, and the results are shown in example 8.5.2.

During a series of tests, the torque at the lathe spindle was applied and measured by a brake. The shaft speed of the motor was measured by a stroboscope, and the electrical input power by a wattmeter. The ratio motor speed : spindle speed was 8 : 1. From the following readings, and using the graph in fig. 8.10, determine the overall efficiency and the mechanical efficiency of the drive at each condition, and plot graphs of efficiencies, input power, and motor speed against output torque. Determine the overall efficiency and the power dissipated in the motor and in the mechanical drive at an output torque of 70 N m.

Test	*Motor speed* (rev/min)	*Input power* (W)	*Output torque* (N m)
1	1493	780	17·5
2	1486	1530	40
3	1481	2340	65
4	1475	3110	85
5	1469	3900	95

[0·54, 600 W, 550 W]

8.7.3 Cars coming off a production line are to have acceleration and braking tests carried out by driving the tractive wheels on a pair of rollers whilst the car is stationary (see fig. 8.14). If the car has a total

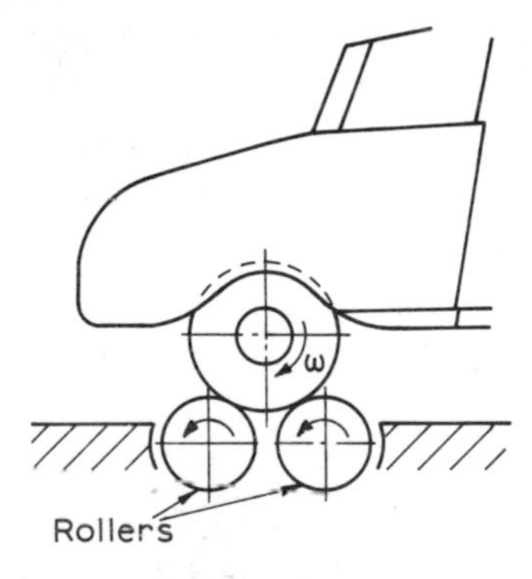

Fig. 8.14

mass of 960 kg, and the moment of inertia of the non-driving wheels is 0·8 kg m^2, calculate

a) the required total moment of inertia of the rollers, if they are 0·25 m diameter and the car wheels are 0·5 m diameter;

b) the rotational velocity of the wheels and of the drum to simulate 100 km/h;

c) the angular deceleration to be applied to the rollers to simulate braking equal to 0·7g.

Discuss the problem of simulating the air drag of the car on the roller, and suggest suitable velocity- and acceleration-measuring devices for the rollers. [30·4 kg m^2, 222·2 rad/s, −55 rad/s^2]

8.7.4 The results of Morse tests at different speeds for the engine of example 8.5.4 are shown in the following table. Using these, calculate the indicated power, friction power, and efficiency values, and plot these against a base of speed, using all the test values. Comment on the shape of the curves obtained.

Engine speed (rev/min)	2000	2500	3000
Spring force with all cylinders firing (N) . .	196·0	189·0	172·5
Spring force with cylinder 1 not firing (N) . .	137·5	131·5	114·5
Spring force with cylinder 2 not firing (N) . .	134·5	128·0	111·5
Spring force with cylinder 3 not firing (N) . .	136·0	129·0	112·5
Spring force with cylinder 4 not firing (N) . .	136·0	129·0	112·5

8.7.5 A test of a reciprocating air compressor gave an indicator diagram similar to that of fig. 8.3, the enclosed area being 9·8 cm^2 and the length 6·2 cm, when using a spring rated at 50 kN/m^2 per mm. Calculate the mean effective pressure, the indicated power expended in compressing the air, and the mechanical efficiency if the input shaft power is 4·54 kW. The cylinder bore is 80 mm, the piston stroke 80 mm, and the shaft speed 700 rev/min. [3·71 kW, 0·82]

8.7.6 Further readings for the centrifugal-pump test of example 8.5.5 are given below.

Spring force (F_s) (newtons)	*Flow rate* ($\dot{Q}$) (m^3/s)	*Suction head* (h_s) (metres of water)	*Delivery head* (h_d) (metres of water)
12·3	$1{\cdot}5 \times 10^{-3}$	1·65	13·0
20·4	$2{\cdot}5 \times 10^{-3}$	1·70	12·4
28·6	$3{\cdot}5 \times 10^{-3}$	0·70	12·0
36·8	$4{\cdot}5 \times 10^{-3}$	0·27	8·4

Calculate the shaft power, the water power, and the efficiency of the pump at these further conditions, and plot these, and those of 8.5.5, against a base of flow rate. Determine from the graph the lift, shaft power, and flow rate at the maximum-efficiency point of the pump.

[15·2 m H_2O, 320 W, $1{\cdot}75 \times 10^{-3}$ m^3/s]

8.7.7 In an electrohydraulic circuit, the electrical input power was found to be 1020 W. At a remote point at the same level as the pump, the flow through a 10 mm bore pipe was 540 ml/s and the static pressure 8·4 bar. Calculate the efficiency of the system at this condition if the fluid has a relative density of 0·88. [0·46]

8.7.8 Show that, for electrical-power measurement using voltmeter and ammeter readings, connected as in fig. 8.15, the error due to the application of the instruments is for (a) $V_{AB}{}^2/Rm$ and for (b) $nV_{AB}{}^2/R$, if V_{AB} is constant, mR is the voltmeter resistance, and nR the ammeter resistance. Verify that these are correct using the values in example 8.5.6.

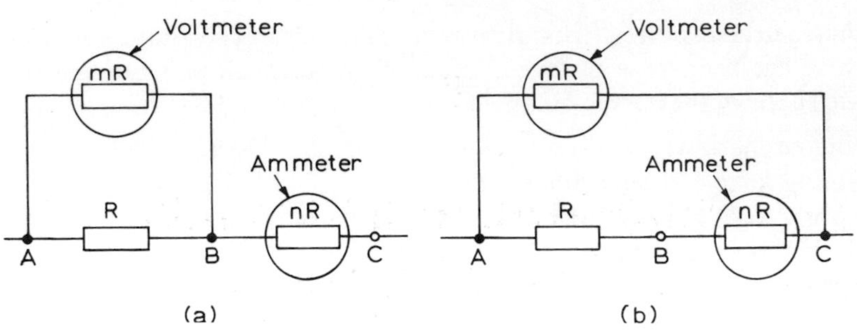
Voltmeter
mR
Ammeter
R
nR
A
B
C
(a)
Voltmeter
mR
Ammeter
R
nR
A
B
C
(b)

Fig. 8.15

9

Liquid-level and Quantity Measurement

9.1 Types of level-measuring devices

Level measurement may be required to give an accurate value of level, as in the maintainance of a supply of liquid at a constant head from a tank, or in the accurate measurement of quantity, such as in aircraft fuel tanks; but more often an approximate value is sufficient, and sometimes only a low-level signal is necessary, so that the contents of a tank may be replenished before it empties. In some cases, for example in a boiler, safety may be dependent on the maintenance of liquid level.

The level may be observed directly by the operator in many cases; in others, a signal which is a function of level may be read locally or transmitted to a remote reading point. Often a level signal may be used to initiate control action, such as opening a valve to introduce more fluid, or switching to another tank.

9.2 Direct methods

For an open tank, an internal scale may be fitted, if convenient, as in fig. 9.1(a). In some cases a hook gauge, as in fig. 9.1(b), may be suitable, for example when timing the discharge of a given quantity of liquid. The conical end of the hook rod makes the instant at which the level is reached easy to determine. For a closed tank or boiler, an external sight-glass may be used, as shown in fig. 9.1(c), but it must be appreciated that, where the vessel is at higher pressure, the glass tube is subjected to this pressure, and suitable metal or armour-plated glass guards and stopcocks are necessary for the protection of personnel.

9.3 Force- and pressure-operated methods

Many indirect methods rely on the force and the pressure caused by gravity on the fluid mass. If the fluid is of uniform density, then the total force exerted is proportional to the volume of liquid, i.e. $F = mg$, and

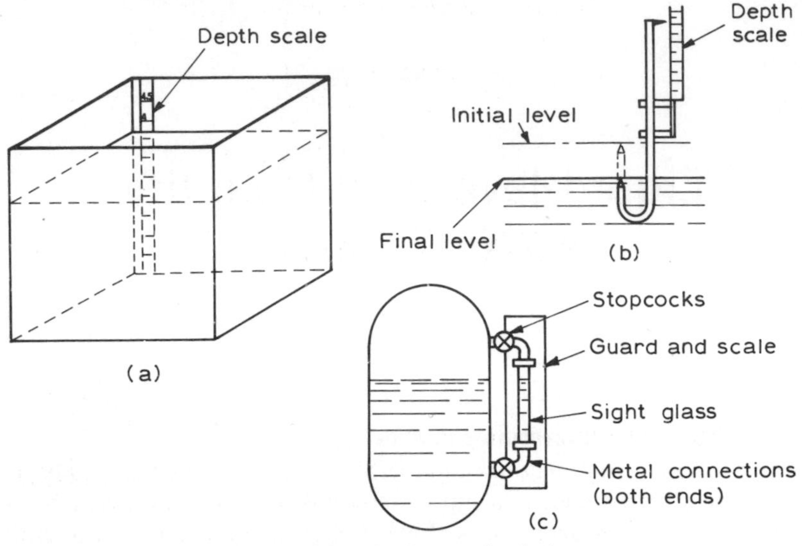

Fig. 9.1 Direct level-measuring devices

the pressure at any point due to the liquid head is proportional to the depth of that point, i.e. $P = \rho g h$.

9.3.1 Force- and mass-measuring methods

In many applications, the measurement of level is a means of measuring the total quantity of fluid, either the volume or the mass. In laboratory experiments, the total mass flow may be measured by collecting the fluid in a tank mounted on platform scales. The measured mass may then be converted to volume.

In industrial applications, it is more convenient to measure the force exerted by gravity on the tank and its contents. This is conveniently done by mounting the tank on load cells of, for example, strain-gauge type or hydraulic type (see Chapter 5), as shown in fig. 9.2. The total force on the base may be designated $\sum F$, then

$$\sum F = mg$$
$$= \rho A h g$$

where A is the *uniform* cross-sectional area, and ρ is the fluid density.

The mass content (m) of the tank is given by

$$m = \rho A h = \sum F/g \qquad 9.1(a)$$

The volume content (V) of the tank is given by

$$V = Ah = \sum F/\rho g \qquad 9.1(b)$$

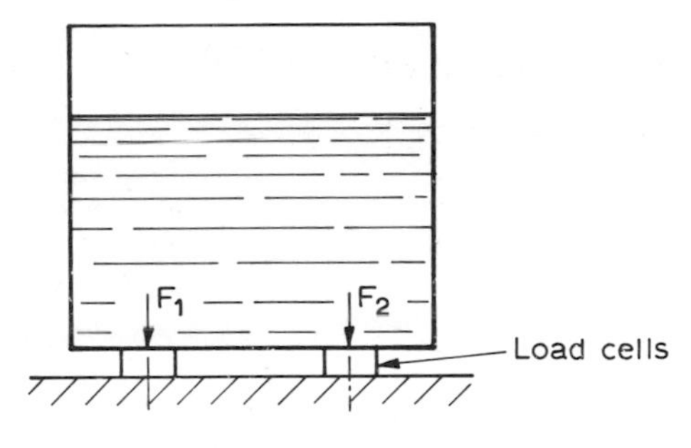

Fig. 9.2

The head (h) is given by

$$h = \sum F/\rho A g \qquad 9.1(c)$$

In each case the required quantity is directly proportional to $\sum F$.

9.3.2 *Pressure methods*

Since the pressure (P) due to a column (h) of fluid of constant density (ρ) is given by $P = \rho g h$, then $h = P/\rho g$. This gives the depth (h) irrespective of whether the vessel has a constant cross-sectional area or not. If the vessel is vented, so that the pressure above the fluid is atmospheric, then a simple U-tube mercury manometer may be used to measure the level. The arrangement is shown in fig. 9.3(a).

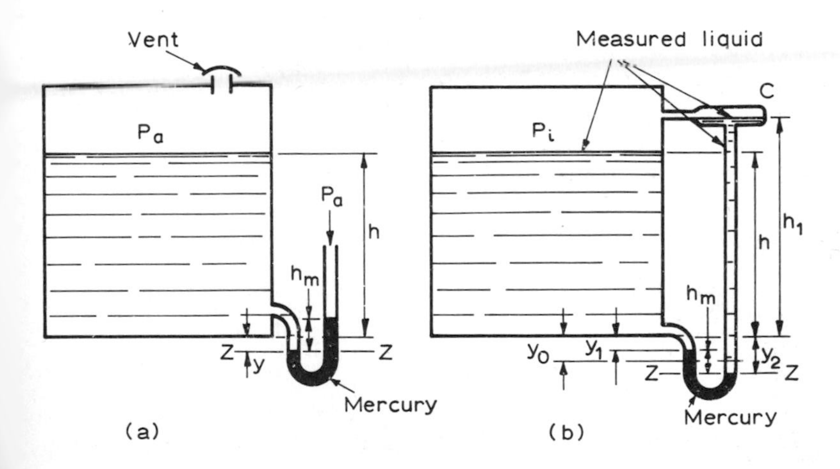

Fig. 9.3 Level measurement by mercury manometer: (a) Vented tank (b) Unvented tank

Equating pressures at level ZZ gives

$$\rho g h + \rho g y + P_a = \rho_m g h_m + P_a$$
$$\therefore \quad \rho h = \rho_m h_m - \rho y$$

where the suffix 'm' denotes mercury, and suffix 'a' denotes atmospheric. If the mercury levels were initially at the same height as the base of the tank, then $y = h_m/2$. Putting this value into the above equation gives

$$\rho h = \rho_m h_m - \rho h_m/2$$
$$\therefore \quad h = \{(\rho_m/\rho) - \tfrac{1}{2}\}h_m \qquad 9.2$$

and $\delta h = K\,\delta h_m$, where $K = \{(\rho_m/\rho) - \tfrac{1}{2}\}$. Hence the mercury scale may be graduated in depth units.

Usually only one leg of the manometer will be calibrated and read, so that the measurements are from a fixed datum. This will, however, reduce the sensitivity by a factor of 2 (see Chapter 5).

If the pressure in the tank is different from the atmospheric pressure, either due to pressurising or to lack of ventilation, then this method will give false readings, unless the other end of the U-tube is connected to the space above the liquid, as shown in fig. 9.3(b). When this is done, liquid will always condense above the right-hand mercury column; hence, to avoid error due to the variable height of this liquid, it is maintained at the full height (h_1) by providing a condensing chamber (C) at the top. Care must be taken to ensure that the column is full, or large error may occur.

For this system, equating pressures at ZZ,

$$\rho g h + \rho g y_1 + \rho_m g h_m + P_i = \rho g h_1 + \rho g y_2 + P_i$$
$$\therefore \quad \rho h + \rho y_1 + \rho_m h_m = \rho h_1 + \rho y_2$$

Since the mercury levels are both y_0 below the tank bottom when $h = h_1$, then

$$y_1 = y_0 - h_m/2$$

and

$$y_2 = y_0 + h_m/2$$

Substituting these values in the above equation gives

$$\rho h + \rho y_0 - \rho h_m/2 + \rho_m h_m = \rho h_1 + \rho y_0 + \rho h_m/2$$

and

$$h = h_1 - (\rho_m/\rho)h_m + h_m$$

Hence

$$h = h_1 - h_m\{(\rho_m/\rho) - 1\} \qquad 9.3$$

and $\delta h = K\,\delta h_m$, where $K = -\{(\rho_m/\rho) - 1\}$. The mercury scale may be graduated in tank-level units using single- or double-leg readings, as in the case of the vented tank.

Remote reading may be obtained for these systems by using a secondary transducer to convert the changes of mercury level (δh_m) into an electrical signal. One method, using the mercury as a conductor, is shown in fig. 9.4. Here the rising level of the mercury above the datum, as h increases, shorts out successive resistors R and increases the current through the ammeter, which is scaled in h units. Other possible systems are photoelectric scanning of the mercury–liquid meniscus, an iron float on the mercury surface in a magnetic field, etc. Although simple in principle, the mercury manometer has its difficulties in practice. The glass-tube system has to be protected, and the pressure it will stand is limited. Also, the secondary transducer for remote signal reading is an added complication.

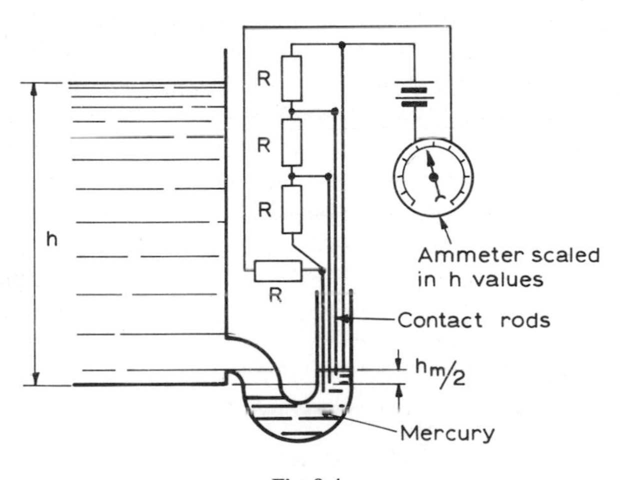

Fig. 9.4

For these reasons, a Bourdon-tube or bellows- or diaphragm-type pressure-gauge may replace the simple U-tube of fig. 9.3, and this is illustrated in fig. 2.5(b). For the closed system, a differential pressure-gauge may be connected as shown in fig. 9.5. For a vented tank, fig. 9.6 shows an air-trap pressure-gauge system at (a) and a diaphragm and pressure-gauge at (b), whilst the system at (c) is a diaphragm whose deflection due to the head (h) of fluid is sensed by an electrical or pneumatic etc. displacement transducer. The level value in the last case may be read remotely at a much greater distance than can be achieved with a capillary-tube system.

A further method of measuring the pressure at the bottom of the tank due to a liquid head is the 'gas-purge' system shown in fig. 9.7. In the case of the vented or open tank shown at (a), the pressure supply

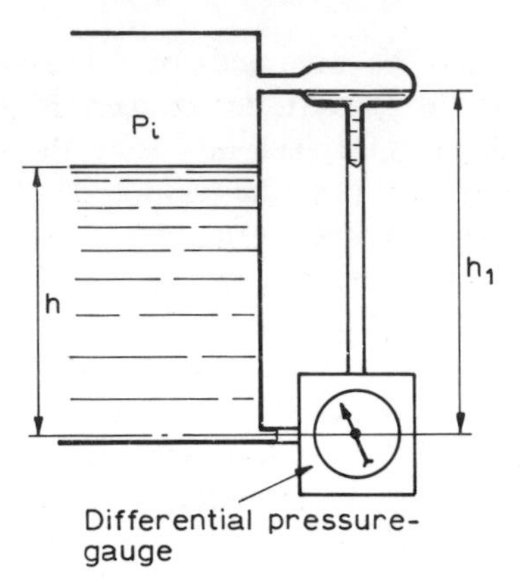

Fig. 9.5

is *just* higher than the pressure due to the maximum head, and a slow discharge of gas occurs as shown. To check that the gas flow is not excessive, a small flow-metering device is fitted. When the flow is small, the pressure at the gauge is virtually the same as the pressure due to the liquid head. If the tank is closed, then a differential pressure-gauge as shown at (b) is necessary. In case (a), $P = \rho gh + P_a$, and h is proportional to the gauge pressure $(P - P_a)$. In (b), $P = \rho gh + P_i$, and h is proportional to the differential pressure $(P - P_i)$. The gas used may be air in many cases, or nitrogen etc. for inflammable liquids. In some cases, liquid is used for the purging instead of gas, especially in the case of slurries or hot viscous liquids.

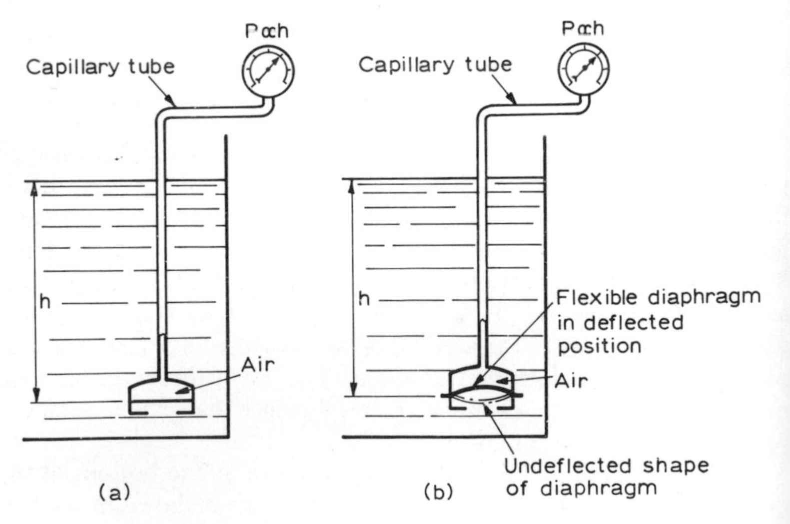

Fig. 9.6 (a) Air-trap pressure-gauge (b) Diaphragm and pressure-gauge

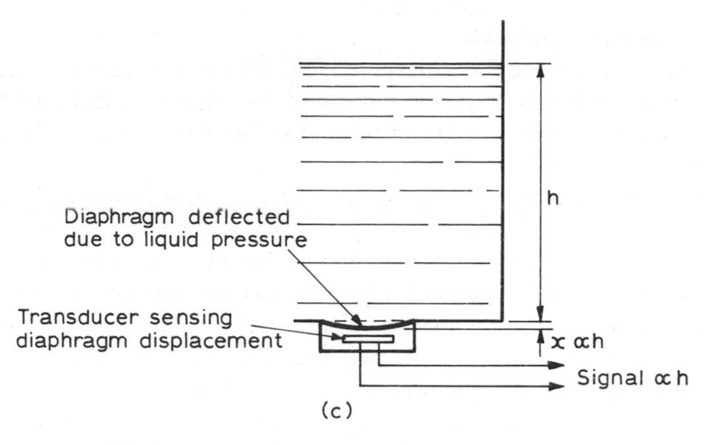

Fig. 9.6(c) Diaphragm and displacement transducer

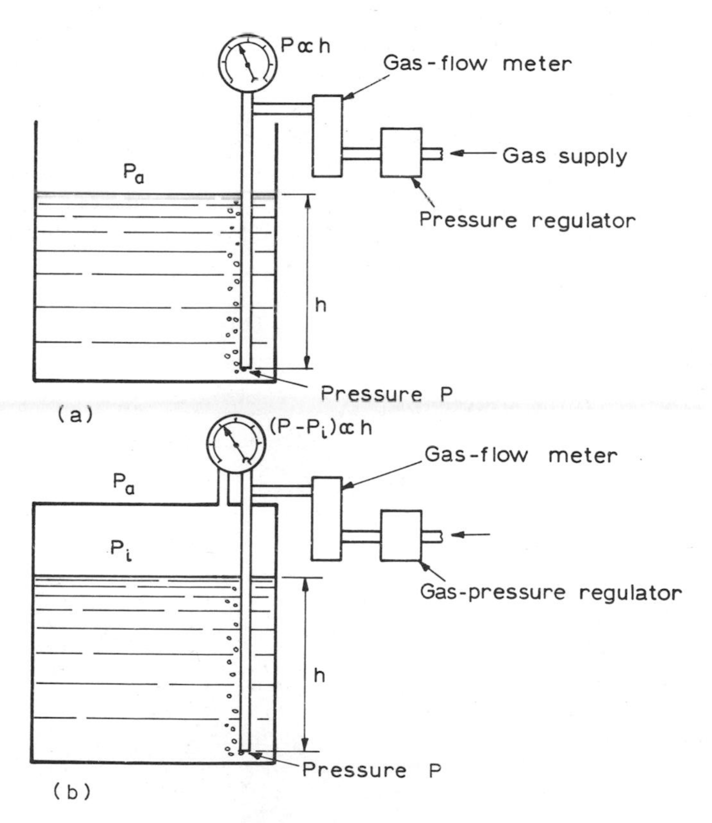

Fig. 9.7 (a) Open or vented tank (b) Unvented tank

9.3.3 Buoyancy methods

The float is a very common level-sensing device, and may be used to indicate or control directly, using a variety of connecting methods such as shown in fig. 9.8(a) and (b), or by means of a secondary transducer as shown in (c).

A 'float' denser than the liquid in which it is immersed may be used to measure level. It is attached rigidly to some point, at which it applies a force proportional to its depth of immersion. The method is illustrated in fig. 9.9. The force due to gravity is constant and equal to mg. An upthrust equal to ρghA exists due to the liquid pressure on the bottom of the float. Hence the net force on the float, and the force applied on the measuring device, is

$$F = mg - \rho ghA$$

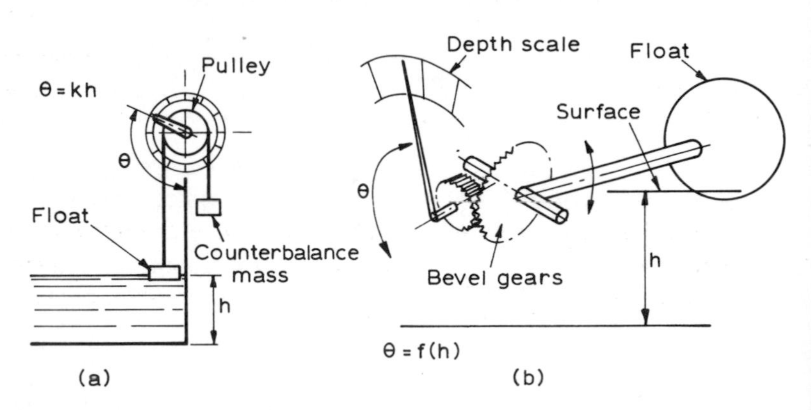

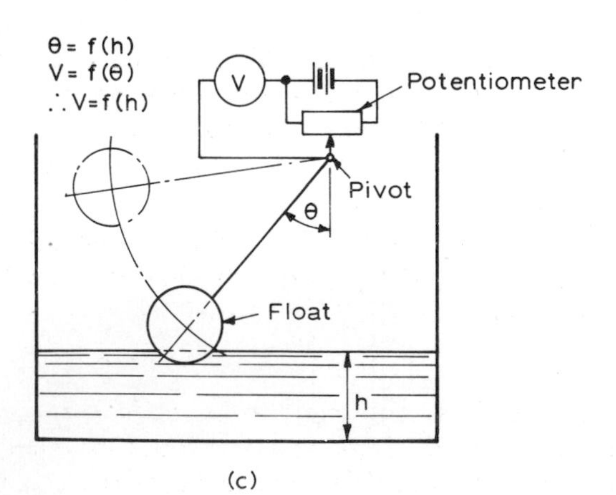

Fig. 9.8

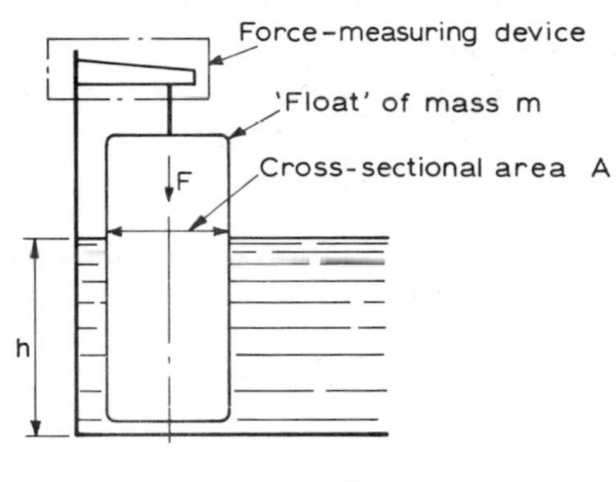

Fig. 9.9

$$\therefore \quad h = (-F + mg)/(\rho g A) \qquad 9.4$$

and $\delta h = K\,\delta F$, where $K = -1/(\rho g A)$. The force-measuring device may give out a mechanical, electrical, or pneumatic signal which is proportional to F, and hence to h.

9.4 Electrical methods

The signals from the sensing methods of sections 9.2 and 9.3 are initially mechanical ones, although secondary transducers may convert them to electrical ones. The following methods are essentially electrical.

9.4.1 Capacitance methods

The electrical capacitance between two electrodes may be varied by altering the distance between them, and also by altering the nature and extent of the material between; this is illustrated in figs 2.20(a) and (b). The system used in level measurement is a capacitance bridge such as that shown in fig. 9.10, where the capacitance C_4 is the value between the central electrode L_1 and the surrounding tubular electrode L_2 in the tank. This capacitance is varied by the height (h) of liquid filling the space between the electrodes. The bridge is balanced initially by varying capacitance C_3, at some datum value of h, when v_0 will be zero. The *amplitude* (V_0) of v_0 is very nearly proportional to h, thus

$$V_0 \approx kh \quad \text{(see section 4.4.2)}$$

Variations of the system exist. In fig. 9.10(b), a system suitable for measuring the level of granular solids and powders is shown. The tank is used as a second electrode and is earthed, as is point D in the bridge.

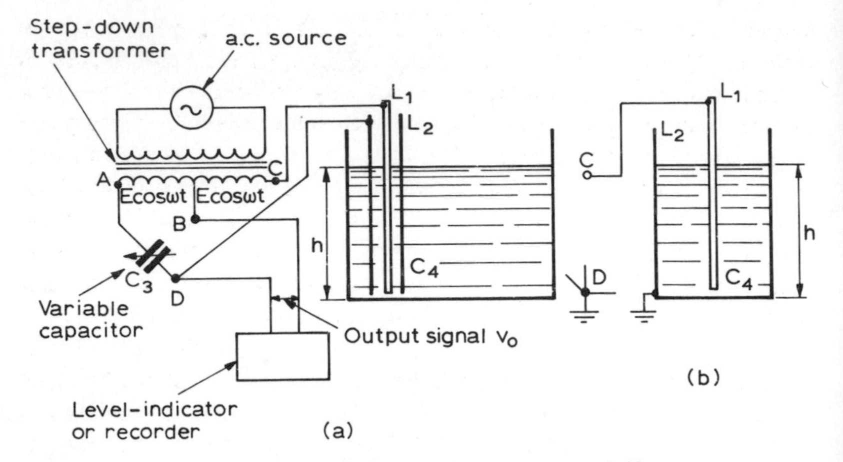

Fig. 9.10 Level measurement by capacitance bridge

The electrodes must be made of a suitable material to match the dielectric characteristics of the measured material, and the system must be calibrated to give reliable measurement.

9.4.2 Photoelectric methods

Photoelectric sensing devices are illustrated in figs 2.26 to 2.28, and are found in several height-measuring devices. The height (h) of the 'float' in a variable-area flow meter such as shown in fig. 7.17(a) is a function of the flow rate ($\dot{Q}$) of fluid through the tube. In one version of this instrument (the 'Flowscan', by G. A. Platon Ltd), the value of h is measured by arranging a photoelectric transmitter–receiver to be traversed by an electric motor over the length (AB) of the scale, as

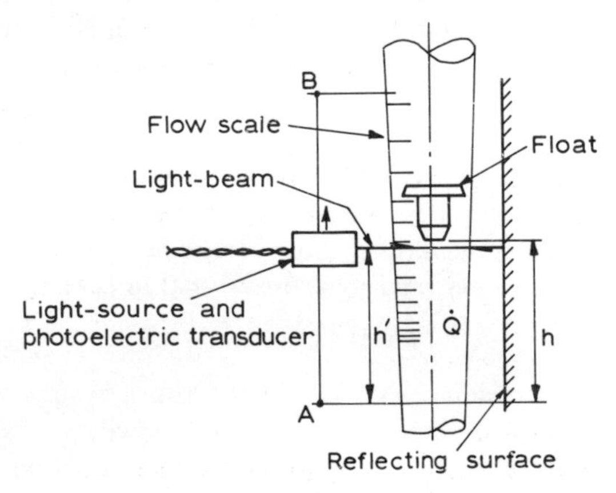

Fig. 9.11 The 'Flowscan' system

shown in fig. 9.11, the reflective surface returning the beam to the receiver. At the same time, the voltage from a variable transformer is varied so that at any moment it is directly proportional to the flow rate represented by a height (h') *on the flow scale*. When the rising beam is interrupted by the bottom of the float, then $h' = h$, and an output circuit is energised. The output-voltage value is a signal of the flow rate ($\dot{Q}$). The cycle time is 2·5 seconds, and hence a voltage pulse proportional to $\dot{Q}$ is produced every 2·5 seconds if the flow rate is constant. This voltage may be held by suitable circuitry until the next pulse, so that the read-out value is corrected about every 2·5 seconds.

Another example of the use of photoelectric pick-ups is shown in fig. 9.12. Light sources at A and C project light beams across the tank towards pick-ups at B and D. The three possible combinations of signal and no signal at B and D will give indications of tank empty, tank full, or an intermediate level. Intermediate beams can add further level signals. The system is suitable for granular solids as well as liquids.

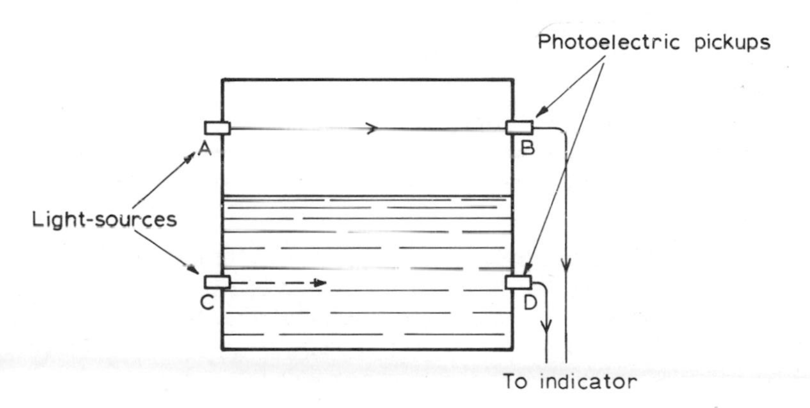

Fig. 9.12 Photoelectric level indication

9.4.3 Ultrasonic method

An ultrasonic transmitter–receiver may be mounted at the top of the tank, projecting a beam downward. This beam is reflected by the fluid surface, back to the receiver. The time between the instants of transmitting and receiving the reflected beam is a measure of the distance travelled by the beam, and hence of the height h. The system is illustrated in fig. 2.14(a).

9.5 Optical methods

A low-level indicator using a single light source is shown in fig. 9.13. The sensing device is a triangular glass prism whose surfaces have been

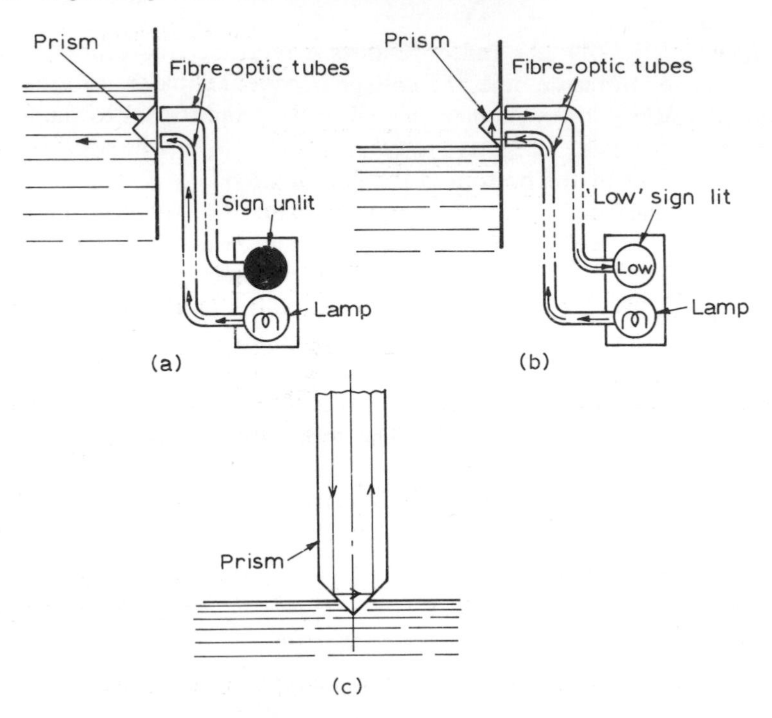

Fig. 9.13 Optical methods of level measurement

specially coated. Light from a source is passed through a fibre-optic tube and into the prism. If the level of liquid is above the prism, the light is not reflected, as shown in (a). When the liquid falls below the prism, the internal critical angle of the surfaces is altered, and the ray is reflected twice through 90°, and returns down a second fibre-optic tube to illuminate a sign or to activate a photoelectric cell. The same principle is used in an 'optical dip-stick' arrangement shown at (c). Here no light is reflected when the prism is immersed.

9.6 Worked examples

9.6.1 To prevent contamination of the mercury manometer for level measurement by the tank contents, sealing chambers and an intermediate sealing fluid are used in the arrangement shown in fig. 9.14. Derive an expression for the liquid head (h) in terms of the cross-sectional area (a) of the manometer tube, the cross-sectional area (A) of the sealing chambers, and the densities ρ, ρ_s, and ρ_m of the measured liquid, the sealing liquid, and the mercury respectively. Calculate the

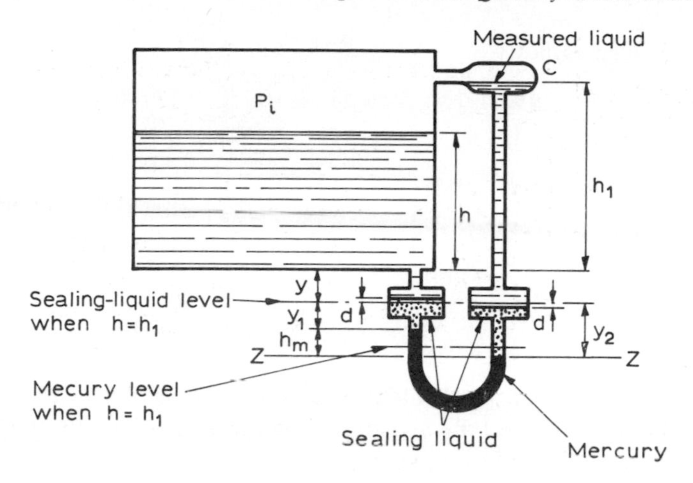

Fig. 9.14

range of movement of the mercury meniscus when h varies from 0 to 6 metres, if $\rho = 0{\cdot}8\ \mathrm{Mg/m^3}$, $\rho_s = 1{\cdot}2\ \mathrm{Mg/m^3}$, $\rho_m = 13{\cdot}6\ \mathrm{Mg/m^3}$, and $a/A = 10^{-2}$.

Equating volume changes in the U-tube and sealing chambers,

$$d \times A = h_m a/2 \qquad \therefore \quad d = \left(\frac{a}{2A}\right)h_m \tag{i}$$

Also,

$$y_2 - y_1 = h_m \tag{ii}$$

Equating pressures at level ZZ,

$$\rho g(h + y - d) + \rho_s g(y_1 + d) + \rho_m g h_m + P_i$$
$$= \rho g(h_1 + y + d) + \rho_s g(y_2 - d) + P_i$$

$$\therefore \quad h - d + (\rho_s/\rho)(y_1 + d) + (\rho_m/\rho)h_m = h_1 + d + (\rho_s/\rho)(y_2 - d)$$

and

$$h = h_1 + 2d + (\rho_s/\rho)(y_2 - y_1 - 2d) - (\rho_m/\rho)h_m$$

Substituting values from equations (i) and (ii),

$$h = h_1 + (a/A)h_m + (\rho_s/\rho)(h_m - (a/A)h_m) - (\rho_m/\rho)h_m$$
$$= h_1 + h_m\{(a/A) + (\rho_s/\rho)(1 - (a/A)) - (\rho_m/\rho)\}$$
$$= h_1 + Kh_m$$

$$\therefore \quad \delta h = K\,\delta h_m$$

where $K = [(a/A) + (\rho_s/\rho)\{1 - (a/A)\} - \rho_m/\rho]$.

For the given values,

$$K = 10^{-2} + (1{\cdot}2/0{\cdot}8)(1 - 10^{-2}) - 13{\cdot}6/0{\cdot}8$$
$$= -15{\cdot}5$$

hence
$$h = 6/(-15{\cdot}5)$$
$$= -0{\cdot}388 \text{ m}$$
$$= -388 \text{ mm}$$

and the movement of one leg of the mercury is 194 mm.

9.6.2 For a level-measuring system as shown in fig. 9.5, calculate the maximum pressure and the maximum differential pressure to which the pressure-gauge is subjected when P_i is 2·5 bar, h_1 is 8 metres, and the liquid density is 1·2 Mg/m^3.

$$\text{Maximum pressure} = P_i + \rho g h_1$$
$$= 2{\cdot}5 \times 10^5 + 1200 \times 9{\cdot}81 \times 8$$
$$= 344{\cdot}2 \text{ kN/m}^2$$

The maximum differential pressure occurs when $h = 0$, then

$$\text{differential pressure} = P_i + \rho g h_1 - P_i$$
$$= \rho g h_1$$
$$= 94{\cdot}2 \text{ kN/m}^2$$

9.6.3 For a liquid of density 820 kg/m^3, calculate the pressure (P) due to a head (h) of 4 metres (a) in an open tank, (b) in a closed tank, when the pressure above the surface is 320 kN/m^2. Hence state the readings on the pressure gauges in fig. 9.7(a) and (b) for these conditions.

a) Pressure at bottom of purge pipe $= \rho g h + P_a$
$$= 820 \times 9{\cdot}81 \times 4 + P_a$$
$$= 32{\cdot}18 \text{ kN/m}^2 + P_a$$
$$= 32{\cdot}18 \text{ kN/m}^2 \text{ gauge pressure}$$

b) Pressure at bottom of purge pipe $= \rho g h + P_i$
$$= 32{\cdot}18 + 320 \text{ kN/m}^2$$
$$= 352{\cdot}18 \text{ kN/m}^2$$

Hence the reading on the gauge in fig. 9.7(a) is 32·18 kN/m^2, and the reading on the gauge in fig. 9.7(b) is 32·18 kN/m^2 *differential* pressure.

9.6.4 The 'float' of a system similar to fig. 9.9 is to measure the level of fluids of a maximum density of 1200 kg/m^3. Its diameter is 0·3 m, and the maximum depth (h) of immersion is 1·5 m, and this will not cover the top of the 'float'. Calculate the minimum mass of the float if no reversal of the force signal is to take place. Calculate the range of the force exerted on the suspension point if the 'float' mass is 150 kg.

Let m be the mass of the float, then, for $F = 0$,

$$mg = \rho g h A$$

and

$$m = \rho h A$$
$$= 1200 \times 1{\cdot}5 \times (\pi/4) \times 0{\cdot}3^2$$
$$\therefore \quad m = 127{\cdot}3 \text{ kg is the minimum mass value}$$

If $m = 150$ kg,

$$\text{the maximum force is } F = 150 \times 9{\cdot}81$$
$$= 1{\cdot}47 \text{ kN, when } h = 0$$
$$\text{the minimum force is } F = 1{\cdot}47 \times 10^3 - \rho g h A$$
$$= 1{\cdot}47 \times 10^3 - 1200 \times 9{\cdot}81 \times 1{\cdot}5 \times (\pi/4) \times 0{\cdot}3^2$$
$$= 0{\cdot}22 \text{ kN, when } h = 1{\cdot}5 \text{ m}$$

Hence the range of force corresponding to a range of h from 0 to 1·5 m is from 1·47 kN down to 0·22 kN.

9.7 Tutorial and practical work

9.7.1 Discuss the use of a well-type manometer in place of the simple U-tube of fig. 9.3(a), stating advantages and disadvantages. Derive an equation for h in terms of h_m and the area ratio for the well and leg.

9.7.2 In the system of fig. 9.3(b), the liquid in the external tube may be at a considerably different temperature from that in the tank. Discuss the error that may arise due to this, and suggest a method of eliminating the effect.

9.7.3 Select suitable methods from those described in this chapter for measuring the level of dry sand in a hopper, giving a signal at a point 100 metres away.

9.7.4 Discuss the difficulties that might arise in the measurement of the level of slurries (i.e. liquids with suspended solids) with the systems described in this chapter.

9.7.5 Discuss errors that might arise due to the use of the level-sensing device of fig. 9.9 in situations (a) of high pressure, (b) of high temperature. Suggest how these might be minimised.

9.7.6 Could the system of fig. 9.12 be made 'fail-safe'; i.e., when the light source at C fails and the level falls below CD, can some indication be given?

9.7.7 Discuss the application of an automatic movement to the optical dip-stick of fig. 9.13(c) so that it always just touches the liquid surface. Sketch a possible system.

9.7.8 Sketch a system for the amplification and display of the very small movement of the level of liquid in the well of a well-type U-tube manometer.

9.7.9 A tank using a pressure-sensitive level-measuring system such as those shown in fig. 9.6 is inadequately vented. Sketch curves showing actual level and indicated level against a base of time when (a) the tank is being filled, (b) the tank is being emptied, in both cases at a constant rate (see Chapter 3).

9.7.10 For the level-measuring system of fig. 9.9, discuss the advantage of having a 'float' whose density is greater than that of water. Would this advantage apply in the system of example 9.8.5, illustrated in fig. 9.17?

9.7.11 Discuss the application of a photoelectric scanning system, such as shown in fig. 9.11, to the measurement of the height of mercury in a manometer and in a thermometer.

9.8 Exercises

9.8.1 Calculate the capacity (V_{max}), in m^3 and in litres, of the trough shown in fig. 9.15. Derive an expression for the height (h) on the internal scale for a given volume (V), and hence calculate the value of h when the trough contains $\frac{1}{2}$ and $\frac{1}{4}$ of its liquid capacity.

[348, 193 mm]

9.8.2 A tank having a mass of 4000 kg is to contain liquid to a total mass of 36 Mg. If it is supported on three interconnected hydraulic load cells, each 150 mm diameter, equally spaced about the centre O, as shown in fig. 9.16, determine the maximum pressure at the gauge, and the change of pressure between the full and empty conditions.

[7·4 MPa, 6·7 MPa]

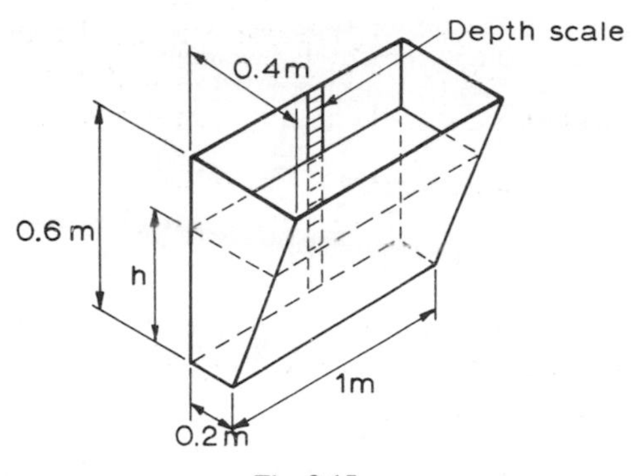

Fig. 9.15

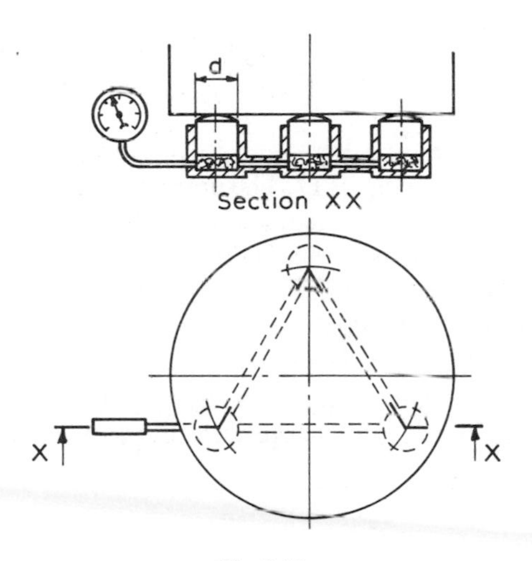

Fig. 9.16

9.8.3 Calculate the sensitivity of a level-measuring system as in fig. 9.3(a) when one leg of the mercury U-tube is calibrated, and the measured fluid has a density of 900 kg/m^3. The relative density of mercury may be taken as 13·6. [−0·0343]

9.8.4 Calculate the sensitivity of the system shown in fig. 9.3(b) if the liquids are water and mercury, and the right-hand leg of the U-tube is calibrated in level units.

If the pressure (P) above the liquid has a maximum value of 200 kN/m^2 absolute, and h_1 is 5 m, calculate the maximum bursting pressure in the U-tube if y_0 is 0·25 m and the bottom of the U-tube is 0·45 m below the tank bottom. [−0·0397, 177 kN/m^2]

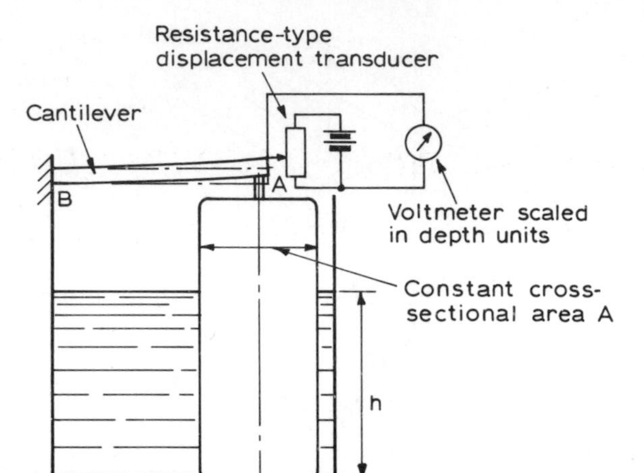

Fig. 9.17

9.8.5 The liquid level in a tank is to be measured using the upthrust from a 'float' as shown in fig. 9.17. The upthrust is to deflect a cantilever AB on which the float is mounted, and the deflection is to be measured by a resistance-type displacement transducer and voltmeter, the sensitivity being 10 volts per millimetre. If the maximum height (h) of the water in the tank is 2 metres, and the cross-sectional area (A) of the float is $0{\cdot}2\ \text{m}^2$, calculate the necessary stiffness of the cantilever in newtons per millimetre if the meter range is 50 V.

What is the maximum percentage error in the measured level due to the float movement? [785 N/mm, $-0{\cdot}25\,\%$]

10

Measurement of Length, Linear Displacement, and Surface Finish

10.1 Length standards

The early standards of length were crude, being based on such variable quantities as the length of the human foot or arm. However, the predecessor of the yard, the 'iron ulna' of Edward I, was the length over the ends of an iron bar, i.e. an *end standard*. It is thought to have differed from the later yard standards by not more than 0·04 inch. Subsequent yard standards were the lengths between transverse lines or dots marked on brass or bronze bars, i.e. *line standards*. The Imperial Standard Yard of 1855 was the length between transverse lines engraved on gold plugs inserted into a 1 inch square bar of metal alloy, the faces of the plugs being on the neutral surface of the bar, as shown in fig. 10.1.

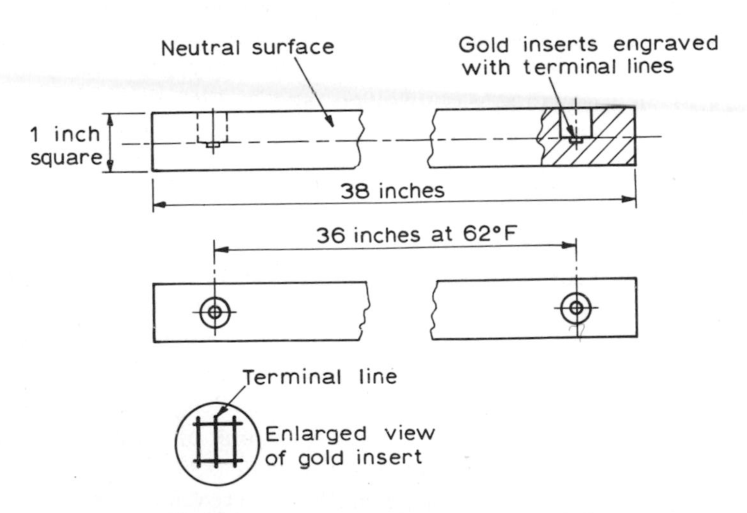

Fig. 10.1 The Imperial Standard Yard of 1855

In 1700 an attempt was made to define a new length unit, the metre, as one ten-millionth part of the earth's meridian passing through the Observatoire in Paris. The surveying work on which this was based was found subsequently to be in error, and the metre was then redefined as the length between the ends of a particular platinum strip at 0 °C. This was replaced in 1889 by the International Prototype Metre, the length between transverse lines marked on the neutral surface of a platinum–iridium bar of cross-section as shown in fig. 10.2.

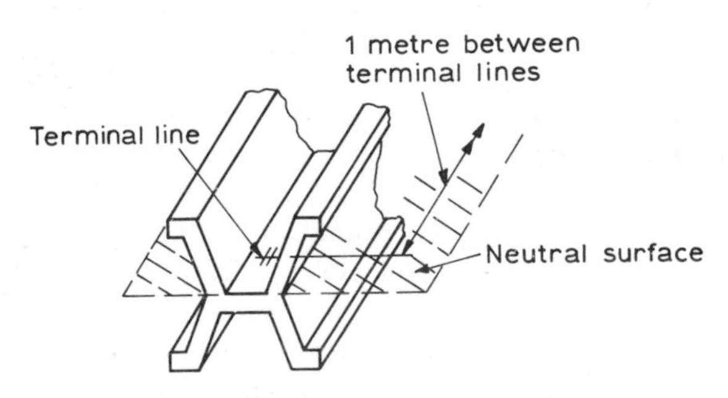

Fig. 10.2 The International Prototype Metre of 1889

The United States Yard was *defined* in 1866 as 3600/3937 of the metre. The International Prototype Metre, as far as is known, has remained a constant length, and since the Imperial Standard Yard has reduced in length continuously, different yard standards have applied in the UK and the USA, with in Canada a different standard again. Whilst these differences have been slight, of the order of five parts in a million, the situation was obviously unsatisfactory, particularly where work of high accuracy was involved, and in 1960 the International Yard, based on the metre, was agreed by several countries, thus:

$$1 \text{ International Yard} = 0{\cdot}9144 \text{ metre.}$$

10.1.1 Light standards of length

It has always been desirable to define length in terms of a standard which is not only constant, but also reproducible anywhere in the world, rather than by reference to some particular and possibly vulnerable piece of metal. Each pure colour of light from a vapourising element has a particular and constant wavelength, at a constant temperature and pressure, and the use of one of these as a length standard was first suggested by J. Babinet in 1829. Starting in 1892, many determinations of the wavelength of the red line of the cadmium spectrum were made, with slightly differing results. With the advent of nuclear physics, pure isotopes of various elements became available, giving very pure, i.e. *monochromatic*, light sources. This has enabled the metre to be now defined in terms of a number of wavelengths of the isotope krypton-86 under carefully controlled conditions (See appendix A), and this provides a standard which is readily reproducible by metrological laboratories anywhere in the world.

10.1.2 Secondary standards of length

In modern technology, end standards are very convenient to manufacture, to calibrate, and to use, and they tend to be used more in precision engineering work than do line standards. The latter, however, have many applications, ranging from the simple rule to the scales used in the beds of some precision machine tools, in conjunction with optical equipment.

From the primary standard, a hierarchy of secondary standards exists. A manufacturer may have slip gauges and end gauges which have been calibrated with reference to the primary standard by the National Physical Laboratory (NPL) or by laboratories approved by the British Calibration Service (BCS) in this country (see section 1.4), or by similar authorities overseas. These master gauges are then used only for comparison with reference gauges which are distributed around the factory. Working gauges of various kinds may be compared with the reference gauges, and hence a direct line exists from the primary standard to the component.

A similar hierarchy of line standards can exist in the form of ruled scales, enabling manufactured scales to be compared with master or reference scales, which in turn are compared with the end standards measured with reference to the light standard.

10.2 Light methods of length measurement

Light is only one of many types of radiated energy: the one that is detected by the eyes. In common with many other radiations, it travels, in a vacuum, at a constant velocity of 299 792·5 km/s. This value is believed to be accurate to within ±0·1 km/s. A common characteristic of these radiations is that they exhibit the properties of a sinusoidal wave, as shown in fig. 10.3. The intensity of the radiation is proportional to the square of the amplitude (a), and the frequency (f) is given by $f = V/\lambda$. The wavelength (λ) determines the characteristics of the radiation, and the position of the visible spectrum in the range is indicated in fig. 10.4, and is seen to extend from about 0·4 μm to 1 μm.

Lamps may be specially constructed to give light of a very narrow spread of wavelengths, and these are referred to as 'monochromatic'

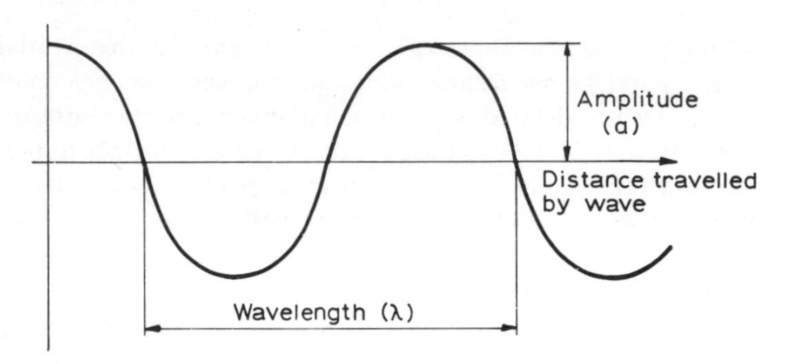

Fig. 10.3

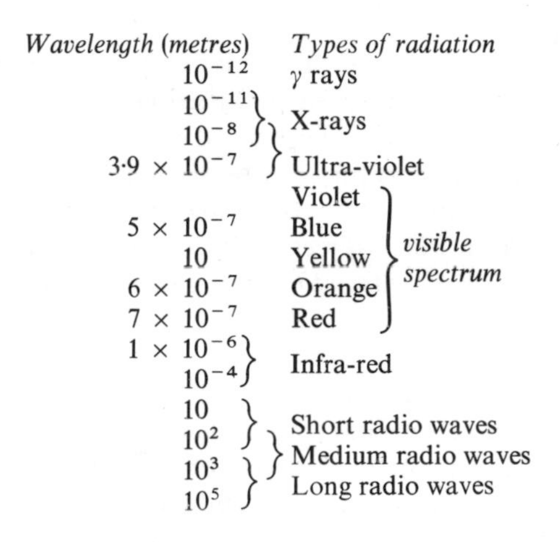

Wavelength (metres)	*Types of radiation*
10^{-12}	γ rays
10^{-11} – 10^{-8}	X-rays
10^{-8} – $3{\cdot}9 \times 10^{-7}$	Ultra-violet
	Violet (*visible spectrum*)
5×10^{-7}	Blue (*visible spectrum*)
10	Yellow (*visible spectrum*)
6×10^{-7}	Orange (*visible spectrum*)
7×10^{-7}	Red (*visible spectrum*)
1×10^{-6} – 10^{-4}	Infra-red
10 – 10^{2}	Short radio waves
10^{2} – 10^{3}	Medium radio waves
10^{3} – 10^{5}	Long radio waves

Fig. 10.4

light sources. For example, the cadmium red line has a wavelength of 0·643 846 96 μm.

10.2.1 Interference of light waves

Light rays from a single monochromatic source are initially all in phase. If rays having travelled different path lengths then recombine, as indicated in fig. 10.5(a), in general their phases will be different, and may give increased or reduced intensity as indicated in fig. 10.5(b). In (i) the rays are seen to recombine in phase, in (ii) antiphased, and in (iii) an intermediate case is shown. Case (i) gives the maximum intensity and (ii) the minimum, i.e. no light. Case (iii) gives intermediate intensity.

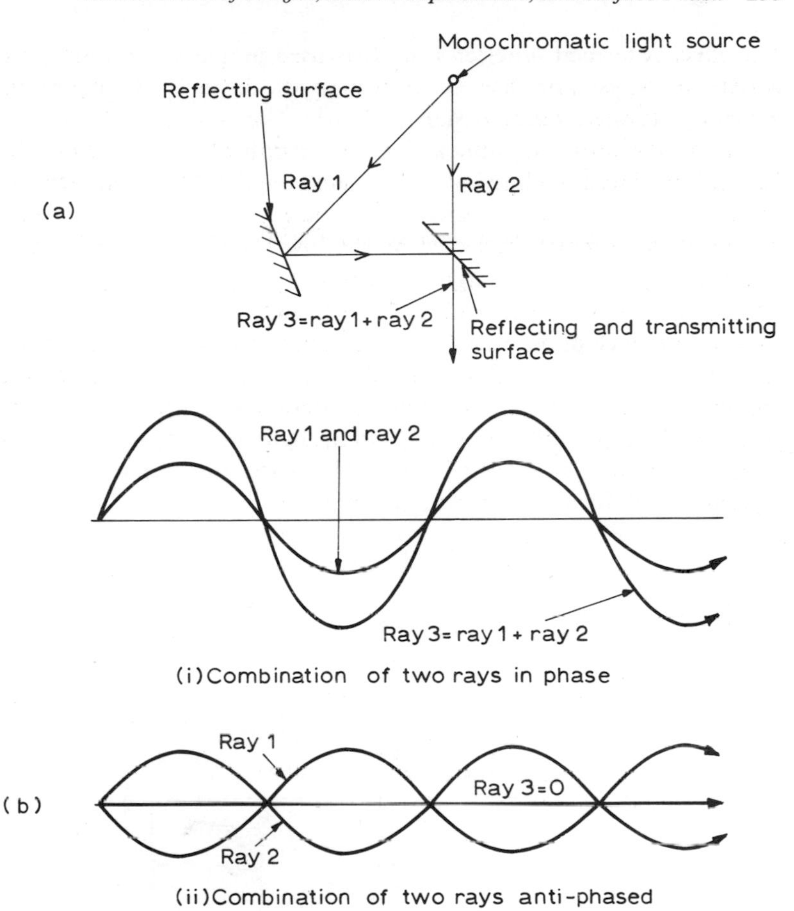

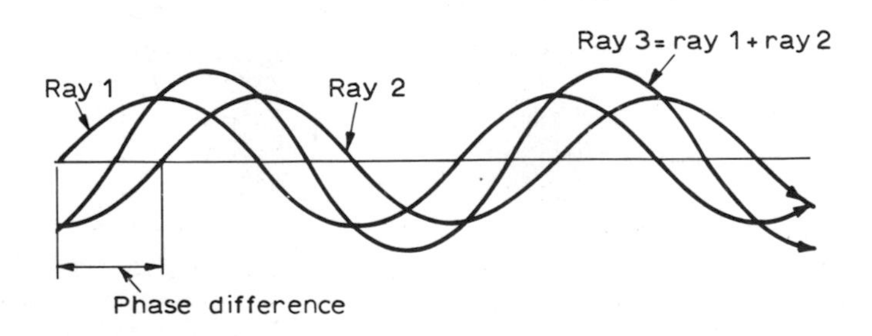

Fig. 10.5 Combination of light rays from a single source

The effect is termed 'interference'. It is used in a large and increasing variety of ways, and has given rise to the branch of dimensional metrology termed '*interferometry*'.

Figure 10.6 shows an 'optical flat', i.e. an optical-glass plate polished flat on both faces, inclined at a *very small* angle (θ) to a flat reflecting surface. Light from a monochromatic source is directed onto the flat. A typical ray is partly reflected by the lower surface of the flat at A, and part of it passes through to be reflected by the workpiece at B. The light rays S–A–eye and S–A–B–C–eye, combining at the eye, have a difference of path length of $\lambda/2$ for the position 1, and similarly at 3 the difference is $3\lambda/2$. In both cases the combining rays are anti-phased, and a dark area is observed. Between these, the position 2 gives a path-length difference of λ, and the light intensity is a maximum, since the combining rays are in phase. Hence the observer sees alternate light and dark bands as shown, i.e. interference *fringes*.

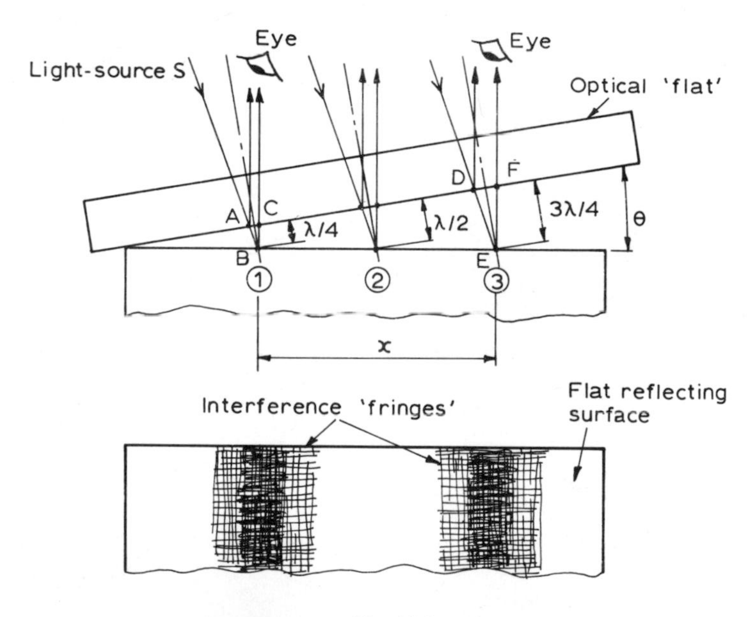

Fig. 10.6

The rate of separation of the surfaces may be calculated if the wavelength (λ) of the light is known, since the separation of the two surfaces increases by $\lambda/2$ in the distance (x) between subsequent fringes.

10.2.2 Length measurement by interference of light

The NPL interferometer, the principle of which is shown in fig. 10.7, will measure the thickness of block gauges up to 100 mm to an accuracy of $\pm 0{\cdot}025\ \mu m$. Two light sources are provided: a mercury-198 isotope lamp and a cadmium hot-cathode discharge lamp. From the former, a violet, a green, and two yellow wavelengths are used, and from the latter the red wavelength, giving light of five accurately known wavelengths. These are separated from the spectrum of the lamp in operation by rotating the constant-deviation prism to a suitable position.

Using each wavelength of light in turn, the fringes due to the baseplate are compared with those from the top of the gauge, and the fraction $f = a/b$ of misalignment of the fringes is measured as shown in fig. 10.7. From fig. 10.6 it can be seen that if the height (h) of the gauge is an exact multiple (N) of $\lambda/2$, then the fringes would coincide exactly. Hence, from the readings the following equations may be written:

$$\begin{aligned} h &= \tfrac{1}{2}(N_1 + f_1)\lambda_1 \\ &= \tfrac{1}{2}(N_2 + f_2)\lambda_2 \\ &= \tfrac{1}{2}(N_3 + f_3)\lambda_3 \\ &= \tfrac{1}{2}(N_4 + f_4)\lambda_4 \\ &= \tfrac{1}{2}(N_5 + f_5)\lambda_5 \end{aligned}$$

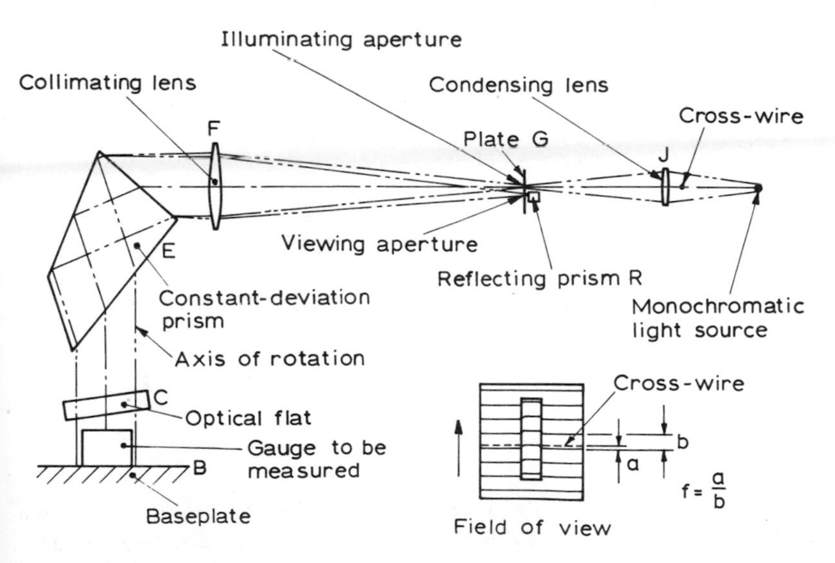

Fig. 10.7 The optical system for the NPL interferometer (0–100 mm)

From the nominal value of h, the N values are estimated and rounded off to integer values. Using these values, h should be the same from each equation; if they are not, then the N values must be increased or decreased until they do agree. To facilitate the work of solution, a special slide-rule is used to give the gauge error directly from the f values.

Other instruments are available using different light sources and wavelengths, but they use the same principle as above. For example, the German Kosters interferometer uses the yellow, green, and red colours of the krypton discharge lamp.

The *laser* (light amplification by stimulated emission of radiation) is becoming increasingly used as a source of coherent and truly monochromatic light. Non-laser light is incoherent in that it is not exactly represented by the waveform of fig. 10.3, but is subject to small random variations. The coherence of the laser beam enables it to be projected in a narrow pencil beam which does not scatter, and it is finding many applications in measurement, including interferometry.

10.3 Length comparison

Many of the measurements made in engineering production are ones of comparison, e.g. between a component and a gauge, or between a component or gauge and a pile of block or slip gauges. Instruments for measuring the difference between one linear dimension and another are called *comparators*. They may operate on fluid-, mechanical-, electrical-, or optical-amplification systems, but in all cases they

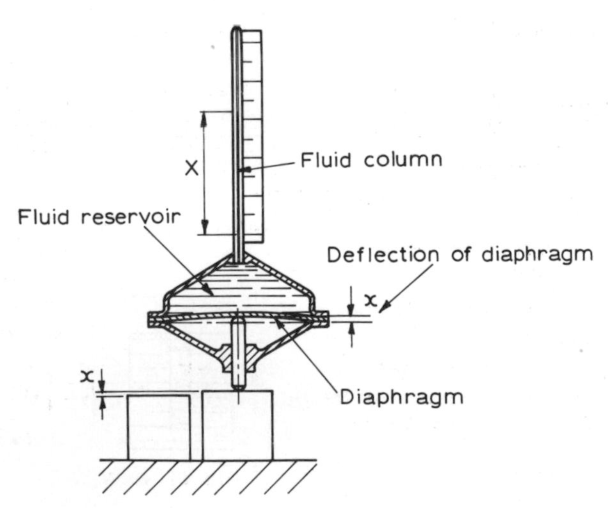

Fig. 10.8 Fluid comparator

provide a high magnification (i.e. high sensitivity). The range of measurement, i.e. difference in dimensions, is small.

A simple fluid comparator is shown in fig. 10.8, where the deflection of a diaphragm displaces liquid from a container into a tube, giving a magnification of X/x. The use of this type is very limited.

10.3.1 Mechanical comparators

Two simple mechanical magnifying elements, the lever and the gear-train, are illustrated in fig. 2.29. They suffer from friction and backlash, and the effects of these are discussed in section 1.3.2. Comparators having very large magnification, which may be 10000 or more, are designed to eliminate friction by using elastic members such as in fig. 1.7(a), (c), and (d) instead of pivots.

An example of this type is the 'reed' comparator shown in fig. 10.9, where a small movement (x) of member A relative to member B produces a large movement (X) of the end of a pointer. There is no friction, and hysteresis is minimised by using suitable steel for the 'reeds'.

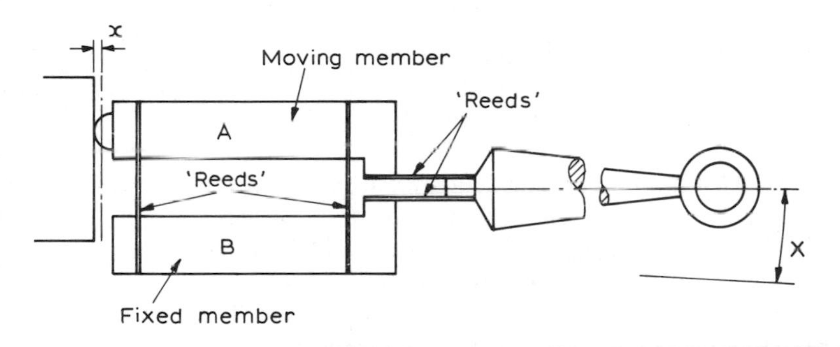

Fig. 10.9 Mechanical system of the Eden-Rolt 'Millionth' comparator

In fig. 10.10 is shown the magnification system of the 'Sigma' mechanical comparator. In this, the small movement (x) of the plunger (P) is allowed by elastic deflection of the two slit-diaphragm supports. The arm AB is attached to the moving side of a cross-strip hinge and magnifies the movement of the plunger at its end B, where a strip-and-drum arrangement causes rotation of the pointer.

A large variety of mechanical comparators have been designed and used, but in recent years they have to some extent been replaced by pneumatic or electric systems.

10.3.2 Pneumatic comparators

Two types of pneumatic displacement transducers are used. Both of

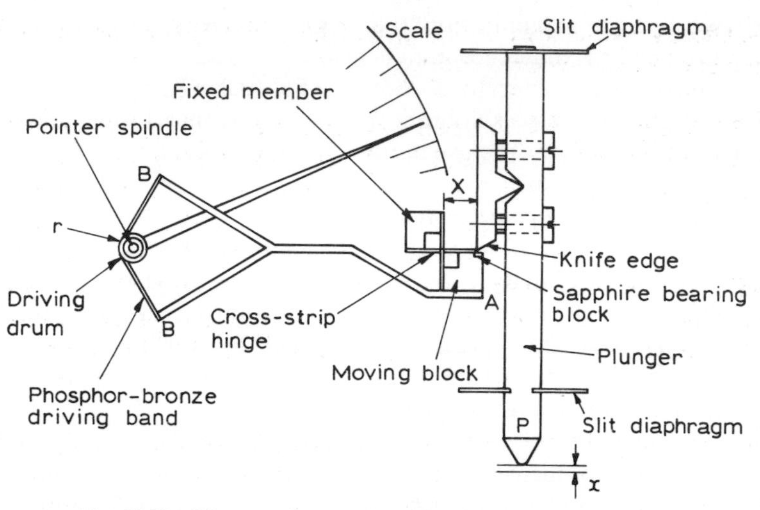

Fig. 10.10 Diagram of movement of Sigma mechanical comparator

these depend on the variation of the flow rate of air through an outlet orifice, the *effective* area of which is varied by the movement of an adjacent wall, as illustrated in fig. 10.11(a).

10.3.2.1 Flow-type pneumatic comparator. It may be shown that, over a limited range, the rate of flow of air ($\dot{Q}$) through the outlet is proportional to the escape area (A_m), which in turn is proportional to the distance (x) of an adjacent wall. Hencc, if a flow meter is used, such as the variable-area type shown in fig. 10.11(b), then $\delta x = K\,\delta\dot{Q}$, and the system forms a displacement–flow transducer. It should be noted that

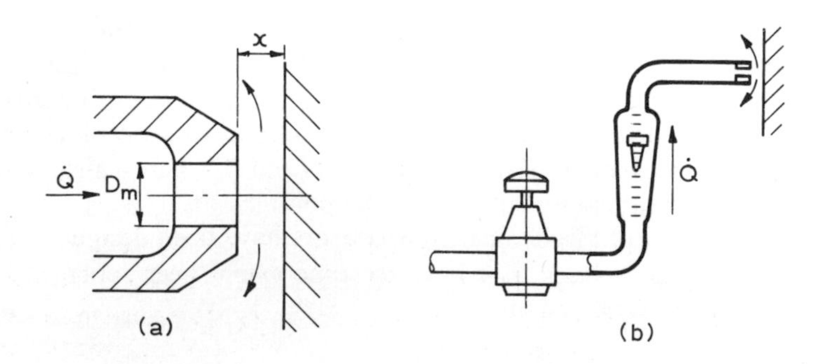

Fig. 10.11

the effective outlet area (A_m) is given by $A_m = \pi D_m x$, where D_m is the diameter of the measuring orifice.

10.3.2.2 Pressure-type pneumatic comparator. This type, illustrated in fig. 10.12 [and also in fig. 2.6(a)] has a fixed orifice (O) of area A_c, which is supplied with air at a constant pressure (P_s). The back-pressure (P_b) between the fixed orifice and the outlet is a function of the effective outlet area (A_m), which again is a function of the displacement (x) of the wall. The arrangement constitutes a displacement–pressure transducer, and is widely used in systems for continuous measurement and/or control, as well as in comparators for dimensional measurement.

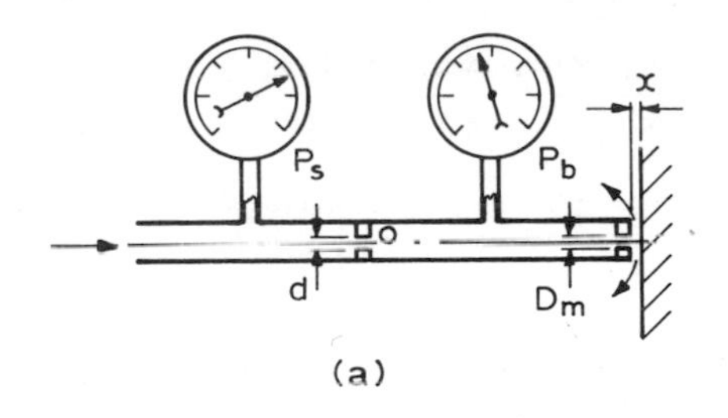

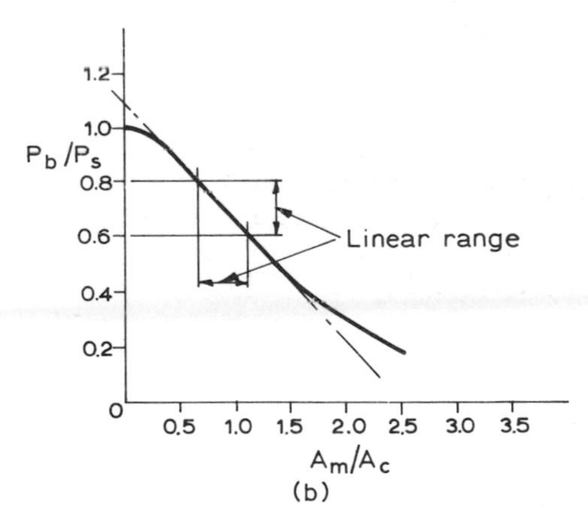

Fig. 10.12

Experiment shows a linear relationship between P_b and x over a limited range of x. Since the supply pressure and the two orifice sizes are possible variables, experimental results may be expressed in the form of dimensionless parameters, i.e. the pressure ratio P_b/P_s and the area ratio A_m/A_c. Practical investigation has shown that, for supply pressures between 15 kPa and 500 kPa, a curve as shown in fig. 10.12(b) is obtained, giving a linear range between values of P_b/P_s extending from 0·6 to 0·8. The tangent to the linear part cuts the P_b/P_s axis at 1·1, and the slope varies slightly, reducing with increasing supply pressure. For the linear range, the relationship may be expressed as

$$(P_b/P_s) = K(A_m/A_c) + b \qquad 10.1$$

where b (= 1·1) and K are constants as discussed above.

The back-pressure P_b may be measured by a Bourdon tube or other elastic-element-type gauge, or by a manometer. If the scale movement of the instrument is δR, then the overall magnification or sensitivity of the system is given by

$$\frac{\text{scale movement }(\delta R)}{\text{change of }x} = \frac{\text{change of }A_m}{\text{change of }x} \times \frac{\text{change of }P_b}{\text{change of }A_m} \times \frac{\text{scale movement }(\delta R)}{\text{change of }P_b}$$

or

$$\frac{\delta R}{\delta x} = \frac{\delta A_m}{\delta x} \times \frac{\delta P_b}{\delta A_m} \times \frac{\delta R}{\delta P_b}$$

or, in the limit,

$$\frac{dR}{dx} = \frac{dA_m}{dx} \times \frac{dP_b}{dA_m} \times \frac{dR}{dP_b} \qquad 10.2$$

where dA_m/dx is the measuring-head sensitivity, dP_b/dA_m is the pneumatic sensitivity, and dR/dP_b is the indicator sensitivity.

Since

$$A_m = \pi D_m x$$

then

$$\frac{dA_m}{dx} = \pi D_m \qquad 10.3$$

Hence the measuring-head sensitivity increases with increasing measuring-orifice size.

Differentiating eqn 10.1 with respect to A_m,

$$\frac{dP_b}{dA_m} = K\frac{P_s}{A_c} \qquad 10.4$$

Hence the pneumatic sensitivity may be raised by raising the supply pressure (P_s) or by decreasing the control-orifice size (A_c). However, A_c depends on A_m, since for linearity, from fig. 10.12, $0{\cdot}6 < (P_b/P_s) < 0{\cdot}8$. Using the mean value of the pressure ratio, $P_b/P_s = 0{\cdot}7$, and substituting in eqn 10.1,

$$0{\cdot}7 = K\frac{A_m(\text{mean})}{A_c} + 1{\cdot}1$$

$$\therefore \quad A_c = -KA_m(\text{mean})/0{\cdot}4$$

Substituting this in eqn 10.4 gives the pneumatic sensitivity:

$$\frac{dP_b}{dA_m} = -\frac{0{\cdot}4P_s}{A_m(\text{mean})} \qquad 10.5$$

Hence, for high pneumatic sensitivity, A_m(mean) must be small.

Examination of eqns 10.1 to 10.5 indicates the manner in which the variables of a given system may be chosen to give either maximum sensitivity, maximum range, or a suitable compromise for a particular measuring system. The linear range of movement (Δx) is small in all cases; hence the system is used to compare the size of one object, usually a manufactured component, with another, either a master component or a slip- or block-gauge pile of known height.

An example of a commercially available instrument is illustrated in diagrammatic form in fig. 10.13. The 'Solex' column gauge incorporates its own pressure regulator which maintains a constant pressure (P_o) before an orifice (G). The back-pressure (P_b), which is varied by the measuring head (S), is measured against P_o by the height (h) of the liquid column, the system forming a well-type manometer.

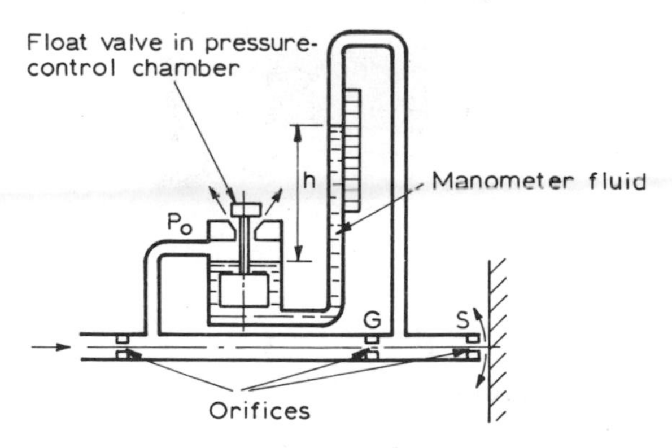

Fig. 10.13 System of Solex column gauge

Advantages are gained by using a pneumatic-bridge circuit analogous to the electrical Wheatstone bridge. The system is shown schematically in fig. 10.14. The advantages are

a) small variations in the supply pressure P_s are compensated for by the differential-pressure measurement;

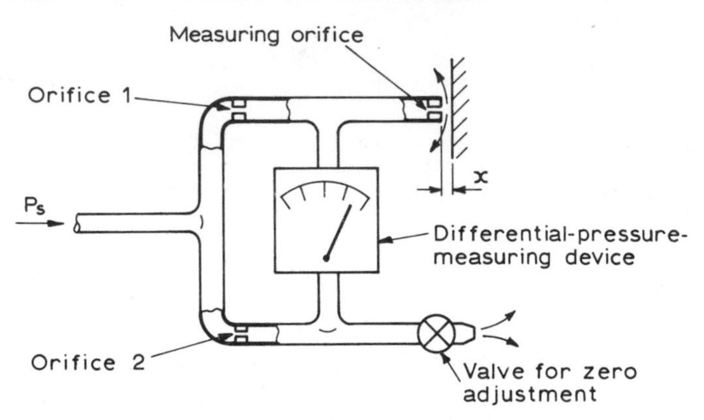

Fig. 10.14

b) the differential pressure may be zeroed, using the adjusting valve, at a chosen x value;
c) the full range of scale may be used on the differential-pressure-measuring device.

A modern instrument using this system is the 'Sigma Dialair' air-gauge unit shown in fig. 10.15; fig. 10.16 shows the system details. The

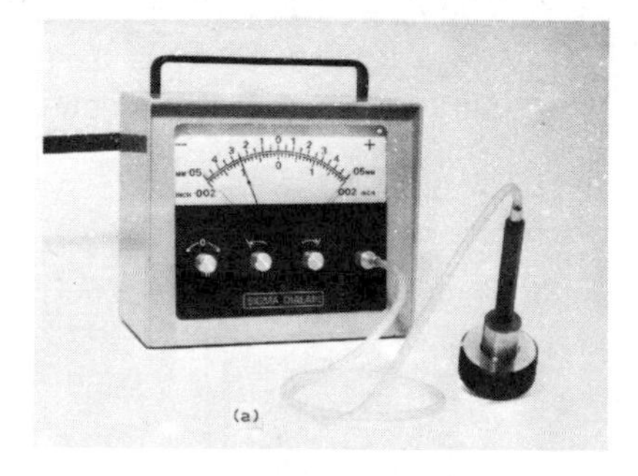

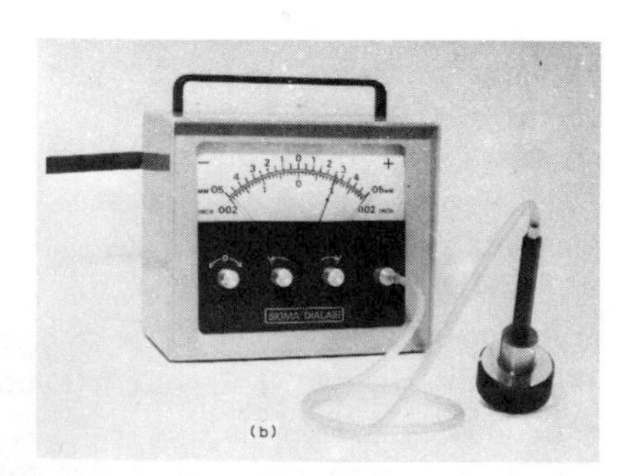

Fig. 10.15 The Sigma Dialair (a) Lower-limit setting (b) Higher-limit setting

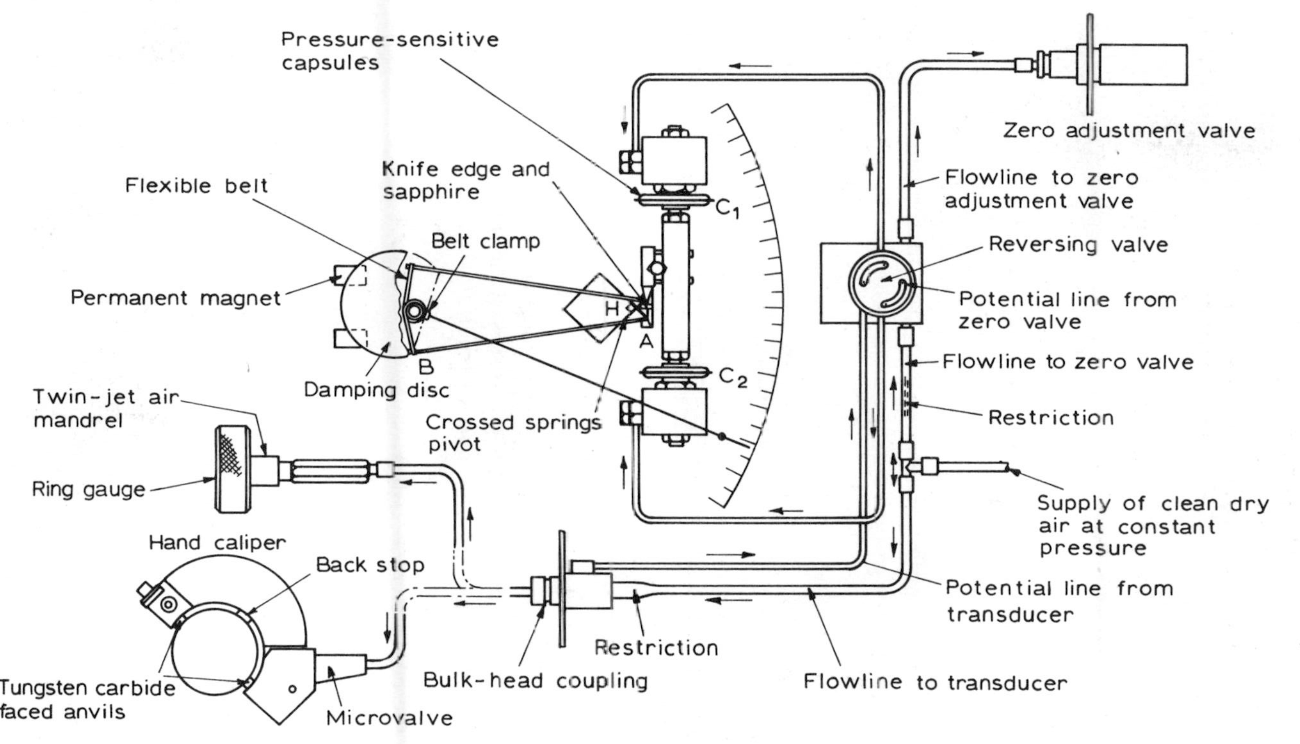

Fig. 10.16 Sigma Dialair indicator circuit

pressure difference across the bridge is sensed by a pair of capsules C_1 and C_2 operating through the knife-edge and sapphire, and causes rotation of the arm AB which is supported on the cross-strip hinge H. The flexible belt rotates the spindle and pointer, the motion of the latter being damped by an eddy-current damper.

The linear range Δx is altered in basic instruments by varying the pressure and area ratios, but is always small. It may be extended, however, by the arrangement shown in fig. 10.17. By means of the parabolic form on the plunger, equal changes of x give equal changes of the effective measuring area, which is the area of the annulus between the orifice and the plunger, and the range (Δx) of the measurement is increased. A similar arrangement is used in the Sigma microvalve units, in the bridge-type circuit shown in fig. 10.16.

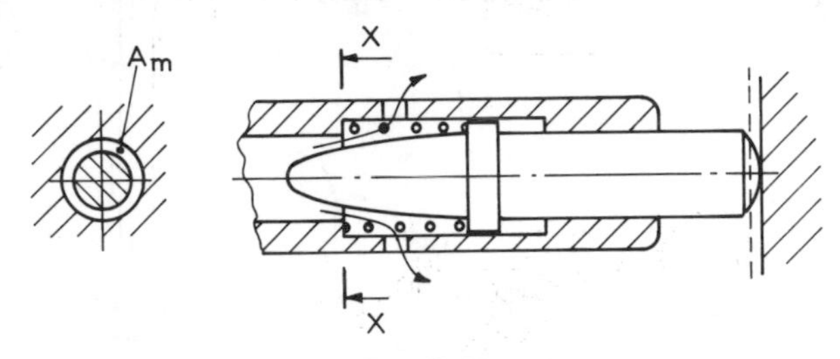

Fig. 10.17

10.3.2.3 Applications of pneumatic gauging. Both the displacement–flow and the displacement–pressure systems lend themselves to the determination of internal and external *diameters*, using air-plug and air-ring gauges as illustrated in fig. 10.18(a) and (b). In these, the diameter of the component is compared with that of an accurate setting ring or plug. Sometimes a pair of rings or plugs is used to set the range of the gauge, as shown in fig. 10.15. In addition, *ovality* and *lobing* may be determined by air-jet gauge arrangements similar to those shown in fig. 10.18(c). Many other geometrical faults may be detected by ingenious arrangements of internal or external measuring jets.

For mass-produced components, several internal or external dimensions may be gauged at the same time. Figure 10.19(a) shows a schematic arrangement of a pressure-sensitive instrument to gauge simultaneously the height of three steps, whilst at (b) is shown a flow-sensitive instrument for the simultaneous gauging of two external diameters. The limits of the tolerance on each dimension can conveniently be indicated by a pair of lines for each dimension measured, as shown. Much more complicated systems are used for the complete gauging of components.

Gauging head	**Application to workpiece**	**Function**
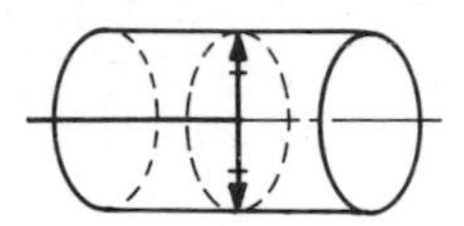 a) Air-plug gauge 2 measuring orifices 1 measuring circuit	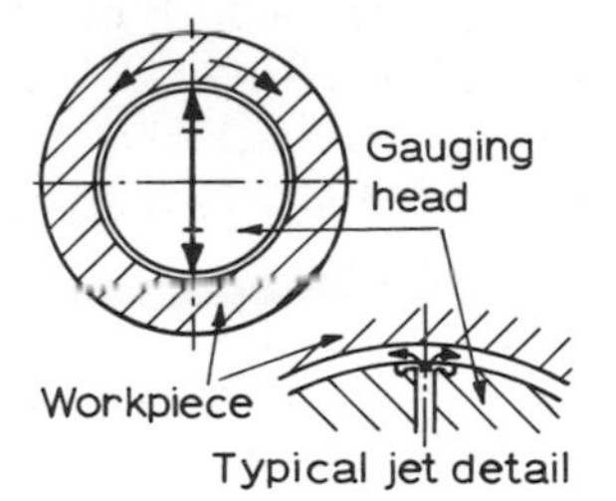	i) Compares internal diameters. ii) Detects ovality by rotation.
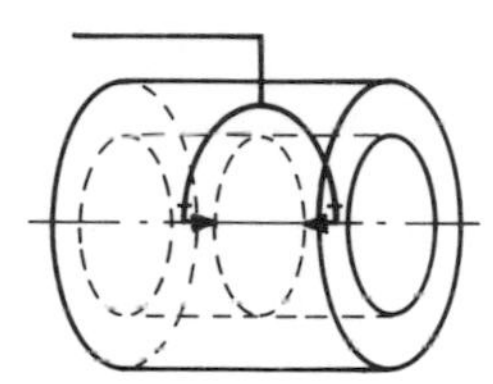 b) Air-ring gauge 2 measuring orifices 1 measuring circuit	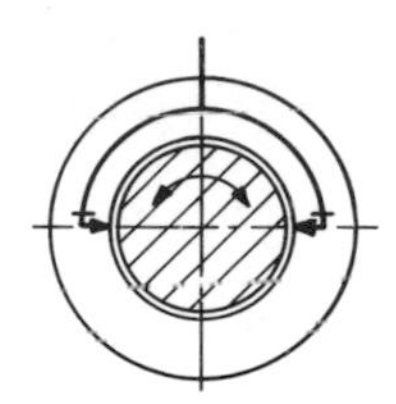	i) Compares external diameters. ii) Detects ovality by rotation.
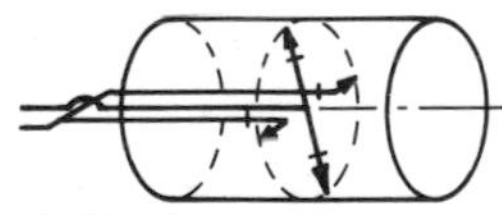 c) Air-plug gauge 4 measuring orifices 2 measuring circuits.	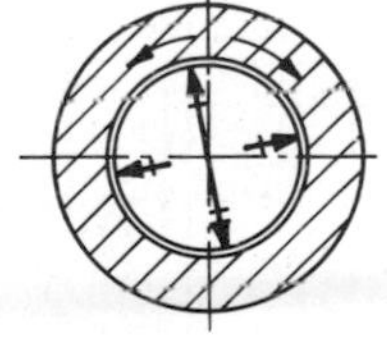	i) Compares external diameters. ii) Detects ovality by different indications on two circuits. iii) Detects other irregularities by rotation.
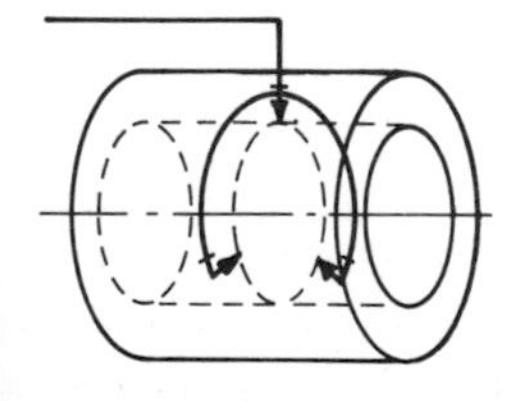 d) Air-ring gauge 3 measuring orifices 1 measuring circuit.	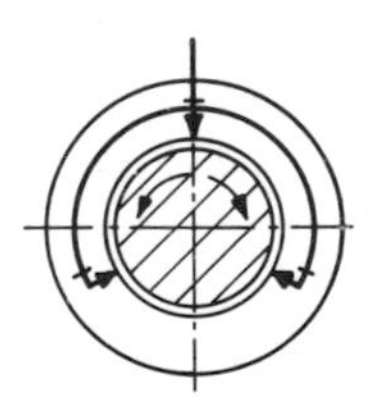	i) Compares external diameters. ii) Detects 3-point lobing by rotation. 

Fig. 10.18 Typical air-gauging-head arrangements for cylindrical components

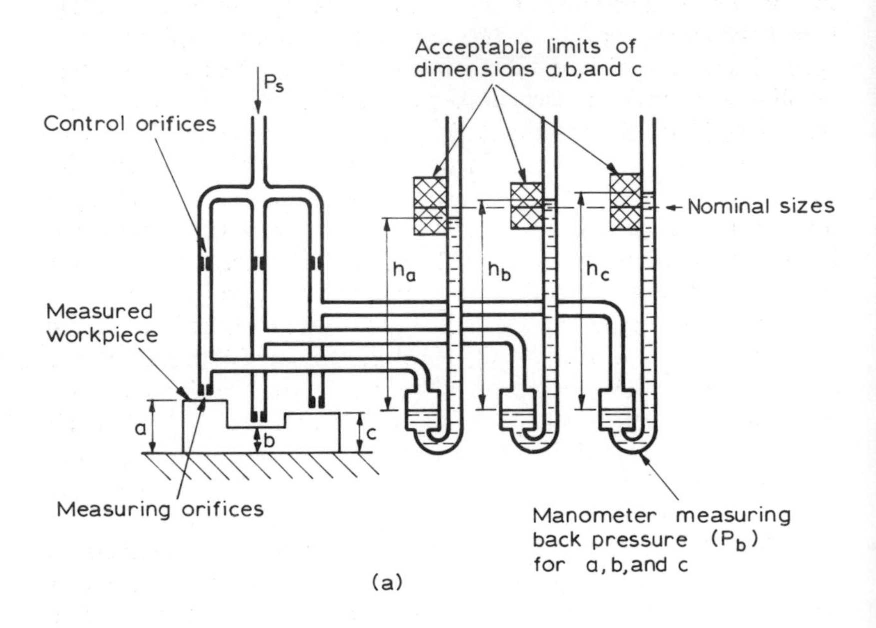

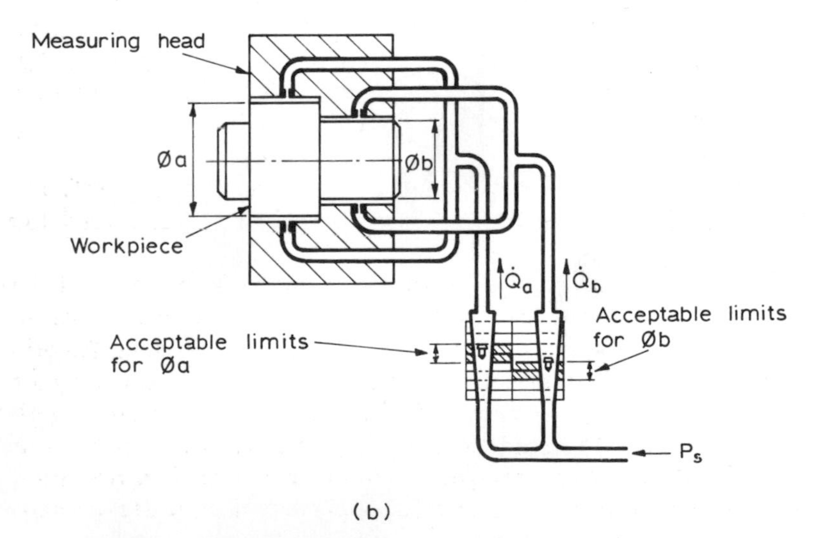

Fig. 10.19 Multiple air-gauging systems: (a) Pressure type (b) Flow type

10.3.3 Electrical comparators

In the device illustrated in fig. 2.23, a steady sinusoidal voltage $v_i = V_i \sin \omega t$ is applied to one coil, and an output-voltage wave $v_o = V_o \sin(\omega t + \phi)$ is induced in the second coil. The output-voltage amplitude depends on the number of turns in the coils and on the reluctance (S) of the iron circuit. This latter is varied considerably by very small changes (δx) in the small air gaps (x) in the iron circuit. This constitutes a displacement–voltage transducer, but the relationship is not linear, and the sensitivity is not very high. These latter characteristics may be improved by using the change of reluctance to cause a change of inductance in coils in two arms of an a.c. bridge.

A typical arrangement is illustrated in fig. 10.20, whose basic circuit is shown in fig. 4.15. Movement (δy) of the plunger (P) causes an iron armature (A) to move nearer to the top iron circuit, thereby increasing the inductance of the upper coil, and away from the lower iron circuit, reducing the inductance of the lower coil. The bridge becomes unbalanced, and a sinusoidal voltage is produced at BD, its amplitude being a function of the input movement δy. In this device, some amplification is obtained in the armature movement, and the sensitivity is doubled since the bridge has two active arms. It can be made to give a very nearly linear displacement–voltage relationship, but the output does not distinguish between movement in one direction from the null point and movement in the opposite direction, unless the phase-angle (ϕ) of the output is detected. The relationships are indicated in fig. 10.20(b). The output voltage may need to be amplified and then demodulated (see ref. 4) before being fed into the indicator, depending on the type of indicator.

As in the case of air-jet sensing, the range of measurement is small, and the difference between a standard length or diameter and the measured distance is obtained.

The main advantage of electrical comparators is the scope of the operations which may be carried out on the output signals. Variation of the range of the transducers is easily arranged by switching; the sum or difference of dimensions may be readily obtained; and switching for control operations, averaging of dimensions, and operation with computers etc. may be carried out. Very versatile measurement and control systems may be arranged. A further advantage is the operation on mains electricity, which is easier to supply than clean air at a constant pressure.

10.3.4 Optical comparators

The use of the optical lever in amplifying small linear and angular

Zero adjustment
Flexible strip hinge
Springs
L_1
V_i
B
V_o
D
A
L_2
P
δy
Workpiece
Amplifier
Demodulator
Indicator

(a)

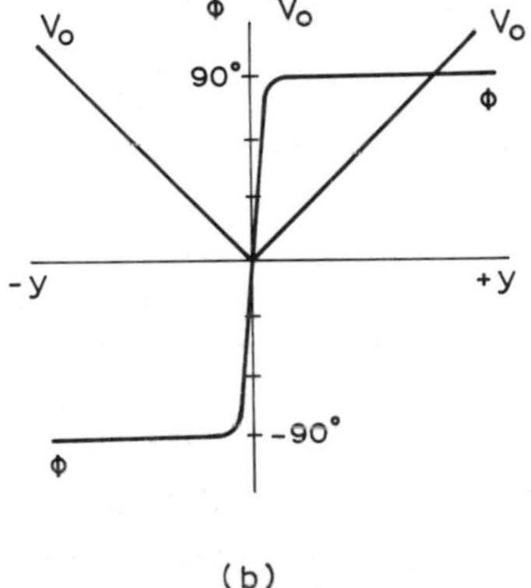

(b)

Fig. 10.20 Variable-reluctance displacement transducer (comparator): (a) System diagram (b) Response diagram

displacements is discussed in section 2.3.3 and illustrated in figs 2.32 and 2.37. Figure 10.21 illustrates a simple comparator for linear dimensions, using this principle. The difference (δy) between two heights is amplified by the lever to give a vertical displacement (δx) and angular displacement ($\delta\theta$) of a pivoted mirror. The reflected ray *smn* is deflected through an angle 2 $\delta\theta$ from the original line *smo*, and gives a reading at δX from the original value. Amplification occurs due to the mechanical- and the optical-lever effects. Commercial instruments

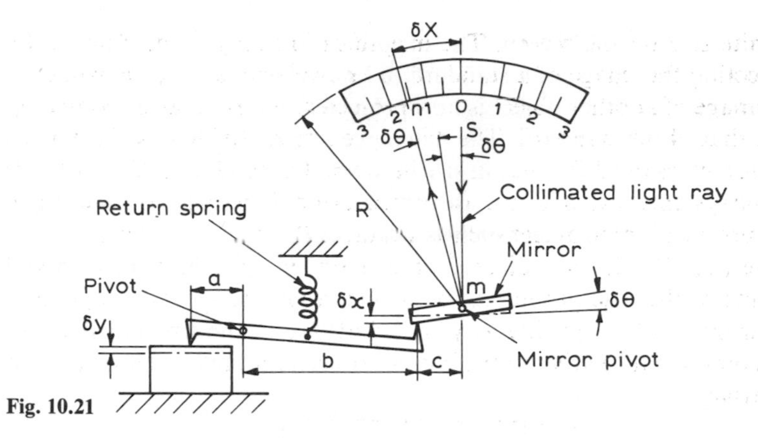

Fig. 10.21

using this kind of system are available (see refs 13 and 16).

The optical calibration and comparison of line standards and of line and end standards may be carried out using a varicty of microscope systems. The microscope has an index line which may be positioned over an engraved line with good precision. To improve this, photoelectric sensing devices are used in more recent instruments. The original need was for transference from the line primary standard to end standards; with the definition of length in terms of light wavelength, the requirement is now reversed, and instruments are now available to compare line standards with end standards, which in turn are compared with the light standards to very high accuracy and precision.

A projector may be used to compare length, and also to compare geometric shape. The principal is shown in fig. 10.22.

An object placed within the projector gives an enlarged image of a

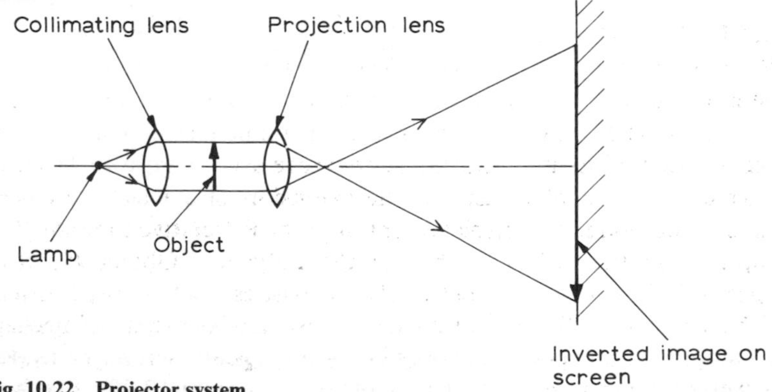

Fig. 10.22 Projector system

definite size on the screen. The magnification may be determined by projecting the image of a standard of known size (e.g. a slip gauge). If the image of another object is now projected, its size may be compared with that of the standard. The difference in size divided by the magnification gives the difference in size between the standard and the object.

The system is extended for use in measuring *shape* in the *shadowgraph*. The profile of some object such as a cam, or the fir-tree root of a turbine blade (fig. 10.23), is projected onto a screen on which an enlarged outline of the true profile has been accurately drawn. The upper and lower limits of the profile may be included, and the size and shape of the component relative to these is easily seen and may be quantified if necessary.

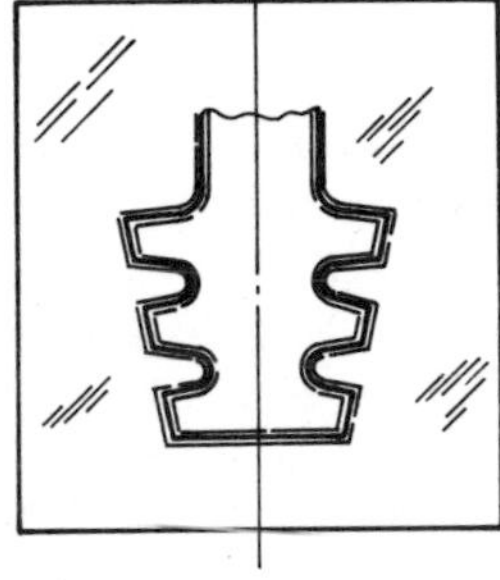

Fig. 10.23 Shadowgraph

10.4 Displacement measurement

Most of the comparators of section 10.3 measure the very small displacement or position of one surface relative to another. The measurement of rather larger displacements may use some of these methods, with a lever or similar system to attenuate the input movement. However, several other methods are also used for the measurement of linear displacements, to varying degrees of accuracy.

10.4.1 General methods

Probably the simplest method of measuring displacement is to observe the movement of a point or marker against a line scale (e.g. as in figs 9.1 and 10.24). For the more accurate measurement of smaller displacements, mechanical amplification systems are used. Figure 10.24 shows a typical method of measuring the extension of a tensile test-piece whilst under load. The displacement of point P_1 relative to point P_2 is amplified in the ratio $(a + b)/a$ at the indicator. Obviously, other types of indicator, having better characteristics, may be used instead of the mechanical dial indicator shown. Also, a wide variety of mechanical amplification systems are possible, such as gears, screws, cams, etc, as well as fluid amplification systems (e.g. as in fig. 2.30). For high

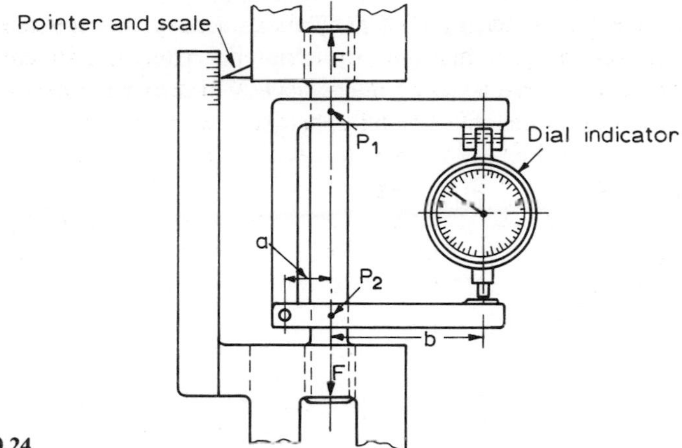

Fig. 10.24

accuracy, such as in the measurement and control of a tool point relative to a datum in precision machine tools, direct methods eliminating such components are preferred. One such method utilises a highly accurate engraved scale in the machine bed, with an optical method of observing the carriage position relative to this.

10.4.2 Electrical methods

A simple and relatively inexpensive displacement–voltage transducer uses the principle of the voltage-dividing potentiometer described in section 4.2.1. The accuracy and sensitivity depend not only on the transducer, but also on the system into which it is connected, and the discrimination is not very good, due to the finite wire size (see fig. 10.25).

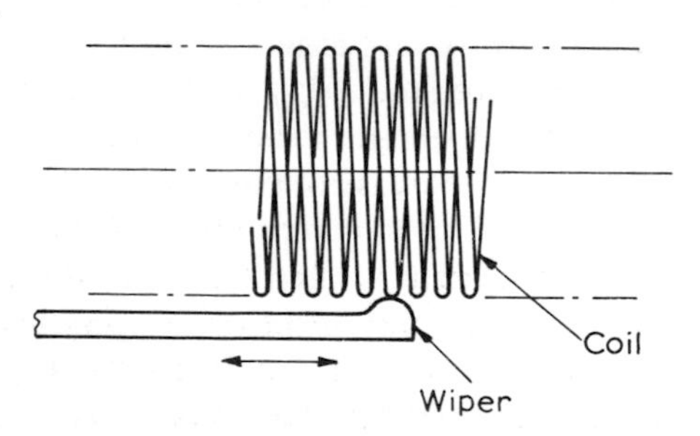

Fig. 10.25

The basic arrangements of non-contacting linear-displacement transducers using variation of capacitance and inductance have been illustrated in figs 2.20(a) and 2.23. Transducers of this type may be used, for example, to measure the axial or radial movement of a rotating shaft, to give warning of excessive vibration or bearing wear. Figure 10.26 shows such a system.

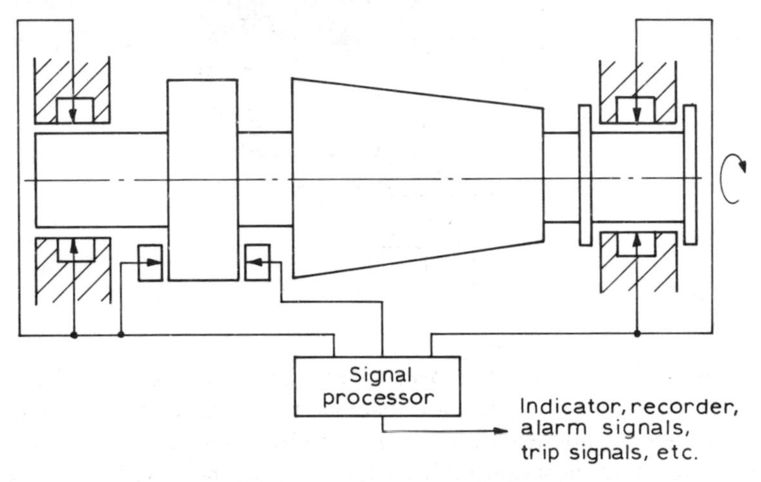

Fig. 10.26 Shaft-displacement monitoring system

A device which is finding increased usefulness is the linear variable-differential transformer (LVDT), illustrated in fig. 10.27. The alternating current in the primary coil A induces a.c. e.m.f.'s in the identical secondary coils B and C, which are coupled to A through the magnetic circuit in the iron core. With the core in the central position, the e.m.f.'s in B and C are of the same amplitude and 180° out of phase with each other, due to the directions of the windings. Since they are connected so that these e.m.f.'s are additive, then $v_0 = 0$. If a displacement (δx) of the core is made, the e.m.f. in one secondary coil is increased and that in the other is reduced, due to the changes of mutual inductance. Now, v_0 has a value, and the amplitude (V_0) of this output is very nearly proportional to δx. To distinguish between positive or negative displacements of the core, the phase of the output relative to the input must be determined, as in the inductance bridge shown in fig. 10.20.

The frequency (ω) of the input is often made very high, so that stray signals from other sources do not affect the output, and also so that low-frequency vibrations of the core do not have any effect on the measurement. A range of these transducers and associated equipment is available in robust form suitable for use in industrial conditions.

The very accurate measurement of the displacement of machine-tool tables etc., of the order of one or two metres, is necessary for measure-

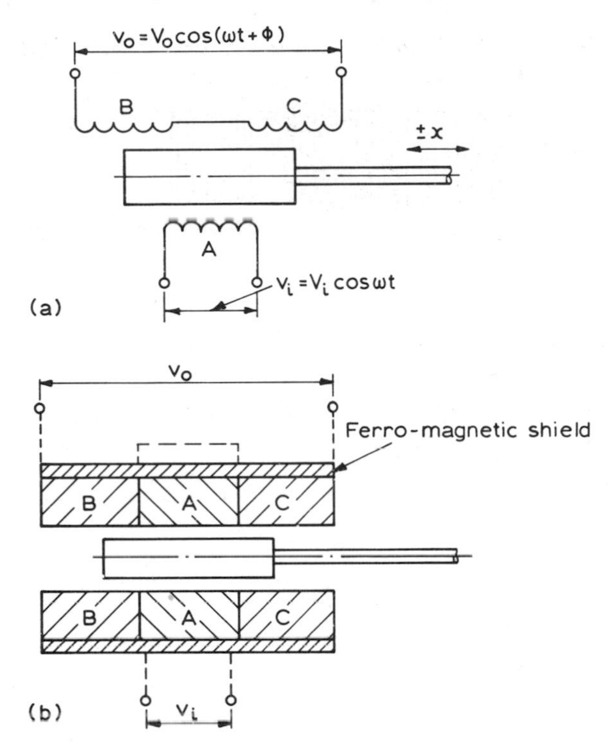

Fig. 10.27 Linear variable-differential transducer: (a) Circuit (b) Cross-section

ment and control purposes. One method is to measure the angular displacement of a leadscrew connected through a nut to the linear motion, and this is described in Chapter 11. The following section describes a more direct method of measuring such linear displacements.

10.4.3 Moiré-fringe methods

If lines are ruled close and parallel on two transparent plates, and they are arranged so that the lines lie at a slight angle to each other, then the light-interference effects will cause light and dark fringes to be formed when light is directed through the pair, as shown in fig. 10.28(a). The distribution of light intensity in the fringes is nearly sinusoidal. When displacement of one plate relative to the other occurs in the x direction, a corresponding but greatly increased movement of the fringes occurs in the y direction. If a light-sensitive transducer is positioned as shown, then it receives an approximately sinusoidal light-intensity–displacement signal [fig. 10.28(b)].

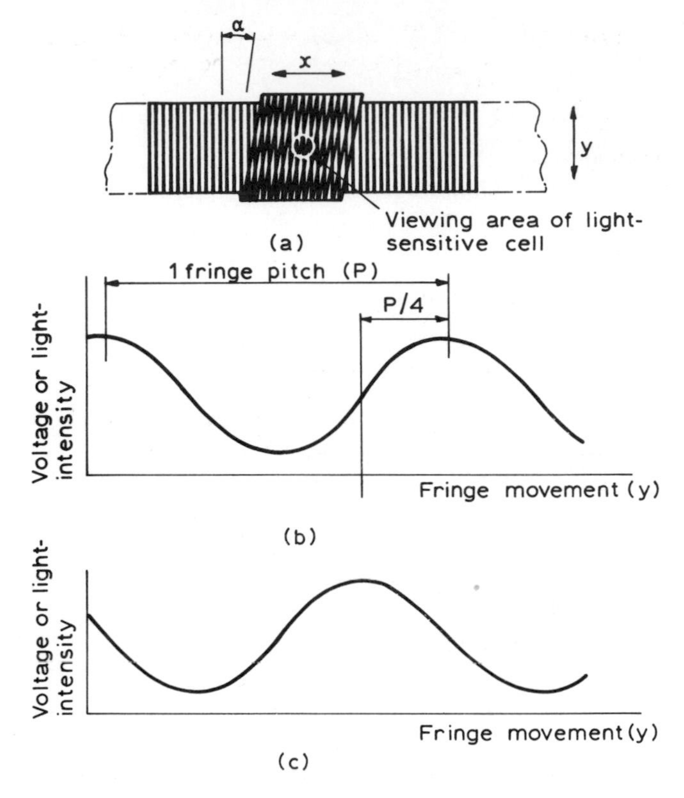

Fig. 10.28 Moiré-fringe displacement transducer (a) Fringe formation and viewing

The distance between two intensity peaks represents a definite movement (δx). The gratings, as they are called, can be manufactured to very close line spacings, of the order of 100/mm, and from a pair of these a relative displacement of about 0·0025 mm can be detected. To determine the direction of motion, a second light-sensitive transducer is added as shown in fig. 10.29, so that its voltage-output curve is at 90° to the first, as in fig. 10.28(c). A phase-sensitive device is then able to detect the direction of motion, and either add or subtract digits at the counter, for each pulse.

The system is usually as shown in fig. 10.29. The lamp, collimating lens, short grating, and light transducers being mounted on the moving part of the machine, and the long grating being fixed on the machine bed. Several variations of this basic system exist and are applied in coordinate positioning in one, two, or three axes, in continuous path control, and in measuring machines.

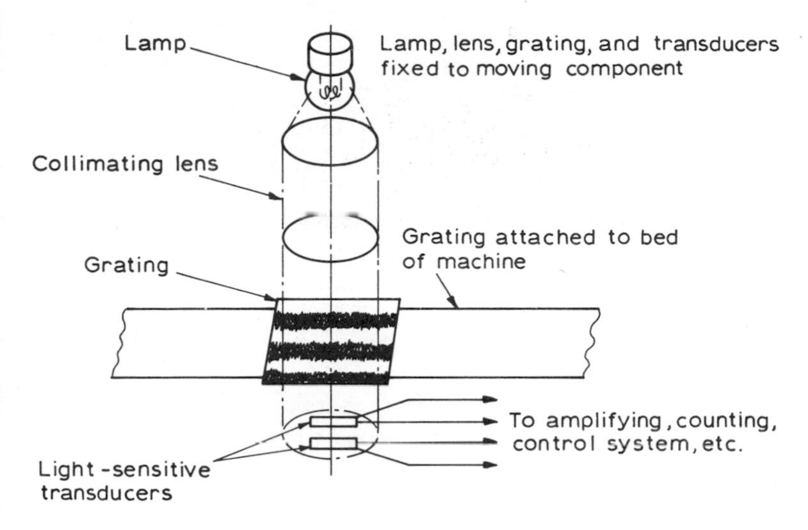

Fig. 10.29

10.5 Surface-finish measurement

No real surface follows a true geometrical shape; a surface which appears to be flat and highly polished will reveal undulations due to the machining and finishing processes to which it has been subjected. The finish is important for many surfaces in a machine, e.g. those used as bearings, and this has often been covered by a note on the drawings, such as 'surface to be smooth'. However, it is desirable to quantify the surface finish, though this presents difficulties since the problem is three-dimensional and, also, machining effects are highly directional. One method widely used is to compare the surface visually and by touch against surface-finish standard specimens, which in skilled hands is a fairly satisfactory method.

10.5.1 Stylus methods

If a straight line across a flat surface, or parallel to the axis of a cylinder, is selected, then the two-dimensional profile along this line may be determined. If a stylus is dragged along the surface, its motion traces out very nearly the profile of the surface, and this may be amplified mechanically as in the Tomlinson surface-meter shown in fig. 10.30, when a magnified trace is obtained on a rotatable smoked-glass screen. The vertical magnification is $\times 100$, and there is no horizontal magnification. If the trace is then projected with a magnification of $\times 50$, then the overall vertical magnification is $\times 5000$ and the overall

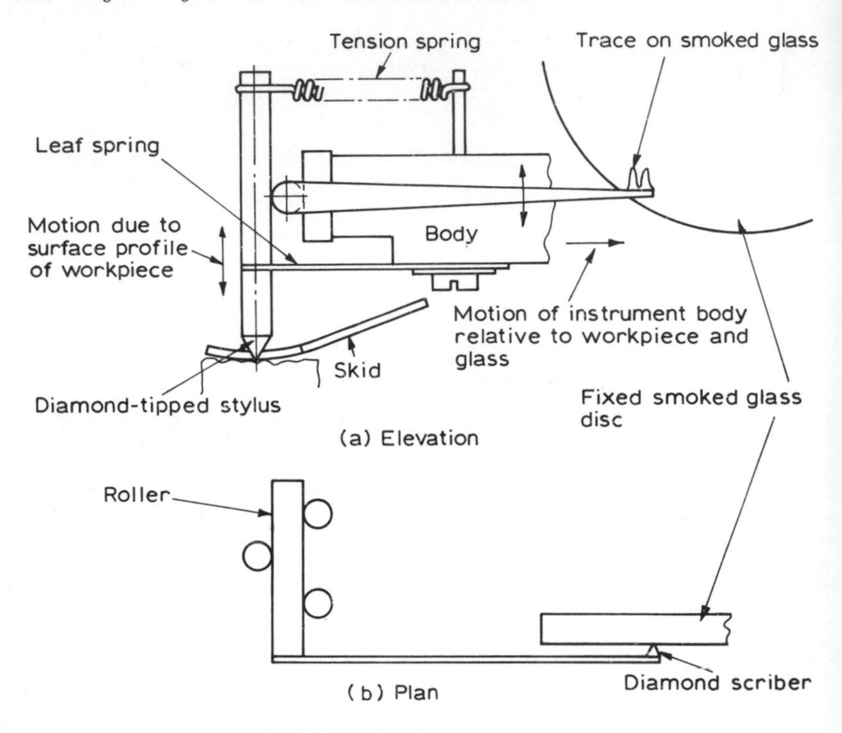

Fig. 10.30 Tomlinson surface meter

horizontal magnification ×50. The device is excellent for the demonstration of surface-finish measurement, but is unsuitable for industrial use.

A more convenient instrument is the Taylor-Hobson 'Talysurf', a system drawing of which is shown in fig. 10.31. The system operates on the inductance-bridge principle, and unbalance of the bridge is caused by variation of the reluctance of the magnetic circuit, due to rocking of the armature as the stylus follows the surface profile. The output is an automatically-drawn trace.

The undulations of the surface may be due to various effects such as tool feed, tool chatter, and vibration of the machine-tool structure and moving parts at various frequencies etc. These may combine to cause a more or less regular pattern which repeats itself over a length, giving *waviness*, as shown in fig. 10.32(a).

A magnified trace may appear as shown in fig. 10.32(b), and the following method is given in BS 1134:part 1:1972, 'Assessment of surface texture', for determining a mean line through the profile and the *arithmetical mean deviation* (R_a) of the profile from the mean line.

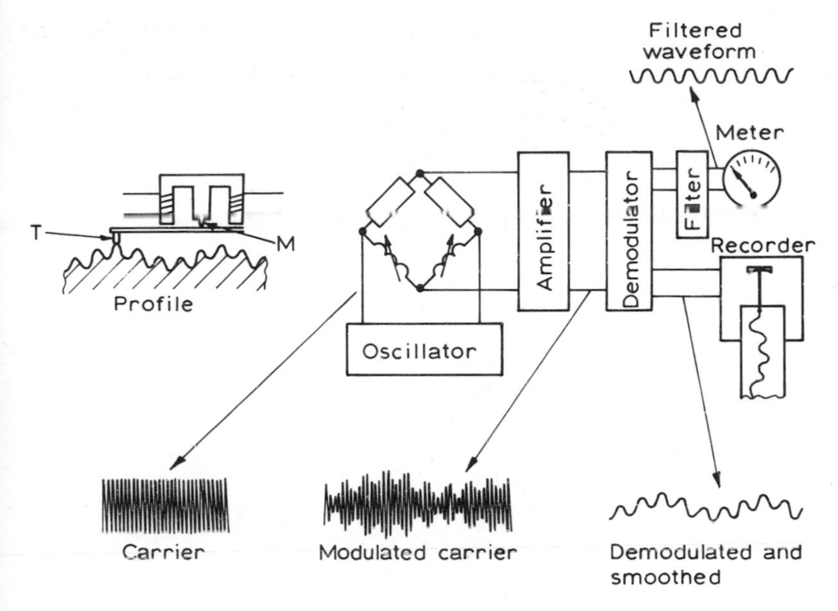

Fig. 10.31 System diagram of the Taylor-Hobson Talysurf

A straight line XX is drawn, generally parallel to the profile over the sampling length (L), and for convenience this may pass through the lowest valley as shown. Where the texture has a distinguishable periodicity, it is essential that the sampling length is chosen to include a whole number of wavelengths.

The area (P) between the profile and the line XX is then determined by measuring ordinates or by using a planimeter. The distance to the mean line is then

$$H_m = P/L$$

The centre line YY can now be drawn parallel to XX. The areas $r_1, r_2, r_3, \ldots$ above, and $s_1, s_2, s_3, \ldots$ below the centre line are then determined. Taking all areas as positive, the arithmetic mean deviation is given by

$$R_a = \frac{\text{sum of areas } r + \text{sum of areas } s}{L} \times \frac{1000}{V_v} \ \mu\text{m}$$

L is expressed in mm, the areas in mm^2, and V_v is the vertical magnification of the trace. Several sampling lengths are taken along the traversing length of the stylus, and the value of R_a for the surface is taken as the mean of these.

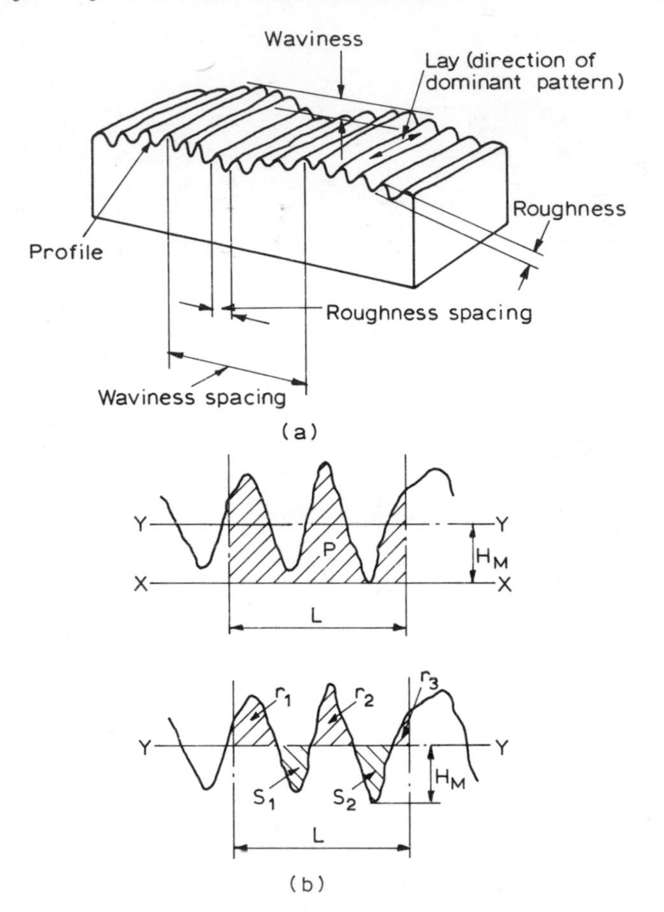

Fig. 10.32 (a) Surface-texture terms (b) Graphical determination of R_a values

The operation is time-consuming and unsuitable for production use. The Talysurf instrument carries out these operations electronically, and gives a meter reading of R_a over a known length of test, in addition to a trace.

It has to be appreciated that surfaces of quite different characteristics can give the same R_a value, and hence the system does not completely fill the requirements for surface-finish evaluation.

10.5.2 Optical-interference method

The formation of interference fringes was described in section 10.2.1. A flat smooth surface viewed through an optical flat, using mono-

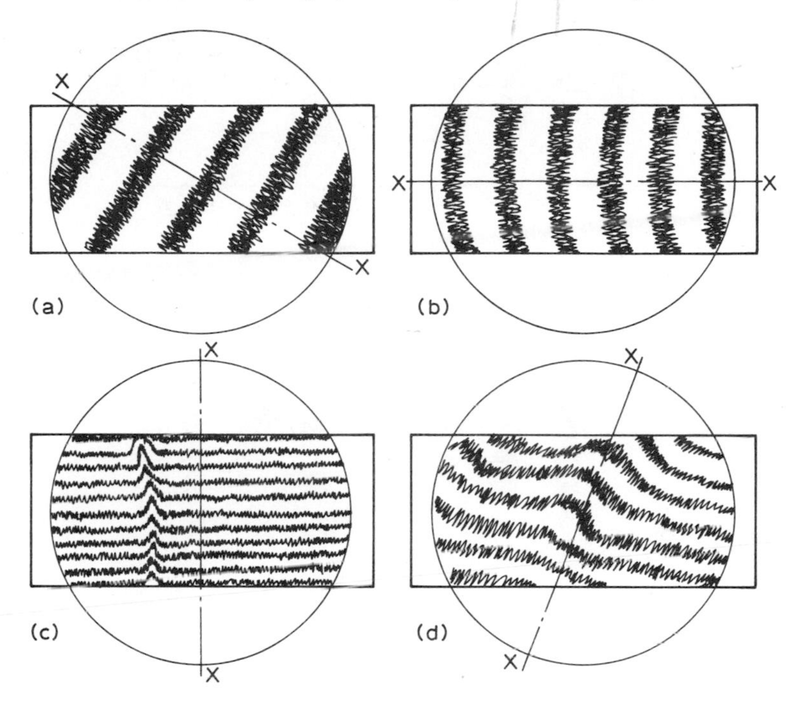

Fig. 10.33 Interference fringe patterns in flatness and surface-finish observations. XX is the trace of the plane containing angle θ (fig. 10.6)

chromatic light, would produce fringes as shown in fig. 10.33(a). A surface which was concave or convex along the centre would give fringes as in (b), whilst one with a scratch on it would appear as in (c). A surface with some general contouring would appear as in (d). The viewing may be carried out using a specially designed surface-finish microscope of relatively low magnification. Whilst it is not possible to determine an R_a value, the appearance of the surface finish of the component may be compared with that of standard specimens. The extent of waviness, concavity, or convexity, or the depth of scratch etc. may be estimated, since, for each fringe spacing, the change of separation of the surface and the optical flat is half of the wavelength of the light used (see example 10.6.7).

10.6 Worked examples

10.6.1 A slip gauge wrung to a flat surface was observed through an optical flat in the light of a cadmium lamp used with a filter, the

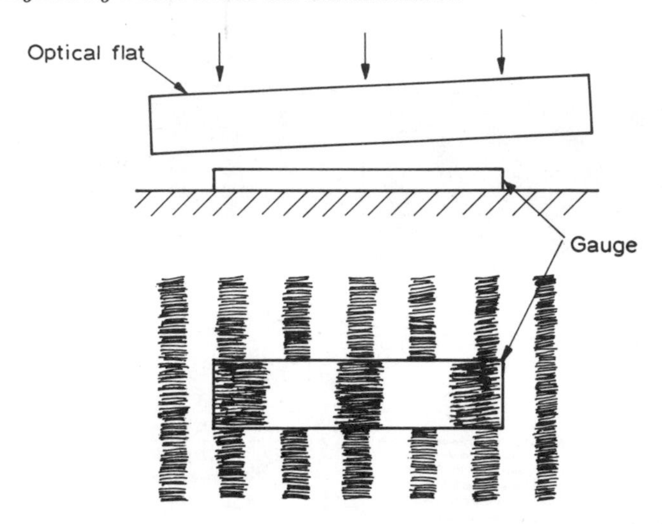

Fig. 10.34

wavelength being 0·509 μm, and fringes as shown in fig. 10.34 were observed. Estimate the change in thickness of the gauge along its length.

There are $2\frac{1}{2}$ fringes along the length of the gauge and $4\frac{1}{2}$ in the corresponding length of the base. The 2 fringes difference represents 2/2 wavelengths change of separation of the top and bottom surfaces of the gauge. Hence the change in thickness of the gauge is $(2/2) \times 0{\cdot}509 = 0{\cdot}5$ μm along its length. (Assuming the accuracy of fringe counting is 1/5 of a fringe, this represents $\lambda/10$, about 0·05 μm; hence the value is given to 1 decimal place.) The test does not indicate the direction of the taper.

10.6.2 Estimate the difference in height between the slip-gauge pile and the component illustrated in fig. 10.35. The cadmium light used has a wavelength of 0·644 μm.

The number of fringes over the distance x is twelve; hence the change of separation of the optical flat and the base over this distance is 12/2 wavelengths.

$$\therefore \quad \text{Difference in height} = \delta y = (12/2) \times 0{\cdot}644$$
$$= 3{\cdot}9\ \mu\text{m}$$

Again, it is not revealed which is higher.

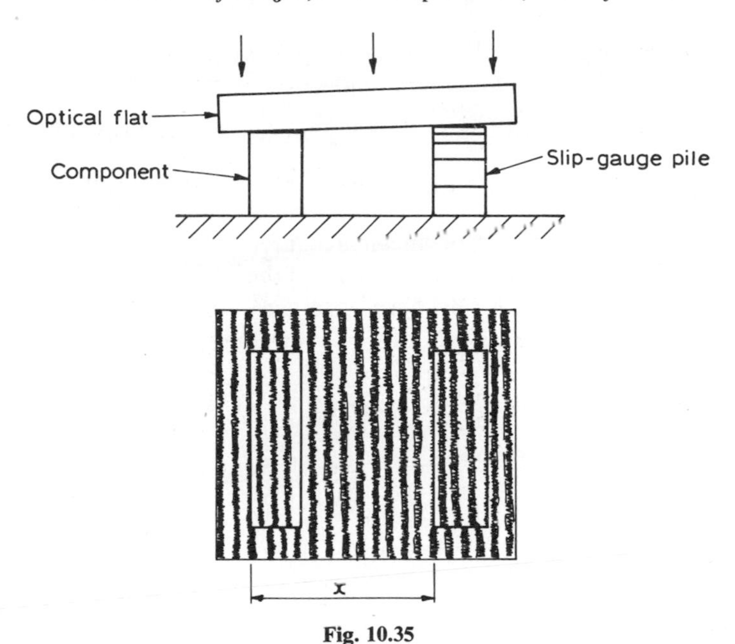

Fig. 10.35

10.6.3 The nominally flat end of a cylindrical component appeared as shown in fig. 10.36 when viewed through an optical flat. What is the shape of the end surface?

Fig. 10.36

The rings are concentric, but crowded more at the outside than the inside, the closer spacings indicating a steeper angle between the surfaces of the component and the flat. It follows that the end of the component is either concave or convex. If it is concave, then, on gently pressing the optical flat in the centre, the lines will become more evenly spaced as the two surface conform more closely to each other. If it is convex, then, on gently rocking the flat, the centre of the fringe circles may be seen to move.

10.6.4 The operation of a pressure-type pneumatic comparator as shown in fig. 10.12 is represented by the equation

$$P_b/P_s = -0{\cdot}5(A_m/A_c) + 1{\cdot}1 \qquad \text{for } 0{\cdot}6 > (P_b/P_s) > 0{\cdot}8$$

The control orifice is 0·5 mm diameter, and the measuring orifice is a 1 mm diameter hole.

Calculate (a) the range of linear measurement; (b) the measuring-head, pneumatic, and overall sensitivities if the back-pressure gauge has a deflection of 25 mm for 1 kN/m² pressure change, and the supply pressure is constant at 200 kN/m² (2 bar) gauge pressure.

a) Substituting $P_b/P_s = 0{\cdot}6$ in the equation, and working in mm units,

$$0{\cdot}6 = -0{\cdot}5 \times \frac{\pi \times 1 \times x_1}{(\pi/4) \times 0{\cdot}5^2} + 1{\cdot}1$$

$$\therefore \quad 0{\cdot}6 = 8x_1 + 1{\cdot}1$$

$$x_1 = 0{\cdot}5/8 = 0{\cdot}062 \text{ mm}$$

Similarly, for $P_b/P_s = 0{\cdot}8$,

$$0{\cdot}8 = -8x_2 + 1{\cdot}1$$

$$x_2 = 0{\cdot}3/8$$

$$= 0{\cdot}037 \text{ mm}$$

Hence the linear range is $x_1 - x_2 = 0{\cdot}062 - 0{\cdot}037$

$$= 0{\cdot}025 \text{ mm}$$

b) From eqn 10.3, the measuring-head sensitivity is

$$\frac{dA_m}{dx} = \pi D_m$$

$$= \pi \times 10^{-3} \text{ m}$$

From eqn 10.5, the pneumatic sensitivity is

$$\frac{dP_b}{dA_m} = -\frac{0{\cdot}4P_s}{A_m(\text{mean})}$$

$$= -\frac{0{\cdot}4 \times 200 \times 10^3}{\pi \times 10^{-3} \times \{(0{\cdot}062 + 0{\cdot}037)/2\} \times 10^{-3}}$$

$$= -\frac{1600}{\pi} \times 10^9 \text{ N/m}^4$$

The pressure-gauge sensitivity is

$$\frac{dR}{dP_b} = \frac{25 \times 10^{-3}}{1000} \frac{m}{N/m^2}$$

By eqn 10.4,

$$\frac{dR}{dx} = \frac{dA_m}{dx} \times \frac{dP_b}{dA_m} \times \frac{dR}{dP_b}$$

$$= \pi \times 10^{-3}\ m \times \left(-\frac{1600}{\pi} \times 10^9\right)\frac{N}{m^4} \times 25 \times 10^{-6}\ \frac{m^3}{N}$$

$$= -40000$$

10.6.5 Show that, for a system as in fig. 10.12(a) with a particular value of K, and $b = 1{\cdot}1$ (eqn 10.1), the pneumatic sensitivity (dP_b/dA_m) does not depend on the measuring-orifice size (D_m).

Taking the mid-point $\bar{x}$ of the linear range, $(P_b/P_s) = 0{\cdot}7$, giving for eqn 10.1

$$0{\cdot}7 = \frac{K\pi D_m \bar{x}}{(\pi/4)d^2} + 1{\cdot}1$$

$$\therefore \quad \bar{x} = (0{\cdot}7 - 1{\cdot}1)\left(\frac{d^2}{4KD_m}\right)$$

$$= B/D_m$$

where B is constant.
The pneumatic sensitivity is given by eqn 10.5:

$$\frac{dP_b}{dA_m} = -\frac{0{\cdot}4P_s}{\pi D_m B/D_m}$$

$$= -\frac{0{\cdot}4P_s}{\pi B}$$

Hence dP_b/dA_m is not affected by D_m if K is constant. However, K does vary slightly with different measuring-orifice sizes. The overall sensitivity is affected, since from eqn 10.3 the measuring-head sensitivity increases with D_m, and referring to eqn 10.2 this is seen to directly affect the overall sensitivity (dR/dx).

10.6.6 Devise an air-jet sensing arrangement to determine (a) the squareness of two faces, (b) the centre distance of two holes.

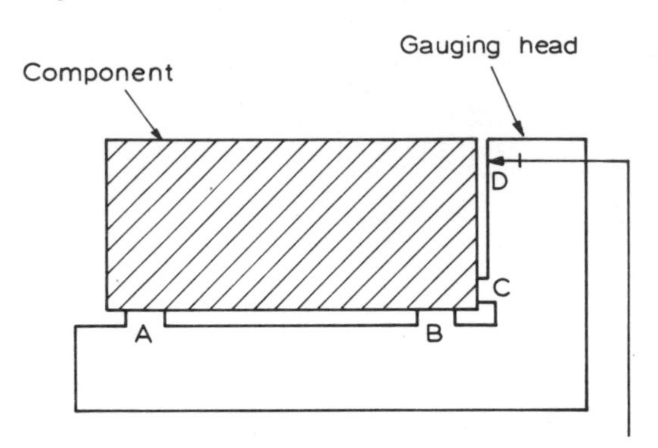

Fig. 10.37

a) Figure 10.37 shows a fixture locating one face at points A and B, and the 90° face at C. At D on the second face, an air jet senses the distance of the face, and this may be compared with that for a square of known accuracy.

b) Figure 10.38 shows a double air-plug gauge with the plugs at nominally the same centre distance as the holes to be measured. The back-pressure or flow-rate values may be read for each of the two measuring systems, with the gauge inserted into two holes whose centre distance is accurately known. The differences in these values compared with those when the gauge is inserted into a component will indicate the difference of the centre distance in a component from the reference value.

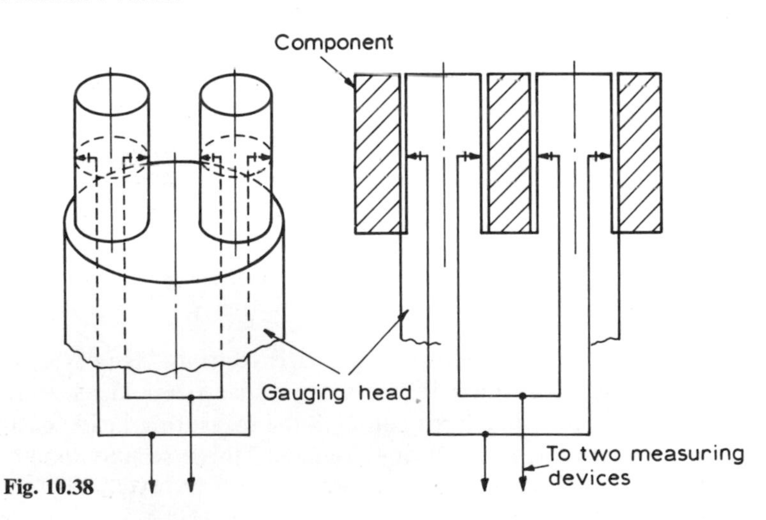

Fig. 10.38

10.6.7 Estimate the depth of the scratch in fig. 10.33(c) if the light used has a wavelength (λ) of 0·509 μm.

Since there are 1 to $1\frac{1}{2}$ fringe displacements at the scratch, this is varying in depth from (1/2) × 0·5 to (3/4) × 0·5, i.e. 0·3 to 0·4 μm.

10.7 Tutorial and practical work

10.7.1 Investigate the history of length units, and produce a written account.

10.7.2 Investigate the various support points for length bars for accurate length measurements. Give your findings in the form of a short report.

10.7.3 Explain why the test in example 10.6.1 does not reveal the direction of the taper of the slip gauge.

10.7.4 In fig. 10.6, refraction at the surfaces of the optical flat has been ignored. What effect, if any, will it have on the fringe pattern?

10.7.5 Explain the significance of the fringe shapes observed on the top faces of the metal discs shown in fig. 10.39. The bottom faces are flat and wrung to a flat base surface. The outer lines represent fringes due to the base.

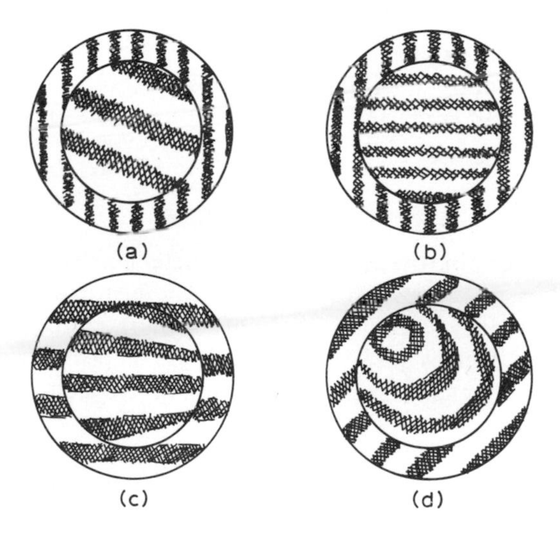

Fig. 10.39

10.7.6 Examine a set of slip gauges, after reading the handling instructions. Referring to the calibration sheet, note the error of each slip from its nominal size and the accuracy of the determination of the sizes. Make up two piles nominally of the same height by wringing together several of the gauges, and compare the heights using (a) a mechanical, electrical, or pneumatic comparator of high magnification and (b) a light-interference method. Discuss any difference in the pile heights which is found, and also any difference in the values found from (a) and (b).

10.7.7 Carry out an accuracy check to BS 1054 on a mechanical, pneumatic, or electronic comparator.

10.7.8 Compare the pneumatic bridge of fig. 10.14 with the electrical bridge of fig. 4.6(a). Draw a diagram of the pneumatic bridge incorporating the reference letters A, B, C, and D corresponding to the electrical bridge.

10.7.9 Discuss the use of one or two master ring or plug gauges in the setting of an air plug or ring gauge for measuring production-component bores or outside diameters.

10.7.10 Sketch air-jet sensing arrangements for determining (a) the variation of a shaft diameter along an axis, (b) the straightness of a bore, (c) the concentricity of two external diameters, and (d) the squareness of a bore to its outside face.

10.7.11 Sketch a suitable arrangement using the pneumatic system of example 10.6.4 to provide a continuous indication of the thickness of metal strip coming out of the rolls, if this is expected to vary between 0·95 mm and 1·05 mm. Describe how you would calibrate such a system.

10.8 Exercises

10.8.1 To determine the difference in diameter between two accurately ground rollers, an optical flat was placed over the top of them as shown in fig. 10.40. When viewed in the light of a sodium source of wavelength 0·593 μm, 10·4 fringes were observed in the length x. Calculate the difference in diameter of the rollers, giving this to a suitable degree of accuracy. [3·1 μm]

10.8.2 Calculate the range and overall sensitivity of the pneumatic comparator of example 10.6.4 if the measuring orifice is changed to 0·5 mm diameter, all other factors being unaltered.
[0·05 mm, −20000]

10.8.3 In a test on an air-jet sensing system, the distance x of the end of a micrometer spindle from the jet was varied, whilst the back-pressure was measured using a precision Bourdon-tube-type pressure gauge (see fig. 10.41). The following values were recorded.

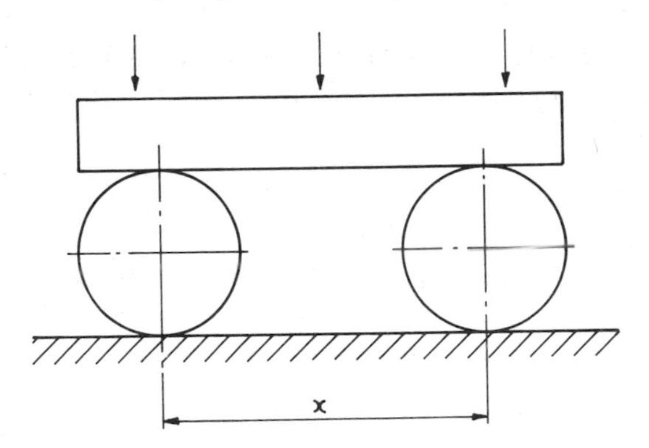

Fig. 10.40

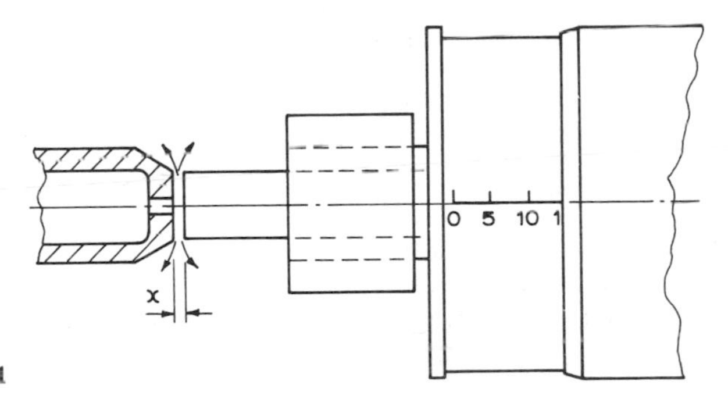

Fig. 10.41

x (mm)	0	0·006	0·018	0·030	0·042	0·064	0·076	0·088
P (bar)	2·70	2·69	2·62	2·45	2·27	2·05	1·86	1·65
x (mm)	0·100	0·112	0·124	0·136				
P (bar)	1·46	1·26	1·10	0·97				

Plot these values, and from the graph (a) determine the range of x over which the pressure versus displacement is linear; (b) check that this linear range applies approximately from $(P_b/P_s) = 0{\cdot}8$ to 0·6; (c) determine the constant K (eqn 10.1) for the system, and check that the constant $b = 1{\cdot}1$ applies, if the diameter (D_m) of the measuring orifice is 1 mm, and the diameter (d) of the control is 0·5 mm; (d) determine the overall sensitivity of the system if the pressure-gauge scale represents 3 bar pressure over a length of 420 mm.

[0·042 to 0·100 mm, −0·40, 2420]

10.8.4 Show that, in the arrangement illustrated in fig. 10.17, the area A_m of the annulus in plane XX is directly proportional to the displacement (x), if the axis of the parabolic cross-section is concentric with the bore at section XX.

11

Measurement of Angle, Alignment, and Angular Displacement

11.1 Angle measurement and standards

The natural unit of angle could be considered to be either (a) the circle, i.e. one rotation, or (b) the radian, i.e. that angle subtended at the centre of a circular arc of length equal to the radius (see appendix A) as shown in fig. 11.1. The radian is used extensively in mathematics and in theoretical work, but it is not a practical unit of measurement in general. The sexagesimal system, dating back to 1000 BC, divides the circle by two diameters into four 90° sectors. Each sector is divided into 90 equal degrees, which are subdivided into 60 equal minutes, again subdivided into sixty equal seconds. Hence the problem of obtaining accurate angle measurement may be considered to be one of accurate division of a 90° angle.

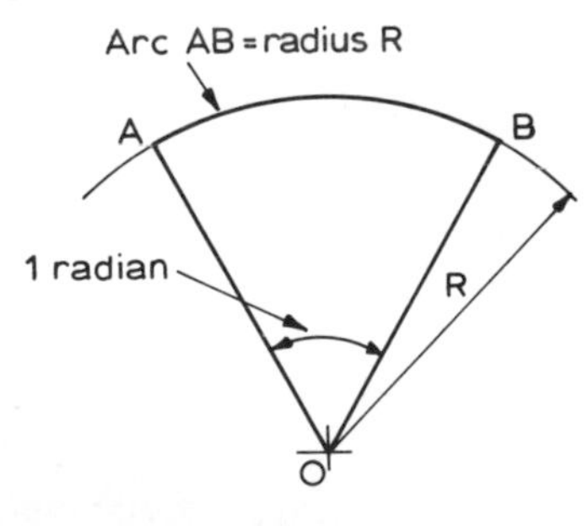

Fig. 11.1

A further way of obtaining an angle standard is through length standards. Very accurate trigonometric ratios of angles are obtained by mathematical methods. Hence the relationships between the sides l, h, and b of a 90° triangle as shown in fig. 11.2 are known for any angle (θ).

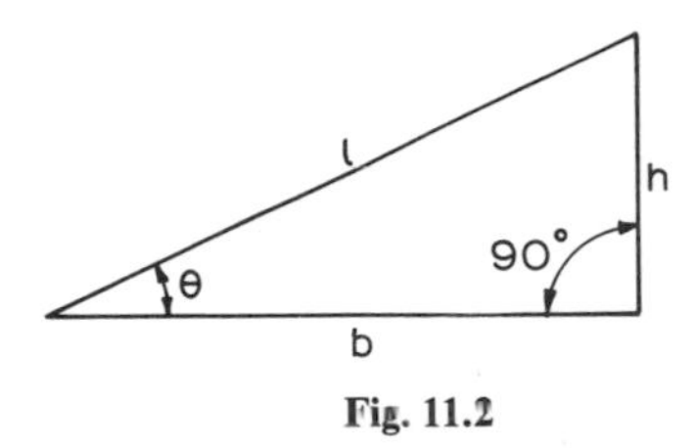

Fig. 11.2

As with length standards, a general division may be made between line standards of angle, such as used on a vernier protractor, and standards measuring angles between faces, in a similar manner to the measurement of length between slip- and block-gauge faces.

11.1.1 Line standards of angle

Protractor scales are divided using the circular dividing engine, a highly developed machine in which metal or glass scales may be ruled directly. Alternatively, in the case of glass, this may be thinly coated with a protective layer through which the lines are ruled; the lines are then etched with hydrofluoric acid. The scales obtained in this way are used in precision instruments such as optical dividing heads and precision clinometers, where the scales are viewed and subdivided using an optical system. These and similar instruments are used as reference standards.

The precision clinometer, reading to an accuracy of 10″, uses a spirit level as a *fiducial indicator*, as indicated in fig. 11.9 for a less precise type of clinometer.

11.1.2 Face standards of angle

Face standards of angle include precision squares and polygons, as shown in fig. 11.3(a), and sets of combination angle gauges, as shown in fig. 11.3(b) and (c). A typical set of combination angle gauges consists of the following:

90° square

27°	9°	3°	1°
27′	9′	3′	1′
27″	9″	3″	

This combination enables any angle to be built up, to a nominal value within $1\frac{1}{2}$ seconds of the required angle. The number of gauges in a set is small, since each gauge may be used in two ways, either to add its angle as in fig. 11.3(b) or to subtract it as in fig. 11.3(c).

Precision polygons may be made from hardened steel or from glass. Their size may range up to 300 mm and the number of faces to 72. They

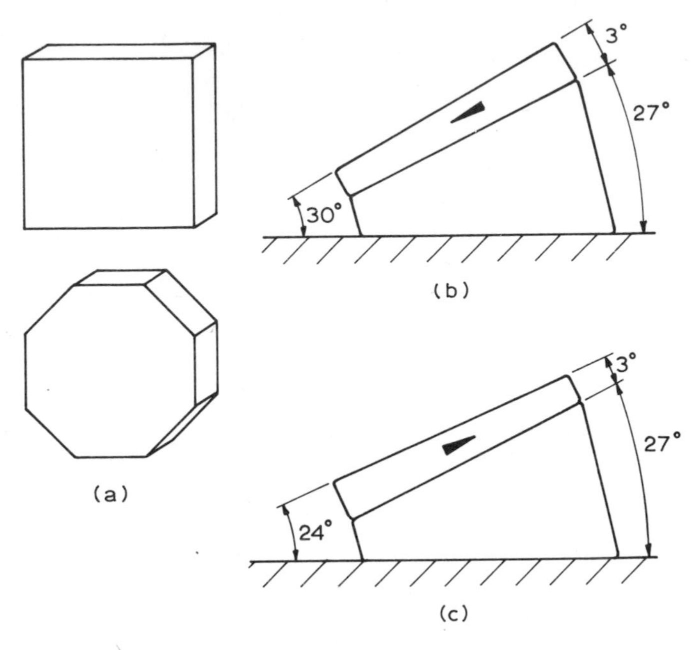

Fig. 11.3

are used in the calibration of equipment such as optical rotary tables etc.

11.1.3 Angles derived from length standards

The high degree of accuracy and precision available for length measurement in the form of slip and block gauges may be utilised for angle measurement using the sine relationship, $\sin\theta$ = (opposite side)/hypotenuse = h/l. Figure 11.4(a) shows a method of using a pair of precision cylinders, together with stacks of slip gauges wrung together to determine h and l and hence obtain the required angle (θ).

Errors in the slip gauges and roller sizes will lead to error in the value of the angle obtained. If δh is the small error in setting h, and δl is the small error in setting l, then the error ($\delta\theta$) in the resulting angle is given by (see appendix C)

$$\delta\theta \approx \pm\left(\frac{\delta h}{h} + \frac{\delta l}{l}\right)\tan\theta \qquad 11.1$$

It is seen from this relationship that, for given fractional errors in the determination of h and l, the error in θ is proportional to $\tan\theta$, and will increase rapidly when θ is greater than 45°.

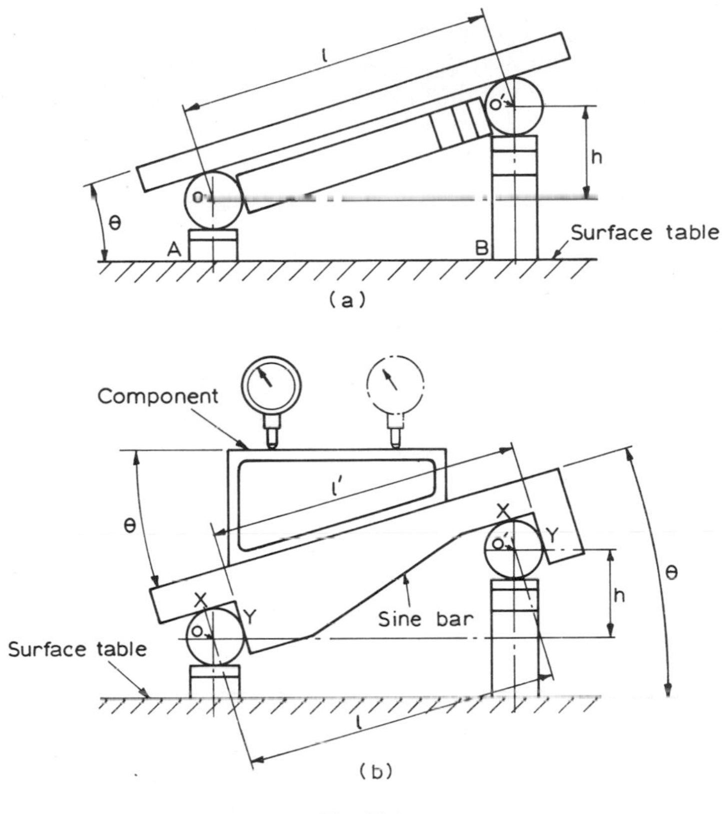

Fig. 11.4

Figure 11.4(b) shows a sine bar, and in this the length l' is accurately known, thus simplifying the procedure. It also illustrates a method of checking the angle of a component by placing it on the bar and then checking its parallelism with the base, either with a dial indicator or with a precision level (see section 11.3.1).

11.2 Angle comparators

The use of comparators for length standards was discussed in section 10.3; in a similar manner, the difference between two nearly identical angles may be determined. This is frequently the difference between the angle of a working standard gauge or instrument and the overall angle of a number of angle gauges wrung together, or the angle between two faces of a standard polygon.

11.2.1 *Optical methods of angle comparison*

Figure 11.5 illustrates the principle of the *auto-collimator*. Light rays passing a target wire are then passed through a collimating lens from which the parallel rays are directed on to a reflecting surface. If the surface is normal to the rays, they return along the same paths, and the image is formed at the target wire, which is in the focal plane of the lens. If the surface is at a small angle (θ) to the normal, then the returning rays are deflected through an angle 2θ, and the image of the target wire is displaced. The displacement of the image of the target wire depends only on the angle (θ) of the reflecting surface, and is not affected by the distance (s). The target wire and its image are viewed through a low-power microscope fitted with a micrometer eyepiece and scale, enabling θ to be read to parts of a second. In some instruments a photoelectric sensor replaces the human eye, giving better precision.

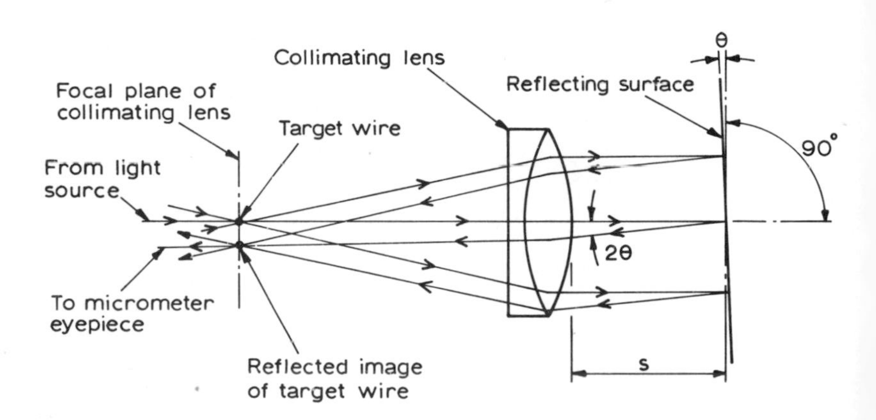

Fig. 11.5 Principle of the autocollimator

The surface on to which the rays are directed must be reflective. A newly machined surface will probably reflect; if it does not, then a parallel slip gauge may be wrung to the surface. Figure 11.6 illustrates the use of the autocollimator to compare the angle of a precision angle-plate with that between two faces of an eight-sided reference polygon. The instrument is firmly fixed in position relative to the surface-plate, and the reading is taken with the polygon in position. The polygon is removed and replaced by the angle-plate, plus slip gauge if necessary, and the new reading is taken. The difference between the two readings must be added to or subtracted from the known angle between the particular pair of polygon faces, to give the angle of the face-plate.

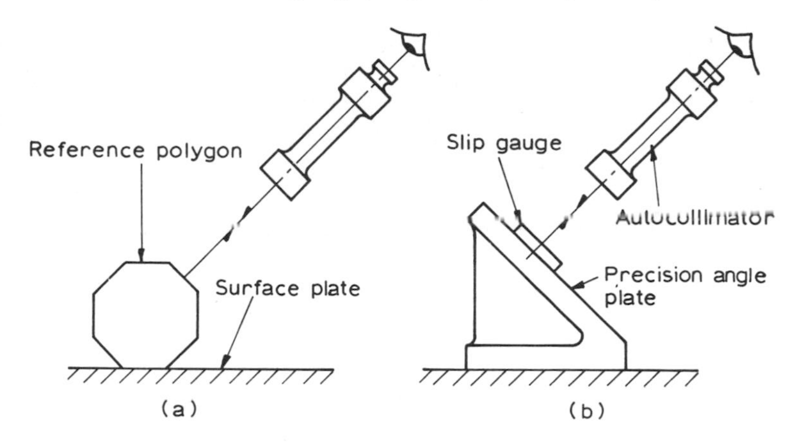

Fig. 11.6

11.2.2 Pneumatic methods of angle comparison

Figure 10.37 shows a method of determining the out-of-squareness of a component by comparing it with a reference square, using a pneumatic gauging head. The method may be used to compare the angle between the faces of a component with that of a reference standard or, with suitable arrangement, the tapers on solids of revolution etc. Figure 11.34 shows a method of comparing the taper of a component with a reference taper, and example 11.5.2 shows a method of calculating the angle difference.

11.3 Measurement of level, alignment, and flatness

Level is defined as a line or plane parallel to a plane tangential to the earth's curvature. It is more readily determined as a line or plane perpendicular to the direction of the earth's gravitational pull, e.g. a line or plane perpendicular to a plumb line.

Alignment is when three or more points are positioned in a straight line, as may be done roughly with the eye or, more accurately, with the aid of optical instruments.

Flatness is the conformity of a surface to a true plane.

11.3.1 Measurement of level

A precision spirit-level (fig. 11.7) consists of a glass *vial*, ground internally to a radius R, enclosing a liquid in which is an air bubble. The hydrostatic pressure due to gravitational attraction will cause the bubble to move to the highest point of the curve. The base AB is horizontal when it is at 90° to CO.

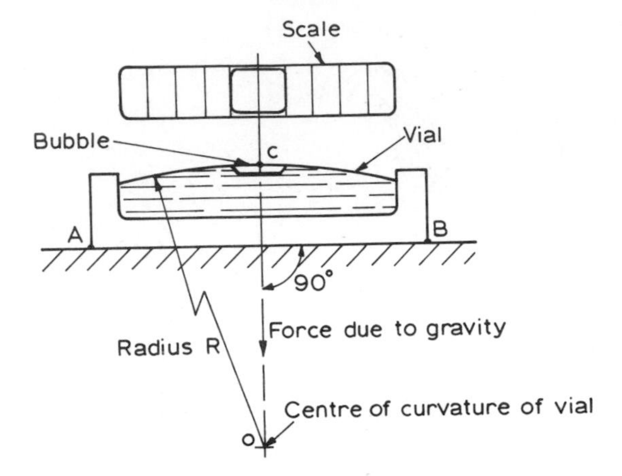

Fig. 11.7

Figure 11.8 shows the geometry of the system. S is the movement of the bubble due to the base AB moving through the angle θ from the level position XX. By similar triangles, since θ is very small,

$$\frac{h}{l} = \frac{s}{R}$$

where (h/l) is the *gradient* of the angle θ, and R is the radius of curvature of the tube in the plane of measurement.

The sensitivity is $s/\theta = s/(h/l) = R$, and is seen to depend only on R. BS 958:1968 gives recommendations regarding base sizes and sensitivities of spirit-levels.

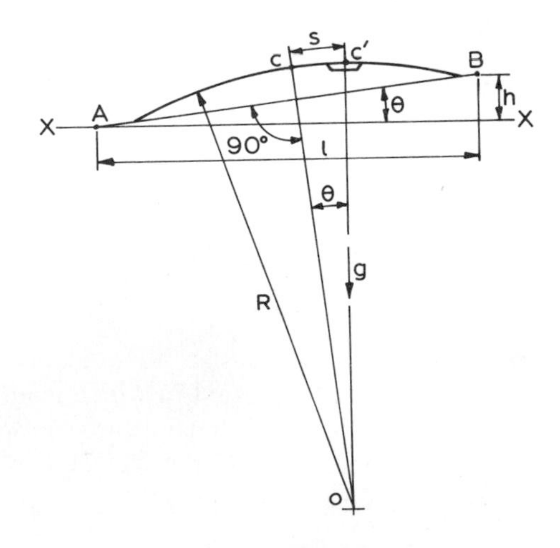

Fig. 11.8

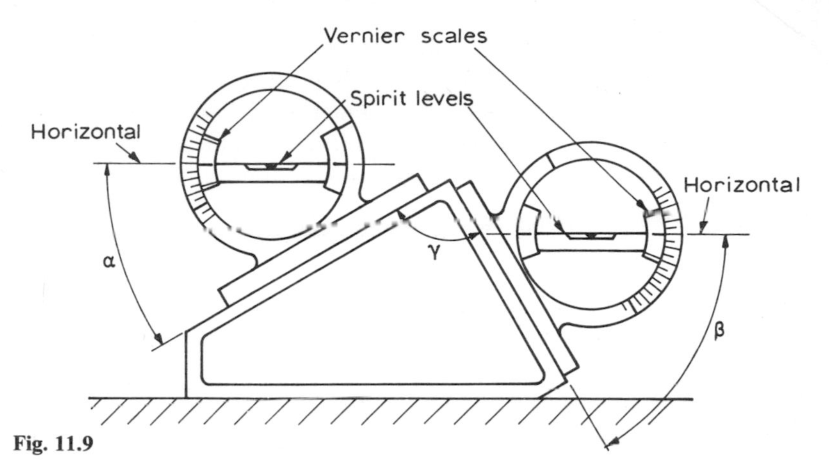

Fig. 11.9

The spirit-level is also used as a fiducial indicator, as shown in fig. 11.9, where the angle γ is found by measuring the angles α and β from the surfaces to the horizontal plane, using a clinometer. Hence

$$\gamma = 180^\circ - (\alpha + \beta)^\circ$$

The principle of the plumb-bob is adapted and refined using modern technology in the 'Talyvel' electronic level, illustrated diagrammatically in fig. 11.10. The pendulum A is supported on fine wires, and is damped by a globule of liquid. The pendulum is positioned between two inductance coils, C_1 and C_2, which form part of an inductance-bridge circuit. The bridge is balanced with the pendulum centrally positioned, when the base AB is horizontal. When the base is inclined, the pendulum moves nearer to one coil and further from the other, unbalancing the bridge. The unbalance signal is amplified and rectified, and is

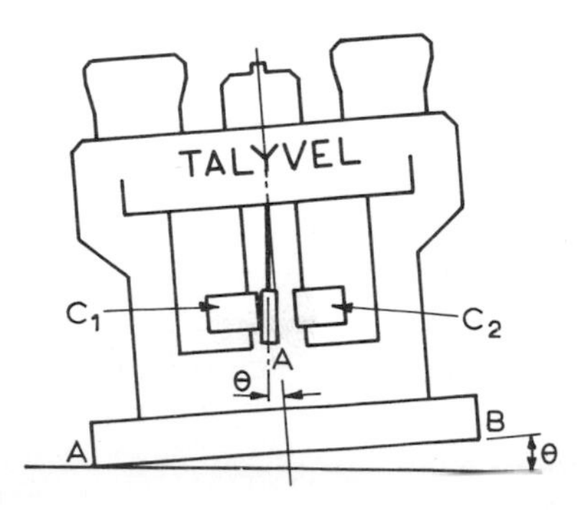

Fig. 11.10 Electronic level

indicated on the meter as an angle in minutes and seconds, or in millimetres per metre. The damping gives a settling of the pointer in about one second, and allows the instrument to be used in the presence of slight vibration. The system has a resolution of one second, a range of $\pm 2\frac{1}{4}$ degrees, and can give remote indication up to 800 metres. A range of accessories allows the instrument to be used in a wide range of applications.

11.3.2 Measurement of straightness and flatness

Straightness and flatness may be defined in mathematical terms, but a practical standard of high accuracy is not so readily available. A ray of light may be taken as a straight line, though its path may be affected by conditions of air temperature, pressure, and humidity. Over a relatively small area, the still surface of a liquid may be taken as a plane, though in fact it is part of a spherical surface.

The inaccuracy of a line or surface may be specified relative to an imaginary straight line or plane, whose position may be defined in several ways: (a) a line through the profile of a surface such that equal areas occur above and below the line, as in surface-texture measurement (section 10.5.1); or (b) a line (or plane) midway between two lines (or planes) touching the extremes of the surface, as in fig. 11.11; or (c) a line (or plane) positioned so that the sum of the squares of the distances from the line (or plane) is minimum (see example 11.5.3).

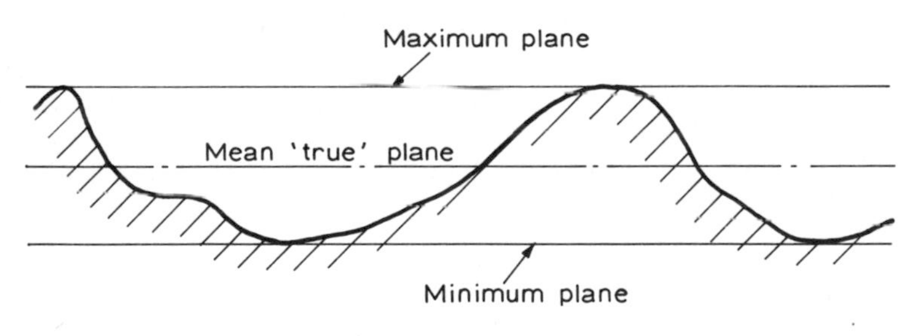

Fig. 11.11

A 'mean true plane' defined as in (b) above is used in the specification of errors of straightness. The least-squares method, however, gives a better datum line or plane, but requires some computation.

A straight line may be established by machining and finishing two straight edges AB and CD together, side by side, as in fig. 11.12(a). They are then placed edge to edge as in (b) and are corrected relative to each other, then reversed as in (c) and again corrected. Testing and correcting side by side and edge to edge is continued until no further

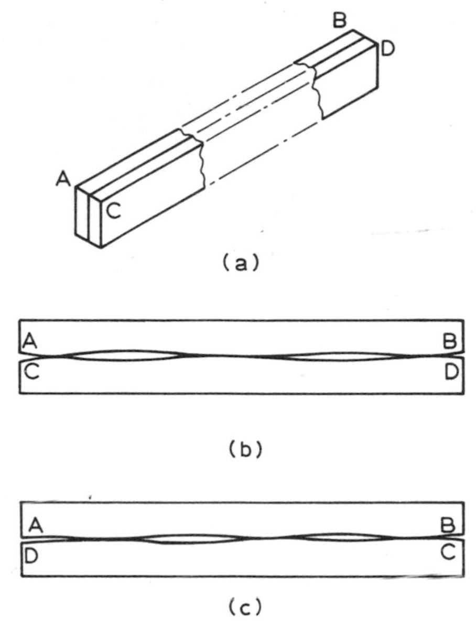

Fig. 11.12

improvement can be obtained. A similar method for surfaces involves scraping three surface-plates until each one appears flat relative to the others.

For comparing a straight edge with another edge or surface, either light methods, marking methods, or transducer methods may be used. White light is transmitted through a gap of 0·002 mm; below this, the longer wavelengths are cut off, until at 0·0005 mm only blue light is seen. The method therefore has very high discrimination. Marking methods use a thin film of marking paste; on holding the surfaces together and making slight lateral relative movement, the surfaces are revealed through the film. The discrimination of this method depends on the thickness of the film and the skill of the operator.

Electrical transducer methods are also available. In the Wayne Kerr system illustrated in fig. 11.13, each non-contacting capacitance transducer can give a signal, through a special feedback amplifier, proportional to the distance from the measured surface, which is normally electroconductive. Two such signals may be fed to sum-and-difference amplifiers to give a variety of indications.

Pneumatic sensing heads may also be used in a similar manner, but manipulation of the output signals of flow rate or pressure is not easy, and these may need to be transduced to electrical quantities.

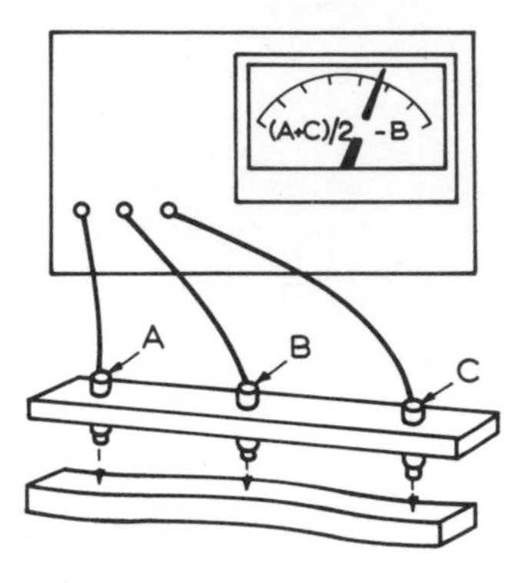

Fig. 11.13

11.3.2.1 Straightness testing. Straight edges may be held in either the horizontal position or the upright position as they are applied to another surface, as shown in fig. 11.14. If used upright, the straight edge should be supported at two points $\frac{2}{9}$ along from each end, if the instrument is more than 1 metre long; this minimises the errors due to bending of the bar under its own weight. In less precise work, the gaps between the two surfaces may be found using feeler gauges. The light method may be employed for more precise work, when a light-box may be used. Alternatively, sensitivity may be increased by using the wedge method shown in fig. 11.15.

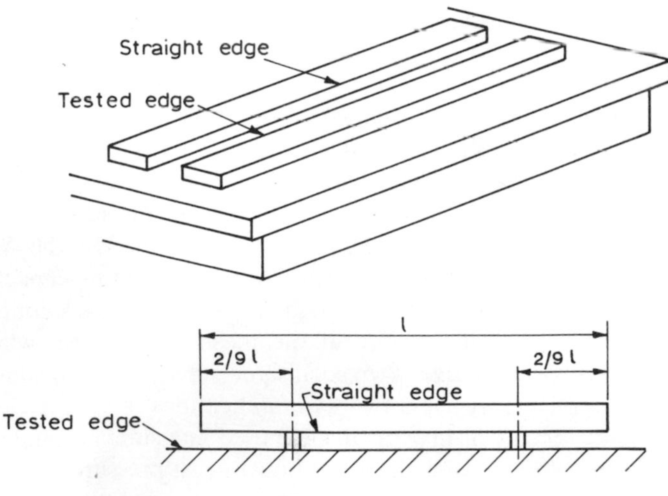

Fig. 11.14

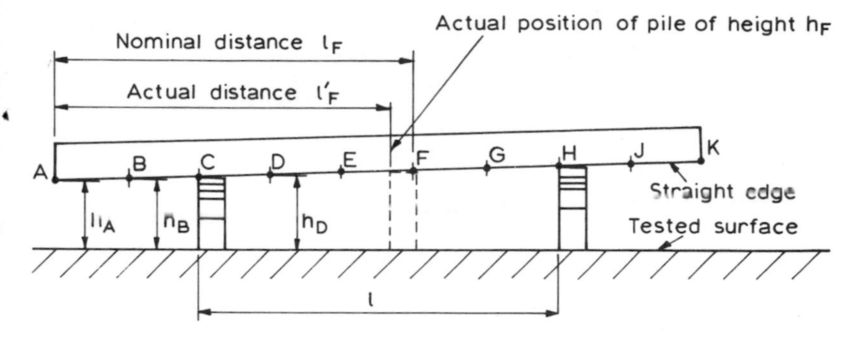

Fig. 11.15

In this method, the theoretical heights h_A, h_B, h_D, etc. are calculated from the slip-gauge-pile heights h_C and h_H and the length l. The profile of the measured surface may be found by either (a) finding pile heights to just fit at the points A, B, D, etc. and noting the difference in height from the theoretical value; or (b) sliding piles of the correct height along until contact is made with the straight edge, when the *actual* distances l_B, l_D, l_E, etc. are recorded.

Another mechanical method uses the beam comparator illustrated in fig. 11.16. The height δh of point C from the line through the points of contact A and B of the feet of the beam is measured, and the instrument is then moved so that the feet are on B and C to measure the height of D relative to BC, etc.

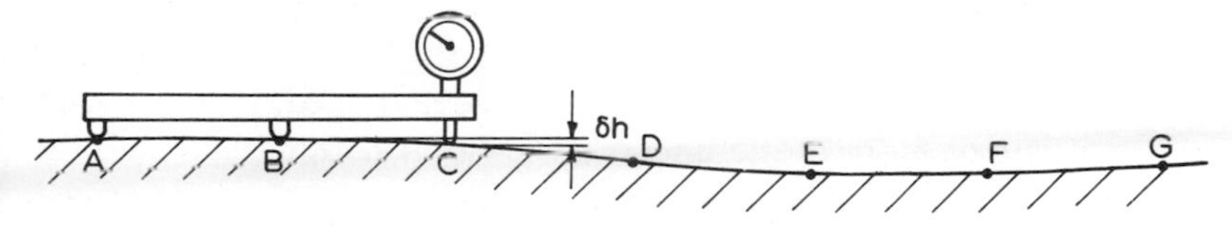

Fig. 11.16 Beam comparator

The precision level, the precision clinometer, or the Talyvel electronic level may be used in a similar method, as shown in fig. 11.17, where the rise or fall relative to either the horizontal plane or to the first pair of points (AB) may be measured.

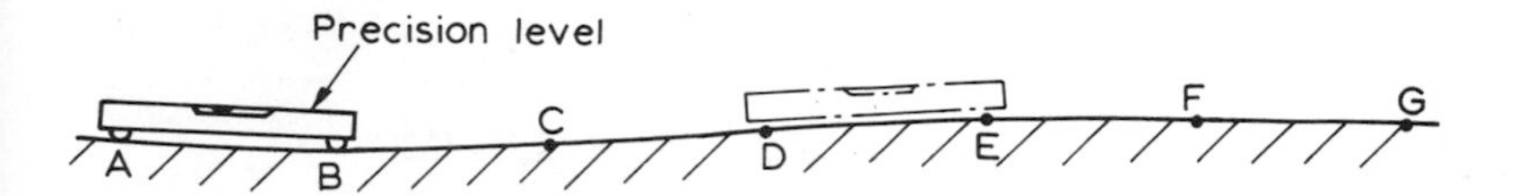

Fig. 11.17

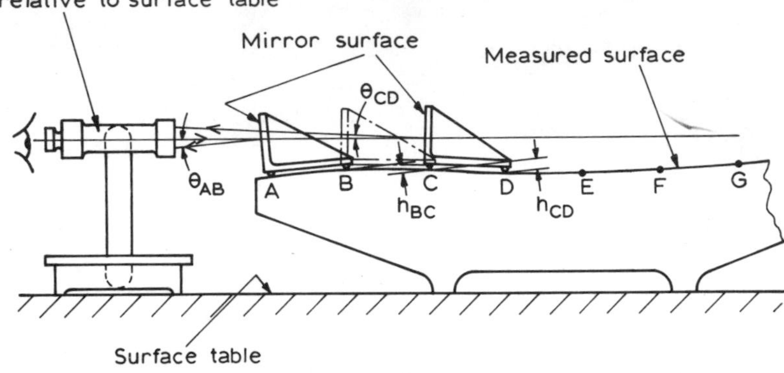

Fig. 11.18

The autocollimator may be used as shown in fig. 11.18. The *change* ($\delta\theta_B$) of the angle of reflection of the rays from the reflector position AB to position BC enables the height (h_{BC}) of point C relative to the straight line through AB to be determined. The change ($\delta\theta_D$) of angle from position AB to position CD gives the height (h_{CD}) of point D from a line through C parallel to AB. Hence the surface profile may be plotted, and errors relative to some datum line may be established as shown in example 11.5.3.

11.3.2.2 Flatness testing. The methods used for straightness testing may be extended to flatness testing by measuring the heights of points marked in a grid pattern as shown in fig. 11.19. The relative heights of points along the grid lines y_0, y_1, etc. are determined, together with those of one or more of the lines x_0, x_1, etc. A datum plane and the deviations from this plane may then be found by the methods described in references 14, 17, and 18.

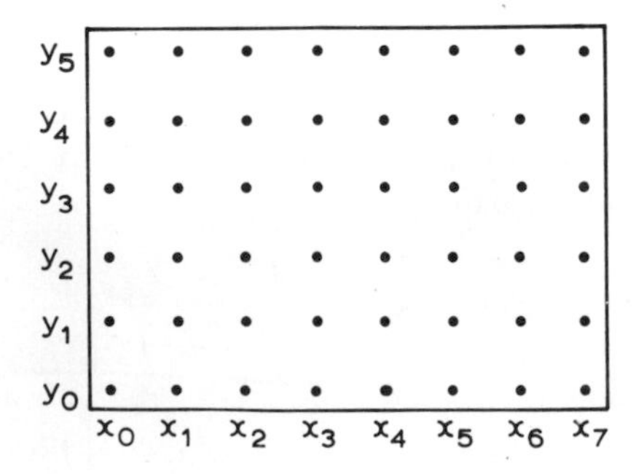

Fig. 11.19

The still surface of a liquid may be used as a reference plane as shown in fig. 11.20, when an electrical contact from the conical pointer to the conducting fluid is used as a fiducial indicator. The method is not often found to be convenient in practice, and, due to the practical difficulties, it is of doubtful accuracy. However, it is used occasionally in special applications.

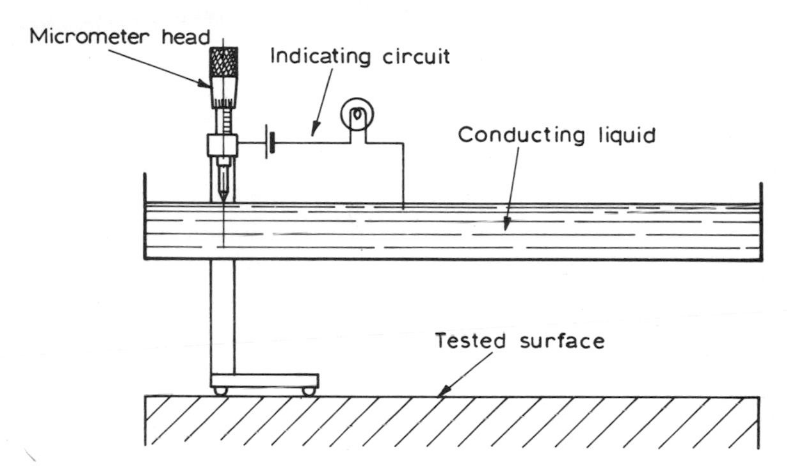

Fig. 11.20

11.3.3 Measurement of alignment

Problems of alignment arise in the manufacture, erection, and checking of structures such as machine-tool frames, aircraft and ships, etc. The alignment of a machine frame of small size may be checked using a straight edge, as shown in fig. 11.21; however, in most cases, optical methods will be preferred.

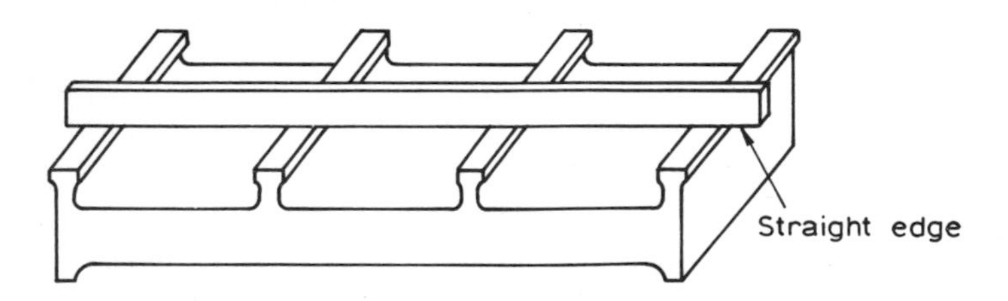

Fig. 11.21

A surveyor's level may be used to check the level and alignment of points, as shown in fig. 11.22. The instrument is adjusted by means of its spirit-level so that its line of sight is horizontal, at a convenient height

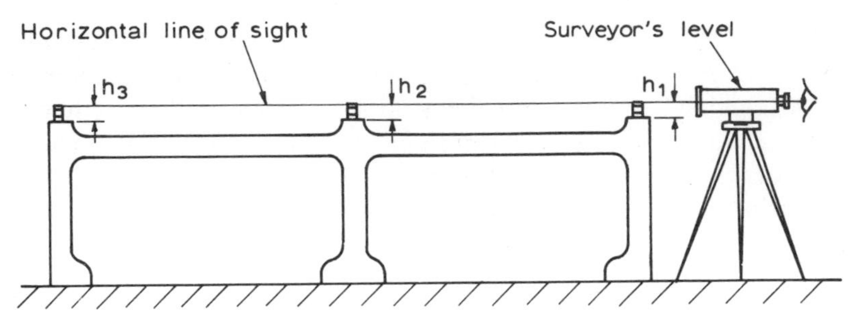

Fig. 11.22

relative to the measured surfaces. Slip-gauge piles are then wrung to each surface in turn, so that, when the level is focused on a particular pile, the top of it coincides with the horizontal cross-wire of the eye-piece. The differences of the heights h_1, h_2, and h_3 give the deviation of the surfaces from a horizontal plane, and enable the deviation from some other straight line to be determined.

The alignment telescope is a more useful tool for the engineer, and is used in conjunction with a collimator unit. Both units are contained in ground casings precisely located relative to the optical axes, enabling them to be mounted in bores by using suitable adaptors.

Figure 11.23 shows the units being used to test the alignment of bores A and B. The units are located in the bores by the adaptors (M). Light from the source (S) illuminates a graticule (G_2) and then passes through a collimating lens (C). With the telescope focused on infinity, the parallel rays from C transmit the graticule image to the telescope, where it is observed relative to the graticule G_1, and the angle θ and its direction is read off. The telescope is then focused on the graticule G_3, the now out-of-focus rays from S and G_2 providing only back lighting, and the displacement (h) of the graticule from the axis of the telescope is observed as horizontal and vertical components. The image of G_2 always appears the same size, but that of G_3 reduces in size as the distance between the units increases, giving a limit to the distance over which the displacement can be effectively found.

A further development in alignment testing involves the use of the laser, which extends the range of measurement. In this system, the collimating unit is replaced by a laser unit giving a narrow parallel light beam which has very little scatter over very large distances. Instead of a telescope, the receiving unit has light-energy sensors which can detect when the receiving unit is coaxial with the laser beam. The movements necessary to achieve this give a measure of the mis-

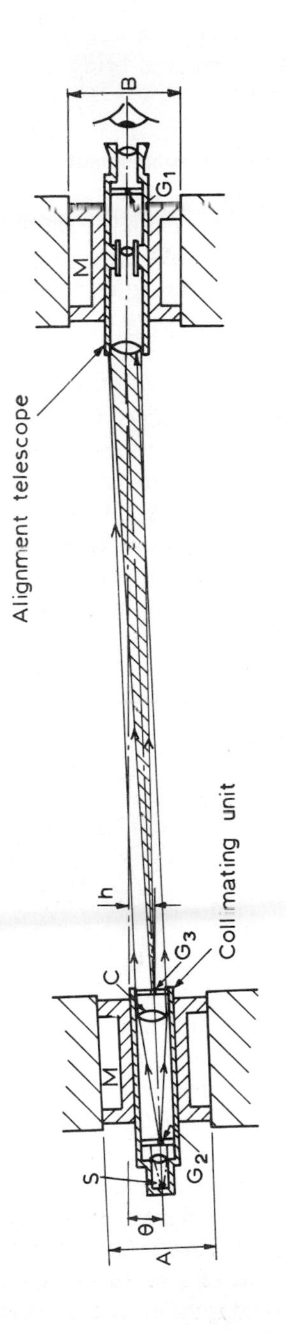

Fig. 11.23

alignment. This method has applications in the alignment of larger structures such as aircraft, ships, and bridges, and in civil-engineering work.

11.4 Measurement of angular displacement

The measurement of angular displacement is required in a large variety of engineering situations, from the small twist of a torsion-test specimen to the angular position of a large radio-telescope, and a large variety of angular-displacement transducers and systems are available.

11.4.1 General methods

The relative angular displacement (θ) of two planes of a torsion test-piece may be measured by the readings of pointers attached to the bar at two points A and B against two fixed protractor scales. A better method is to fix a scale to one point and a pointer to the other, as shown for torsion measurement in fig. 11.24, to give a direct indication of relative angular displacement. For the very small elastic twist of shorter bars, more amplification of the movement is necessary. In many applications, the nearly straight-line relative displacement of two arms of larger radius is used to measure θ by means of a linear-displacement transducer.

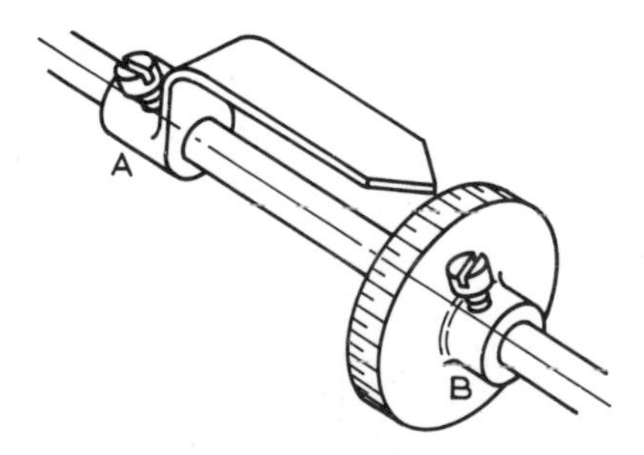

Fig. 11.24

Figure 11.25 illustrates the use of air-jet sensing to measure small angular displacements in this way. For small angles, $x = R\theta$, where θ is in radians; hence $\delta\theta = \delta x/R$ and, since $\delta x = K\,\delta P$ over a small range, $\delta\theta = \delta P \times (K/R)$ for this range.

Optical methods, with their potentiality for large amplifications, are useful in many applications. Figure 2.32 illustrates a method of measuring the angular displacement of a beam using a light ray and mirror, and the method is readily adapted to the measurement of small twist values.

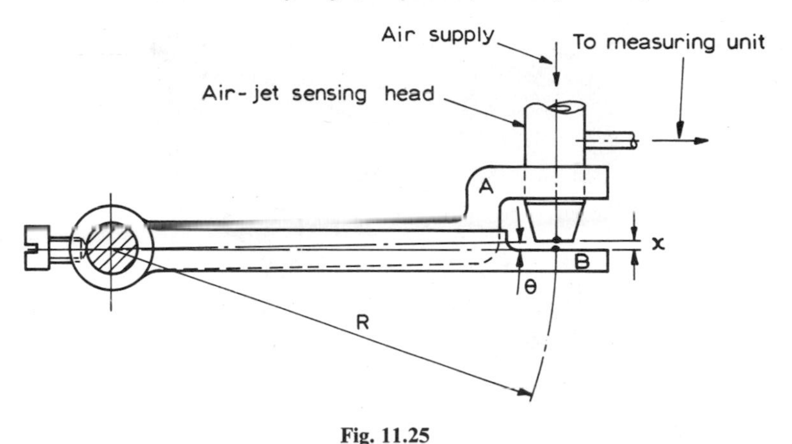

Fig. 11.25

11.4.2 *Electrical methods–analogue*

Electrical effects provide the greatest variety of methods for angular-displacement measurement, some of which are capable of measuring rapidly changing angles such as occur in vibration (see Chapter 14).

The voltage-dividing potentiometer of section 4.2.1 is adapted as indicated in fig. 4.4 for the measurement of angular displacement. Using a resistance wire wound on a single ring, i.e. a toroidal coil, displacements up to nearly 360° may be measured. If the wire is wound on a helical former, as in fig. 11.26, displacements up to many complete turns may be measured. The wiper has to follow a helical path also, and this may be provided by, for example, a screw of the same pitch as the helix, on the end of the spindle.

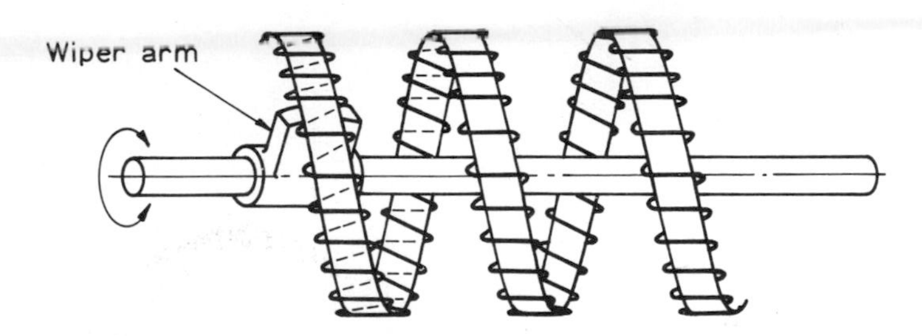

Fig. 11.26 Helical potentiometer

A useful type of angular-displacement transducer uses the variation of the capacitance between two parallel sets of metal plates as their overlap varies due to rotation [see fig. 2.20(a)]. A typical unit used as an angle transmitter in instruments is shown in fig. 11.27. The

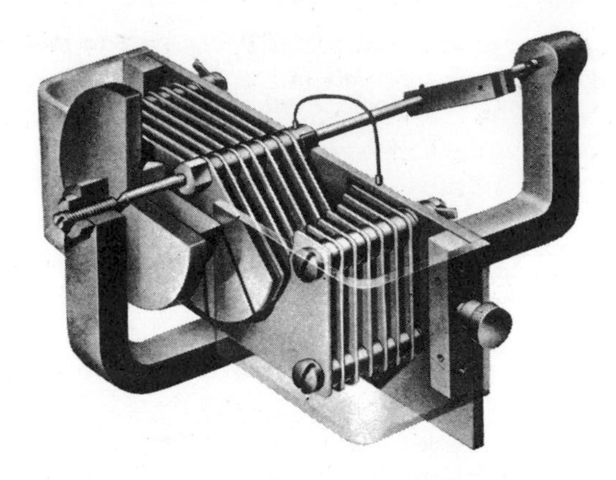

Fig. 11.27 The Sigma transmitter

capacitance variation unbalances an a.c. bridge (see fig. 4.14). A typical system arrangement is shown in fig. 11.28. In this, both arms of the bridge are active; the capacitance of one increases as the other decreases, as rotation occurs, thereby doubling the sensitivity of the transducer plus bridge. Angular-displacement transducers of this kind are made to high standards of accuracy and precision and, having infinite resolution, for use in high-grade instrument applications such as aircraft navigational systems.

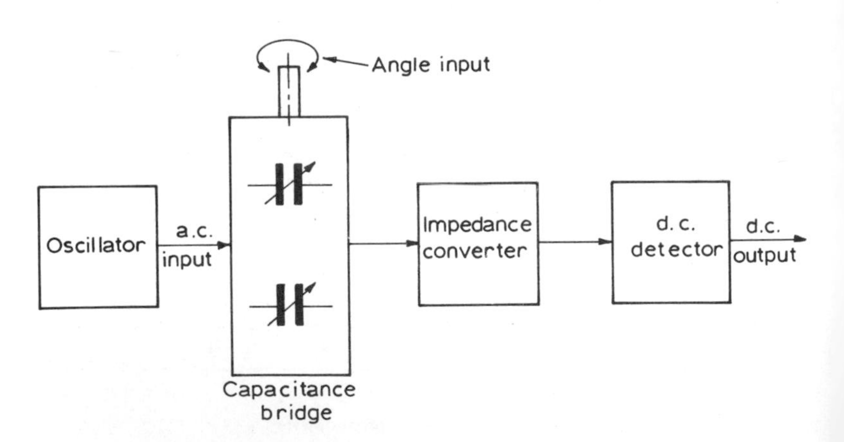

Fig. 11.28

In both the potentiometric and capacitive angle transducers, the output is normally required to be linear with input, e.g. $\delta\theta = K\,\delta R$ or $\delta\theta = K\,\delta V_{max}$. However, these devices may be modified so that the output is proportional to a function of θ, e.g. $\sin\theta = KR$ or $\cos\theta = KV_{max}$ etc., to suit particular applications. The capacitive type has the advantage of being stepless, unlike the potentiometric type, and has low operating torque and negligible wear.

Many angle transducers use the variation of reluctance and inductance in various arrangements to unbalance an a.c. bridge. The device illustrated in fig. 11.29 has a rotor (R) and a stator (S) made of magnetic material. The coils A, B, C, and D are wound on pole pieces on the stator, and are connected into a bridge configuration energised across UT by an a.c. potential $v_i = V\cos\omega t$. When $\theta = 0$, the bridge is balanced, giving $v_0 = 0$ and $I_1 = I_2$, and the flux Φ at each pole is the same. For a small angular displacement (θ) of the rotor, the reluctance at the gaps A and D is increased, whilst that at B and C is reduced, altering the flux values. It may be shown (ref. 4) that the output p.d. across VW is given by $v_0 = K\theta \times V\cos\omega t$.

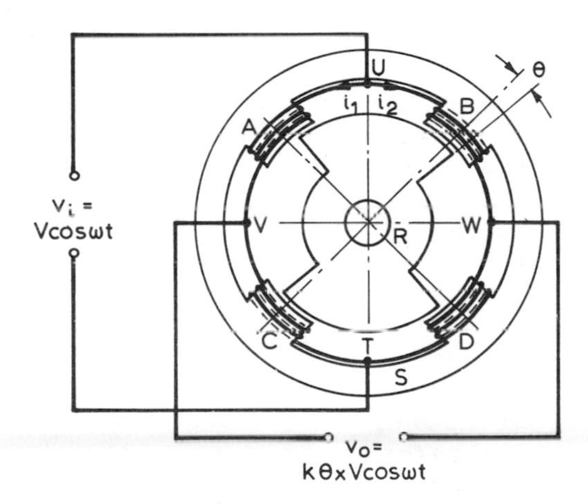

Fig. 11.29

Larger displacements may be measured using the system shown schematically in fig. 11.30. It consists of a transmitter and receiver, both of which have the general configuration of a six-pole electric motor. The distribution of flux induced in the transmitter stator windings depends on the angular position of its rotor (R_1), and this distribution is reproduced in the stator windings of the receiver, tending to turn its rotor (R_2) to the same angle. When $\theta_R = \theta_T$, v_0 is zero. The value of v_0 is a function of $\theta_R - \theta_T$, and the system may be used in a variety of ways for absolute or differential angle measurement, and in automatic control systems. Further examples of the many angle

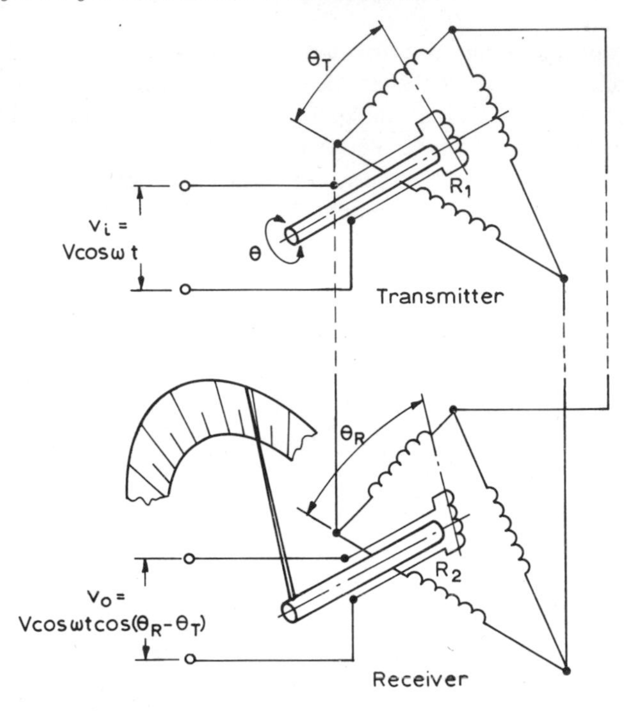

Fig. 11.30

transducers using changes of reluctance and inductance are shown in references 2, 4, and 5.

The linear variable-differential transducer described in section 10.4.2 is also adapted to the measurement of angular displacement.

11.4.3 Electrical methods–digital

The methods of angular-displacement measurement described in section 11.4.2 are all of analogue type, giving a voltage or voltage amplitude as a measure of displacement. In many systems a digital output is preferred, often for the reason that this may be readily handled by digital computers. A basic idea is shown in fig. 11.31, which shows a slotted metal disc making and breaking an electrical circuit to an electromechanical counter. The signal the counter receives is either (a) current flowing or (b) no current flowing, and movement at the plunger from slot to land to slot will actuate the counter to move one digit. If 100 slots are cut equally round the periphery, then the counter will indicate angular movement for any number of rotations, with a

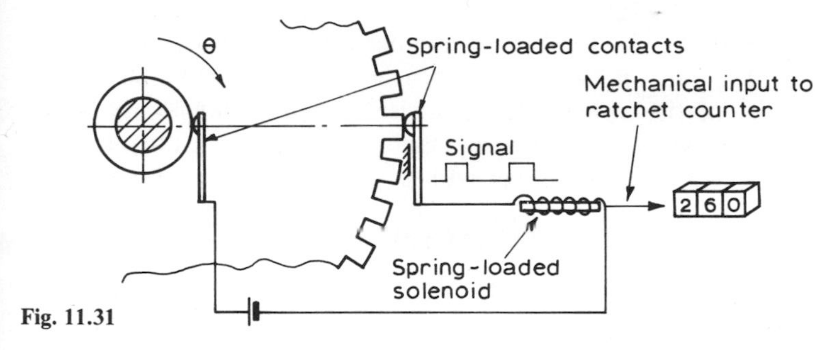

Fig. 11.31

discrimination of 0·01 of a revolution. However, in this crude form, it will not count back when the direction of rotation is reversed.

In electrical systems, it is convenient to have 'OFF' referred to as 0 and 'ON' referred to as 1, and to use these in a binary code to indicate number. Thus the binary number 1011 is the equivalent of the denary number $1 \times 2^3 + 0 \times 2^2 + 1 \times 2^1 + 1 \times 2^0 = 8 + 0 + 2 + 1 = 11$ (which is $1 \times 10^1 + 1 \times 10^0$). Although this appears to be a more awkward way of using the denary number 11, it is much more suitable for processing in digital systems. The binary-coded disc shown in fig. 11.32 has four brushes, B_1 to B_4, and is divided into 16 sectors, numbered 0 to 15. The shaded areas are conducting, maintained at a given potential, and the white areas are non-conducting. A p.d. on brush B_1 indicates a 1 in the 2^0 column, on brush B_2 a 1 in the 2^1 column, and so on.

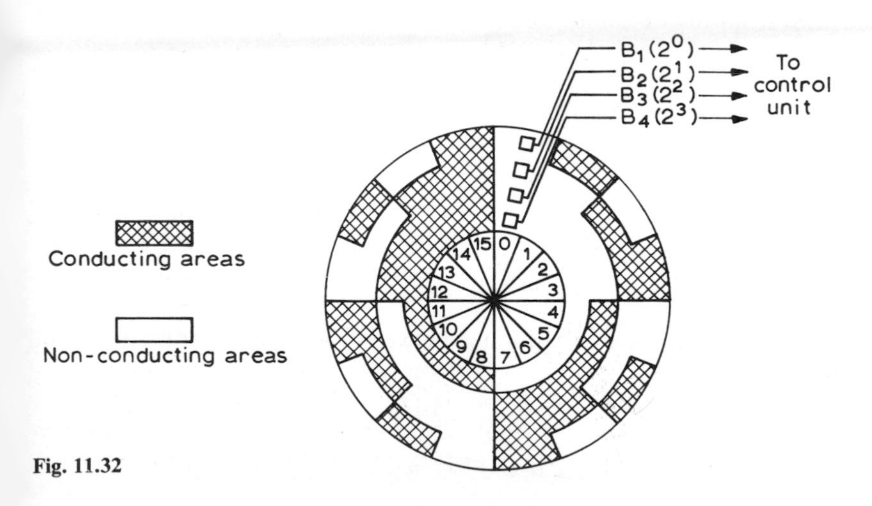

Fig. 11.32

It is seen that a unique binary number is given by the presence or absence of a p.d. on each of the brushes for any particular sector, irrespective of the approach, clockwise or anticlockwise, provided that all the brushes are making contact in that sector. Unfortunately, when one brush makes contact with a sector before the others, as is bound to happen during disc rotation, a mixture of the codes of the two sectors is signalled on the digital circuits, giving a false indication of sector. This difficulty is overcome in two ways.

a) Two brushes may be used for each binary digit. When one set of brushes is over the transition line from one sector to the next, the brushes are disconnected by additional circuitry, and the second set, in the middle of the sector, is operative.

b) An alternative number system, known as the cyclic code or Gray code, may be used, in which successive numbers differ from each other by only one digit column, and the error cannot be greater than one sector. References 2, 12, and 40 discuss digital transducers in more detail.

The Moiré-fringe method of measuring linear displacement using diffraction gratings, described in section 10.4.3, may also be adapted to measure angular displacement. In this application the gratings are ruled radially.

11.5 Worked examples

11.5.1 A twelve-sided reference polygon was calibrated by mounting it on a rotary table and viewing adjacent faces with two autocollimators whose axes were positioned at angles very nearly 30° apart, as shown in fig. 11.33. The readings of the angular displacement R_1 and R_2 of the faces from the nominal positions were recorded for each of the twelve corners thus:

Corner	A	B	C	D	E	F
R_1 (sec)	+ 4·2	−16·3	−11·3	−10·0	−22·4	−9·3
R_2 (sec)	+ 6·0	0·0	0·0	+2·0	−1·0	+10·0

Corner	G	H	J	K	L	M
R_1 (sec)	−11·8	−11·8	−15·2	0·0	−10·5	−8·5
R_2 (sec)	0·0	0·0	−3·0	0·0	0·0	0·0

Determine the error in the angle between each pair of adjacent faces.

It is necessary to determine the value of the angle θ.

Angle A $= \theta + R_{2A} - R_{1A}$, angle B $= \theta + R_{2B} - R_{1B}$, etc.

hence $\quad A + B + \ldots M = 360° = 12\theta + \sum R_2 - \sum R_1$

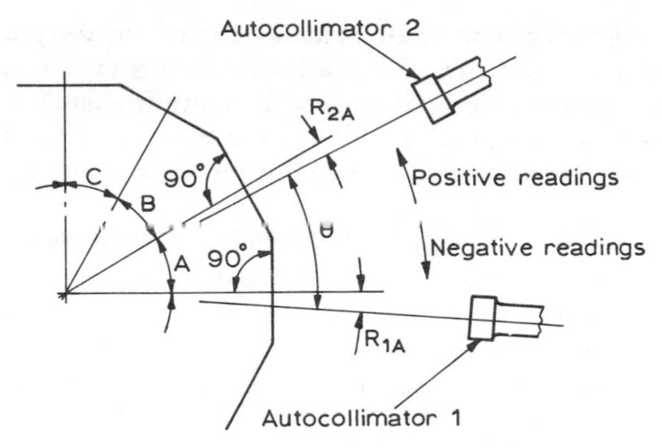

Fig. 11.33

and
$$\theta = (360° + \sum R_1 - \sum R_2)/12$$
$$= 30° + (\sum R_1 - \sum R_2)/12$$

Tabulating the above values to find $\sum R_1$ and $\sum R_2$,

Corner	R_1 (sec)	R_2 (sec)	*Error* $= -11{\cdot}4 - R_1 + R_2$ (sec)	
A	4·2	6·0	−9·6	
B	−16·3	0·0		4·9
C	−11·3	0·0	−0·1	
D	−10·0	2·0		0·6
E	−22·4	−1·0		10·0
F	−9·3	10·0		7·9
G	−11·8	0·0		0·4
H	−11·8	0·0		0·4
J	−15·2	−3·0		0·8
K	0·0	0·0	−11·4	
L	−10·5	0·0	−0·9	
M	−8·5	0·0	−2·9	
	$\sum R_1 =$ −122·9	$\sum R_2 =$ 14·0	−24·9	25·0

Hence
$$\theta = 30° + (-122{\cdot}9'' - 14{\cdot}0'')/12$$
$$= 30° - 11{\cdot}4''$$
$$\therefore \text{ angle A} = 30° - 11{\cdot}4'' + 6{\cdot}0'' - 4{\cdot}2''$$
$$= 30° - 9{\cdot}6''$$

The error values are tabulated in columns 4 and 5, the maximum error being −11·4″ at corner K. The check additions of columns 4 and 5 show a discrepancy of 0·1 sec, this being due to rounding off in the calculation for θ.

The error values are high for a reference polygon, see 11.6.4.

11.5.2 The angle comparator of fig. 11.34 uses the back-pressure pneumatic sensing system with manometer indication, and has a sensitivity (K_1) of 4000, i.e. 40 mm movement of the liquid surface occurs for a difference of 1/100 mm in diameter between the component and the master gauge used in setting the system at diameters d_1 and d_2.

Determine the angle error ($\delta\theta$) of the component when the angle (α) between the levels in the manometers is 30°, and also the approximate angle sensitivity (K_2) of the system.

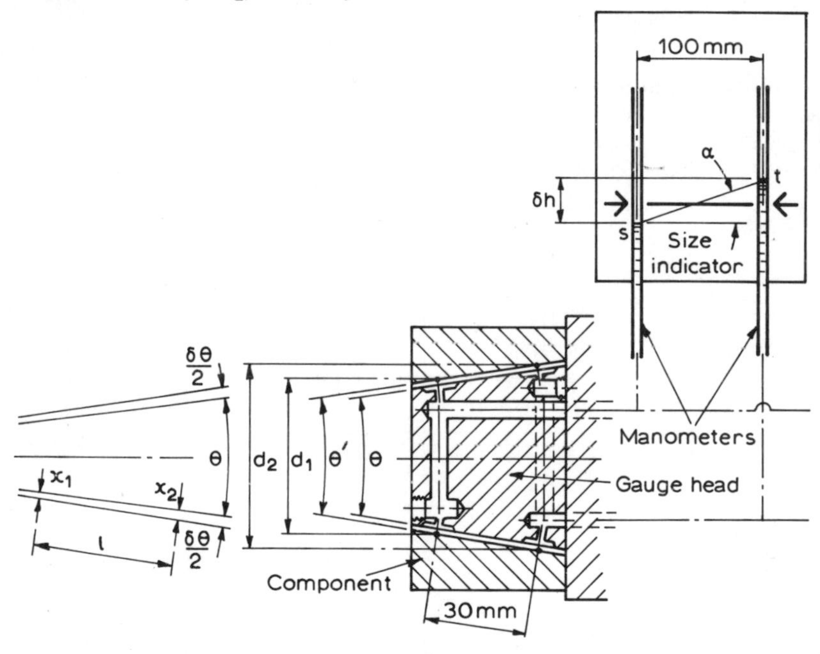

Fig. 11.34

$$\delta h = 100 \tan 30°$$
$$= 57{\cdot}74 \text{ mm}$$

When $d_2 - d_1$ is the same as on the master gauge, $\delta h = 0$. The *error* $[\delta(d_2 - d_1)]$ in the *difference* between d_1 and d_2 is indicated by δh.

Hence $\delta(d_2 - d_1) = \delta h/K_1$. For small values of θ, the difference (x) between the nominal diameter and the actual diameter is very nearly the same as the radial difference, hence

$$\begin{aligned}\delta\theta/2 &= (x_2 - x_1)/l \\ &= \delta(d_2 - d_1)/2l \\ \therefore\quad \delta\theta &= \delta(d_2 - d_1)/l \\ &= \delta h/K_1 l \\ &= 57{\cdot}74/(4000 \times 30) \\ &= 0{\cdot}00048 \text{ rad} \\ &\approx 1{\cdot}7 \text{ minutes}\end{aligned}$$

$$\text{Angle sensitivity } (K_2) = \frac{\text{output angle } (\alpha)}{\text{input angle error } (\delta\theta)} = \frac{30 \times 60}{1{\cdot}7} \approx 1059$$

$$\text{or angle sensitivity } (K_2') = \frac{\delta h}{\delta\theta} \approx \frac{57{\cdot}74}{1{\cdot}7}$$

$$\approx 34 \text{ mm/sec}$$

11.5.3 A straightness test on a measuring-machine bed was made using an autocollimator and reflector as shown in fig. 11.18. Each minute of arc increase in the angle observed corresponds to a rise of 25·4 μm of the front of the reflector relative to the rear. From the angle readings tabulated in column 2 below, construct a profile graph of the

Points	*Angle reading* θ (minutes)	$\delta\theta = \theta - \theta_{AB}$ (min)	*Rise* δy (μm)	*Total rise y relative to AB* (μm)
A	—	—	—	
B	30·83	—	—	
C	31·14	+0·31	+7·87	+7·87
D	31·24	+0·41	+10·40	+18·27
E	31·08	+0·25	+6·35	+24·62
F	31·10	+0·27	+6·86	+31·48
G	31·23	+0·40	+10·17	+41·65
H	31·12	+0·29	+7·36	+49·01
J	31·35	+0·52	+13·20	+62·21
K	31·48	+0·65	+16·50	+78·71
L	31·47	+0·64	+16·25	+94·96
M	31·42	+0·59	+15·00	+109·96
				$\sum y =$ +518·74

surface relative to the initial points AB, and state the maximum deviation from a line through the end points. Determine the least-squares line, show it on the graph, and state the maximum deviation from it.

From the graph, fig. 11.35, of the deviation of the surface from the AB reference line, it is seen that the maximum deviation from a line joining the ends is -21 μm at point H, and the surface is higher at the ends than at the middle. However, it may be preferred to give the deviation from some mean line through the surface.

The least-squares method (see ref. 17) gives a line from which the sum of the squares of the errors is minimum. The equation of the line is given by $y = mx + c$, and it passes through the mean point $\bar{x}$, $\bar{y}$. The slope is given by

$$m = \sum (x_m y_m) / \sum (x_m)^2$$

where $x_m = x - \bar{x}$, and $y_m = y - \bar{y}$.

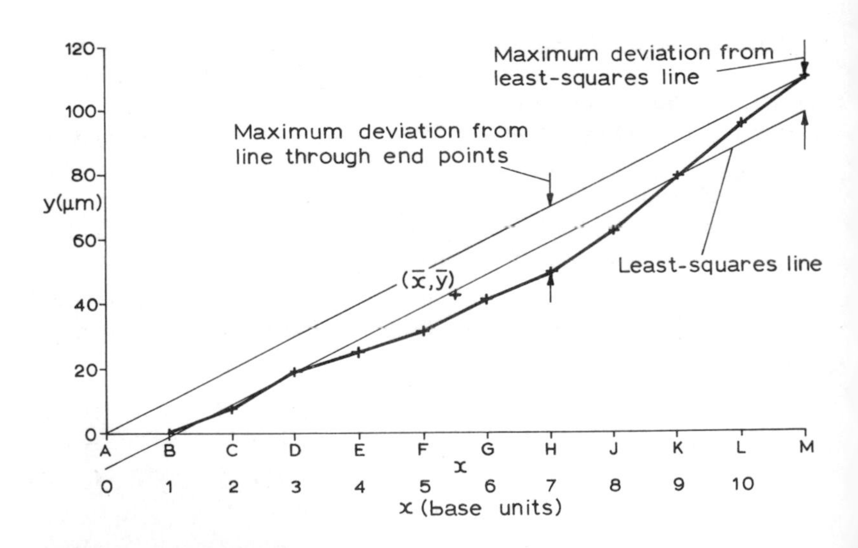

Fig. 11.35

Taking the length of the reflector base as unity, $\bar{x} = 5{\cdot}5$ units.

$$\bar{y} = \sum y/N = 518{\cdot}74/12 = 43{\cdot}26\ \mu\text{m}$$

Reference	$x_m = x - \bar{x}$		$y_m = y - \bar{y}$		$x_m y_m$		x_m^2
	+	−	+	−	+	−	
A		5·5		43·26	237·9		30·25
B		4·5		43·26	194·7		20·25
C		3·5		35·39	123·9		12·25
D		2·5		24·99	62·5		6·25
E		1·5		18·64	28·0		2·25
F		0·5		11·78	5·9		0·25
G	0·5			1·61		0·8	0·25
H	1·5		5·75		8·6		2·25
J	2·5		18·95		47·4		6·25
K	3·5		35·45		124·1		12·25
L	4·5		51·70		232·7		20·25
M	5·5		66·70		366·9		30·25
					1432·6	0·8	
					$\sum (x_m y_m) = 1431{\cdot}8$		$\sum (x_m)^2 = 143{\cdot}0$

Hence $m = 1432/143 = 10{\cdot}01$ μm/base unit.

The least-squares line passes through the point $x = 12$, $y = 43{\cdot}26 + 5{\cdot}5 \times 10{\cdot}01 = 98{\cdot}3$ μm.

From the graph, it is seen that the maximum deviation from the least-squares line is +12 μm at point M, with a similar deviation at A.

11.6 Tutorial and practical work

11.6.1 Examine the line-standard measuring instruments available to you, e.g. vernier bevel protractors, clinometers, etc., and compare their precision and accuracy from available information and observation.

11.6.2 For the sine-bar arrangement shown in fig. 11.4(b), discuss the possible effects of (a) lack of circularity of the cylinders, (b) lack of alignment of the surfaces XX, (c) lack of parallelism of the surfaces YY, (d) inaccuracy of the 90° angles between the faces XY.

11.6.3 In the method of angle comparison illustrated in fig. 11.6, describe how you could ensure that no movement of the auto-collimator occurred relative to the surface-plate between taking readings of angle for the reference polygon and the angle-plate.

11.6.4 In the calibration of a polygon as in example 11.5.1, discuss the effect of (a) the centre of the polygon not coinciding with the axis of the table, (b) the intersection of the axes of the autocollimators not coinciding with the axis of rotation or with the centre of the polygon.

11.6.5 The angle values in the calibration of the polygon in example 11.5.1 were observed during a student experiment. Discuss possible reasons for error in the measured values, which may have lead to error values of angle higher than the actual ones.

11.6.6 For the pneumatic angle comparator of example 11.5.2 (fig. 11.34), construct a shaded diagram on graph paper, indicating the limits of the line *st* through the liquid surfaces on the indicator manometers, if the mean diameter of the conical bore of the component has limits of $\pm 0{\cdot}01$ mm and the angle has limits $\pm 10'$.

11.6.7 For the pneumatic angle comparator of example 11.5.2, investigate the error in angle measurement due to using the approximation $x_2 - x_1 \approx (d_2 - d_1)/2$ for components having larger θ.

11.6.8 Sketch a system using air-jet sensing to measure the straightness and thickness of bars 600 mm long, and describe how it would be used.

11.6.9 Adjustment is usually provided on a spirit-level by a fine-pitch screw which moves one end of the vial relative to the base. Describe a procedure using a precision surface-plate to adjust the instrument to a correct zero.

11.6.10 Discuss the reason for taking the differences of angle from the *initial angle* when using the autocollimator for straightness and flatness testing as in example 11.5.3, and compare the method with the tabulations required when using the spirit-level and beam-comparator instruments.

11.6.11 The rails alongside a water tank for ship-model testing are to be parallel to the water surface. If they are 200 m long, suggest a method of measuring the correct height of points along the rails *before* the tank is filled with water.

11.6.12 Describe a stroboscopic method of measuring the twist of a rotating shaft.

11.6.13 Suggest a method of making the counting of the digital angle transducer of fig. 11.31 reversible.

11.6.14 Write down the binary numbers from 0 to 15, and check that these are represented by the sector contacts shown in fig. 11.32. How many sectors would be possible using five, six, or seven brushes?

11.6.15 Determine the greatest angular-displacement error which could be indicated on the binary-coded disc of fig. 11.32 due to brushes not all contacting in the same sector.

11.6.16 Discuss the use of angle transducers in conjunction with a leadscrew to indicate precisely the position of a machine-tool carriage.

11.6.17 Suggest methods of calibrating angular-displacement transducers with reference to line and end standards of angle.

11.6.18 Draw the bridge-circuit diagrams for the systems shown in figs 11.28 and 11.29.

11.6.19 Sketch an arrangement for using an LVDT to measure small angular displacements.

11.6.20 Sketch an optical method of measuring relative twist during a torsion test on a metal bar.

11.7 Exercises

11.7.1 Using eqn 11.1, construct a graph showing the actual error ($\delta\theta$) in the angle setting of a sine bar where the slant length (l) is 200 mm and the accuracy of the determination of this and of the vertical height (h) is $\pm 0{\cdot}01\%$. Calculate values at 10° intervals from 0 to 60°, and state your conclusions.

11.7.2 The straightness of a surface along a line was tested using a straight edge by the wedge method shown in fig. 11.15. The slip-gauge piles supporting the straight edge at C and H were 10·250 and 10·300 mm respectively, and it was found that a pile 10·232 mm high would just pass under point A. The calculated slip-gauge piles were found to just touch the straight edge at the following positions.

Point	B	D	E	F	G	J.	K
Distance from A (mm)	40	120	160	200	240	320	360
Distance of contact point (mm) from nominal position	−4	+4	+8	+9	+6	−6	−8

Determine the sensitivity of the arrangement, and construct a graph of distance between surfaces against distance from A. Find the maximum deviation from (a) a line through CH, (b) a line through AK.

[4000, −0·003 mm at K, +0·005 mm at F]

11.7.3 BS 958 : 1968 recommends spirit-levels of 200 mm base length, with one scale division indicating a change of level of 0·0025 mm in 100 mm, and of 500 mm base length indicating a change of 0·005 mm per 100 mm per scale division.

Calculate the radius of curvature (R) of the vial in each case, if the length between scale divisions along the curvature is 3 mm, and also the sensitivity in terms of seconds of arc per scale division. If the reading of the bubble position can be made to a fifth of a division, determine the discrimination of each type.

[120 m, 60 m, 5·2″, 10·3″, 0·001 mm, 0·002 mm]

11.7.4 A spirit-level having a base length of 200 mm was used as shown in fig. 11.17 to measure the straightness of a machine bed. The following readings of rise and fall were recorded.

	A	B	C	D	E
rise (mm)	0	+0·020	+0·017	+0·000	−0·015
	F	G	H	J	K
	+0·025	−0·010	+0·022	+0·005	−0·017

Plot the cumulative values against distance along the bed, and hence determine the position and magnitude of the maximum deviation from a straight line joining the two end points. [+0·02 mm at J]

11.7.5 A pneumatic angle-transducer similar to that shown in fig. 11.25 uses the sensing head of example 10.6.4, the radius (R) to the jet being 25 mm. Calculate (a) the range of measurement (θ) in minutes of arc, (b) the sensitivity in N/m^2 per minute of arc.

[3·43 minutes, −11 630 N/m^2 minute]

12

Strain and Stress Measurement

12.1 Strain and stress measurement

Direct, i.e. tensile or compressive, strain may be regarded as the change of length (δx) due to an applied force. *Unit* strain (ϵ), which is commonly referred to as strain, is the change of length per unit of *original* length (x_0).

i.e.

$$\epsilon = \delta x / x_0 \qquad 12.1$$

However, in some work on the properties of materials, particularly for larger strain values, a *natural* or *logarithmic* strain (ϵ') is used, defined as

$$d\epsilon' = dx/x \qquad 12.2$$

where x is the *instantaneous* length.

$$\therefore \quad \epsilon' = \int_{x_0}^{x} dx/x = \ln(x/x_0) = \ln\left(\frac{x_0 + \delta x}{x_0}\right) \qquad 12.3$$

Direct stress (σ) and shear stress (τ) are defined as force per unit area of material. Their values may be found by photoelastic methods, using models made from translucent materials having particular stress–optical properties. For other materials, stress is usually calculated from measured extension, compression, deformation, or strain values, using theoretical stress–strain relationships.

12.1.1 Stress–strain relationships

Where simple tensile or compressive loading is applied to a member, the stress may be calculated from the observed strain using the modulus of elasticity (E) value for the material, provided its proportional-limit value has not been exceeded. Thus direct stress = direct strain × modulus of elasticity

or $$\sigma = \epsilon \times E \qquad 12.4$$

However, attention must be paid to the method of applying the force to the material, and also to changes of section which occur, since these may alter the pattern of stress and strain considerably. In fig. 12.1(a), the tensile stress in the bar will be uniform across the section, over the gauge-length AB, i.e. the shaded area, *only* if the pins and holes by which the force is applied to the bar are far enough away. In (b), the tensile stress across the section YY will vary sharply across the section. The mean stress is found from $\sigma_{YY} = F/A_{YY}$, but the stress at the wall of the hole may be about three times this value, (see examples 12.5.4 and 12.5.6). The effect is referred to as *stress concentration*.

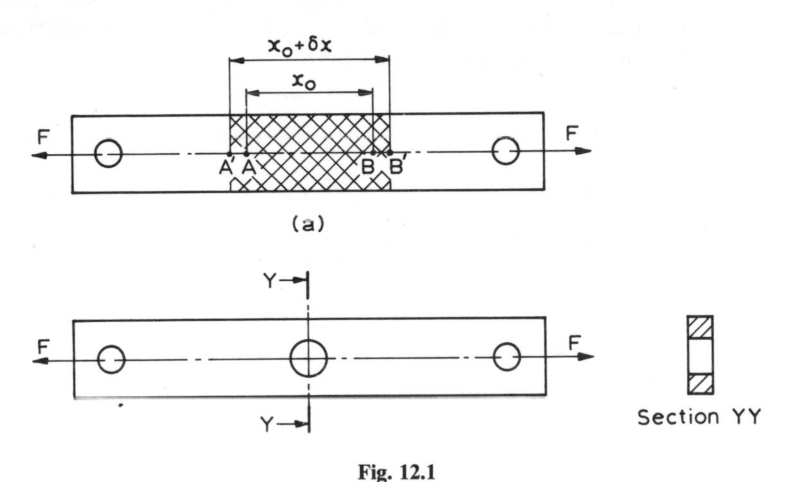

Fig. 12.1

In addition to the strain (ϵ_x) occurring in the direction of the applied force (F_x), strains (ϵ_y, ϵ_z) are found to occur in the lateral directions as shown in fig. 12.2.

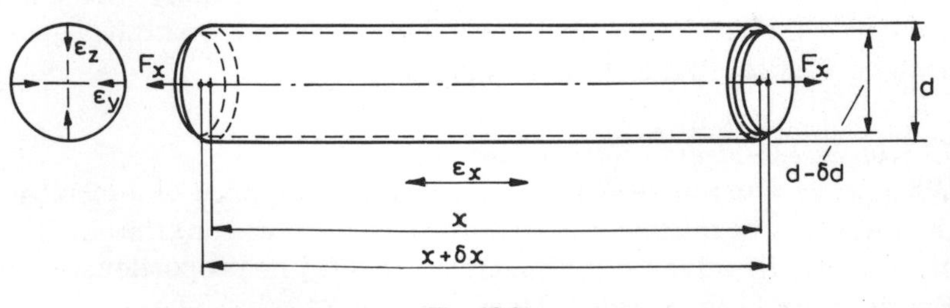

Fig. 12.2

The strains are related thus:

$$\epsilon_y = -\nu\epsilon_x, \ \epsilon_z = -\nu\epsilon_x \qquad 12.5$$

where ν is Poisson's ratio, approximately 0·3 for metals.

When forces are applied to a body in two or three mutually perpendicular directions, or when a couple is applied in addition to a force or forces, then a two- or three-dimensional stress–strain field results. However, strain-measuring instruments are usually mounted on a *free* surface, i.e. one that has no stress normal to it, and hence that surface is part of a one- or two-dimensional stress–strain field. A one-dimensional system is shown in figs 12.1 and 12.2, and for this system $\sigma_x = \epsilon_x \times E$ for strain within the proportional limit.

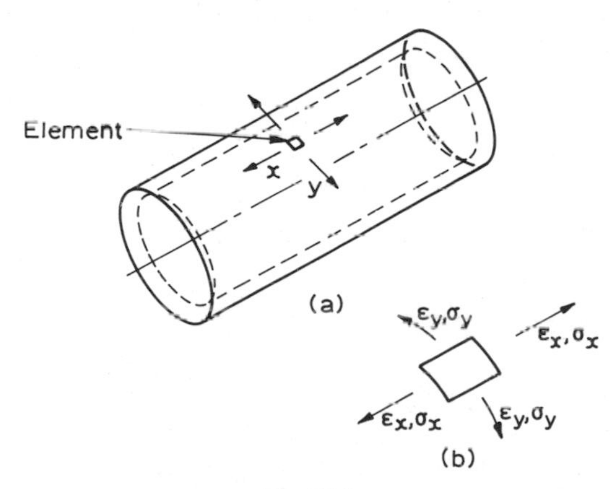

Fig. 12.3

Figure 12.3(a) shows a cylindrical vessel subjected to internal pressure. In (b) are shown the stresses on a very small element of metal at the surface. The hoop stress is σ_y and the longitudinal stress, caused by pressure on the end plates, is σ_x. The longitudinal stress (σ_x) causes a lateral strain ($-\nu\sigma_x/E$) in the y direction, and the hoop stress (σ_y) has a similar effect in the x direction; hence

$$\epsilon_x = (\sigma_x - \nu\sigma_y)/E \quad \text{and} \quad \epsilon_y = (\sigma_y - \nu\sigma_x)/E \qquad 12.6$$

Solving these as simultaneous equations gives

$$\sigma_x = E(\epsilon_x + \nu\epsilon_y)/(1 - \nu^2)$$
$$\sigma_y = E(\epsilon_y + \nu\epsilon_x)/(1 - \nu^2) \qquad 12.7$$

Hence, if ϵ_x and ϵ_y are measured, σ_x and σ_y may be calculated.

If, in addition to the internal pressure, a torque is applied to the cylinder, then the stress system applied to an element at the surface of the cylinder is as shown in fig. 12.4(a), where τ_{xy} is the shear stress caused by the torque, and τ_{xy}' is the complementary shear stress balancing it. This is the general case of two-dimensional stress. It may be shown that in two directions (1 and 2) at θ and $\theta + 90°$ to the reference direction, as shown in (b), the stresses σ_1 and σ_2 are a maximum and a minimum respectively, and the shear stresses in these two directions are zero.

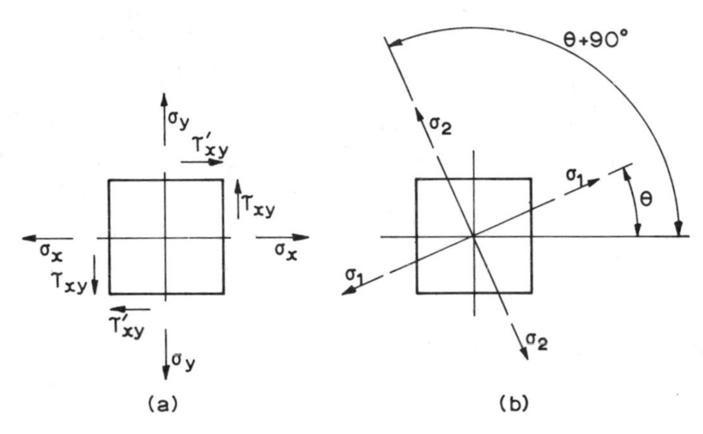

Fig. 12.4

These are the *principal stresses*, and it may be shown that maximum and minimum strains, i.e. *principal strains*, also occur in the same directions. Also, bisecting these directions are two directions in which the shear strains are maximum. It should be noted that, in fig. 12.3, σ_x and σ_y are the principal stresses (σ_1 and σ_2).

The foregoing is an extremely brief account of the *elastic* stress–strain relationships, which may be investigated further in any strength-of-materials textbook. It is included here to show that the relationships in the two-dimensional stress system are complex, and that care should be exercised in interpreting the results of strain measurements. The main implication is that, unless the directions of the principal stresses and strains are known, then, to obtain their magnitude and direction by strain-gauges alone, the strain must be measured in at least three directions. Alternatively, the directions of the principal stresses may be found by a brittle-lacquer test (see section 12.2.3), and a pair of gauges at 90° to each other may be used in these directions at chosen points.

12.2 Strain measurement

In the simple system of fig. 12.1(a), the strain at each of the points A and B is expected to be the same, i.e. the strain is uniform. For the cantilever of fig. 12.5, the bending moment varies linearly from 0 at $x = 0$ to maximum at the built-in end, and for a constant section the strain on the top surface varies in the same manner, as shown, and $\epsilon_A \neq \epsilon_B$. A strain-measuring device contacting at A and B, or in contact along the length between A and B, would measure some intermediate value between the point values ϵ_A and ϵ_B. In many applications, the strain is varying continuously from point to point, and hence a device for *strain* measurement should operate over as small a length or area as possible, unless the strain is known to be constant.

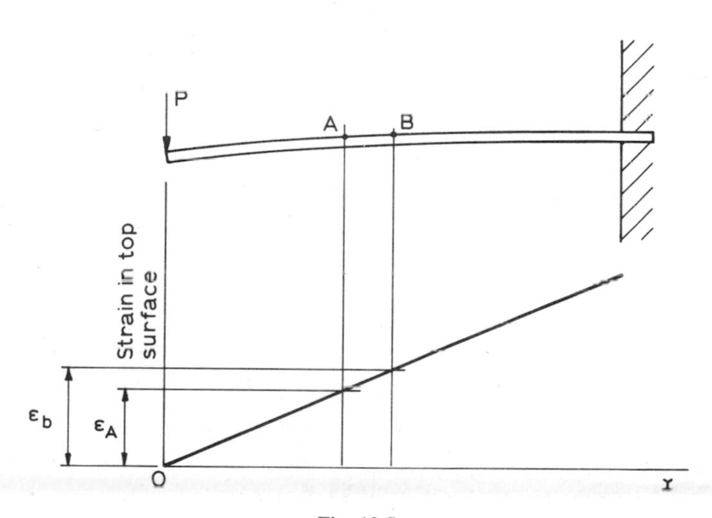

Fig. 12.5

12.2.1 Mechanical and optical strain-gauges

A wide variety of mechanical devices are available to measure the change of length of a tensile test-piece, or the change of length between points on a beam etc. The gauge-length of these may vary from 1 mm or less to more than 1 m, according to the scale of the measured object. With the shorter gauge-lengths, considerable magnification is necessary to provide a suitable signal, though with larger gauge-lengths this is not always necessary, and is the main reason for their use. One ingenious device, the Scratch Strain-gauge (Prewitt Associates, Lexington, Kentucky) uses the strain to rotate a small brass disc, on which it is recorded directly. The device may be left secured to a structure such as a bridge or aircraft frame, to record the strain pattern over a period.

Of the types requiring amplification, mechanical, pneumatic, and optical devices are used in wide variety, and may be scaled in length units or in strain units. The Huggenberger gauge shown in fig. 12.6 is one of a range using lever amplification. A typical gauge-length is 25 mm, though smaller ones may be obtained. As lever L pivots about P_1 due to change of the gauge-length, the top end of the lever has an amplified movement. This is transmitted to the pointer lever, pivoted at P_4, by the cross-link C, pivoted at P_2 and P_3, giving further amplification. The scale reading is divided by a calibration constant to give the strain value.

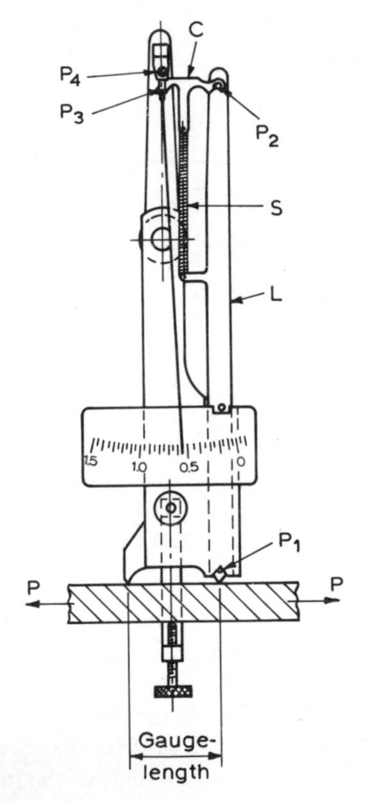

Fig. 12.6 Huggenberger extensometer

In general, any of the small-displacement-measuring devices described in Chapter 10 may be applied to strain measurement. An example is the air-jet sensing system shown in fig. 12.7. In this, the two

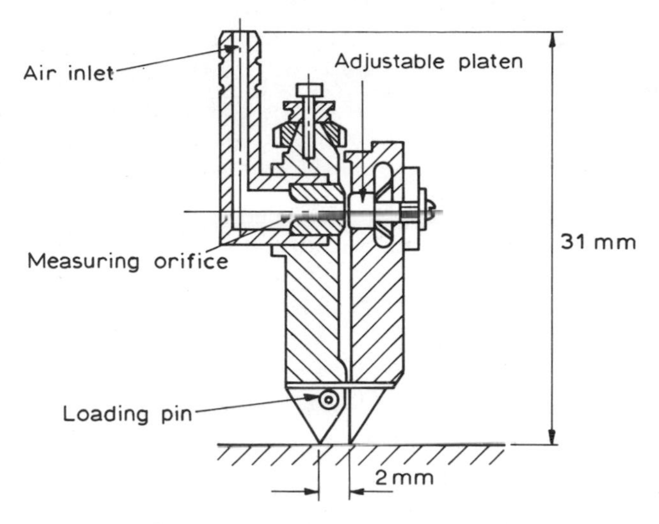

Fig. 12.7 STNC **pneumatic extensometer**

knife-edges arc held to the measured surface by applying a suitable force to the loading pin. Change in length of the 2 mm gauge-length gives a change in the distance between the orifice and platen, amplified by rotation about the spring hinge. This change is sensed by a change of the air back-pressure or flow rate (see section 10.3.2). The system has high sensitivity and precision.

12.2.2 *Electrical strain-gauges*

Small changes of resistance, inductance, capacitance, reluctance, permeability, etc. are all used in various applications to measure strain, usually in conjunction with bridge circuits to give a suitably magnified signal. The piezoelectric effect is also used in some specialist applications.

12.2.2.1 Electrical-resistance strain-gauges. When an electrical conductor is strained, it is found that a change of its resistance occurs, and the phenomenon is used to measure strain. The resistance of a conductor of uniform cross-section A, length l, and uniform resistivity ρ is given by

$$R = \rho l/A$$

Taking logarithms,

$$\ln R = \ln \rho + \ln l - \ln A$$

Differentiating,

$$\frac{\mathrm{d}R}{R} = \frac{\mathrm{d}\rho}{\rho} + \frac{\mathrm{d}l}{l} - \frac{\mathrm{d}A}{A}$$

But

$$\frac{\mathrm{d}A}{A} = -2\nu\frac{\mathrm{d}l}{l}$$

where ν is Poisson's ratio, and substituting this in the above equation gives

$$\frac{\mathrm{d}R}{R} = \frac{\mathrm{d}\rho}{\rho} + \frac{\mathrm{d}l}{l} + 2\nu\frac{\mathrm{d}l}{l}.$$

Hence

$$\frac{\mathrm{d}R}{R} = \frac{\mathrm{d}l}{l}(1 + 2\nu) + \frac{\mathrm{d}\rho}{\rho}$$

Dividing through by $\mathrm{d}l/l$, and writing as differentials,

$$F = \frac{\delta R/R}{\delta l/l} = 1 + 2\nu + \frac{\delta\rho/\rho}{\delta l/l} \qquad 12.8$$

where F is the *strain-sensitivity factor* or, if applied to a strain-gauge, the *gauge factor*. Since the value of ν for metals is about 0·3, the first two terms give a value of around 1·6. However, F may vary from −12·0 for nickel to +5·1 for an iridium(5 %)/platinum(95 %) alloy. A common value of F in commercial gauges is 2. Hence the fractional resistance change is not due solely to length and cross-sectional-area changes, but is also affected by a change of the resistivity (ρ) of the material with strain.

A variety of metals are available for use in strain-gauges, and the relevant properties of a number of these are shown in fig. 12.8. An important factor is the stability of a gauge material. Although high sensitivity is desirable, some of the metals having this property are unstable in use, and are utilised only in specialist applications. In many of the materials, the change of resistance due to temperature change is high enough for a change of a few degrees to give a false indication of strain as big as that expected due to force application, unless compensating methods are used.

Strain-gauges may be of either unbonded or bonded type. In the unbonded types, fine wire is stretched out between two or more points which may form part of a rigid base which itself is strained (fig. 12.9). Alternatively, the wire may be connected between points forming part of a transducer assembly, for example in instruments for measuring force, torque, pressure, acceleration, etc. Movement apart of the points A and B causes tensile strain in the resistance wire, and the change in

Material	Composition	Gauge factor F	Electrical resistivity $\rho(\mu\Omega m)$	Temperature coefficient of resistance $\alpha(/°C)$	Temperature coefficient of linear expansion $\lambda(/°C)$	Characteristics
Advance	Cu 58 %, Ni 42 %	2·1	0·49	0·000020	0·000016	Up to 400 C. Commonly used. Low value of α. Easily worked.
Iridium/platinum	Ir 5 %, Pt 95 %	5·1	0·24	0·001300	0·000009	Used for resistance thermometers
Iso-elastic	Ni 36 %, Cr 8 %, Mo 0·5 %, Fe bal.	3·5	1·12	0·000175	0·000007	Used for dynamic strain measurements
Karma	Ni 75 %, Cr 20 %, Fe, Al	2·1	1·50	0·000020		Similar to Advance
Manganin	Ni 4 %, Mn 12 %, Cu 84 %	0·47	0·41	−0·000006		
Nichrome V	Ni 80 %, Cr 20 %	2·5	1·10	0·000100	0·000012	Used for dynamic high-temperature work up to 1200°C
Nickel		−12·0	0·08	0·005000	0·000013	
Steel					0·000012	
Duralumin	Al 95 %, Cu 4 %, Mn, Mg, Si				0·000023	

Fig. 12.8 Properties of electrical-resistance strain-gauge materials and others

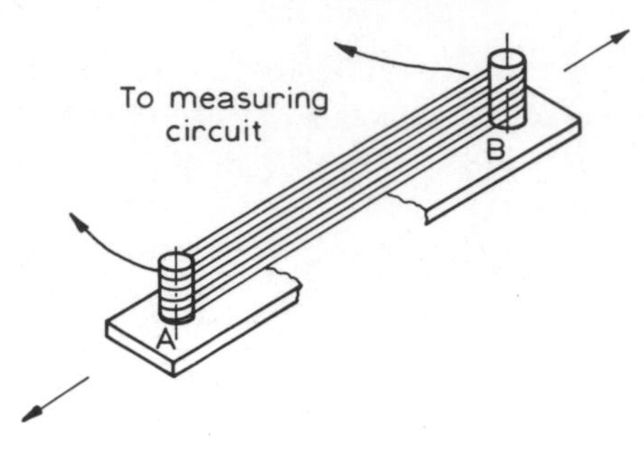

Fig. 12.9

its resistance is measured by a suitable circuit, to give a signal which is a function of strain.

The bonded gauge is a suitably shaped piece of the resistance metal which is bonded close to the surface whose strain is to be measured. Typical arrangements are shown in fig. 12.10. At (a), the exploded view shows a thin circular wire, typically 0·025 mm in diameter, shaped into a grid pattern; this is then cemented between suitable thin sheets of an insulating material, typical ones being strong paper or plastic. In (b), instead of wire, the grid is made from thin metal foil etched to the form shown; this has several advantages over the wire type, and is widely used. The assembled gauge is then bonded to the surface with a thin layer of suitable adhesive, and is finally waterproofed with a thin layer of suitable wax or lacquer. The grid should now sustain very nearly the same strain as the material to which it is bonded, both in tension and in compression.

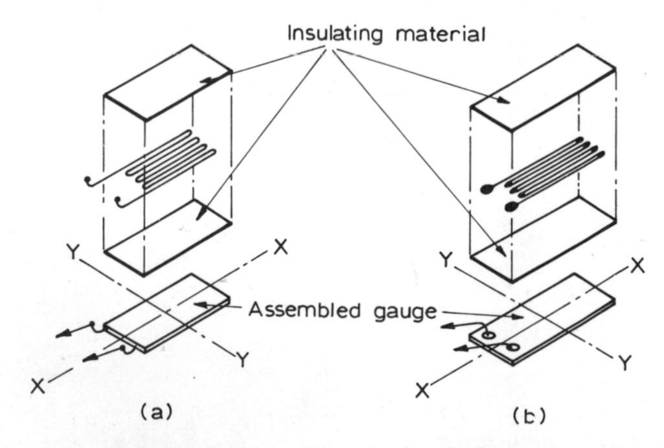

Fig. 12.10 Electrical-resistance strain-gauges: (a) Wire gauge (b) Foil gauge

The gauge is most sensitive to the strain along its axial direction XX, but the ends are in the YY direction, and strain always occurs in this direction, due to the Poisson's-ratio effect. This causes change of resistance, and may lead to an error of 2 % in the measured strain when using the wire bonded gauge. However, in the foil gauge the thickened ends reduce this *cross-sensitivity* effect to virtually zero.

Some special types of foil gauge are self-temperature-compensating, but, when using normal gauges, temperature compensation should always be obtained by suitable arrangement of the system. Figure 12.11 shows a simple null-type Wheatstone bridge, similar to that of fig. 4.6(a), where two strain-gauges having very nearly the same resistance values are used. One, referred to as the *active* gauge, is the measuring one, fixed to the strained member. The other, the *dummy* gauge, is for temperature compensation, and is fixed to a piece of the same material as the active gauge and is kept at the same temperature.

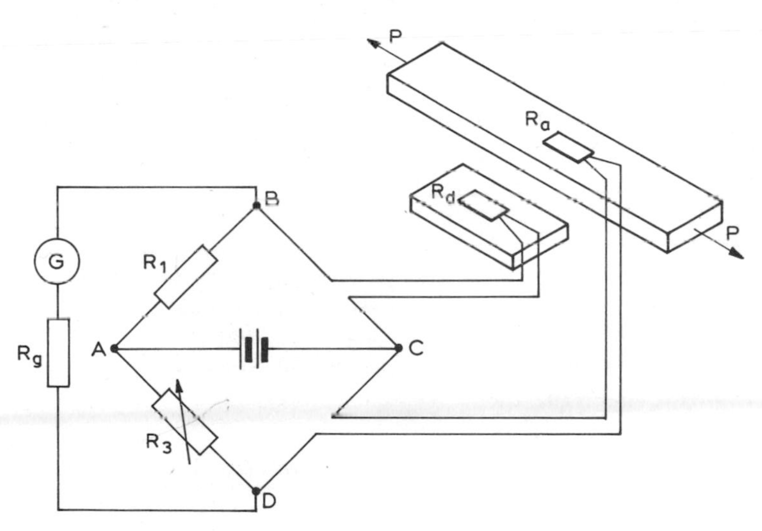

Fig. 12.11

With no force applied to the bar, the bridge is balanced initially by adjusting R_3, then

$$\frac{R_1}{R_3} = \frac{R_d}{R_a} \quad \text{(see section 4.3.1)} \qquad 12.9$$

If the temperature changes by $\delta\theta$, then, due to the different expansion rates of the gauge and the metal to which it is bonded, the gauge suffers a strain of magnitude $(\alpha - \alpha_g)\delta\theta$, where α_g and α are the temperature coefficients of linear expansion for the gauge metal and the base metal respectively. Also, the resistivity of the gauge metal changes with temperature (see example 12.5.2), and in connection with this it should be noted that the current through a gauge has a heating effect, and should not be excessive. Further, a change of strain sensitivity may also occur due to temperature change. These effects will lead to changes of resistance $\delta R_a'$ and $\delta R_d'$ in the gauges, which will be virtually identical.

Then
$$\frac{R_1}{R_3} = \frac{R_d + \delta R_d'}{R_a + \delta R_a'}$$

and
$$\frac{R_1}{R_3}(R_a + \delta R_a') = R_d + \delta R_d'$$

Substituting $(R_1/R_3)R_a = R_d$ from eqn 12.9 gives

$$R_d + \frac{R_1}{R_3}\,\delta R_a' = R_d + \delta R_d'$$

The bridge is seen to remain balanced, since $R_1/R_3 = 1$ and $\delta R_a' = \delta R_d'$, and the bridge is temperature-compensated as long as both gauges are at the same temperature.

Hence, if δR_a is the change of resistance of the active gauge due to the application of force P, and δR_3 is the change required in resistance R_3 to again obtain balance, then, from eqn 12.9,

$$\delta R_a = \left(\frac{R_d}{R_1}\right)\delta R_3 \qquad 12.10$$

Let $\epsilon(= \delta l/l)$ be the strain in the gauge, and, since the gauge constant is $F = (\delta R_a/R_a)/(\delta l/l)$, then

$$F\epsilon = \frac{\delta R_a}{R_a} = \frac{R_d}{R_1} \times \frac{\delta R_3}{R_a} \quad \text{from eqn 12.10}$$

and
$$\epsilon = \left(\frac{1}{FR_1}\right)\delta R_3 \qquad 12.11$$

When the strain due to bending is to be measured for a symmetrical member such as the cantilever of fig. 12.12, then the bridge sensitivity may be doubled by using a two-active-arm bridge. Two active gauges are used, the one on the outer surface having positive strain applied to it, and that on the inner surface negative strain. This gives an increase in the resistance of one arm and a decrease in the adjacent arm, both effects unbalancing the bridge. For the voltage-sensitive bridge of section 4.3.2, eqn. 4.8 becomes

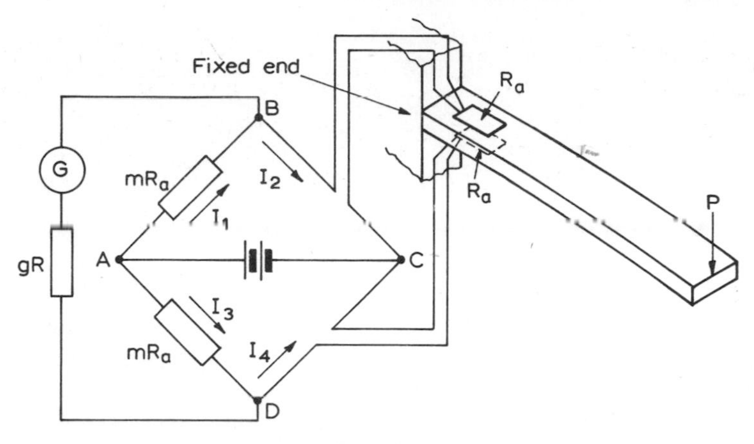

Fig. 12.12

$$I_1 = I_2 = V_{AC}/\{mR + R(1 - q)\}$$

where $$q = \delta R_a/R_a$$

Substituting this and eqn 4.9 in eqn 4.7, and putting $n = 1$,

$$V_{BD} = V_{AC}\left(\frac{mR}{mR + R(1 - q)} - \frac{mR}{mR + R(1 + q)}\right)$$

$$= V_{AC}\frac{2mq}{(m + 1 + q)(m + 1 - q)}$$

$$\approx \frac{2mV_{AC}}{(m + 1)^2}\left(\frac{\delta R_a}{R_a}\right) \qquad 12.12$$

which for a given strain value doubles the output voltage V_{BD} of the single-active-arm bridge, eqn 4.11.

Since $$F\epsilon = \delta R_a/R_a$$

then $$\epsilon = V_{BD}\frac{(m + 1)^2}{2mV_{AC}} \qquad 12.13$$

Considering the balanced condition, the same changes of temperature of each gauge will give the same changes of resistance in each gauge, and of the same sign, hence the analysis applied to the dummy-compensated bridge applies, and the null-type bridge and the voltage- and current-sensitive bridges are compensated for temperature when two adjacent arms are active, if $n = 1$.

It may also be shown that, with two active arms as above in 'push-pull', the change of the variable resistance of the null-type bridge to rebalance is doubled, and the output current of the current-sensitive bridge is doubled for the same strain value as compared with the single-active-arm bridge. Voltage- or current-sensitive bridges may have four active arms, thereby quadrupling the sensitivity, and the application of these bridges is discussed in section 12.3.1.

If the directions of principal strain and stress are not known, then clusters of three or more strain-gauges are used. These are known as *rosettes*, and typical forms are shown in fig. 12.13. The determination of the magnitudes and directions of the principal strains and stresses from the readings on the rosette gauges is carried out either by a graphical method or by using mathematical relationships (see refs 8, 20, 23, 26).

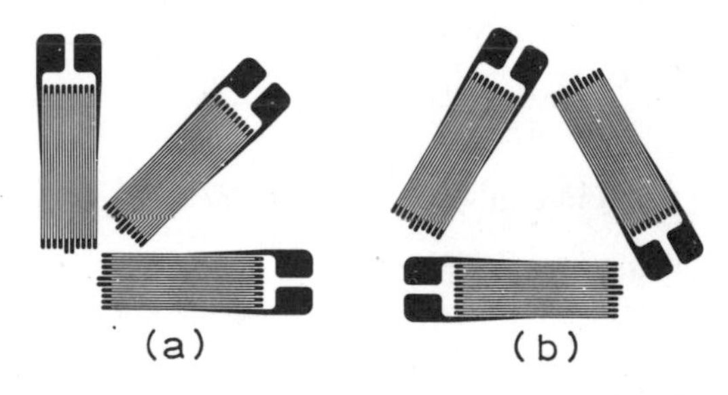

Fig. 12.13 Strain-gauge rosettes: (a) 45°/90° rosette (b) Delta rosette

A.C. excitation of resistance-strain-gauge bridges is commonly used, for the reasons given in section 4.4. Using the amplitude of the a.c. output across the points BD, the equations derived for the d.c. Wheatstone bridge still apply.

Calibration of bonded-type gauges used for strain measurement must be carried out with them bonded to the measured component. The method usually adopted is to introduce a precision resistance (R_p) into the bridge in parallel with the gauge to be calibrated, as illustrated in fig. 12.14. When the switch (S) is closed, the resistance changes from R_a to $R_aR_p/(R_a + R_p)$. The change of resistance is

$$\delta R_a = R_a - R_aR_p/(R_a + R_p) = R_a^2/(R_a + R_p)$$

$$\therefore \quad \frac{\delta R_a}{R_a} = \frac{R_a}{R_a + R_p} \qquad 12.14$$

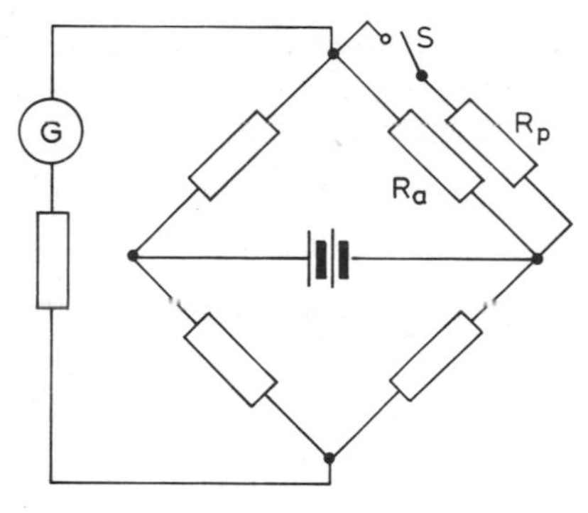

Fig. 12.14

From eqn 12.11,

$$\epsilon = \frac{1}{F}\frac{\delta R_a}{R_a}$$

$$= \frac{1}{F}\left(\frac{R_a}{R_a + R_p}\right) \qquad 12.15$$

12.2.2.2 Other electrical strain-gauges. The electrical-resistance strain-gauge has in recent years become an important tool of the design and development engineer. It is widely used for both static and dynamic measurements. However, difficulty arises due to drift of the output signal at higher temperature values, and 550 °C is about the limit for static strain measurements. In dynamic strain measurement, drift is not so important, since the change of strain over the cycle of loading is measured, and, using suitable metals, gauges may be used at much higher temperatures.

A recently developed gauge (fig. 12.15) uses change of capacitance to measure static or dynamic strain in the temperature-range −260 °C to 650 °C or more. The gauge, of overall dimensions approximately 25 × 4 × 4 mm, is micro-spot-welded to the measured material at the ends of the gauge-length (G). An increase of gauge-length causes different changes of the arch heights x_1 and x_2, causing an increase in the gap (z) between two flat electrodes, thereby reducing their capacitance. Neither the mechanical amplification ($\delta z/\delta G$) nor the capacitance–gap relationship ($\delta C/\delta z$) is linear, but the non-linearities tend to cancel each other, and the relationship shown in fig. 12.15(c) is given as typical for the gauge. The gauge material may in some cases be matched for thermal expansion with the measured material, to prevent

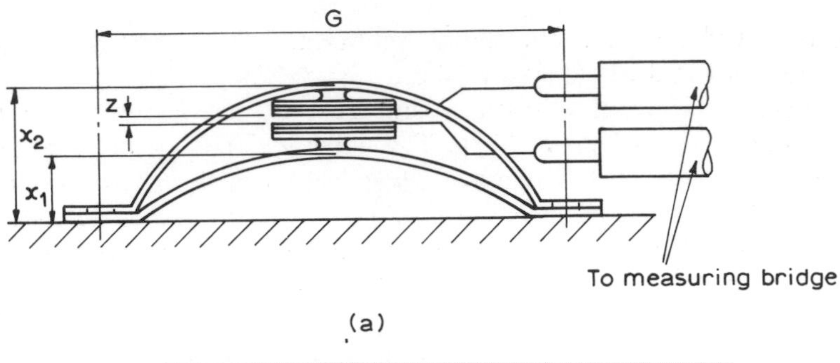

(a)

(b)

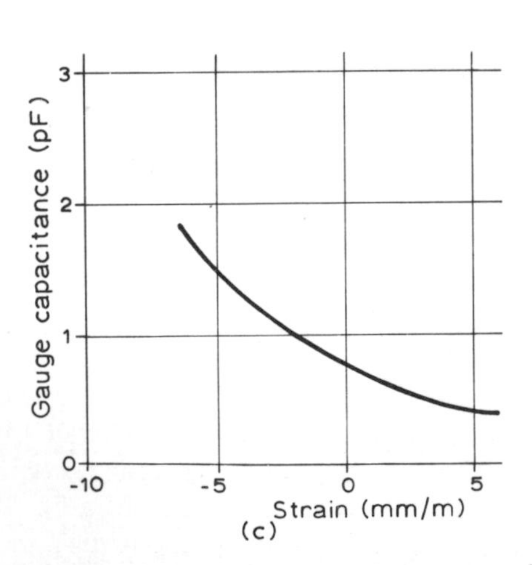

(c)

Fig. 12.15 CERL-Planar capacitance strain-gauge

apparent strain due to differential expansion. Strain sensitivity is high; hysteresis and drift are relatively low. (See ref. 24.)

Semiconductor strain-gauges have also been developed, and an example is shown in fig. 12.16. The gauge, which is 11 × 7 mm or smaller, is bonded to the measured structure, and changes of strain cause changes of current in the material. A typical gauge has a sensitivity of −100, with a maximum strain of 0·003 and a temperature-range of −10 to +70 °C.

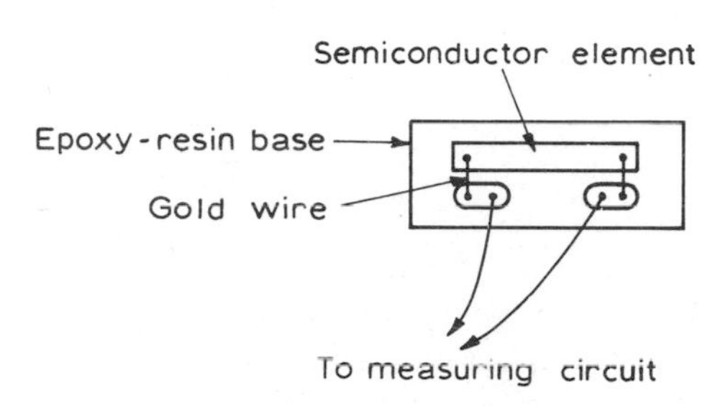

Fig. 12.16 Semiconductor strain-gauge

A wafer of piezoelectric crystal may be bonded to a surface, and will suffer an electric charge when strained. The sensitivity of such an arrangement is high, several hundred times that of an electrical-resistance gauge; however, stability is poor, and the method is suitable mainly for dynamic strain measurement. Gauges should be calibrated dynamically.

A few strain transducers are based on the magnetostrictive effect (refs 5, 7). In this, the permeability of a magnetic material is altered when subjected to strain, giving a change of reluctance in a magnetic circuit. This may be used in an arrangement basically similar to that shown in fig. 2.23, but with a change of permeability due to strain as the input, instead of a change of reluctance due to a varying air gap.

A further type of strain-gauge using the change of resistance is illustrated in fig. 2.16(b). The gauge consists of carbon granules cemented into a matrix and bonded to the measured surface. Strain of the surface causes an increase of the contact pressure between granules and a change of contact resistance. Sensitivity is high, but the stability is poor, and this type of gauge is used only in special applications.

A recently introduced strain transducer consists of a fine wire stretched inside a tube. The initial tensioning produces a natural frequency of transverse vibration (f_1). Strain of the tube causes increase or decrease of the wire tension and a change of frequency to f_2. The wire is vibrated by electrical means, and the frequency of vibration is picked up by an electrical coil. The change of strain is given by

$$\delta\epsilon = K(f_1^2 - f_2^2) \qquad 12.16$$

The transducer units are robust and may be buried in concrete or built into foundations etc. They are particularly suitable for the long-term measurement of slowly varying strains.

12.2.3 Strain-sensitive lacquers

When carrying out a tensile test on a specimen which has been normalised after machining, the brittle oxide film is seen to crack and fall off as the metal strains past the yield point, and this effect has been developed for strain measurement. Special lacquers have been developed which, when painted on the surface, will crack in tension below the yield point of metals, at strain values in the region of 0·0005 to 0·0015. Cracks appear in a direction normal to the direction of the principal stress/strain, and the field covered by the cracks at any moment indicates the extent of the region which has passed the *threshold strain*, i.e. the particular strain at which the lacquer cracks.

The method does not lend itself to the accurate evaluation of strain, but is valuable in giving a visual indication of strain distribution and direction, particularly if used with gradually increasing loading. Some degree of accuracy, however, may be obtained by applying the lacquer to both the component and to a cantilever strip, taking care that each is subjected to the same treatment, i.e. having the same lacquer thickness, and being subjected to the same temperature and humidity. When subjected to an end force, producing a bending moment increasing with distance, the lacquer on the cantilever will crack up to the point where the threshold strain has applied, and a similar strain will apply on the component at the point when the lacquer cracks. In the test shown in fig. 12.17, the threshold strain or stress may be calculated from the bending moment P_1x on the section QQ. The values of the force (P_2) on the test bar when, for example, cracks first appear near the hole and when cracks reach the edges of the strip, enable the stress distribution across the section to be determined. The method can be used only for tensile stresses in the lacquer, and to measure compressive stresses in a component it must first be loaded to give the stress, then lacquered and the loading removed, when the lacquer is stressed in tension.

Details of lacquers are given in refs 20 and 22.

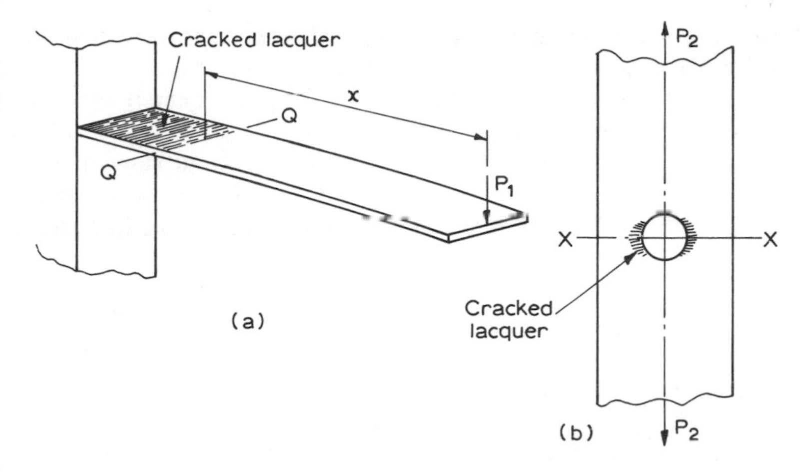

Fig. 12.17 Brittle-lacquer test: (a) Calibration strip (b) Test bar

12.3 Applications of strain-gauging

Strain measurement is not usually carried out for its own sake, but either as a means of determining maximum stress values or, in specialised transducers, to measure force, pressure, acceleration, torque, etc.

12.3.1 ***Stress measurement***

Where large areas of uniform stress and strain exist, the 'extensometer' type of strain-gauge, working on a larger gauge-length, may be used. An application of this type is illustrated in fig. 1.13 and is discussed in section 1.3.5. The 'Demec' gauge is an example of the type, and consists of a beam with a fixed point at one end and a movable point at the other end operating a dial gauge through a one-to-one ratio bell-crank lever. A reference beam has recesses accepting the two gauge-points for zeroing the indicator. A setting bar has two points for setting two metal buttons with centre holes for the gauge-points as they are bonded to the measured material. Readings are taken of the distance apart of the button centres, relative to the reference beam, before loading and subsequently at any time after loading. This system has the advantage that no delicate instrument is left exposed to environmental hazards.

As discussed in section 12.1, when the stress system is biaxial, the stress–strain relationship is complex. In many engineering structures and machines, components are subjected to such loading, and mathematical evaluation of stresses is frequently difficult and sometimes impossible. Recourse must then be made to practical measurement, and the initial problem is to determine at what point or points the strain value is to be measured, and then to determine at what angle the strain is to be measured.

The brittle-lacquer technique of section 12.2.3 is extremely useful for initial investigation of the strain field. If the component is coated all over and is subjected to gradually increasing load, in the same manner as in service, the crack propagation will indicate the most highly stressed areas and the direction of the higher tensile principal stress. Strain-gauges may then be positioned at these points for further tests.

The highest strains often occur at rapid changes of section, frequently in corners, and strain-gauges of very small length are necessary for mounting in such places. Electrical-resistance gauges are available in lengths down to about 1 mm, and mechanical types to about 2 mm; however, the electrical type is easier to mount, and if necessary the electrical measurements for static strains may be made quite simply in the field by using battery-powered bridge circuits and indicators.

The resistance gauge and bridge are convenient in some cases for separating bending and direct stresses. The cantilever of fig. 12.18, subjected to the end force *P* at an angle, has a combination of direct and bending stress. However, if the two gauges are connected into a bridge as in fig. 12.12, the direct component of stress, being of the same magnitude and direction at both gauges, cancels out. The system is therefore responsive to bending stress only.

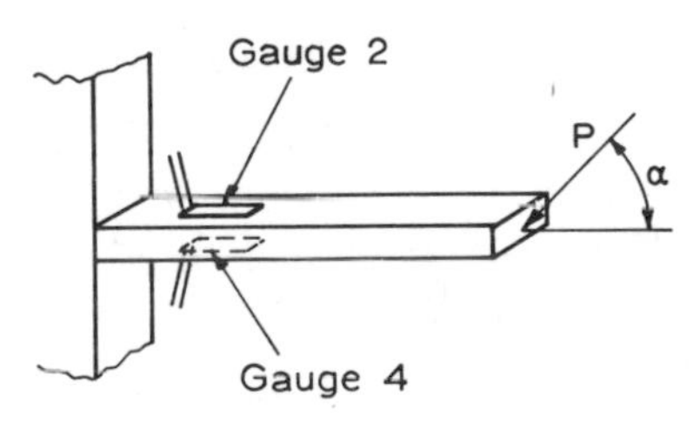

Fig. 12.18

For a circular shaft subjected to pure torsion, as shown in fig. 12.19(a), the shear stress τ_{xy} is balanced by the complementary shear stress τ_{xy}' as shown in (b). It may be shown that direct, i.e. tensile and compressive, stresses occur at 45° to the axial direction as shown in (c), and the magnitudes of these are the same as the shear-stress values (τ_{xy}). Also, the shear-stress value in each of these directions is zero. Hence it is convenient to place two gauges in these principal directions as indicated in (c), and to use them in push-pull fashion in a two-active-arm bridge to determine τ_{xy}. Special gauges are in fact produced for this purpose, and a typical one is illustrated in fig. 12.22(a). Other

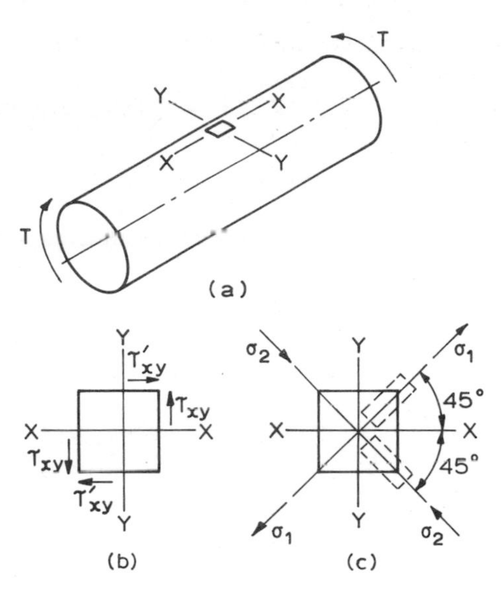

Fig. 12.19

arrangements of two and four active gauges may be used which will eliminate or reduce the effects of axial force and bending moment, when only shear stress is to be measured (see ref. 8).

If the principal directions are to be determined by rosettes, then the gauge sizes should be as short as possible so that the minimum change of direction occurs in their length. In some cases, gauges are superimposed one on top of the other to obtain as near to point values as possible. However, with very small gauges, the current may have to be made very small to prevent temperature rises.

12.3.2 The electrical-resistance strain-gauge as a general transducer

The electrical-resistance strain-gauge is of great value as a primary transducer for the measurement of a wide range of quantities. Figure 12.20 shows a pair of gauges in an arrangement whereby their output is a function of the very small displacement (x) of a plunger.

Many load cells for measuring forces use this type of gauge, an example being shown in fig. 5.38, capable of measuring both tensile and compressive force. Another typical arrangement, illustrated in fig. 12.21, consists of a number of strain-gauges bonded to a solid or hollow pillar. At (a), two gauges diametrically opposite, numbers 1 and 4, are placed axially, and, since they are in *opposite* sides of the bridge, their output is additive. The temperature-compensating gauges (2 and 3) are bonded to a separate piece of the same material. In (b), the

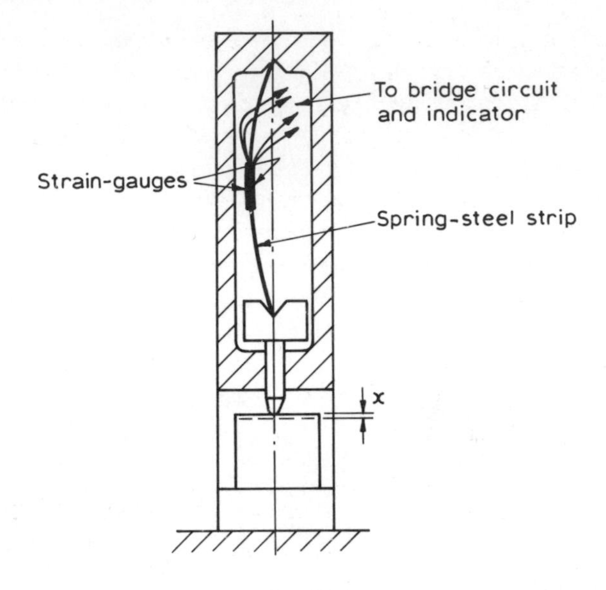

Fig. 12.20

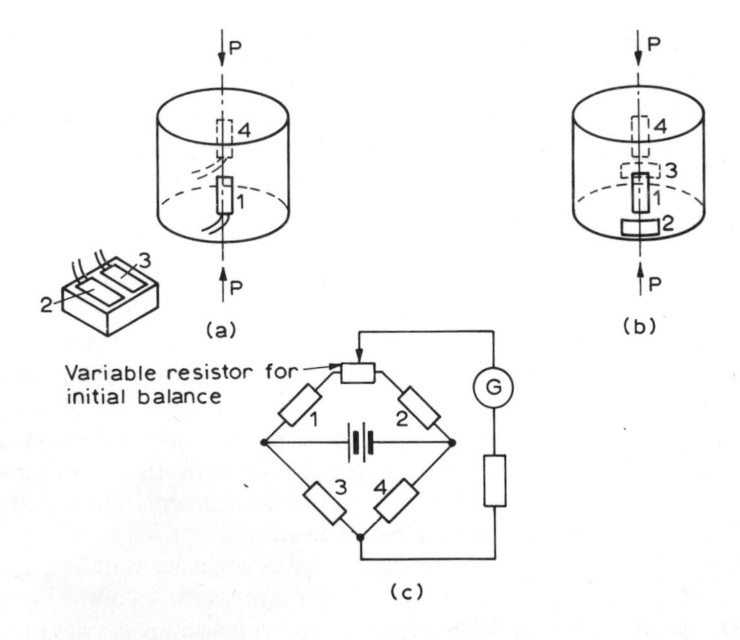

Fig. 12.21 Strain-gauge load cell

compensating gauges are bonded on to the cylinder in a transverse direction, and they suffer a positive strain due to the Poisson's-ratio effect. Since they are in the bridge arms *adjacent* to the axial gauges, their output is additive to gauges 1 and 4, which measure negative strain; hence system (b) is $(1 + \nu)$ times more sensitive than system (a), and $2(1 + \nu)$ times more sensitive than a single active gauge.

Diaphragms are used to measure force and pressure as illustrated in figs 5.5(c) and 5.27(c), using a strain-sensitive transducer. Special foil gauges have been developed for this purpose, and two are shown in fig. 12.22(b) and (c). The double spiral is sensitive to tangential strain, and the Element Redshaw to both tangential and radial strain in a two-active-arm bridge. They provide much greater sensitivity than standard gauges.

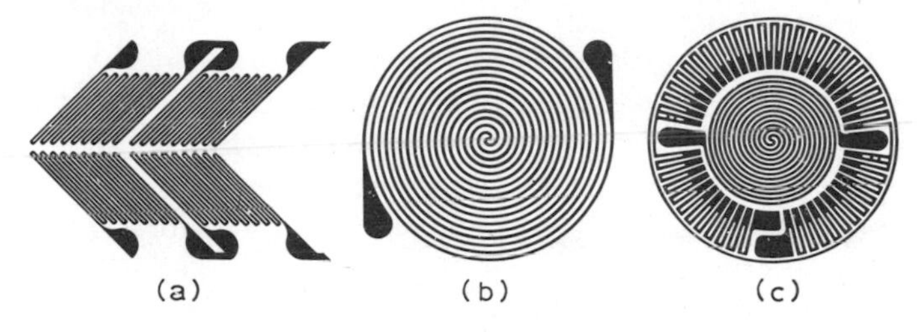

Fig. 12.22 Special-purpose resistance strain-gauges (a) Torque gauge (b) Double spiral (c) Element Redshaw

In the applications discussed in this section, calibration is usually carried out directly in units of bridge-output or indicator units per unit input of deflection, torque, force, pressure, etc., the strain value not being of direct significance.

12.4 Photoelastic determination of stress

The photoelastic method depends on the stress–optical properties of a number of translucent materials such as glass, Araldite, Perspex, etc. which exhibit the property of *double refraction* or *birefringence* when subjected to stress. This causes a retardation of light waves passing through the material, proportional to the stress at any particular point. In suitable optical systems, this can be made to produce a *fringe* pattern similar to those obtained in dimensional interferometry (section 10.2), and the relationship may be expressed thus:

$$\sigma_1 - \sigma_2 = nF/t \qquad 12.17$$

where σ_1 and σ_2 are the principal stress values, n is the number of wavelengths retardation (i.e. number of fringes from a zero stress point), F is the material fringe value [stress/(n/unit thickness)], and t is the material thickness.

F is determined for a particular material by a calibration test.

Using this relationship, and counting n fringes from a zero-stress point on the fringe pattern, the value of $(\sigma_1 - \sigma_2)$ may be determined for points in the material. Modification of the optical system then gives *isoclinic* patterns from which the locus of points having the same principal stresses may be observed. With further computation, the complete two-dimensional stress field for the model may be determined. The method requires a knowledge of stress theory, and the brief treatment here is intended only to give an appreciation of the role of the method in engineering.

12.4.1 Photoelastic models

Two-dimensional models are made by casting or by machining from sheet material, and are subjected to the appropriate force system in a *transmission polariscope*, in which polarised light is transmitted through the model, the resulting fringe pattern being viewed on a screen, or photographed. Typical patterns obtained during a calibration test and a component test are shown in fig. 12.23.

The method may be extended to three-dimensional models by means of the *frozen-stress* technique. In this, the force system is applied whilst the model is hot, and is maintained until it is cold. The stress pattern is found to remain in the material. The model is then sliced to give two-dimensional models which are viewed without further loading. From these two-dimensional stress patterns, the three-dimensional pattern may be built up.

12.4.2 Photoelastic coatings

In this technique, a thin but even layer of the birefringent material is cast or bonded to the surface of the tested component. The component or the bonding agent must be reflective, and, with a suitable optical system, reflected fringe patterns may be viewed. Very little change of stress is caused by the coating if the component is metal, because of the large difference in their elastic-modulus values. For the same reason, stresses beyond the elastic limit of the metal may be observed.

References 20 and 25 are recommended for further reading.

12.5 Worked examples

12.5.1 Two mechanical strain-gauges of short gauge-length were mounted at right angles to each other on the outer surface of a spherical vessel subjected to internal gas pressure only. Both readings were

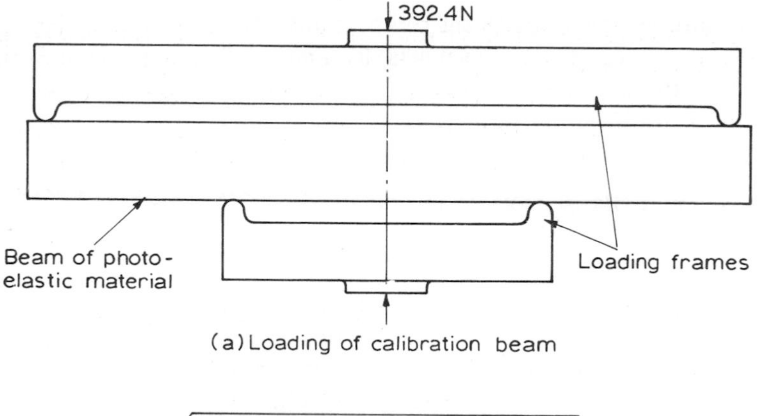

(a) Loading of calibration beam

(b) Fringe pattern observed for centre of calibration beam

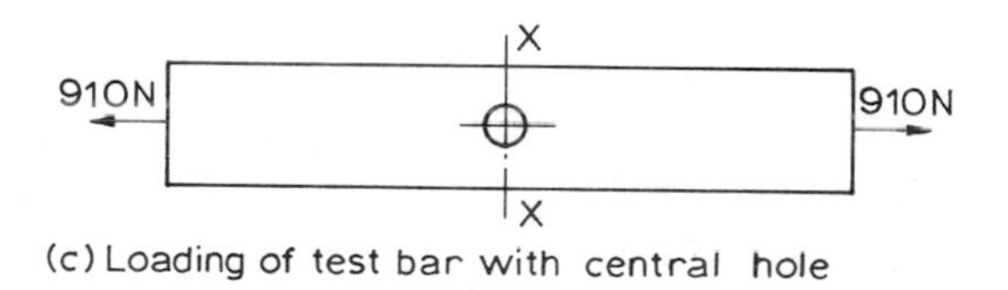

(c) Loading of test bar with central hole

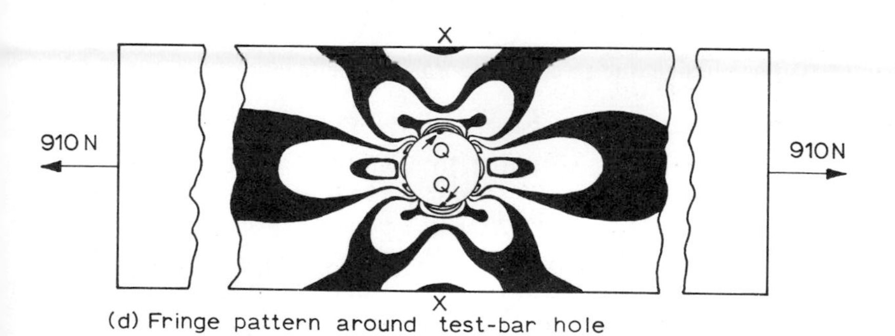

(d) Fringe pattern around test-bar hole

Fig. 12.23 Photoelastic test

0·000 63 tensile strain. Determine the stress in the surface of the metal if Poisson's ratio is 0·30 and the modulus of elasticity is 200 GN/m^2.

Since any section through the centre point of the sphere has identical size and loading, the stress value is the same in all directions tangential to the surface.

From eqn 12.7,

$$\sigma_x = E(\epsilon_x + \nu\epsilon_y)/(1 - \nu^2) = E\epsilon(1 + \nu)/(1 + \nu)(1 - \nu) = E\epsilon/(1 - \nu)$$

$$= 200 \times 10^9 \times 0{\cdot}00063/(1 - 0{\cdot}3)$$

$$= 180 \text{ MN/m}^2 \quad \text{or} \quad 180 \text{ N/mm}^2$$

12.5.2 A single electrical-resistance strain-gauge of resistance 120 Ω and sensitivity (F) of 2·0 is bonded to steel having an elastic-limit stress of 400 MN/m^2 and modulus of elasticity 200 GN/m^2. Calculate the change of its resistance (a) due to a change of stress equal to $\frac{1}{10}$ of the elastic range, (b) due to a change of temperature of 20 °C, if the material is Advance alloy.

Using values from fig. 12.8 where necessary,

a) change of stress $= 400 \times 10^6/10 = 40 \times 10^6$ N/m^2

change of strain $= \sigma/E$

$$= 40 \times 10^6/200 \times 10^9$$

$$= 200 \times 10^{-6}$$

From eqn 12.8,

$$\delta R = \epsilon F R$$

$$= 200 \times 10^{-6} \times 2{\cdot}0 \times 120$$

$$= 0{\cdot}048\ \Omega$$

b) (i) The change due to the temperature coefficient of resistance (α) is

$$\delta R = R\alpha\,\delta T$$

$$= 120 \times 20 \times 10^{-6} \times 20$$

$$= 0{\cdot}048\ \Omega$$

b) (ii) The strain due to differential expansion of the gauge metal and the steel is

$$\epsilon = (\lambda_s - \lambda_g)\delta T$$

$$= (12 - 16) \times 10^{-6} \times 20$$

$$\delta R = \epsilon F R$$

$$= -80 \times 10^{-6} \times 2 \times 120$$

$$= -0{\cdot}019\ \Omega$$

The changes of resistance due to temperature change are seen to be of the same order as the changes due to the measured strains likely to be encountered, and if uncompensated will lead to unacceptable errors. The change due to differential expansion partly offsets the change due to coefficient of resistivity in this example, but the effects are additive with some metal combinations.

12.5.3 Two electrical-resistance strain-gauges are bonded to a Duralumin cantilever and connected into a bridge circuit as shown in fig. 12.12. Each gauge has a resistance of 100 Ω, a sensitivity (F) of 2·1, and $m = 1$, $V_{AC} = 4$ V. If a force P is applied until the mean stress at the gauges is ± 200 MN/m^2, calculate
a) the change of resistance in arm 3 to restore balance in a null-type bridge,
b) the p.d. across BD if gR is 10 kΩ,
c) the current I_G if gR is 400 Ω,
d) the current through the strain-gauges.

The modulus of elasticity (E) for Duralumin is 70 GN/m^2.

a) Using eqn 12.11, but with a factor of 2 for two active arms,

$$\delta R_3 = 2\epsilon F R = 2\sigma F R/E$$

$$= 2 \times 200 \times 10^6 \times 2{\cdot}1 \times 100/70 \times 10^9$$

$$= 1{\cdot}2\ \Omega \text{ (a reduction of resistance in this case)}$$

b) This is a voltage-sensitive bridge. Using eqn 12.12 and putting $m = 1$,

$$\text{output p.d.} = V_{BD} = \frac{2V_{AC}}{4}\left(\frac{\delta R_a}{R_a}\right) = \frac{V_{AC}}{2}\cdot\frac{F\sigma}{E}$$

$$= \frac{4 \times 2{\cdot}1 \times 200 \times 10^6}{2 \times 70 \times 10^9}$$

$$= 12 \text{ mV}$$

c) With the relatively low value of resistance in the indicator circuit, this is a current-sensitive bridge. Using eqn 4.15 with a factor of 2 for two active arms, and putting $m = 1, n = 1, g = 4$,

$$\text{output current} = I_G = \frac{2V_{AC}q}{R(4 \times 4 + 4)} = \frac{V_{AC}}{10R}\left(\frac{\delta R_a}{R_a}\right) = \frac{V_{AC}}{10R}\cdot\frac{F\sigma}{E}$$

$$= \frac{4 \times 2{\cdot}1 \times 200 \times 10^6}{10 \times 100 \times 70 \times 10^9} = 24\ \mu\text{A}$$

d) Gauge current $I = V_{AC}/2R = 4/(2 \times 100)$

$= 20$ mA

12.5.4 In a brittle-lacquer test as shown in fig. 12.17, the strip used in the calibration cantilever was 20×4 mm, and a force (P_1) of 30 N caused cracking of the lacquer of the top surface up to a distance (x) of 256 mm. A long strip 30×6 mm with a single central hole 10 mm in diameter was tested in tension, and the lacquer immediately to the side of the hole showed cracks when the axial force (P_2) was 6·2 kN. Estimate the static stress-concentration factor at the hole.

$$\text{Stress in top surface of cantilever at QQ} = \sigma = My/I$$

$$= F \times y \times 12/bd^3$$

$$= \frac{30 \times 256 \times 2 \times 12}{20 \times 4^3}$$

$$= 144 \text{ N/mm}^2$$

The same stress is expected to apply when the lacquer cracks around the hole in the test strip.

$$\text{Nominal stress at section XX} = P_2/\text{area of metal}$$

$$= 6{\cdot}2 \times 10^3/(20 \times 6)$$

$$= 51{\cdot}7 \text{ N/mm}^2$$

$$\text{Static stress-concentration factor} = (\text{actual stress})/(\text{nominal stress})$$

$$= 144/52$$

$$= 2{\cdot}8 \text{ approx.}$$

12.5.5 A shaft is to transmit power up to 44 kW at a constant speed of 1400 rev/min, and it is proposed that the torque be sensed by a pair of gauges of the type shown in fig. 12.22(a) bonded to a specially machined portion of the shaft. The gauges are to be connected in push-pull in an equi-armed voltage-sensitive bridge, the output of which is to be calibrated in power units. If the maximum strain value of the gauges is 0·0015, their resistance 120 Ω, and gauge factor 2·1, calculate

a) the diameter of steel shaft to which they should be bonded, if its modulus of elasticity is 200×10^9 N/m^2,

b) the output voltage at full power if the bridge energising voltage is 6 V,

c) the sensitivity of the system in volts/kilowatt.

a) Angular velocity $= \omega = \dfrac{2\pi N}{60} = \dfrac{2 \times 22}{7} \times \dfrac{1400}{60} = \dfrac{440}{3}$ rad/s

Power $= T\omega$,

$$\therefore \quad \text{torque } T = 44000 \times \frac{3}{440} = 300 \text{ Nm}$$

Maximum strain value is 0·0015,

$$\begin{aligned}\therefore \quad \text{maximum direct stress} &= 0{\cdot}0015 \times E \\ &= 0{\cdot}0015 \times 200 \times 10^9 \\ &= 300 \text{ MN/m}^2\end{aligned}$$

For pure torsion, the maximum shear stress τ has the same value.

From the shaft equation, $\dfrac{T}{I_p} = \dfrac{\tau}{r}$

where T = applied torque, $I_p = \pi d^4/32$, and τ = shear stress at outside radius r.

$$\therefore \quad 300 \times 10^3 \times \frac{32}{\pi d^4} = \frac{300}{1} \times \frac{2}{d}$$

$$d = 17{\cdot}2 \text{ mm}$$

b) Using eqn 12.12,

$$\begin{aligned}V_{BD} &= \frac{V_{AC}}{2}\left(\frac{\delta R_a}{R_a}\right) = \frac{V_{AC}}{2}.F\epsilon \\ &= \frac{6}{2} \times 2{\cdot}1 \times 0{\cdot}0015 \\ &= 9{\cdot}45 \text{ mV}\end{aligned}$$

c) Overall sensitivity $= 9{\cdot}45/44$
$= 0{\cdot}21$ mV/kW ($= 0{\cdot}21$ μV/W)

This strain-gauge system on a rotating shaft necessitates high-quality slip rings to transfer the signal to the stationary member, and for this type of application other systems of torque measurement (e.g. those described in Chapter 5) are usually preferred.

12.5.6 A calibration beam of uniform cross-section 20 × 6 mm made from a photoelastic material was subjected in a polariscope to symmetrical loading as shown in fig. 12.23(a). As loading progressed to the final value shown, six fringes were observed to travel from the top and bottom edges of the central portion to the centre of the section as shown in (b). Determine the stress represented by one fringe order for the material. Hence determine the stress and the static stress-concentration factor for the hole in the test strip of the same material in (c), if an axial load of 510 N produces ten fringes past the hole edge at QQ as indicated in (d). The strip is 18 mm wide by 6 mm thick, and has a 6 mm diameter hole.

For the central portion, the bending moment is uniform and equal to $50 \times 392{\cdot}4/2 = 9810$ N mm.

From $\sigma/y = M/I$, where M is the applied bending moment, σ is the direct stress at y from the neutral axis, and I is the second moment of area about the neutral axis $= bd^3/12$,

$$\sigma = \pm\frac{My}{I} = \frac{9810 \times 10 \times 12}{6 \times 20^3} = 24{\cdot}5 \text{ N/mm}^2$$

Hence, referring to eqn 12.17, $F/t = 24{\cdot}5/6 = 4{\cdot}1$ N/mm^2 per fringe.

In the test shown in fig. 12.23(b), since ten fringes have originated from the points Q, then the stress at these points is

$$\sigma_1 = 4{\cdot}1 \times 10 = 41 \text{ N/mm}^2$$

since σ_2 normal to the free edge at Q $= 0$.

The nominal stress at section XX is

$$\sigma_n = F/A = 981/72$$
$$= 13{\cdot}6 \text{ N/mm}^2$$

Stress-concentration factor $= \sigma_1/\sigma_n = 41/13{\cdot}6 = 3{\cdot}0$ approx., and is of the same order as that found by the brittle-lacquer test of example 12.5.4.

12.6 Tutorial and practical work

12.6.1 The strain values across the face of a test specimen of rectangular cross-section 30×6 mm due to combined bending and tension are to be measured. Suggest a suitable type of strain-gauge, and design a test specimen with an offset loading such that no stress concentration occurs in the test length, and both tensile and compressive strain occurs across the section.

12.6.2 Sketch the stress system applied to an element in the outer surface of a hollow cylindrical member subjected to (a) tension and torsion, (b) bending and torsion, (c) internal pressure and bending. In which of these cases would strain-gauge rosettes need to be used?

12.6.3 Referring to section 12.2.2.1, show that, if A is the area perpendicular to a direction l, then $\mathrm{d}A/A = -2\nu\,\mathrm{d}l/l$.

12.6.4 Discuss the use of unbonded wire resistance strain-gauges in applications requiring the measurement of (a) tensile and compressive strains, (b) compressive strain only.

12.6.5 Show that, for a null-type Wheatstone bridge, if an increase (δR) in the resistance of one arm and a decrease of the same magnitude in an adjacent arm occurs, then a change of resistance of $2\delta R$ is required in a third arm to restore balance. What is the effect of a resistance change $+\delta R$ in two opposite sides of such a bridge?

12.6.6 Referring to fig. 4.9, show that for two identical strain-gauges operating in push-pull in adjacent arms of a current-sensitive bridge, when the fractional resistance changes ($q = \delta R/R$) are of the same magnitude but opposite sign, then

$$I_G \approx 2V_{AC}mq/\{2m(m + 1) + g(m + 1)^2\}$$

12.6.7 Discuss, and compare the differences between, measurement of static and dynamic strain, both at normal and extreme temperatures.

12.6.8 Discuss the effect of connecting the two strain-gauges in fig. 12.19 into opposite sides of the bridge circuit. Sketch methods of attaching gauges to a shaft so that they are sensitive to (a) torque but not axial force, (b) bending but not torque or axial force.

12.6.9 It is required to determine the stresses in the chassis of a heavy goods vehicle whilst in operation. Suggest different ways of doing this, and discuss their advantages and disadvantages.

12.6.10 Discuss the relative importance of maintaining a constant voltage or voltage amplitude across a null-type, a voltage-sensitive, and a current-sensitive bridge.

12.6.11 Scaling fig. 12.6, calculate the approximate values of magnification, i.e. (pointer movement)/(change of gauge length), and strain sensitivity, i.e. (scale reading)/strain, for that Huggenberger extensometer. Describe a method of calibrating the instrument.

12.7 Exercises

12.7.1 A tensile test was carried out on a long steel bar of rectangular cross-section, to which one Huggenberger extensometer was attached in the axial direction, and a second nearby in the transverse direction, both near the middle of its length. The readings were as follows:

Axial gauge	0	0·20	0·40	0·60	0·80	1·00	1·20	1·40
Transverse gauge	0	−0·05	−0·11	−0·17	−0·22	−0·27	−0·33	−0·38

The sensitivity in scale units per unit strain of the axial gauge was 1028, and that of the transverse gauge 1025. Plot axial strain against transverse strain, and determine the value of Poisson's ratio for the material. [0·28]

12.7.2 Two strain-gauges attached to the surface of a cylindrical pressure vessel, one in the axial and one in the circumferential direction, gave strain values of 0·000 18 and 0·000 72 respectively. Calculate the hoop and longitudinal stress values if the cylinder is of steel having a modulus of elasticity 200 GN/m^2 and Poisson's ratio 0·29.
[84 MN/m^2, 170 MN/m^2]

12.7.3 Calculate the fractional change (q) of resistance of a strain-gauge made of nichrome-V material (a) due to stress of 200 MN/m^2 in the metal to which it is bonded; (b) for a temperature change of 100 °C (i) when bonded to steel, (ii) when bonded to Duralumin. E for steel is 200 GN/m^2; E for Duralumin is 70 GN/m^2.
[$2{\cdot}5 \times 10^3$, 10×10^{-3}, $2{\cdot}86 \times 10^{-3}$, $11{\cdot}1 \times 10^{-3}$]

12.7.4 For an electrical-resistance strain-gauge of sensitivity (F) equal to 2·0 and resistance 100 Ω, determine the value of the shunt resistance to calibrate for a strain of (a) 0·0001, (b) 0·0005, (c) 0·001.
[499 900 Ω, 99 900 Ω, 49 900 Ω]

12.7.5 A load cell consists of a solid cylinder of steel 40 mm diameter with four strain-gauges bonded to it and connected into the four arms of a voltage-sensitive bridge, the arrangement being as shown in fig. 12.21(b) and (c). If the gauges are each of 100 Ω resistance and gauge factor 2·1, and the bridge energising voltage is 6 V, determine the sensitivity of the cell in volts per kilonewton. E for steel is 200 GN/m^2; ν for steel is 0·29. [32·4 μV/kN]

13

Temperature Measurement

13.1 Temperature and standards

Temperature is probably the most frequently measured variable, and one can readily quote such diverse examples as measurement of the temperature of a domestic oven, of the surface of a far distant star, of parts of the human body, or of a red-hot billet of metal being rolled. We can sense temperature roughly by touch, or by receiving visible or invisible radiation from a hot body; however, temperature is not readily defined in simple terms.

A substance has internal energy partly due to the motion of its molecules, which may be oscillatory in the case of a solid or liquid, or of random velocity with collisions in the case of a liquid or gas. This energy is manifested in the temperature of the body. If two bodies at different temperatures are in contact, or in view of each other, a transfer of internal energy from the one at higher temperature to the other at lower temperature occurs, this transferred energy being referred to as 'heat'. Changes of the temperature of a substance bring about a wide variety of effects which may be categorised as physical, chemical, electrical, and optical, and many of these are used as transducing effects in temperature-sensing devices.

13.1.1 The thermodynamic temperature scale

A temperature scale may be defined in thermodynamic terms. From the second law of thermodynamics, the efficiency of a heat engine working between temperatures T_1 and T_2 is $\eta = 1 - Q_2/Q_1$, where Q_1 is the heat given up by the *source* at temperature T_1, and Q_2 is the heat received by the *sink* at temperature T_2. For a hypothetical engine working on the ideal Carnot cycle, the efficiency is $\eta = 1 - T_2/T_1$, and so in this case $Q_2/Q_1 = T_2/T_1$. Hence $T_2 = T_1Q_2/Q_1$ or $T = T_0Q/Q_0$, and a temperature T may be *defined* in terms of heat quantities Q and Q_0 to and from this hypothetical engine, and some datum temperature point T_0. The unit of thermodynamic temperature so defined is the kelvin (K). At the very low pressure of 611·2 N/m^2, ice, water, and steam can co-exist in equilibrium, and this state is referred to as the *triple point of water*. The kelvin is defined as 1/273·16 of the temperature of the triple point of water, i.e. it is a temperature interval.

The use of an engine is impractable for this purpose, and thermodynamic temperatures are realised by using the characteristic gas equation, which applies to perfect gases, thus:

$$PV = mRT \qquad 13.1$$

where P is the absolute pressure, V is the volume of a mass m, R is the characteristic gas constant of the gas, and T is the thermodynamic temperature.

Hence for a given mass of perfect gas confined in a fixed volume, its pressure is proportional to its temperature. No real gases are perfect, but constant-volume thermometers are constructed using near-perfect gases such as helium, hydrogen, and nitrogen, incorporating elaborate methods of correction. In this way, thermodynamic temperature values are established by measurement scientists over a range of approximately 20 K to 1000 K, and are compared with values in the practical temperature scale.

13.1.2 The International Practical Temperature Scale

Although the thermodynamic temperature scale is fundamental and necessary, a more readily available means of reference and calibration is required. This is provided in the International Practical Temperature Scale (IPTS), the latest publication of this being in 1968 (ref. 32). The scale has eleven fixed points, including the temperatures of the triple points of hydrogen, oxygen, and water; the boiling points of some pure liquids; and the freezing points of some pure metals. The boiling- and freezing-points, with one exception, are at the standard pressure, 101·325 kN/m^2. The freezing-point values may be identified by plotting curves during slow cooling, when 'arrest points' are obtained as indicated in fig. 13.1.

The IPTS (1968) defines the Celsius temperature (θ) as

$$\theta = T - T_0 \qquad 13.2$$

where T is the thermodynamic temperature in kelvin units, and $T_0 = 273{\cdot}15$ K.

Having specified fixed points, it is necessary to specify also the method of measuring temperature values between the fixed points. That this

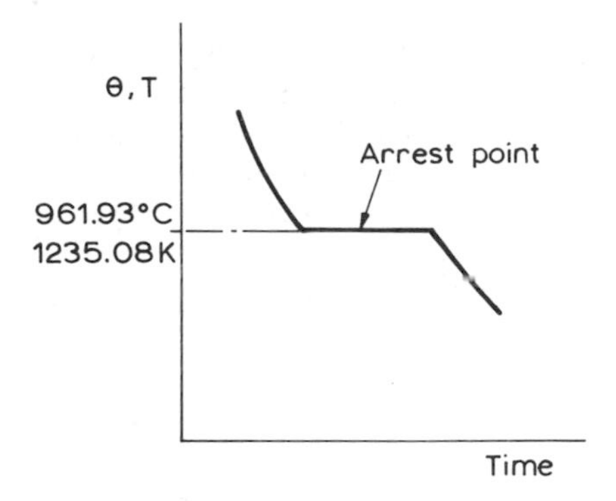

Fig. 13.1 Arrest point for pure silver

is in fact necessary may be demonstrated by careful measurement of a series of temperature values between reproducible values at say 0 °C and 100 °C using several different methods, e.g. (i) the volumetric expansion of mercury, (ii) the constant-volume pressure increase of hydrogen, (iii) the change of resistance of an electrical conductor, etc. It is found that at some intermediate temperature, when one of the measured changes has altered by x% of its range for 100 degrees, then the others will have changed by slightly different proportions of their range. Some methods give a more nearly linear response than others when their performance is compared with the thermodynamic scale; however, to measure intermediate temperature values with good repeatability, the stability of the method is of more importance than its linearity. For this reason, the platinum resistance thermometer is specified in the IPTS for interpolation in the range 13·81 K to 630·74 °C. In the range 13·81 K to 0 °C, the resistance–temperature relationship is defined by a reference function and specified deviation equations. From 0 °C to 630·74 °C, two polynomial equations define the resistance–temperature relationship. The constants in these equations are found by making resistance measurements at specified fixed points.

In the range 630·74 °C to 1064·43 °C, the Celsius temperature (θ) is defined by the equation

$$E = a + b\theta + c\theta^2 \qquad 13.3$$

where E is the e.m.f. of a standard thermocouple of platinum–10% rhodium/platinum. The constants a, b, and c are calculated from the values of E at 630·74 °C ± 0·2 °C, as determined by a platinum resistance thermometer, and at the freezing-points of silver and gold.

Above 1064·43 °C, the thermodynamic temperature (T) is defined by an equation relating the radiated energies at temperatures T and

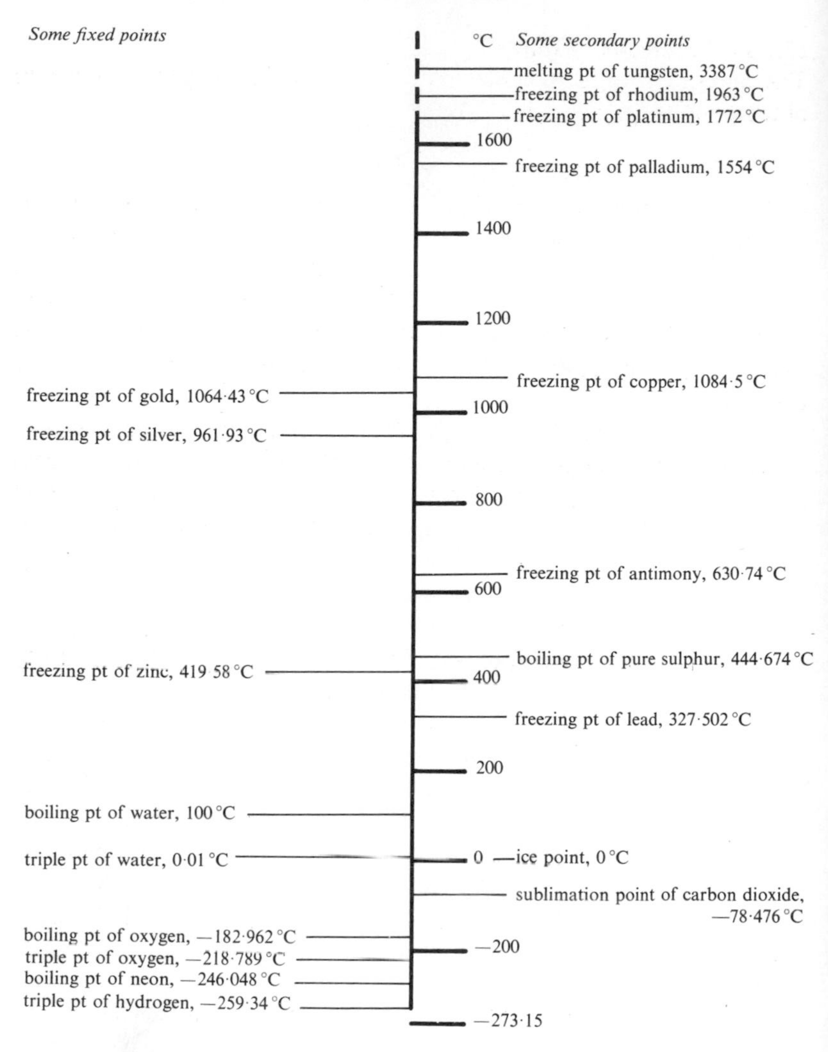

Fig. 13.2 Some points of the International Practical Temperature Scale, 1968

T(Au), the latter being the gold-point temperature, at the same wavelength of radiation, for black-body radiators (see section 13.8).

13.2 Expansion thermometers

Most solids and liquids expand when they undergo an increase in temperature. The direct observation of the increase of size, or of the signal from a secondary transducer detecting it, is used to indicate temperature in many thermometers.

13.2.1 Expansion of solids

The change (δl) of the original length (l) of a solid due to a change ($\delta\theta$) of temperature is given by

$$\delta l = l\alpha\, \delta\theta \qquad 13.4$$

where α is the coefficient of linear expansion, usually taken as constant over a particular temperature range.

The expansion of a single metal is widely used in the rod-type thermostat, shown diagrammatically in fig. 2.9(a), but not usually in indicating thermometers. It should be noted that, as shown in this figure, the expansion of the frame or support for the rod must also be taken into account.

The bonding together of two strips of material having different expansion rates to form a bimetal strip, as shown in fig. 2.9(b), causes bending of the strip when subjected to temperature change. Referring to fig. 13.3, let the initial straight length of the bimetal strip be l_0 at

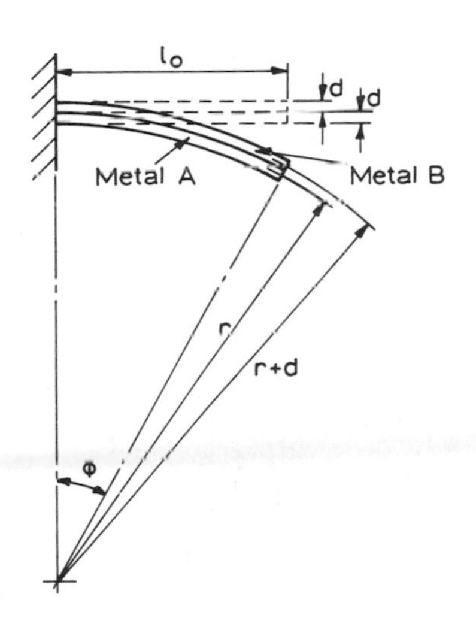

Fig. 13.3 Deflection of a bimetal strip

temperature 0 °C, and let α_A and α_B be the linear expansion coefficients of the materials A and B respectively, where $\alpha_A < \alpha_B$. If the strip is assumed to bend in a circular arc when subjected to a temperature θ, then

$$\frac{r + d}{r} = \frac{\text{expanded length of strip B}}{\text{expanded length of strip A}}$$

$$= \frac{l_0(1 + \alpha_B\theta)}{l_0(1 + \alpha_A\theta)}$$

$$r = d(1 + \alpha_A\theta)/\theta(\alpha_B - \alpha_A) \qquad 13.5$$

If Invar is used for strip A, then α_A is virtually zero, and the equation becomes

$$r = d/\theta\alpha_B \qquad 13.6$$

It is seen that for thinner strips, r is smaller; i.e. more bending occurs. Example 13.10.1 shows a deflection calculation for a bimetal strip. Reference 28 shows relative-deflection-versus-temperature curves for some commerical bimetals. To increase the sensitivity, bimetals are coiled in helical form, see fig. 13.4, when a change of temperature produces a twisting of one end relative to the other. This may give a direct reading of temperature with a pointer on the scale, or may operate a voltage-dividing potentiometer etc.

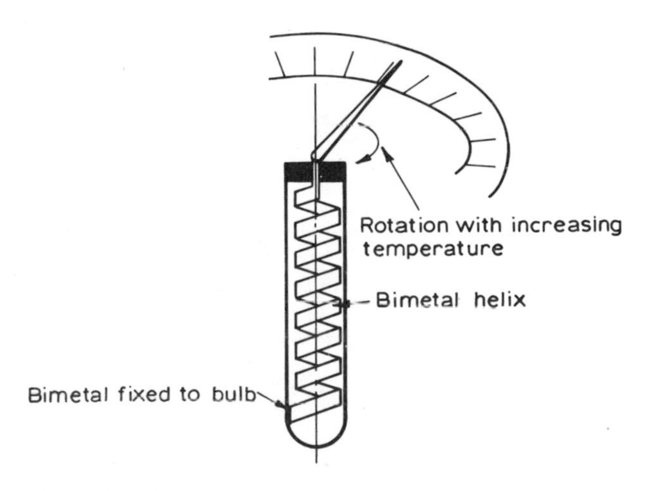

Fig. 13.4 Bimetal thermometer

13.2.2 Expansion of liquids

The change (δV_l) of the volume (V_l) of a liquid due to a change ($\delta\theta$) of its temperature is given by:

$$\delta V_l = V_l\beta_l\,\delta\theta \qquad 13.7$$

where β_l is the coefficient of volumetric expansion of the liquid, taken as an average value over a temperature range.

The container will also have a volume change (δV_c), and, if its temperature change is again $\delta\theta$, then

$$\delta V_c = V_c \beta_c \, \delta\theta$$

where β_c is the volumetric expansion coefficient of the container.

But $V_l = V_c$, hence the apparent volume change (δV) is

$$\delta V = \delta V_l - \delta V_c$$

$$\therefore \quad \delta V = (\beta_l - \beta_c) V \, \delta\theta \qquad 13.8$$

The simple liquid-filled glass-bulb and capillary-tube thermometer is widely used because of its repeatibility, reliability, and cheapness. The preferred liquid is mercury over its range from freezing at −35 °C to vapourising at 375 °C, extended to 510 °C by pressurising the capillary tube with an inert gas such as nitrogen. Other liquids extend the range down to −200 °C, as illustrated in fig. 13.5(a).

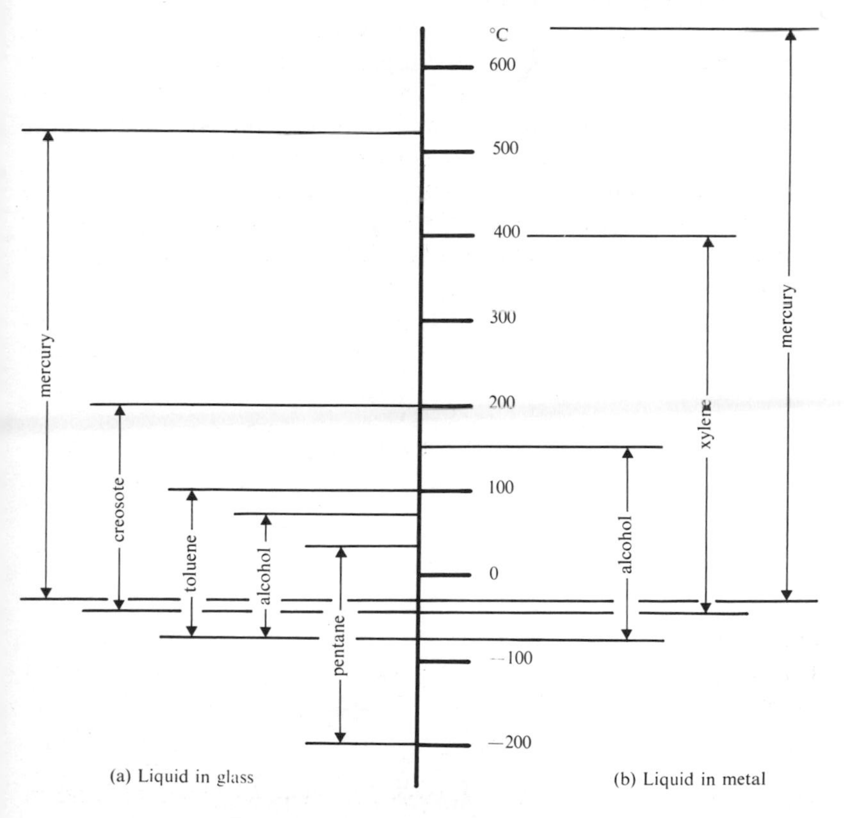

Fig. 13.5 Thermometer filling-liquid ranges

Although glass thermometers are protected at either the bulb or tube end, or both, by metal sheaths, they are too delicate for general adoption in industry. Also, it is not always convenient or possible to read a thermometer near to the measurement point. The liquid-in-metal type, illustrated in fig. 2.10(b), is a more robust instrument, and can be arranged for remote reading. The metal bulb is connected to a metal capillary tube, connected at its outer end to a pressure-sensing instrument such as a Bourdon-tube gauge. The increase in volume of the liquid exerts a pressure in the Bourdon tube, tending to open it and giving a reading on the scale, which is calibrated in temperature units. The system is usually filled under pressure. Mercury is a common filling liquid and, because of its corrosive action on many metals, is always used with stainless-steel materials. Other filling liquids and their temperature ranges are shown in fig. 13.5(b).

13.3 Perfect-gas thermometers

The ideal-gas equation, eqn 13.1, shows that, for a given mass of gas, the temperature is proportional to the pressure if the volume is constant, and proportional to the volume if the pressure is constant. It is much simpler to contain a gas in a constant volume and measure the pressure than to measure the volume at constant pressure; hence constant-volume gas thermometers are common, whilst constant-pressure ones are rare. The simple laboratory type is illustrated in fig. 2.11(a). The industrial type illustrated in fig. 13.6 is filled usually with nitrogen, or for lower temperatures hydrogen, the overall range covered being about −120 °C to 300 °C. At the higher temperatures, diffusion of the filler gas through the metal walls is excessive, the loss of gas leading to loss of calibration.

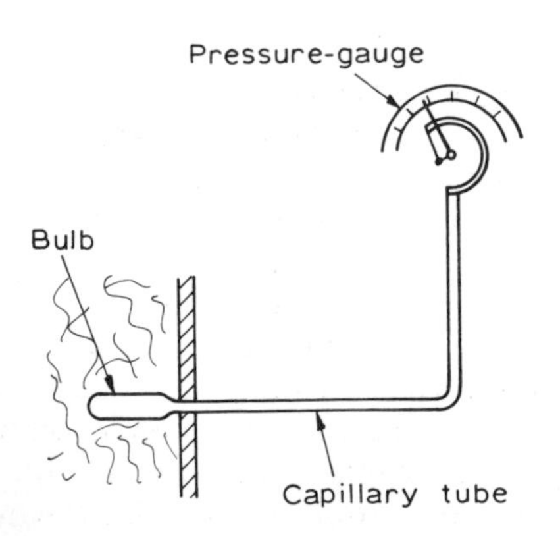

Fig. 13.6 Gas thermometer

13.4 Chemical- and physical-change methods

13.4.1 Temperature-sensitive coatings

Many engineers know the method of determining the tempering temperature of tools and components by observing the colour of the oxides at the surface. The blue colour of many carbon-steel springs is due to their being tempered at a temperature of about 300 °C. The tempering colour changes from pale yellow at 220 °C to dark blue at 300 °C, as the oxide film increases in thickness whilst the temperature of an object is steadily increased. However, the method is not precise, as the oxide film can build up at a constant temperature if it is maintained for a period.

Chemical and physical changes in some substances cause distinct colour changes at fairly well-defined temperature values. Paints and crayons incorporating these are available commercially which give a positive colour indication of temperature in the range from 40 °C to 1400 °C. The values available vary in steps of around 5 °C at the lower end to 20 °C or more at the higher end, with precision of some 5%. Some of these are available with two, three, or four different temperature changes.

The colour changes are in general non-reversible, and, if narrow strips of material of different change-temperature values are painted or crayon-marked on a component such as a piston face or a turbine blade, a plot of the surface-temperature distribution may be built up. The colour changes are time-dependent to some extent, and many temperature values are based on 30 minute heating times. Shorter periods at higher temperatures and longer periods at lower temperatures can produce the same change. BS 1041 : part 3 gives some details of the method.

13.4.2 Change-of-phase methods

Substances can exist in solid, liquid, and gaseous phases. Most of the fixed points of the IPTS are the phase-change points of pure substances; however, mixtures of substances can be contrived to melt at predictable temperature values, and these are available in the form of pyrometric cones about 60 mm high. When a cone bends over so that its tip touches the base level, the stated temperature has been reached, provided the heating has been steady. The cones, made of mixtures of aluminium silicate (china clay), magnesium silicate (talc), sodium aluminium silicate (feldspar), silica (quartz), iron oxide, boric acid, calcium carbonate, etc., are available with collapsing temperatures between 600 °C and 2000 °C. As with the indicating paints, the collapsing temperature is time-dependent, mainly due to the poor conductivity of the cones, and precision is only about 10 °C. By placing a number of cones of different collapsing temperatures in a furnace adjacent to the heat-treated components, the component temperature may be judged from the number of collapsed cones. Similar devices covering the same temperature range are made in the form of pellets and bars (see BS 1041 : part 3).

The phase change from liquid to vapour* is dependent on pressure and temperature. If a vessel contains part liquid and part vapour from the liquid, the quantities of each will adjust until the pressure of the vapour corresponds to the temperature of the liquid–vapour meniscus and is not affected by the temperature distribution in the liquid or the vapour. A vapour-pressure thermometer working on this principle is illustrated in fig. 2.11(b). As shown, the pressure–temperature relationship is not linear; however, this is not a serious disadvantage since scales may be calibrated accordingly, or some compensation may be arranged in the pressure-gauge linkage design. If all the liquid vapourises

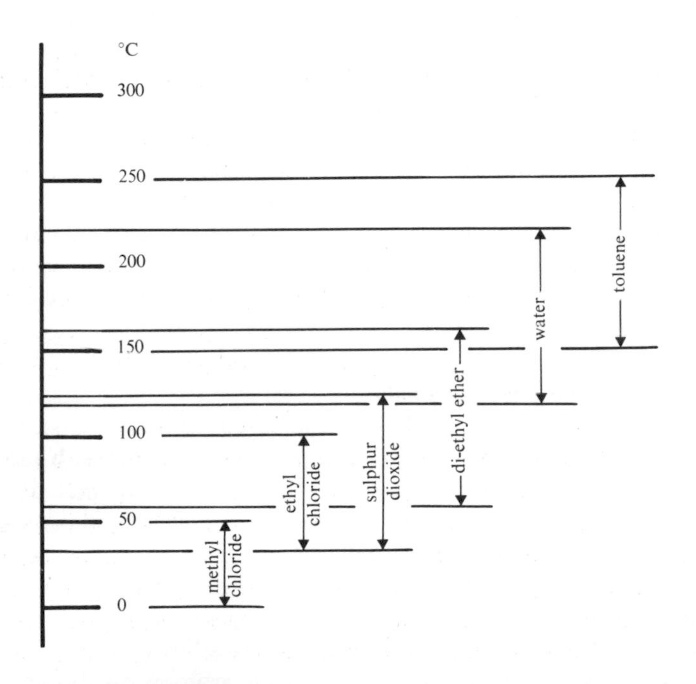

Fig. 13.7 Filling liquids and ranges for vapour-pressure thermometers

* A vapour is a gas near to its pressure and temperature of generation from the liquid.

or condenses, filling the system with a single phase, then the vapour pressure–temperature relationship no longer applies, and the thermometer does not read correctly; hence charging must be carefully controlled.

A number of filling liquids are available, giving instrument ranges of about 100°C between 0°C to +250°C, as shown in fig. 13.7. The instrument is cheaper to make than the liquid-in-metal type, and does not suffer from error due to the ambient temperature of the capillary. However, difficulties arise in the measurement of near-ambient temperatures (see section 13.5.1.2).

13.5 Errors in non-electrical thermometers

Errors may be considered in two categories:

a) *static errors*, i.e. errors in reading steady temperature (the measured fluid may, however, be moving); and

b) *dynamic errors*, which occur when following fluctuating temperatures, or when the thermometer is brought suddenly into contact with the measured temperature.

13.5.1 Static errors

Static errors may occur in thermometers due to the general effects discussed in section 1.3. In addition, the following may cause error.

13.5.1.1 Immersion effects. Any thermometer may cause heat flow (Q) through the capillary or sheath, as shown in fig. 1.12. The energy lost may reduce the measured temperature significantly.

In liquid-filled thermometers, errors due to partial immersion of the bulb may be large, the temperature indicated being in rough proportion to the amount of the bulb immersed. In vapour-pressure thermometers, the temperature of the liquid–vapour meniscus determines the vapour pressure in the system, and incomplete immersion may not lead to error.

13.5.1.2 The effects of ambient temperature. In the liquid-filled thermometer, a change of temperature of the capillary or the pressure spring may lead to error if the liquid and the capillary material have different expansion coefficients. Although the effect is minimised by making the bulb volume large and that of the capillary and pressure spring small, this is not sufficient for high-quality instruments. Compensation is provided in some cases by using a small wire thread through the capillary bore, the materials being chosen so that the volume expansion of the space matches that of the liquid. Another method, though more expensive, is to use a dummy capillary with a second

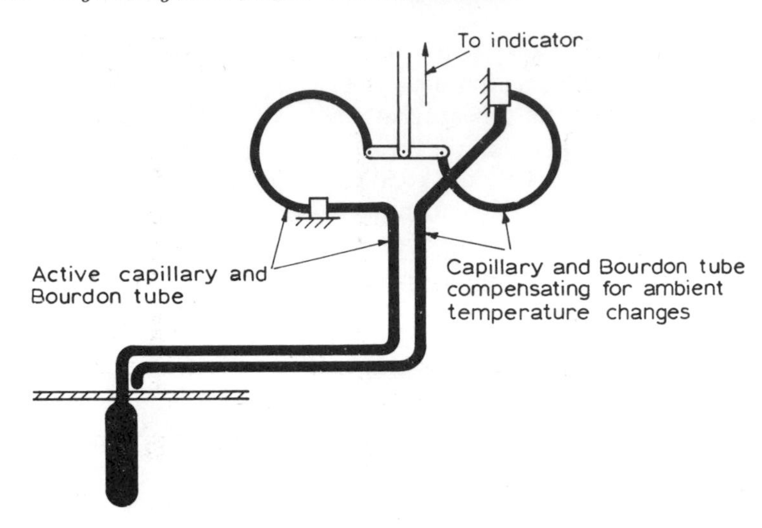

Fig. 13.8 Ambient temperature compensation, liquid-in-metal thermometers

pressure spring as shown in fig. 13.8. Thirdly, a bimetal compensating strip may be used in the pointer linkage.

The compensating-capillary method is not completely satisfactory in the case of gas thermometers, because of the high expansion rate of the gas and the equalising of the pressure in the active capillary and not in the compensating one. However, since the range is large, changes of ambient temperature are not so significant, and the use of a bulb of large volume relative to the capillary volume is usually sufficient.

Ambient-temperature effects can be troublesome in vapour-pressure thermometers under some conditions. With the measured temperature at a lower value than the tube and the pressure-gauge, the tube and gauge will be filled with vapour. When the bulb temperature is higher than ambient temperature, vapour will condense in the tube and spring. The conditions are illustrated in fig. 13.9(a) and (b). Unsteady operation will result as vapourising or condensing conditions occur in the capillary, as the bulb temperature passes through the ambient-temperature value. The effect may be avoided either by fixing the range of measurement of an instrument completely above or below ambient, or by using a second filling liquid which will not vapourise in the range, as shown in fig. 13.9(c).

13.5.1.3 Head errors. The head of liquid (h) in an upright expansion-type thermometer (see fig. 13.10) will exert a fluid pressure on the bulb, causing it to expand, and causing an apparent lowering of temperature.

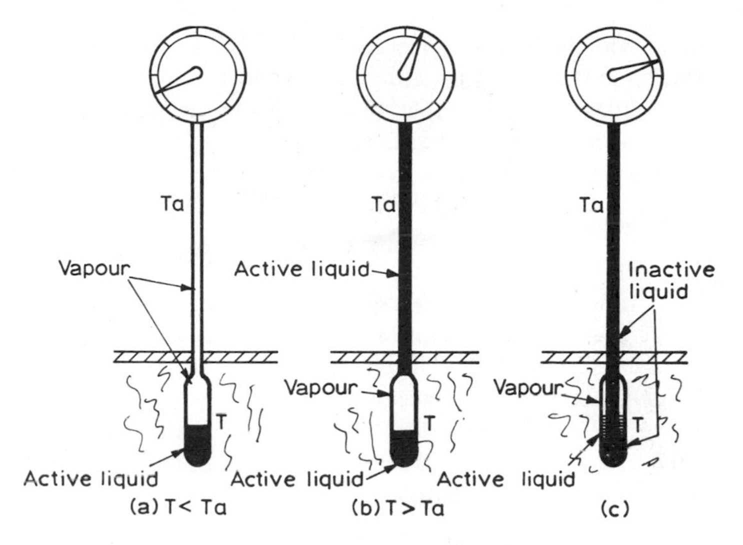

Fig. 13.9 Vapour-pressure thermometers

This will not apply if the axis is horizontal, and will give the opposite effect if the instrument is inverted. However, in systems filled under high pressure, the change due to the head will be a very small proportion of the total pressure, and error is negligible.

The head due to a liquid-filled capillary will exert either a positive or a negative pressure at the pressure-measuring instrument, as may be seen from figs 13.8, 13.9, and 13.10. Again, the error due to this becomes relatively small when the filling pressure is high. If the thermometer is

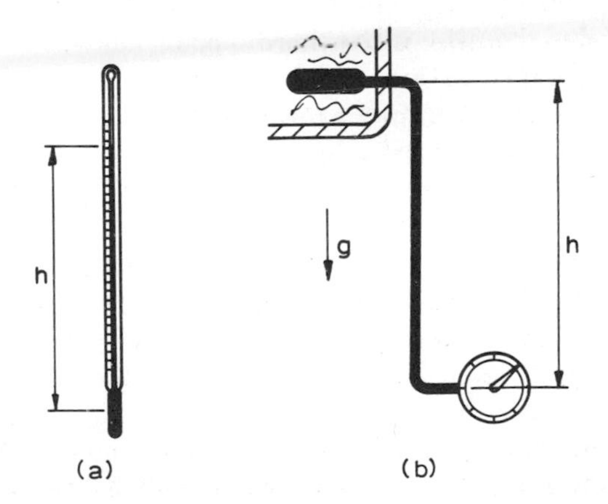

Fig. 13.10

calibrated with the relative positions of the bulb and pressure-gauge the same as in the operating situation, no error due to head occurs in the liquid-filled type. Example 13.10.2 shows the magnitude of the error in a vapour-pressure thermometer.

In vapour-pressure thermometers, the filling pressure is low, and the pressure due to liquid head may be a higher proportion of the total pressure, giving appreciable temperature error in some cases.

13.5.1.4 Ambient-pressure errors. The pressure spring usually measures gauge pressure, i.e. pressure difference from atmospheric; hence the indicated temperature will change due to atmospheric pressure change, giving error. With systems filled at high pressures, the error is negligible; but, as with head errors, the vapour-pressure thermometer is affected to a greater extent. However, the cost of measuring absolute pressure is usually prohibitive.

13.5.1.5 Ageing. Glass, after heating to a high temperature during manufacture, does not contract to its original length immediately when cooled, but continues to contract over a very long period. To reduce this effect, thermometers are annealed during manufacture for several days at a temperature above the maximum they will measure, and are then cooled slowly over several days. In connection with this, thermometers should be subjected to high temperatures for short periods only (see BS 1041: part 2, section 2.1: 1969).

13.5.2 Dynamic errors

Dynamic errors arise in thermometers due to the heat-transfer and heat-capacity effects of the sensing element, since heat must be supplied to raise the temperature of the thermometer fluid, and this must be transferred from the working fluid. Step inputs to a thermometer occur when it is suddenly placed in the measured fluid. Fluctuating temperatures occur in ovens, process fluids, etc. due to various causes, and these sometimes approximate to ramp and sinusoidal form. The characteristic equations of thermometers are basically first-order, though the addition of a well round the bulb tends to give an over-damped second-order effect. The response of first- and second-order systems to step, ramp, and sinusoidal inputs is discussed in Chapter 3, and example 3.5.5 gives an example of the calculation of the dynamic lag of a temperature-measuring system. (See also question 3.6.7.)

The response of a temperature-measuring system may be affected by the mode of heat transfer to the outer surface of the sensing element. This will be mainly by conduction or convection in the thermometers discussed so far, since radiation effects are not predominant below about 600 °C. A thermometer bulb in a still fluid will have a lower heat-input rate than if it is immersed in a moving fluid, and the velocity of the fluid can have a considerable effect (see ref. 29). At very high velocities, a further effect can occur as the fluid is brought to rest adjacent to a thermometer and its kinetic energy is converted to internal energy. The effect is discussed in section 13.6.3.1.

13.6 Temperature measurement by the thermo-electric effect

If two electrical conductors of dissimilar materials are joined to form a circuit as shown in fig. 13.11, it is found that, when the two junctions are at different temperatures (θ_1 and θ_2), small e.m.f.'s (e_1 and e_2) are generated at the junctions, and the algebraic sum of these causes a current (i) to circulate. This is the *Seebeck effect*, discovered in 1822 by Thomas Seebeck, who listed 35 metals in order of their relative thermo-electric e.m.f.'s. The resultant e.m.f. is the same for any particular pair of metals with junctions at particular temperatures, and is not affected by the size of the conductors, the areas in contact, or the method of joining them.

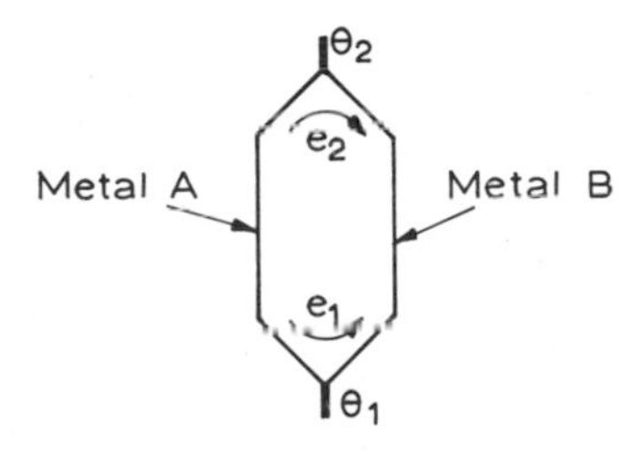

Fig. 13.11 Thermocouple

Later work indicated that the thermo-electric e.m.f.'s were due to separate effects, thus:

Peltier effect Peltier observed that heat flow occurred to the junction at the higher temperature and from that at the lower temperature. The quantity of electricity is related to the heat flow.

Thomson effect Thomson predicted and confirmed that, in addition to the Peltier e.m.f., a further e.m.f. occurred in each material of a thermocouple due to the temperature gradient between its ends, when it forms part of a thermocouple circuit.

From these two effects it is shown (see ref. 29) that, using some simplifying assumptions, the resultant e.m.f. of a thermocouple is

$$e = a(T_1 - T_2) + b(T_1^2 - T_2^2) \qquad 13.9$$

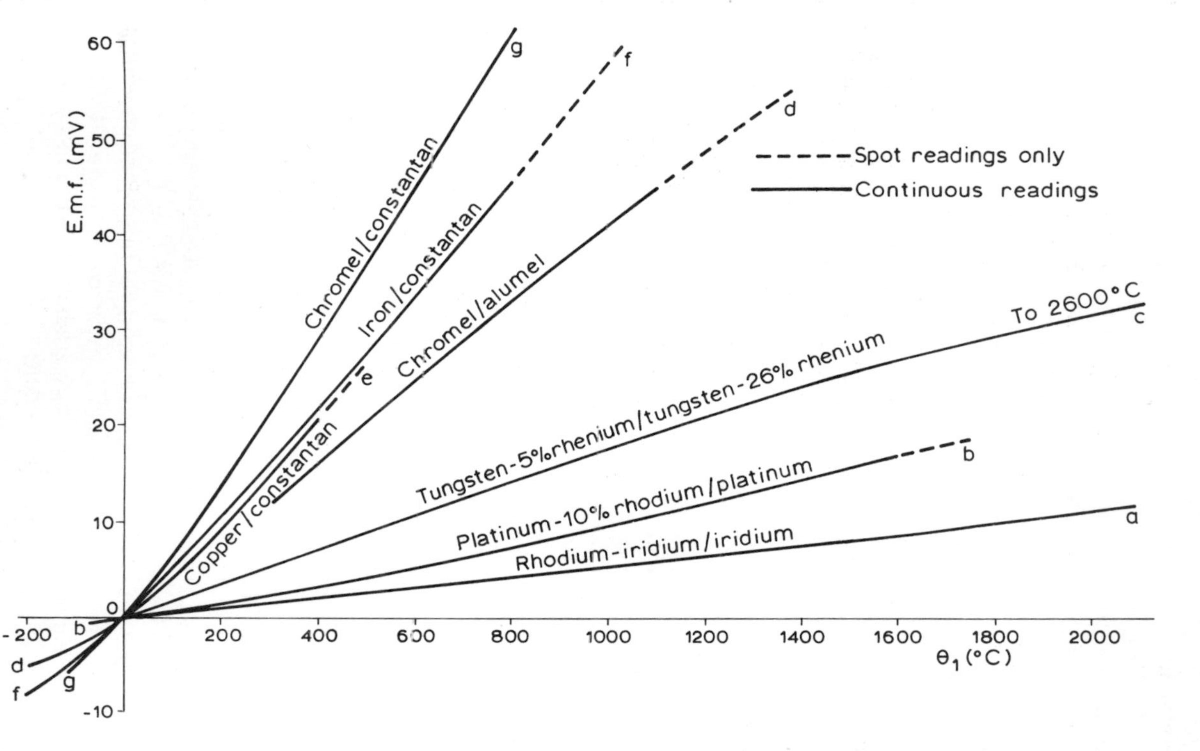

Fig. 13.12 (a) Approximate e.m.f./θ_1 values for $\theta_2 = 0°C$ (data from BS 1041 : part 4 : 1966 and BS 4937 : parts 1, 3, and 4 : 1973)

	Base-metal couples				Rare-metal couples		
Positive wire	Copper	Iron	Chromel 90% Cr, 10% Ni	Chromel	Platinum 90% rhodium 10%	Tungsten 95% rhenium 5%	Rhodium iridium
Negative wire	Constantan 40% Ni, 60% Cu approx.	Constantan	Alumel 94% Ni, 2% Al, + Si and Mn	Constantan	Platinum	Tungsten 72% rhenium 26%	Iridium
Temperature range (°C)	−250 to +400	−200 to +850	−200 to +1100	−200 to +850	0 to +1400	0 to +2600	0 to +2100
Spot maximum (°C)	500	1100	1300	1100	1650		
Characteristics	Resists oxidising and reducing atmospheres up to 350°C. Requires protection from acid fumes.	Low cost. Corrodes in the presence of moisture, oxygen, and sulphur-bearing gases. Suitable for reducing atmospheres.	Resistant to oxidising but not to reducing atmospheres. Susceptible to attack by carbon-bearing gases, sulphur, and cyanide fumes.	Suitable for oxidising but not for reducing atmospheres, carbon-bearing gases, and cyanide fumes. High e.m.f.	Low e.m.f. Good resistance to oxidising atmospheres, poor with reducing atmospheres. Calibration is affected by metallic vapours and contact with metallic oxides.	For use in non-oxidising atmospheres only. The 5% rhenium arm is brittle at room temperatures	Similar to platinum/rhodium–platinum.

Fig. 13.12(b) Industrial thermocouple ranges and characteristics

where T_1 and T_2 are kelvin temperatures, and a and b are virtual constants. The e.m.f. of many couples is found to approximate to this quadratic relationship, and curves for several are shown in fig. 13.12(a), when one junction, the *reference junction*, is held to 0°C and the temperature of the other junction, the measuring junction, is the variable temperature. Values of e.m.f. are tabulated against temperature for various couple materials in **BS** 1041: part 4, 'Code for temperature measurement – thermocouples', and in **BS** 4937 'International thermocouple reference tables'.

The thermocouple circuit must normally be broken to introduce a galvanometer-type instrument to measure the current or a known e.m.f. to balance the generated e.m.f. In doing this, additional materials and junctions are introduced which may be at various temperatures, and the following laws summarise the effects of these.

Law of intermediate metals If one or both of the junctions of a thermocouple are opened, and one or more metals are interposed, the resultant e.m.f. is not altered provided that all the new junctions are at the same temperature as the original junction between which they are positioned.

Law of intermediate temperatures The e.m.f. of a thermocouple with junctions at θ_1 and θ_3 is the algebraic sum of the e.m.f.'s of two couples of the same materials with junctions at θ_1 and θ_2, and θ_2 and θ_3 respectively.

The first law applied to the circuit of fig. 13.13(a) means that if the reference junction at a temperature θ_r is replaced by a galvanometer, with junctions at temperatures θ_a, θ_b, θ_c, and θ_d, the resultant e.m.f. will be unaltered if all of these are the same as the original reference-junction temperature (θ_r).

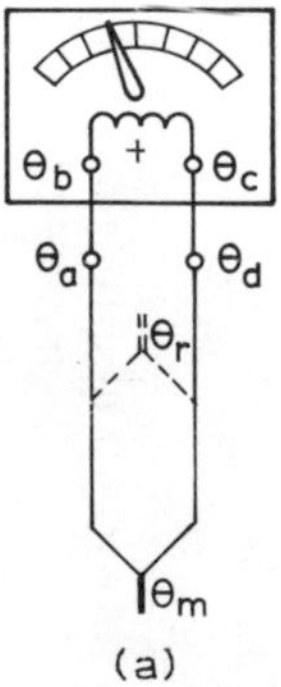

(a)

Fig. 13.13(a) Application of the law of intermediate metals

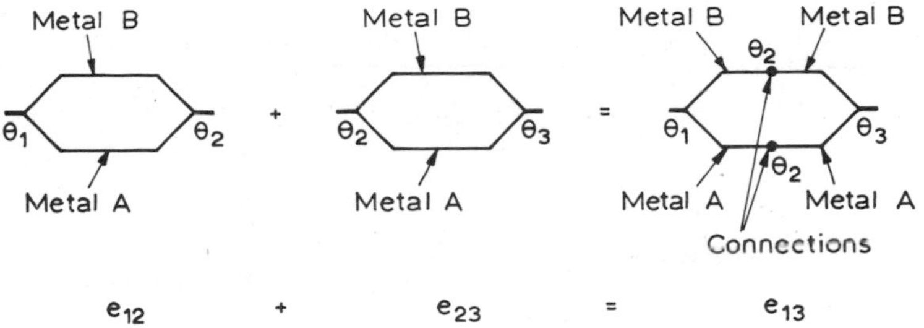

Fig. 13.13(b) The law of intermediate temperatures

The law of intermediate temperatures is illustrated in fig. 13.13(b), where it is seen that a pair of intermediate junctions between materials having the same thermo-electric values, if maintained at the same temperature values (θ_2) as each other, has no effect on the resultant e.m.f. due to junctions at temperatures θ_1 and θ_3.

13.6.1 Thermocouple applications

Thermocouple materials are divided into two categories, i.e. rare-metal types using platinum, rhodium, iridium, etc., and base-metal types. The requirements in many industrial applications are high output e.m.f., resistance to chemical change when in contact with working fluids, stability of e.m.f., mechanical strength in their temperature range, and cheapness. The table of fig. 13.12(b) summarises some of the properties of thermocouples. The standard platinum couple, which defines temperatures in the range from 630·74 °C to 1064·43 °C, is seen to have a wider range of usefulness than this.

Since the e.m.f. generated does not depend on the size of the wires, the smallness of these is limited only by their mechanical strength. However, care must be taken that the wires are not under strain, which can lead to error due to change of resistance. Some hot-junction arrangements are shown in fig. 13.14. At (a) is a simple arrangement consisting of a pair of wires electrically insulated by ceramic sleeves, the wire ends being twisted together or welded. The junctions in (b), (c), and (d) are more robust types, having the wires insulated by mineral material and protected by a metal outer casing which prevents the ingress of damaging hot gases or damage due to pressure and force. The miniaturised types of junction shown in (e) and (f) were originally intended for medical use, the former for injection into muscle tissue, the latter for suturing to internal surfaces, but both find applications in engineering. Other types of hot junction have shaped or pointed ends for insertion into soft objects or for contact with a surface. Miniature

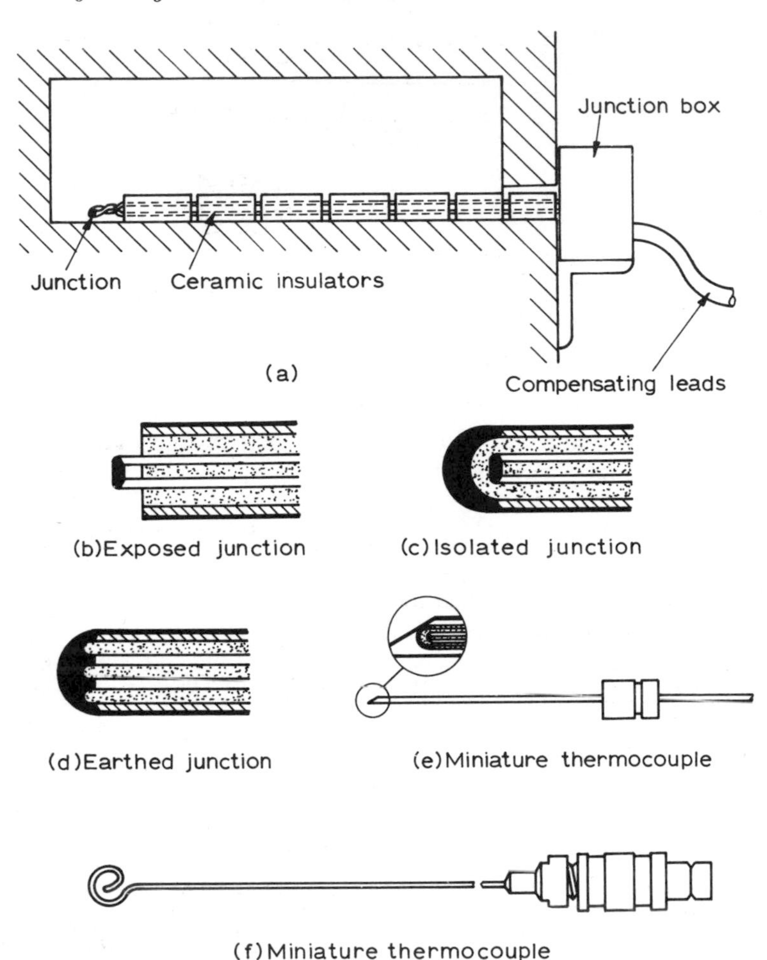

Fig. 13.14 Industrial measuring junctions

types may be buried in solid materials for the measurement of internal temperatures.

Some applications are arduous. For example, in the measurement of molten-metal temperatures in excess of 1600 °C, a dip-and-remove technique with a rare-metal couple is used. Difficulty arises in providing a protective sheath for this which will withstand the molten-metal temperature. Expendable sheathed tips for the couple are used to provide some economy in the use of the rare metals (see ref. 34).

For good repeatability of temperature measurement, the reference junction(s) must be maintained at a constant temperature. This may be

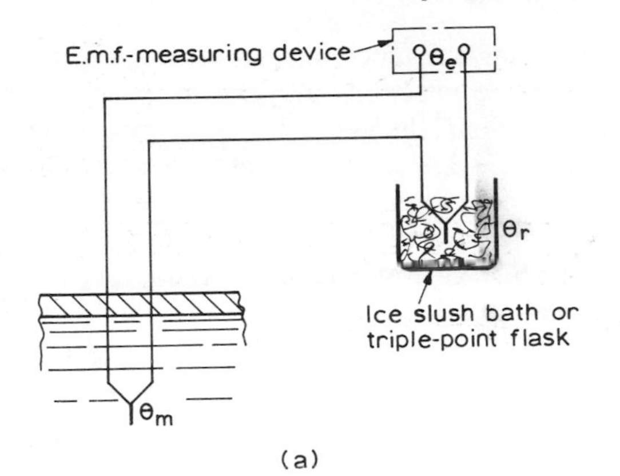

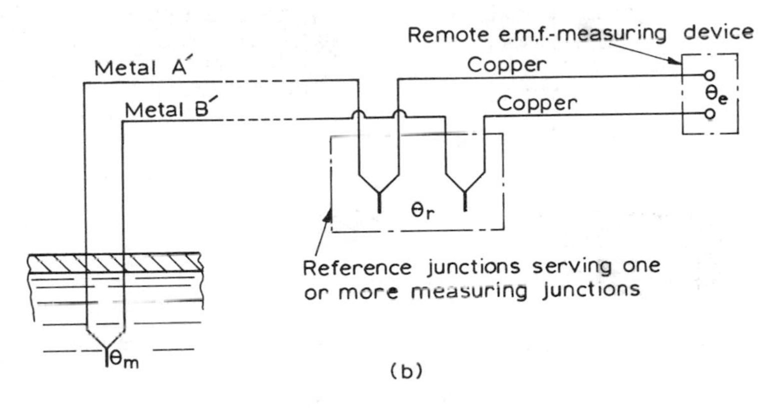

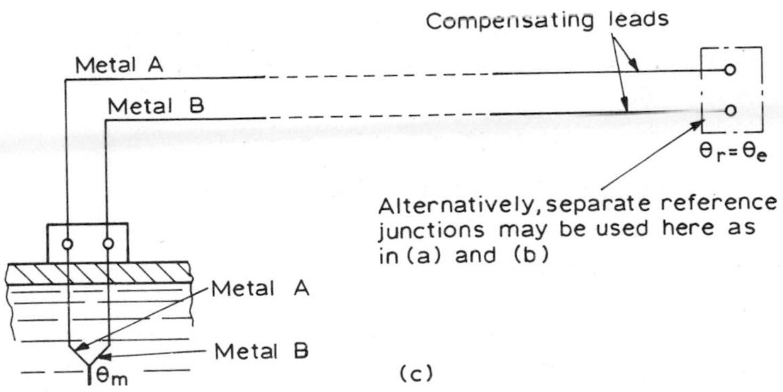

Fig. 13.15 Thermocouple connections

done by immersing the junction(s) in an ice–water mixture, and, in laboratory work, connections may be made as in fig. 13.15(a). In industrial applications, the use of ice–water mixtures is not attractive, and alternatives are automatic temperature-controlled containers or, in less precise work, room temperature, away from the high or low temperature being measured, as shown in (b).

To avoid the high cost of long leads made from the more expensive materials, *compensating leads* may be used from the head of the sensing element, as shown in (c). These have the same thermo-electric e.m.f. as the hot-junction metals, but they have inferior high-temperature properties. From the law of intermediate temperatures, if both of the connections in the junction box are at the same temperature as each other, even though this changes, then the intermediate junctions here have no effect. In (a), (b), and (c), if all the connections at the e.m.f.-measuring device are at the same temperature, then by the law of intermediate metals the junctions here have no effect on the resultant e.m.f. In using a system of thermocouples, a common reference-temperature point may be maintained to which all the compensating leads are taken, whilst the indicators may be at a considerable distance from both these and the measuring junctions.

The resultant e.m.f. at a particular temperature may be increased by multiplying the number of hot and reference junctions as shown in fig. 13.16. In the example shown, the number of measuring junctions, and hence the e.m.f., is multiplied by three. The arrangement is known as a 'thermopile', and is used as a sensitive sensing device in radiation pyrometers. If the thermocouples in this arrangement are at different temperatures, θ_1, θ_2, and θ_3, then the resultant e.m.f. is a measure of the mean value. The materials used are frequently silver–bismuth or chromel–constantan. Further thermocouple arrangements are shown in refs 2 and 11.

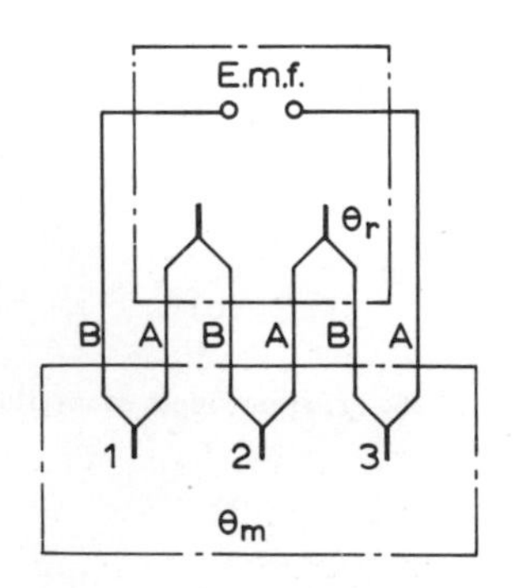

Fig. 13.16 Thermopile

13.6.2 Thermocouple-e.m.f. measurement

The small current circulating in a thermocouple circuit due to the resultant e.m.f. (e) may be measured by a d'Arsonval galvanometer-type instrument (see section 2.4.2) of high resistance. The meter responds to current (i) thus:

$$i = e/(R_m + R_e) \qquad 13.10$$

where R_m is the meter resistance and R_e is the external-circuit resistance.

For good repeatability, R_e and R_m must remain at the same value as at calibration, and, if a meter is changed, the calibration should be checked.

The use of transistors and integrated-circuit methods has led to the development of compact and cheaper d.c. amplifiers, many using digital voltage indicators. These may be made with high input resistance, giving R_m much greater than R_e, and making changes of the latter insignificant when applied to a thermocouple.

An excellent method of measuring the e.m.f. is provided by the d.c. voltage-balancing potentiometer, discussed in section 4.2.2 and illustrated in fig. 4.5(b). The thermocouple measuring points are connected to points E and F, and are connected to the circuit by connecting F to S after standardising the slide-wire with the standard cell (SC). The position of the slider at C_2 for no galvanometer (G) current gives the couple e.m.f. The junctions in the potentiometer may constitute the reference junction of the couple, or a separate reference junction may bc used.

Industrial potentiometric indicators and recorders are usually automatically standardised and balanced. A secondary transducer senses the magnitude and direction of the small unbalance current and initiates control action to move the slider to eliminate it, using a servo-motor or similar device, to the null-balance position. A typical arrangement is shown in fig. 13.17.

13.6.3 Errors in thermocouple systems

13.6.3.1 Static errors. In the galvanometer-type instrument, a swamp resistance (R_s) is usually placed in series with the galvanometer resistance (R_g) as shown in fig. 13.18. The resistance of these, and of the external-circuit couples, wires, and the compensating leads, if used, may vary due to their temperature being different from those at calibration. Varying resistance may occur across the measuring junction,

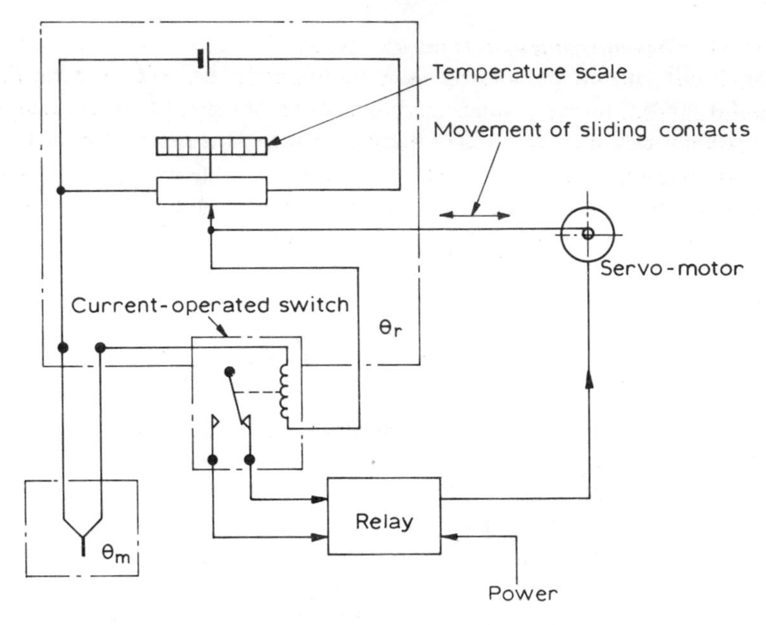

Fig. 13.17 Automatically balanced thermocouple potentiometer

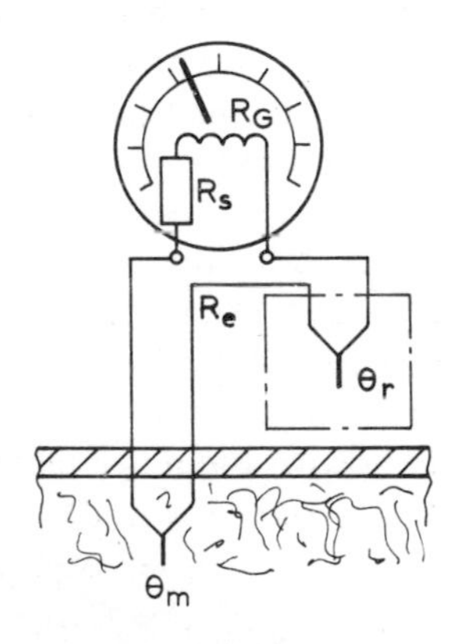

Fig. 13.18

depending on how it is joined, and also across the remaining junctions of the circuit.

The effect of varying temperatures of the meter junctions on the resultant thermal e.m.f. has been discussed in section 13.6.2. These should all be at the same value, but if this is different from the temperature at calibration then error will occur. One method of compensation for this is the use of a bimetal strip, such as in fig. 13.19, to move the 'fixed' ends of the hair-springs of the galvanometer, so correcting the reading as the reference-junction temperature changes. A further method of compensation, using a Wheatstone bridge, is described in ref. 9. Other possible causes of error are leakage currents and induced e.m.f.'s due to nearby electric circuits, and electrolytic effects due to moisture at the junctions of dissimilar metals.

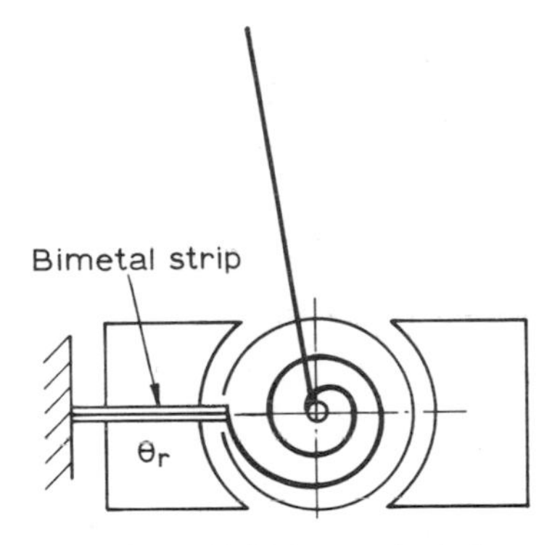

Fig. 13.19

The potentiometer with null balance is inherently more accurate than the galvanometer-type instrument. It eliminates errors due to resistance changes, and the instrument may be frequently standardised for voltage with the built-in standard cell. Compensation for reference-junction temperature change may be made by connecting two further resistances R_A and R_B to the basic potentiometer (fig. 13.20). If R_A is made of nickel, which has a high positive temperature coefficient of resistance, and the other resistances are made of manganin, having virtually zero coefficient, then the increase of resistance of R_A as the ambient temperature increases may be arranged to balance the loss of thermo-electric e.m.f.

At 600 °C and above, radiation effects may lead to the hot junction of the couple radiating a significant proportion of the heat it receives to the cooler walls of a pipe or furnace. In a gas stream, a radiation shield may be positioned round the thermocouple as shown in fig. 13.21(a). The shield may also be used when the walls are considerably hotter than the measured gas temperature. In a furnace where the gas velocity is low, a suction pyrometer may be used. In this, the hot gas is continuously drawn out of the furnace past the thermocouple. Correction figures may be applied for radiation losses such as these under known conditions (see ref. 30).

When gases or vapours carry suspended liquid droplets, the measured temperature is invariably that of the liquid. Probes have

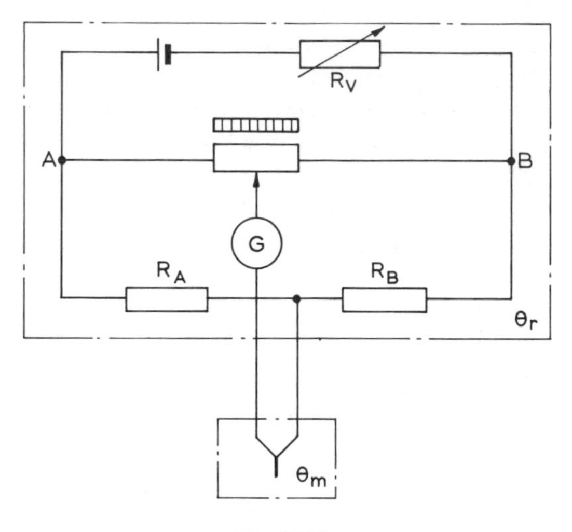

Fig. 13.20

been designed to separate the two phases and measure only the gas temperature (see ref. 2).

Gases at higher velocities are brought to rest locally at the thermocouple, due to its intrusion into the stream. The kinetic energy is transformed into internal energy, resulting in a local rise in temperature at the probe. The resulting temperature is the *stagnation temperature* (T_s). The steady-flow energy equation of the gas is written

$$mc_pT_1 + mV_1^2/2 = mc_pT_2 + mV_2^2/2$$

where m is the mass of gas, c_p is its specific heat at constant pressure, V_1 is the gas-stream velocity, $= V$, V_2 is the velocity at the probe, $= 0$, T_1 is the temperature of the gas stream, $= T_g$, and T_2 is the stagnation temperature at the probe, $= T_s$.

Hence $$T_g = T_s - V^2/2c_p \qquad 13.11$$

A total-temperature probe such as shown in fig. 13.21(b) will measure T_s by bringing the gas-stream velocity virtually to zero at the hot junction. The velocity term is found to be significant at gas velocities above about 50 m/s.

13.6.3.2 Dynamic errors. The dynamic response of thermocouples depends on the response of the measuring junction and the response of the e.m.f.- or current-measuring instruments. The hot-junction response depends on its heat capacity and the rate of heat transfer to

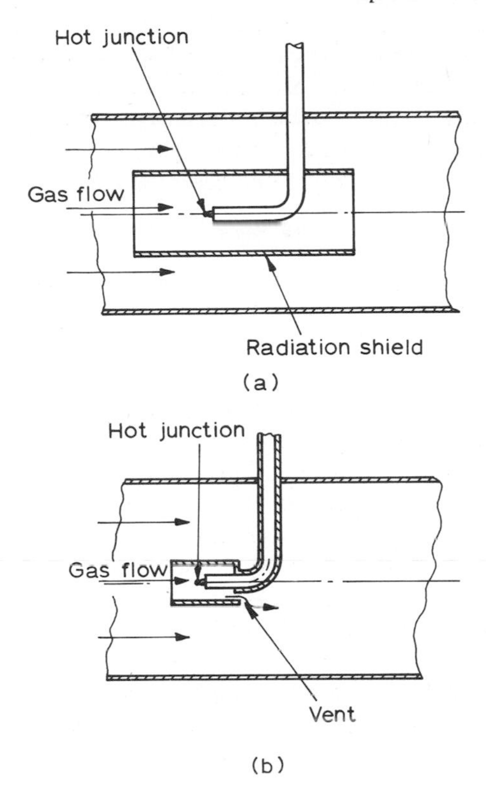

Fig. 13.21 (a) Thermocouple radiation shield (b) Total-head temperature probe

and from the junction. Thin wires and a bare junction give a high heat-transfer rate in, a low heat capacity, and a low heat-transfer rate out along the wires, and hence a quick response. In the more robust probes, the response is inevitably slower, and is similar to that of liquid- and gas-filled bulbs.

The response of the galvanometer instrument is typically faster than that of the measuring probe, but the response of a system is slower than the part of it having the slowest response. The self-balancing potentiometers have typically a longer response time than the galvanometer, and the shortest response time may be expected from the digital voltmeter-type instrument. The overall response characteristics determine the lag and the temperature error in following fluctuating temperatures (see example 3.5.5).

13.7 Temperature measurement by resistance change

The electrical resistance of most metals increases with temperature, exceptions being carbon, which decreases, and manganin, which decreases very slightly (see fig. 12.8). The resistance of electrolytes, semiconductors, and insulators decreases with increasing temperature.

13.7.1 Electrical-resistance thermometers

Platinum, which may be manufactured in very pure form having a stable resistance–temperature relationship, is widely used in industrial and laboratory applications. Its temperature range extends higher than the range in which it is used to define Celsius temperatures (see section 13.1.2). Copper and nickel are also used in industrial resistance thermometers. Curves of relative resistance values at different temperature values for these metals are shown in fig. 13.22. Nickel is seen to be more sensitive and copper more linear when compared with platinum, but both have disadvantages.

Resistance-type temperature bulbs use sensing elements in the form of wires or foil. Thin films deposited on insulating surfaces are also used for temperature sensing. In the wire type, the arrangement is commonly a helical coil wound as a double wire to avoid inductive effect. The laboratory type is often wound on a crossed mica former and enclosed in a Pyrex tube as shown in 13.23(a). The tube may be evacuated or

Temperature °C	*Resistance* (Ω)		
	Platinum	*Nickel*	*Copper*
−200	0·18	0·23	
−100	0·60	0·54	
−50			0·79
0	1·00	1·00	1·00
50			1·22
100	1·39	1·60	1·44
150			1·67
200	1·77	2·34	1·90
250			2·14
300	2·14	3·23	
350		3·80	
400	2·49		
500	2·83		
600	3·20		

Fig. 13.22 Relative resistance/temperature values for resistance-element metals (data from BS 1041 : part 3 : 1960)

filled with inert gas to protect the platinum. The industrial type is typically as shown in (b), the former being of grooved ceramic and the wire being protected by a glass coating or by a stainless-steel tube.

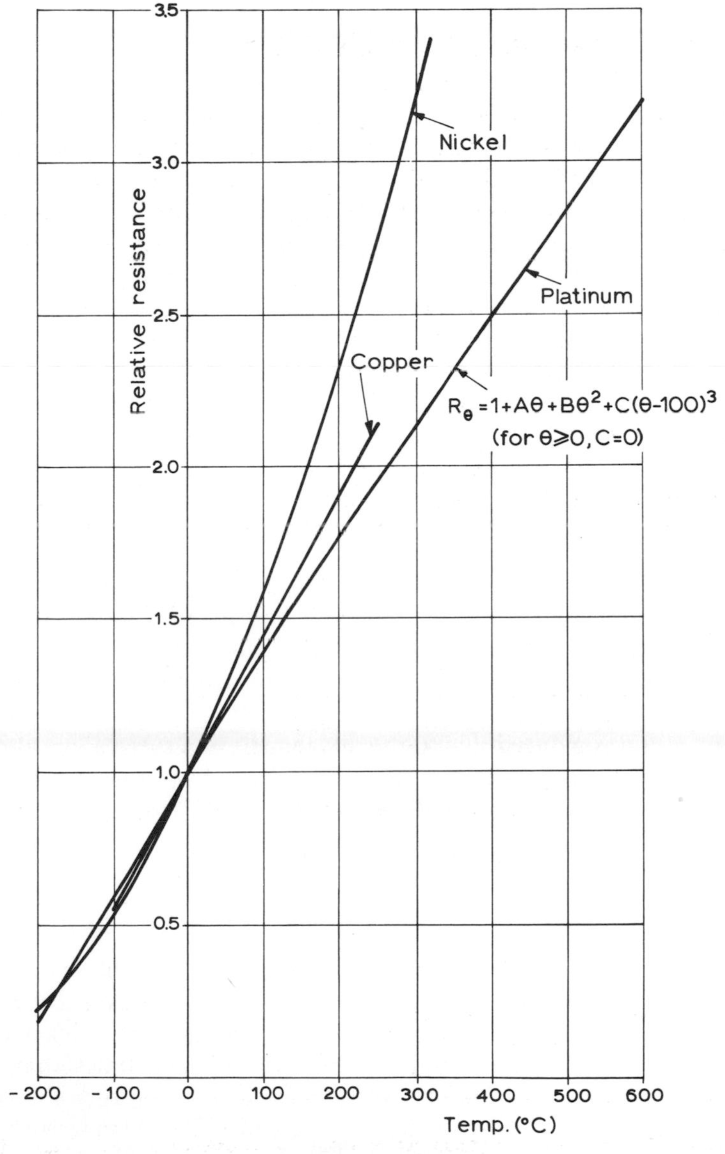

Fig. 13.22 *(cont'd)*

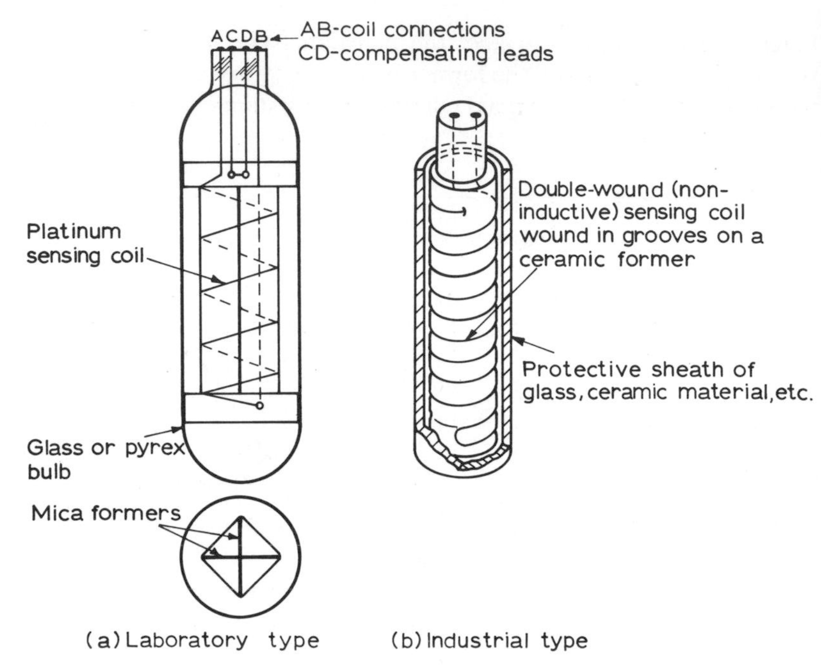

Fig. 13.23 Resistance thermometers

Resistance elements are also available as thin etched grids of metal foil, of similar shape to foil-type strain-gauges. They are constructed of platinum and may be bonded to a plastics backing for attachment to a surface. They may be open-faced or coated, and have a fast response compared with the bulb types. Strain-gauge measuring circuits may be used to measure the resistance changes; however, care must be taken that no strain is induced due to the attachment to a surface, or, alternatively, in some applications the strains from two sensors may be arranged to cancel out their resistance changes (see Chapter 12). Thin-film sensors have an extremely rapid response, and find applications in the aero-space industry.

13.7.2 Thermistors

Thermistors consist of small pieces of ceramic material made by sintering mixtures of metallic oxides of manganese, nickel, copper, cobalt, iron, etc. They have large negative temperature coefficients of resistance, which are non-linear but typically ten times that of platinum or nickel. Curves for some commercial thermistors and the forms of typical transducers are shown in fig. 13.24. Their high resistance enables them to be used in very small elements in the form of beads,

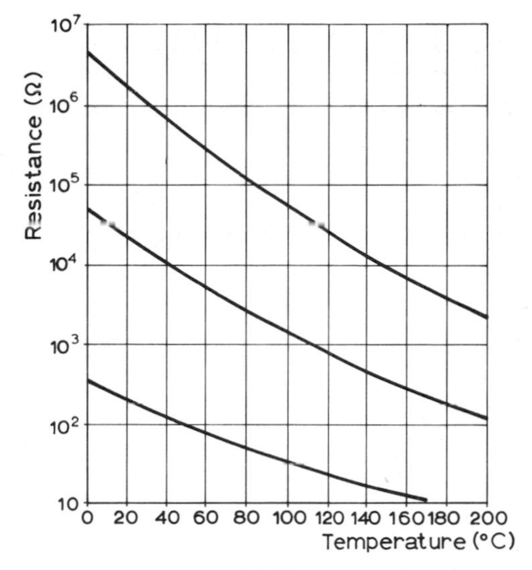

Fig. 13.24(a) Typical *R/θ* curves for thermistors

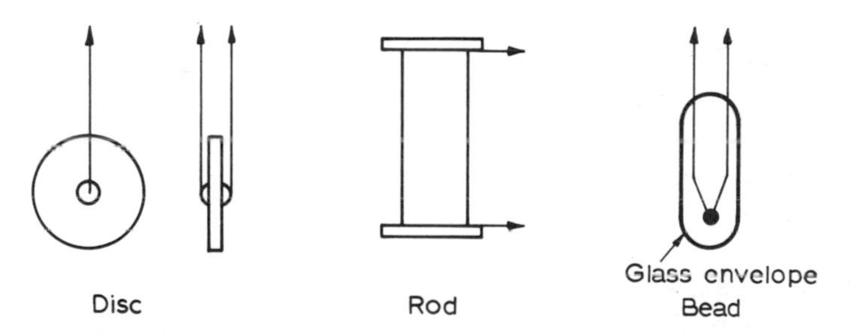

Fig. 13.24(b) Typical thermistor forms

discs, and rods. They can have quick response times in the smaller forms, and their high sensitivity makes them particularly useful in temperature-control systems.

Their temperature–resistance relationship may be expressed in the form

$$R_T = R_0 a e^{b/T} \qquad 13.12$$

where R_0 is the resistance at the ice point, R_T is the resistance at the absolute temperature T(K), and a and b are constants over a small range of temperature.

13.7.3 Resistance-measuring circuits

The simplest resistance-measuring circuit is provided by the null-balance Wheatstone bridge illustrated in fig. 4.6(a), with an automatically balanced type in (b). However, the temperature gradient of the lead wires from the sensing resistance to the measuring bridge will cause a change of resistance of the measuring arm of the bridge not due to the measured temperature. Many methods of compensating for this change have been devised and used (see refs 2, 9, and 29); section 4.3.4 discusses temperature compensation of the leads using the four-lead system shown in fig. 4.10, and gives the bridge conditions under which this is achieved.

Voltage- or current-sensitive bridges using either a.c. or d.c. activation may be employed, giving direct readings which are not quite linear with resistance change. However, the resistance change of metal conductor sensors is also not linear with temperature change, and the two effects may be arranged to balance out to some extent and so enable the use of linear temperature scales on the indicating instrument.

Thermistors have much greater non-linearity of the resistance–temperature relationship than have conductor elements. If Wheatstone-bridge-type circuits are used, the output is not linear, and the system must be calibrated over the range. A half-bridge incorporating a sensing thermistor may be used as described in ref. 2 to produce an output voltage linear with temperature, when the energising e.m.f.'s and bridge resistances have particular values depending on the thermistor constants and the temperature for null balance of the bridge. Since the sensitivity of thermistors is high, some ten times that of platinum, no lead compensation is usually necessary.

13.8 Radiation methods of temperature measurement

The temperature-sensing methods previously described in this chapter have all involved inserting some kind of sensor into the zone of the measured temperature. The top limit of temperature is reached when the inserted material fails in some way, such as melting or oxidising of the bulb or couple material, unstable changes of resistance, etc. The top temperature limit has until fairly recently been about 1600 °C, using platinum–10 % rhodium/platinum thermocouples. Demands from the metallurgical and the aero-space industries have led to the development of rare-metal couples of tungsten–molybdenum, rhenium, iridium, rhodium, etc., so that temperatures up to some 2800°C may be measured. However, in most industrial applications, temperatures above 1600 °C are measured by the radiation from the hot body, and indeed radiation pyrometers of various types are used in the range from room temperatures up to about 5000 °C.

A body at a temperature T radiates energy, the rate being a function of the temperature and also of the type of surface the body has. Highly polished bodies emit, and also absorb, a relatively small amount of energy, and it is easily demonstrated that a matt black body emits much more energy at the same temperature. The ideal emitting surface is the *black-body* or *full radiator*, and the energy (q_b) radiated per unit area from this is given by the Stefan–Boltzmann law, thus:

$$q_b = \sigma T^4 \qquad 13.13(a)$$

where σ is the Stefan–Boltzmann constant, $= 56{\cdot}7 \times 10^{-9}$ W/m^{2}K^4, and T is the kelvin temperature of the surface.

Prevost's theory of exchange states that, for two black bodies in sight, each will radiate energy to the other, and hence the net energy transfer per unit area from 1 to 2 is given by

$$q_b = \sigma(T_1{}^4 - T_2{}^4)$$

where T_1 and T_2 are the surface temperatures, and $T_1 > T_2$.

If T_1 is much higher than T_2, for example when T_2 is an instrument casing at room temperature and T_1 is more than about 600 °C, then the T_2 term is insignificant. In this case, and considering a *grey body*, i.e. a theoretical surface whose *emissivity* is less than a full radiator's, then

$$q = \epsilon\sigma T_1{}^4 \qquad 13.13(b)$$

where ϵ is the emissivity $= q/q_b$.

Where a hot body is totally enclosed by walls *at the same temperature*, then both walls and body radiate and absorb heat at the same rate. If a small hole is made in the container, this area will behave as a full radiator, since rays leaving the enclosure will have been reflected many times. This situation applies very nearly to a furnace with a *small* hole through which a radiation pyrometer may be sighted, either on the body, as shown in fig. 13.25, or on the ceiling of the furnace. In this case, eqn 13.13(a) applies very nearly. Where the temperature of an

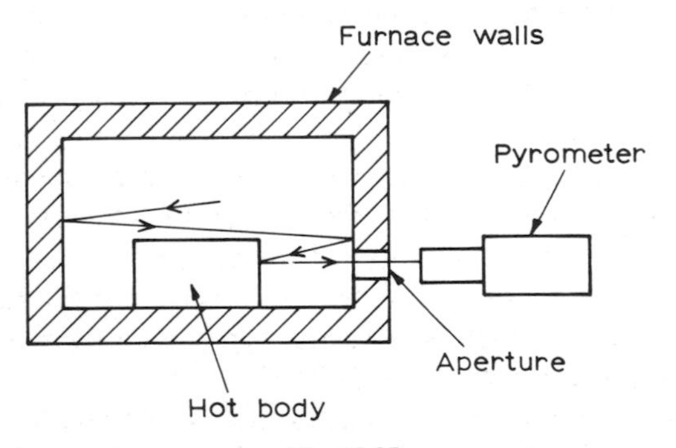

Fig. 13.25

object in the open is measured, due regard must be made to the emissivity of the surface, and eqn 13.13(b) applies.

The energy is radiated over a range of frequencies of the electromagnetic spectrum, the distribution for any particular wavelength (λ) being given by Planck's radiation law, thus:

$$q_{b\lambda} = c_1/\{\lambda^5(e^{c_2/\lambda T} - 1)\} \qquad 13.14$$

where $q_{b\lambda}$ is the energy radiated at wavelength λ, and c_1 and c_2 are constants.

Energy-distribution curves calculated from this equation are shown in fig. 13.26 for three temperature values, and the small visible band of the range is indicated. It is seen that, if a vertical line is drawn at a particular frequency value, then the radiated energy has a particular intensity at each temperature. If a vertical band representing a range of frequencies is drawn, it is seen that the energy radiated at a particular temperature is represented by the area in the band under that temperature curve. The values shown are for full radiators; curves for grey bodies are reduced by the factor ϵ, and actual surfaces follow a less

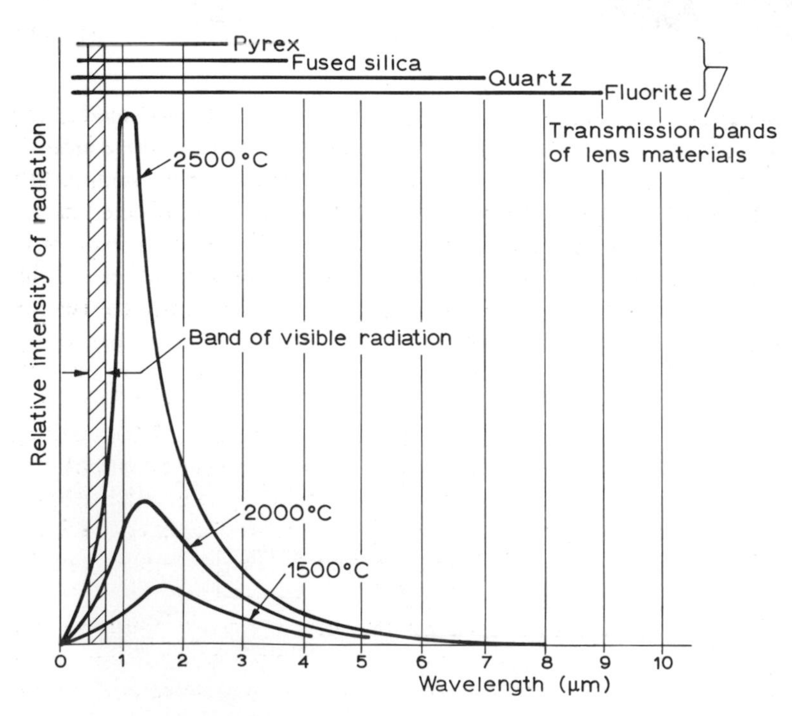

Fig. 13.26 Distribution of radiant energy from a full radiator

	Temperature °C						
	600	800	1000	1200	1400	1600	1800
Iron, unoxidised	0·2			0·37			
Iron, oxidised	0·85			0·89			
Iron, oxidised, smooth	0·96	0·92					
Molten cast iron					0·29	0·29	
Molten steel						0·28	0·28
Nickel, oxidised				0·75	0·75		
Fireclay			0·75	0·52	0·45		
Silica bricks			0·62	0·54	0·46		
Alumina bricks			0·30	0·23	0·19		

Fig. 13.27 Approximate values of emissivity (ϵ)

regular distribution. The emissivity (ϵ) of different types of surface also varies at different wavelengths, and some typical values are shown in fig. 13.27. For temperature measurements, either a wide or a narrow band of radiation may be used.

13.8.1 Wide-band pyrometers

The so-called *total-radiation pyrometer* has some means of directing the radiation from the measured surface on to a temperature-sensing element. This may be done by means of a tube, usually with diaphragms to prevent reflections, or by a parabolic reflector, as shown in fig. 13.28(a), or more usually by a lens system as shown in (b). Where lenses are used, the transmissability of the glass determines the range of the frequencies passing through. The ranges of some lens glasses are shown in fig. 13.26.

The radiated energy absorbed by the sensing element and its mounting causes a rise of temperature, which leads to heat flow from these to cooler surroundings. A steady temperature of the sensing element occurs when the net heat flow is zero. The sensing element may be a resistance element of conductor or thermistor type, or a thermopile. The response time of the instrument will depend on the heat capacity of the sensor, its area, and the heat flow from it, but it is typically of the order of one second.

The radiation from the sighting hole of the furnace must 'fill' the instrument, since radiation from another source will indicate partly another temperature. The cone of 'sight' of a pyrometer is given for each instrument, such that increasing areas must be viewed if the

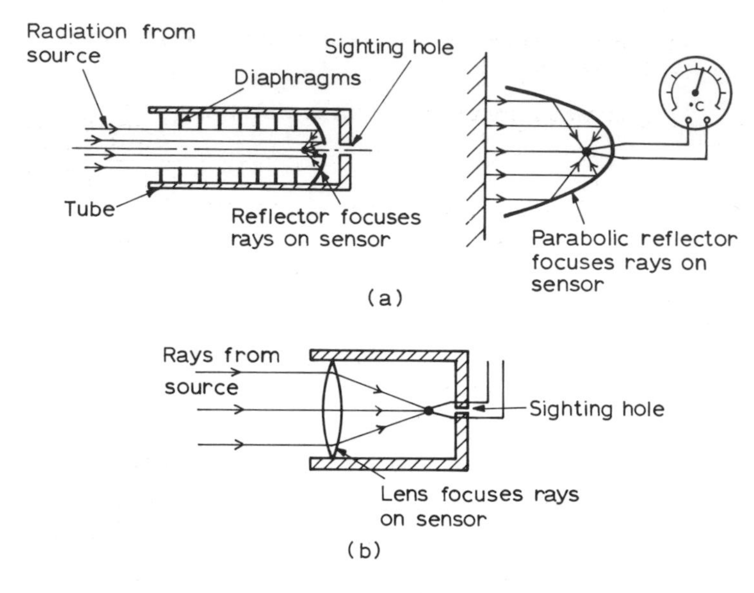

Fig. 13.28

distance between the instrument and the source is greater. Many instruments are of fixed focus, and hence have a limited range of distance from the source.

13.8.2 Narrow-band pyrometers

A widely used instrument in this category is the *disappearing-filament* pyrometer, shown schematically in fig. 13.29. The filament is heated electrically to a temperature such that its colour matches that of the radiation from the measured source, and hence becomes invisible to the operator. If the filament temperature is too high, it appears bright against the source, and vice versa as shown. The ammeter is scaled to read the temperature of the filament, which should be the temperature of a full-radiator source against which its intensity is matched. The filter restricts the light passing through to the eye to a very narrow band at 0·65 μm wavelength; this value is particularly suitable, since the intensity changes of bright metal surfaces with change of temperature are largest at this wavelength. The absorption screen cuts down the radiation passing to the filament and allows the temperature range to be increased. The range of the instrument may be up to 3000 °C.

A further type of instrument is responsive to radiation in the visible and near infra-red range. The sensing element is usually a photo-electric element, or in some cases a sensitive resistive element known as

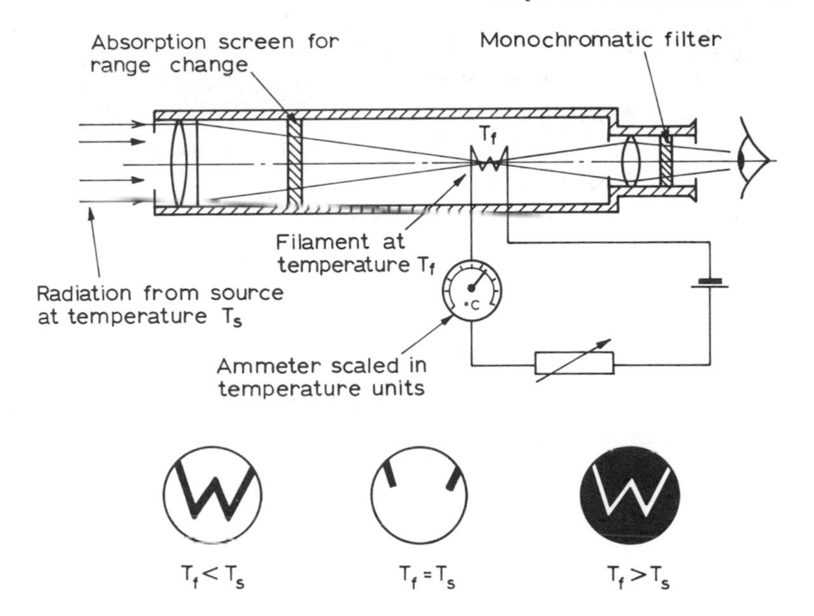

Fig. 13.29 Disappearing-filament pyrometer

a *bolometer*. The photo-sensitive element may be of the photo-voltaic type shown in fig. 2.26, a photo-conductive cell (fig. 2.27), a photo-emissive cell [fig. 2.28(a)], or a semiconductor-type element such as that shown in fig. 2.28(b). These are responsive to varying wavebands of radiation, which may be further narrowed using filters. They have a very rapid response, of the order of a few milliseconds. The errors caused by variation of the emissivity of metals due to oxidation are less at lower wavelengths, and hence this type of instrument, together with the disappearing-filament pyrometer, is inherently more precise than the total-radiation types.

In some versions of the infra-red pyrometer, the radiation from a standard-temperature source such as a tungsten strip lamp (see section 13.9.2) is viewed by the photocell sensor alternately with the radiation from the measured source. The standard source temperature is automatically adjusted by a control system until the two radiation intensities are matched, giving the measured temperature value in terms of, for example, the current through the standard lamp.

13.9 Calibration of thermometers and pyrometers

The calibration of temperature-measuring instruments must be traceable to the primary standard, which is the thermodynamic temperature scale of the SI. However, the temperatures of the defining fixed-point conditions of the International Practical Temperature Scale are known in terms of thermodynamic temperature within a small uncertainty, typically 0·01 K at the boiling-point of oxygen and 0·2 K at the freezing-point of gold, and for engineering purposes the IPTS (1968) may be taken as the reference scale.

13.9.1 Determination of point values

The defining temperature point of the thermodynamic scale, 273·16 K at the triple point of water, may be obtained using a triple-point cell and flask as shown in fig. 13.30. During manufacture, the cell is evacuated, partly filled with pure water, then sealed at the triple-point pressure (611·2 N/m^2). After following a freezing procedure, the equilibrium temperature may be maintained for a week or more if the cell is kept well packed with ice in the flask. Thermometers of various types are placed in the centre tube for calibration to an accuracy within $\pm 0{\cdot}0001\,°C$.

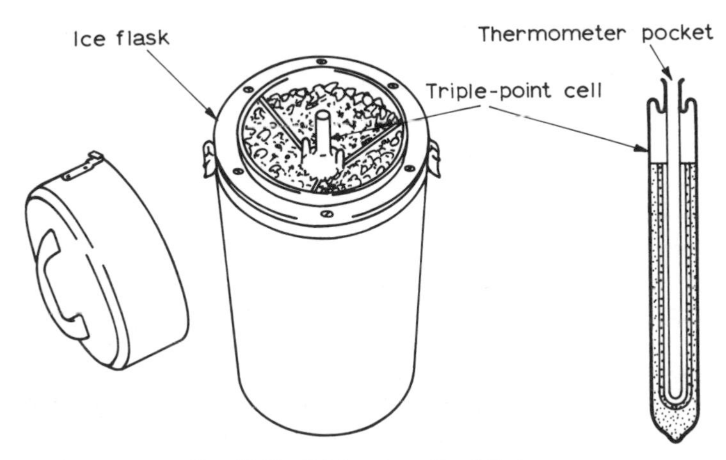

Fig. 13.30 Triple-point cell and flask

Boiling- and freezing-point apparatus is described and illustrated in ref. 33. For boiling-points, a small quantity of the pure liquid is heated electrically in an insulated metal container as shown in fig. 13.31. One or more radiation shields are interposed between the thermometer pockets and the container; without these, significant error can occur in the calibrated temperature. The pressure has a significant effect on the boiling-point temperature, and, for precise measurement, the apparatus is connected to a large reservoir maintained precisely at the standard atmospheric pressure.

Freezing-point temperatures are much less affected by pressure

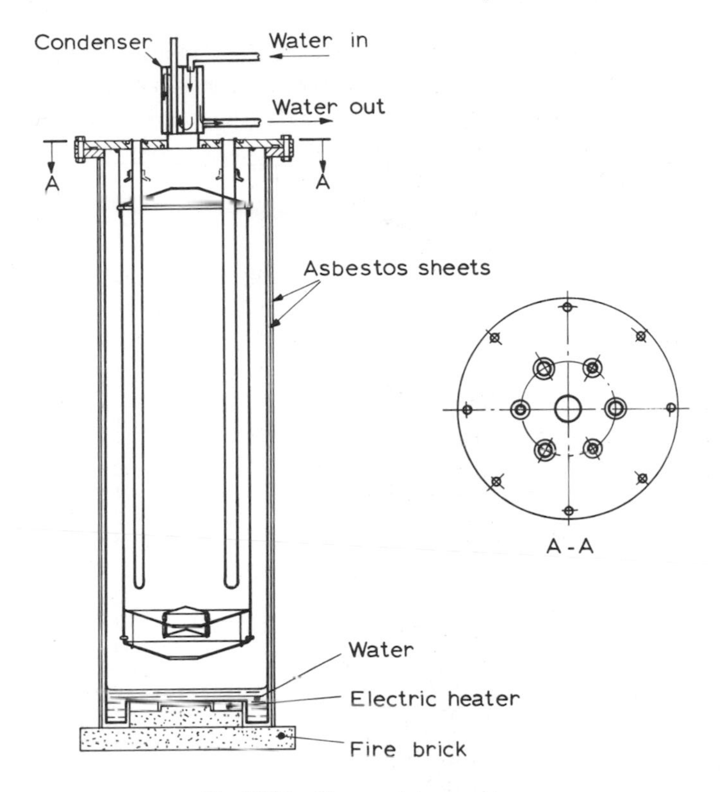

Fig. 13.31 Steam-point apparatus

variations, but are more susceptible to the effects of impurities in the materials. A typical freezing-point-of-metals apparatus is the tubular furnace shown in fig. 13.32. The depth of immersion of the thermometer in the central pocket should be sufficient to overcome conduction effects, and the metal should be heated in a uniform temperature zone. This is obtained by using a large length/diameter ratio, preferably more than 20. The temperature is raised to just above the melting-point, and the assembly is then allowed to cool slowly. A plot of the thermometer output against time reveals the arrest point and its corresponding output value for the temperature point.

A useful method for the point calibration of thermocouples uses the melting of a small quantity of pure metal fixed between the hot-junction wires. This assembly and a second junction are heated together slowly in the furnace. At the melting-temperature of the metal, the e.m.f. of the test couple remains stationary, and then falls as it breaks, whilst the e.m.f. of the second couple rises at a constant rate.

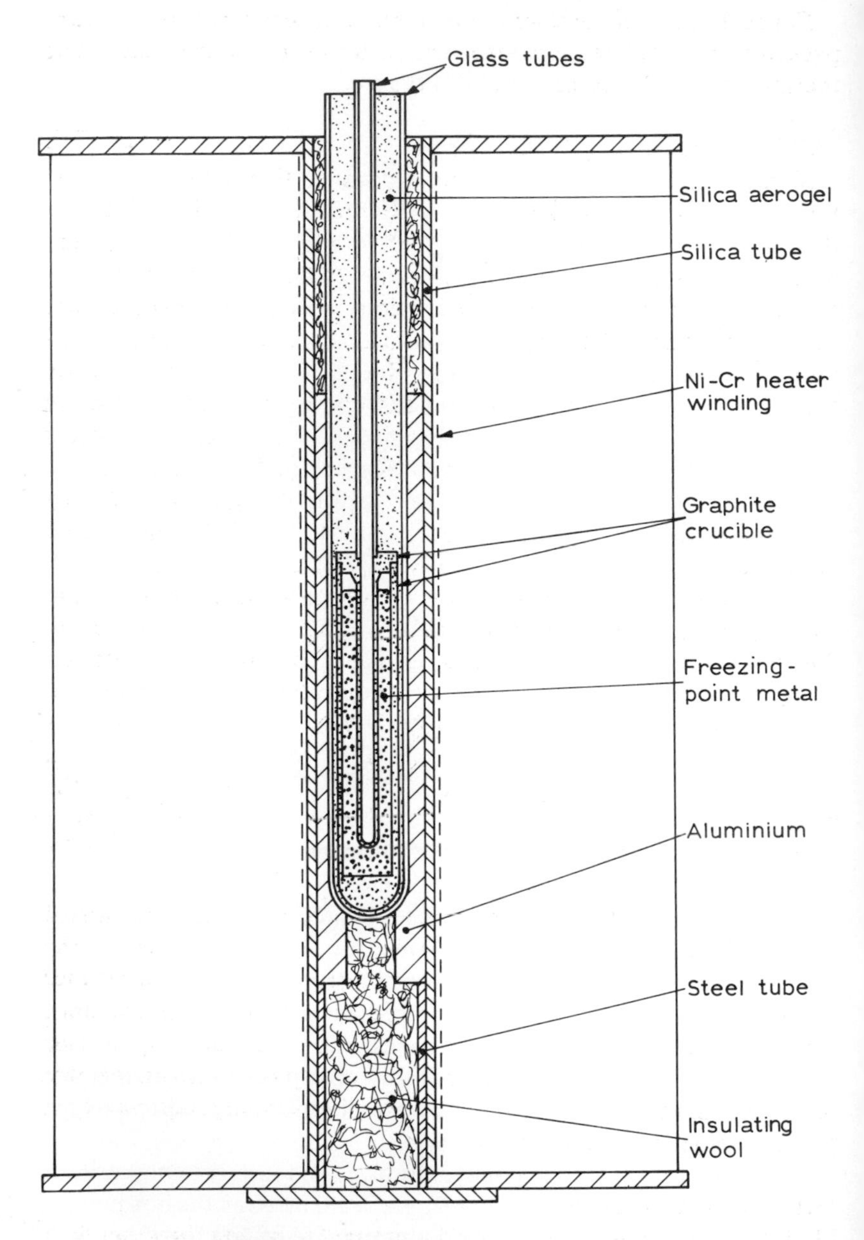

Fig. 13.32 Freezing-point-of-zinc apparatus

For realisation of the ice-point temperature, shavings of pure ice are packed in an insulated container with a drain to remove liquid. The thermometer is placed in a cavity in the ice.

13.9.2 Comparison of thermometers

Defining instruments, i.e. platinum resistance thermometers for the range 13·81 K to 630·74°C, and platinum–10% rhodium/platinum thermocouples for the range 630·74°C to 1064·43°C, are calibrated using the IPTS (1968) interpolation formulae, using values of resistance or e.m.f. measured at given fixed points. Secondary standards are calibrated against these by the National Physical Laboratory and other laboratories approved by the British Calibration Service, by comparison methods. Working standards are calibrated against the secondary standards. The fixed points and secondary reference points of the IPTS (1968) are available for point checks on all instruments.

Comparison in the middle range of temperature is made by immersing the standard instrument with others in a heated or cooled and stirred liquid bath. The liquids are typically iso-pentane (−150°C to 0°C), water (0°C to 100°C), oils (80°C to 300°C), and a mixture of sodium and potassium nitrate (200°C to 600°C). Heating is carried out by an external or immersion-type electric element, cooling by circulating a refrigerated fluid, and the temperature is controlled to a constant value.

Above about 600°C, salt baths are difficult to use, and comparisons are made in the tubular-type furnace, the range of which may be within the range 100°C to a top limit of about 1850°C. The accuracy of the furnace method is inferior to the liquid bath, particularly at lower temperatures, where heat transfer is by conduction rather than by radiation.

Calibration of disappearing-filament optical pyrometers is carried out using tungsten strip lamps. These latter are calibrated at the National Physical Laboratory in terms of current and luminance temperature, the temperature scale being derived from the standard photoelectric pyrometer using a wavelength of about 0·66 μm (see ref. 33). The filter of the disappearing-filament pyrometer allows this same wavelength to pass. Other methods of calibrating radiation-type pyrometers are also described in ref. 33.

13.10 Worked examples

13.10.1 A bimetal-strip element has one end fixed and the other free, the length of the cantilever being 40 mm. The thickness of each metal is 1 mm, and the element is initially straight at 20°C. Calculate the movement of the free end in a perpendicular direction from the initial line when the temperature is 180 °C, if one metal is Invar and the other is a nickel–chrome–iron alloy with a linear expansion coefficient (α) of $12{\cdot}5 \times 10^{-6}$/°C.

By eqn 13.6, since α for Invar = 0,

$$r = d/\theta\alpha_b$$

$$= \frac{1}{160 \times 12{\cdot}5 \times 10^{-6}} = \frac{10^6}{2000} = 500 \text{ mm}$$

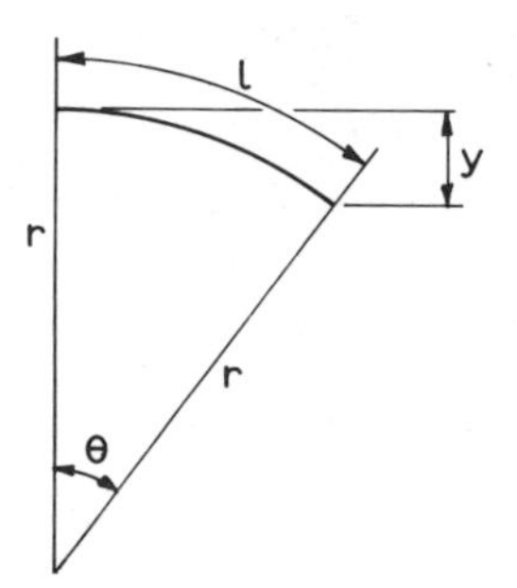

Fig. 13.33

$$\text{From fig. 13.33,} \quad \phi = \frac{l}{r} = \frac{40}{500} = 0{\cdot}080 \text{ rad}$$

$$= 4°35'$$

$$y = r - r\cos\phi = r(1 - \cos\phi)$$

$$= 500(1 - 0{\cdot}996\,81)$$

$$= 1{\cdot}60 \text{ mm}$$

13.10.2 A vapour-pressure thermometer filled with pure water is positioned with the Bourdon-tube pressure-gauge 5 m below the bulb. Using tables of thermodynamic properties, calculate the error in °C due to the head effect when the bulb is at a temperature of 200 °C, if this has not been taken into account in calibration.

The capillary will be full of water under the given conditions, the pressure due to this being

$$P_h = \rho g h = 1000 \times 9{\cdot}81 \times 5$$

$$= 49{\cdot}050 \text{ kN/m}^2$$

The vapour pressure is the pressure of saturated steam at this temperature, which from steam tables is 1555 kN/m^2. Hence the pressure at the gauge will be 1555 + 49 kN/m² = 1604 kN/m².

From the tables, the saturation temperature at 1600 kN/m² is 201·4 °C and the saturation temperature at 1700 kN/m² is 204·3 °C; hence at 1604 kN/m² the saturation temperature is

$$\theta_s = 201{\cdot}4 + \frac{1604 - 1600}{1700 - 1600}(204{\cdot}3 - 201{\cdot}4)$$

$$= 201{\cdot}4 + \frac{4}{100} \times 2{\cdot}9$$

$$= 201{\cdot}5\,°\text{C}$$

The thermometer will read approximately 1·5 °C high because of the head effect.

13.10.3 A chromel–alumel hot junction is connected directly to a potentiometer. This is at a temperature of 20 °C, and gives a reading of 27·07 mV. Determine the measured temperature, assuming the couple conforms to the values given in BS 4937: part 4: 1973.

The measured e.m.f. is the algebraic sum of those at the hot and reference junctions, the latter being those at the potentiometer, i.e. $(e_m - e_r)$. But the required value is $(e_m - e_0)$, i.e. with the second junction at 0 °C, so that standard tables may be referred to for the temperature.

But $$(e_m - e_r) = (e_m - e_0) - (e_r - e_0)$$

and $$(e_m - e_0) = (e_m - e_r) + (e_r - e_0)$$

From the tables, $(e_r - e_0)$ for 20 °C is 0·80 mV, and, since the measured e.m.f. $(e_m - e_r)$ is 27·07 mV, then

$$(e_m - e_0) = 27{\cdot}07 + 0{\cdot}80$$

$$= 28{\cdot}87\ \text{mV}$$

From the tables, this is seen to correspond to a temperature of 670 °C.

13.10.4 A thermocouple circuit as shown in fig. 13.18 uses chromel–alumel couples which give an e.m.f. of 33·3 V when θ_m = 800 °C and θ_r = 0 °C. The resistance of the instrument coil (R_g) is 50 Ω, and 0·1 mA gives full-scale deflection. The resistance (R_e) of the junctions and leads is 12 Ω.

Calculate (a) the value (R_s) of the series resistance if a temperature of 800 °C is to give full-scale deflection, (b) the approximate error due to a rise (δR_e) of the external resistance equal to 1 Ω, (c) the approximate error due to a rise of 10 °C in the copper coil of the galvanometer if it has a temperature coefficient of resistivity (α) of 0·00426/°C. (The series resistance (R_s) is assumed to have a zero temperature coefficient.)

a)
$$e = i(R_g + R_s + R_e)$$
$$\therefore \quad 33{\cdot}3 \times 10^{-3} = 0{\cdot}1 \times 10^{-3}(50 + R_s + 12)$$
$$R_s = 333 - 50 - 12$$
$$= 271\ \Omega$$

b)
$$i' = \text{current with increased external resistance}$$
$$= e/(R_g + R_s + R_e + \delta R_e)$$
$$= 33{\cdot}3/(50 + 271 + 12 + 1) = 33{\cdot}3/334$$
$$= 0{\cdot}09970\ \text{mA}$$

The approximate temperature error is $-0{\cdot}0003 \times 800/0{\cdot}1 = -2{\cdot}4$°C
c) The resistance change of the galvanometer coil is

$$\delta R = R_g \times \alpha \times \delta T$$
$$= 50 \times 0{\cdot}00426 \times 10$$
$$= +2{\cdot}13\ \Omega$$

The result is again a low temperature indication by approximately $2{\cdot}4 \times 2{\cdot}13$ or about -5°C

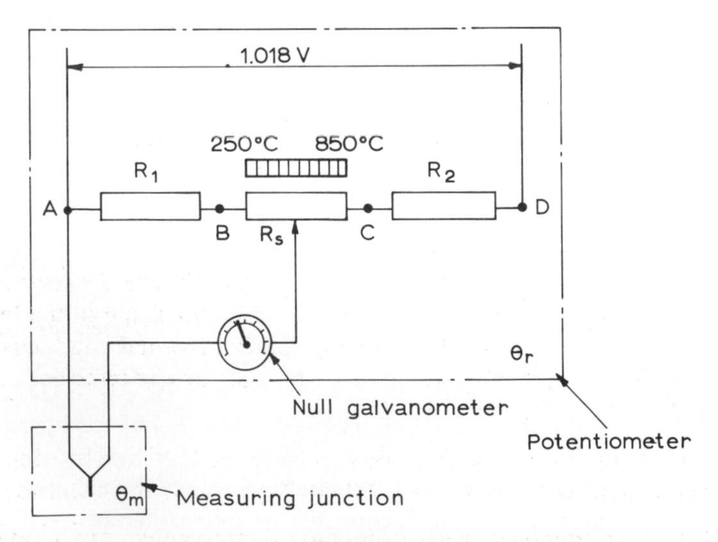

Fig. 13.34

13.10.5 The e.m.f. of an iron–constantan thermocouple is to be measured by the potentiometer shown in fig. 13.34. A potential difference of 1·0186 V is applied over the points AD and is standardised using a standard Weston cell. The current through the resistors is to be 2 mA, and the range of temperature measurement is to be from 250 °C to 850 °C. Calculate the value of the resistances R_1, R_s, and R_2 for an ambient temperature $\theta_r = 20\,°C$.

From BS 4937: part 3: 1973, the e.m.f. of the iron–constantan couple with the reference junction at 0 °C is

$$\text{e.m.f. at } 20\,°C = 1{\cdot}019 \text{ mV}$$
$$\text{e.m.f. at } 250\,°C = 13{\cdot}553 \text{ mV}$$
$$\text{e.m.f. at } 850\,°C = 48{\cdot}716 \text{ mV}$$

The resultant e.m.f.'s at the θ_m values given, for $\theta_r = 20\,°C$, are

$$\text{at } 250\,°C,\ e_1 = 13{\cdot}553 - 1{\cdot}019 = 12{\cdot}534 \text{ mV}$$
$$\text{at } 850\,°C,\ e_2 = 48{\cdot}716 - 1{\cdot}019 = 47{\cdot}697 \text{ mV}$$

$\therefore$ the p.d. over the points BC must be

$$47{\cdot}697 - 12{\cdot}534 = 35{\cdot}163 \text{ mV}$$

If the current (i) = 2 mA, then

$$R_S = 35{\cdot}163/2 = 17{\cdot}581\ \Omega$$

and

$$R_1 = V_{AB}/i = 12{\cdot}534/2 = 6{\cdot}267\ \Omega$$

Also

$$V_{AD} = i(R_1 + R_S + R_2)$$
$$\therefore\ R_1 + R_S + R_2 = 1018{\cdot}6/2 = 509{\cdot}3$$

and

$$R_2 = 509{\cdot}3 - (17{\cdot}581 + 6{\cdot}267)$$
$$= 485{\cdot}5\ \Omega$$

13.11 Tutorial and practical work

13.11.1 Discuss the effect on the sensitivity and the speed of response of a liquid-in-glass thermometer of (a) the size and shape of the bulb, (b) the size of the capillary bore, (c) the density, thermal conductivity, and specific heat of the filling liquid, (d) the thickness, specific heat, and thermal conductivity of the glass.

13.11.2 A liquid-in-glass thermometer is seen to 'dip' by a few degrees before rising, when thrust into a hot fluid. Explain why this effect occurs.

13.11.3 Two identical mercury-in-glass thermometers are suddenly immersed in water at 90 °C. In one case the water is still; in the other it is circulated by a stirrer. Compare the transient and steady-state responses of the two thermometers, giving reasons for the differences.

13.11.4 Discuss the reasons for filling liquid- and gas-type thermometers at high pressures, but not vapour-pressure thermometers.

13.11.5 Sketch a system to measure the temperature of the tip of a lathe tool, whilst cutting, by measuring the thermo-electric e.m.f. between the dissimilar metals of the tool and workpiece. Suggest also a method of calibrating the system.

13.11.6 Discuss the effects on the accuracy and/or response of the instrument due to heat transfer through (a) the capillary of a bulb-type thermometer, (b) the wires from the hot junction of a thermocouple.

13.11.7 Discuss the difference between standardising and calibrating a thermocouple-plus-potentiometer instrument.

13.11.8 Explain how the change (δR_A) of the resistance R_A in fig. 13.20 can compensate for the variation of the reference temperature θ_r.

13.11.9 Describe a method of determining the increase of temperature of the field winding of an electric generator during operation.

13.11.10 Discuss methods of measuring the surface temperature of (a) a metal, (b) a non-metal.

13.11.11 Discuss the use of the metal-foil grid-type temperature sensors, in a Wheatstone-bridge circuit, for the rapid measurement of temperature difference. Suggest applications where their use would be particularly suitable.

13.11.12 Using manufacturers' literature, compare the stated accuracies of thermometers and pyrometers over comparable ranges.

13.11.13 Discuss the application errors that might arise due to the use of thermometers and pyrometers.

13.11.14 Is error likely to arise in resistance thermometers due to thermo-electric e.m.f.'s at junctions of dissimilar metals in the circuit?

13.11.15 Discuss the static and dynamic errors that might arise in electrical-type thermometers and pyrometers, indicating their likely magnitudes.

13.11.16 Sketch a liquid-in-metal thermometer with a suitable electrical transmitting system to read at a distance of 200 m.

13.11.17 The current passing through a resistance-type temperature sensor has a heating effect (i^2R). Discuss the possible effects of this on the accuracy of measurements made with metal sensing elements and with thermistors, taking account of their temperature–resistance relationships.

13.11.18 Sketch a system suitable for measuring the *average* temperature of the exhaust gases of a turbine engine in the exhaust cone (at about 600 °C).

13.11.19 Would changes of the volume of the bulb due to changes of temperature and pressure of the fluid in which it is immersed lead to significant error in the temperature measurement in the case of (a) a liquid-in-metal thermometer, (b) a vapour-pressure thermometer?

13.11.20 For a particular furnace at one temperature, determine the repeatability of your own temperature readings using a disappearing-filament pyrometer, and compare these with the values and repeatability of other operators' readings.

13.12 Exercises

13.12.1 A gas thermometer is filled with nitrogen, and the system is adjusted to a pressure of 5 MPa when the bulb is at the triple-point-of-water temperature. Calculate the system pressure at −100 °C and +400 °C. [3·17 and 12·3 MN/m^2]

13.12.2 A rod-type thermostat similar to that shown in fig. 2.9(a) consists of a rod of material A enclosed in a tube of material B. If the effective length of the rod and tube assembly is 200 mm, calculate the displacement (x) of the free end of A for a temperature change of 10 °C, if the linear expansion coefficients (α) are $\alpha_A = 10 \times 10^{-6}$/°C, $\alpha_B = 20 \times 10^{-6}$/°C.

Show that, for the same temperature change, the deflection (y) of the bimetal thermostat of example 13.10.1 would be about four times this value, and discuss the applications of the two types of instrument. [0·02 mm, ≈0·08 mm]

13.12.3 A liquid-filled thermometer has a bulb volume eight times that of the capillary plus Bourdon tube. What is the error due to an increase of ambient temperature of 15 °C from the temperature at calibration, if no compensating device for this effect is fitted? [+1·88°C]

13.12.4 Determine the effect on a vapour-pressure thermometer filled with water of a drop in ambient pressure of 2·5 kN/m^2 at the

Bourdon-tube gauge when the measured temperature is 120 °C.

[The reading will be ≈ 0·4°C high.]

13.12.5 The e.m.f. of a platinum–10% rhodium/platinum thermocouple with an ice-point reference temperature was measured using a precision potentiometer under carefully controlled conditions, and was found to be 5·538 mV at the antimony point, 9·131 mV at the silver point, and 10·336 mV at the gold point. By interpolation from the e.m.f. values given in BS 4937: part 1: 1972, show that these differences will lead to errors of −1·34, −1·51, and +0·17 °C respectively unless the e.m.f. errors are allowed for in calibration.

13.12.6 A thermistor has a resistance of 3980 Ω at the ice point and 794 Ω at 50 °C. If eqn 13.12 applies, calculate the values of the constants a and b, and calculate the range of resistance to be measured if the instrument is to measure temperatures in the range 40 °C to 100 °C.

[$a = 300 \times 10^{-6}$, $b = 2845$, resistance range ≈ 1050 Ω to 239 Ω]

13.12.7 Calculate the maximum velocity (V) of air immediately upstream of the thermocouple probe of fig. 13.21(b) if the error due to this in measuring the temperature of the moving gas is not to exceed 3°C. The specific heat at constant pressure for air is $c_p = 1005$ J/kg K.

[78 m/s]

14

Vibration Measurement

14.1 Vibration and its measurement

14.1.1 Types of vibration

Vibration is the repeated cyclic oscillation of a system consisting of matter which may be in solid, liquid, or gaseous form. It is termed *natural vibration* when the frequency of oscillation is that, or those, determined by the magnitude and distribution of mass, stiffness, and damping in the system, and there is no sustained exciting force. Hence, if a force is applied to deflect part of the system from its equilibrium position and is then removed, the system will oscillate at a natural frequency, if the damping ratio (ζ) is less than 1, until the energy put into the system is dissipated (see section 3.2.2 and fig. 3.5).

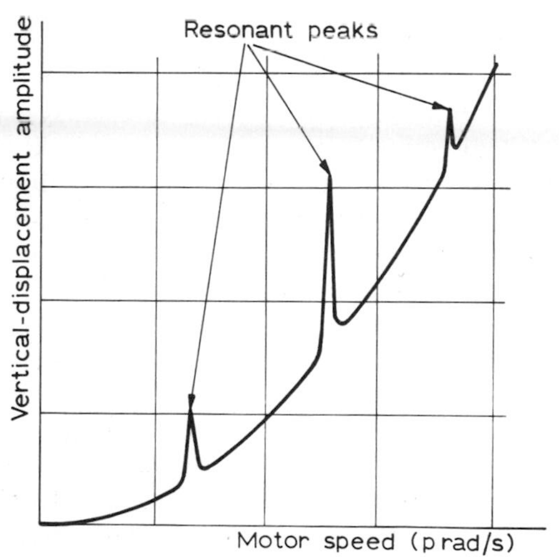

Fig. 14.1

If a vibration is imposed on a system by a periodic force, then the system is said to have *forced vibration*. If the (circular) frequency (p) of the forcing is the same as, or near to, a natural frequency (ω_n), then the amplitude of the oscillation increases as shown in fig. 3.11(a), and the condition is referred to as '*resonance*'. Figure 14.1 also illustrates this effect, showing the amplitude of the vertical component of displacement due to vibration of an electric-motor mounting bracket. The amplitude is seen in general to increase as the square of the motor speed, since centripetal force $= p^2 \times$ radius of mass centre, but at certain speeds the amplitude increases sharply. These resonant peaks occur when the frequency of the out-of-balance force of the motor coincides with a natural frequency of vibration of the bracket, i.e. when $p/\omega_n = 1$.

A further category of vibration is referred to as *self-excited*, although in this case the exciting force is external, and the reaction of the system to the force sustains the vibration. An example is the 'flutter' of an aerofoil section due to a gas stream.

14.1.2 *Modes of vibration*

Vibrating systems may be classified into two types: the lumped system and the continuous system. The simplest lumped system, shown in fig. 14.2(a), consists of a single mass restrained by some kind of elastic member, that shown being a coil spring. The mass has a natural frequency of oscillation in the YY direction given by $f_n = \sqrt{(k/m)}/2\pi$, or $\omega_n = \sqrt{(k/m)}$. It may also have other types of vibration, e.g. torsional oscillation about the YY axis and pendulum vibration. These are the different *modes* of vibration. The two-mass lumped system of (b) can vibrate in the XX direction in two characteristic modes: either both masses travelling the same way, or the two travelling in opposite directions, at any given instant. Each of these modes has a different natural frequency, which may be calculated from the m and k values, taking damping into account if it is a known quantity.

Continuous systems, such as the bar in fig. 14.2(c), may vibrate in an infinite number of different modes, each of different frequency, in the XY plane as shown. Similar vibrations at again different frequencies may occur in the ZY plane. Longitudinal vibrations can occur along the OY axis. It will be appreciated that, for objects of complex shape, many vibration modes are possible, each with a different frequency. The points such as O, A, B, C, D, etc. on the bar of fig. 14.2(c) which remain stationary in any particular mode of vibration are termed *nodes*.

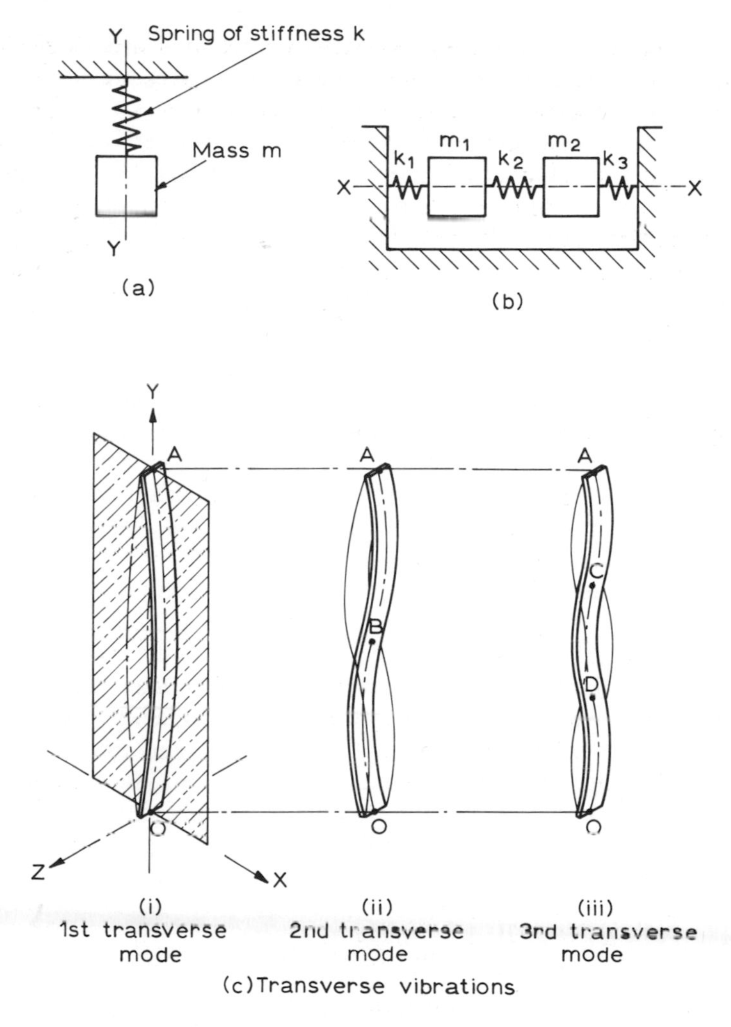

Fig. 14.2

14.1.3 Vibration measurement

Natural vibrations may be initiated by a chance force such as a hammer blow, and they usually die down at a rate depending on the amount of damping, if the initiating force is not repeated. Forced vibration is caused by the application of forces such as those due to unbalanced reciprocating and rotating masses in engines, inaccuracy of meshing gears, or intermittent cutting as in a milling operation. Care should be taken to minimise such effects in design, manufacture, and operation of machines and structures. Self-excited vibrations are frequently caused by gas streams, and often affect aerofoil surfaces such as wings, compressor and turbine blades, fans, impellers, etc.

Vibration causes a fluctuation of stress in a solid material which may lead to premature failure by fatigue. Larger amplitudes of displacement cause larger acceleration and stress values. Resonant and self-excited vibrations can lead to large amplitudes, particularly if damping is small. A spectacular failure caused by wind-induced 'self-excited' vibrations was the destruction of the suspension bridge over the Tacoma Narrows in the USA. Much knowledge of vibration has been gained in recent years, and methods of solution of vibration problems by computer have been developed, but recourse has frequently to be made to the measurement of vibration characteristics by tests during development, either on the full-scale machine or on a model. Vibration monitoring is carried out in service on such important machines as power-station turbines and generators, to give early warning of conditions capable of leading to eventual destruction.

The quantities required to be measured in vibrating systems are displacement, velocity, acceleration, and stress, usually only the peak or amplitude values being required. Frequencies are measured, and the modes of vibration at particular frequency values may be observed. Displacement, velocity, and acceleration are related to each other, since velocity is the rate of change of displacement, and acceleration is the rate of change of velocity. Hence each quantity may be obtained by differentiating or integrating, for example electrically, one of the other quantities which has been measured.

14.1.4 Simple harmonic vibration

For simple harmonic vibrations (SHM), the relationships are

$$\text{displacement} = x = A\cos\omega t$$

where A is the amplitude (maximum value) of x and ω is the circular frequency of the vibration

$$\text{velocity} = V = \mathrm{D}x = -\omega A\sin\omega t$$

$$\begin{aligned}\text{acceleration} = a = \mathrm{D}^2x &= -\omega^2 A\cos\omega t\\ &= -\omega^2 x\end{aligned}$$

From these expressions for the instantaneous values of displacement, velocity, and acceleration, the maximum values are seen to be

$$\text{maximum displacement} = A \qquad 14.1$$

$$\text{maximum velocity} = \hat{V} = \omega A \qquad 14.2$$

$$\text{maximum acceleration} = \hat{a} = \omega^2 A \qquad 14.3$$

For simple harmonic vibrations, the restoring force when a part is deflected from the equilibrium position must be proportional to the displacement, and in many systems this applies. Hence many natural and self-excited vibrations are SHM.

14.1.5 Overtones, harmonics, and frequency spectrum

In some systems, more than one mode of vibration may occur simultaneously, and a resulting vibration may be the sum of two or more separate ones. The frequencies higher than the first or fundamental one are known as *overtones*. If they are simple multiples of the basic frequency, such as occur in string and wind musical instruments, they are also known as *harmonics*. For the bar of fig. 14.2(c), the frequency of transverse vibration in mode 'n' is n^2 times the fundamental fre-

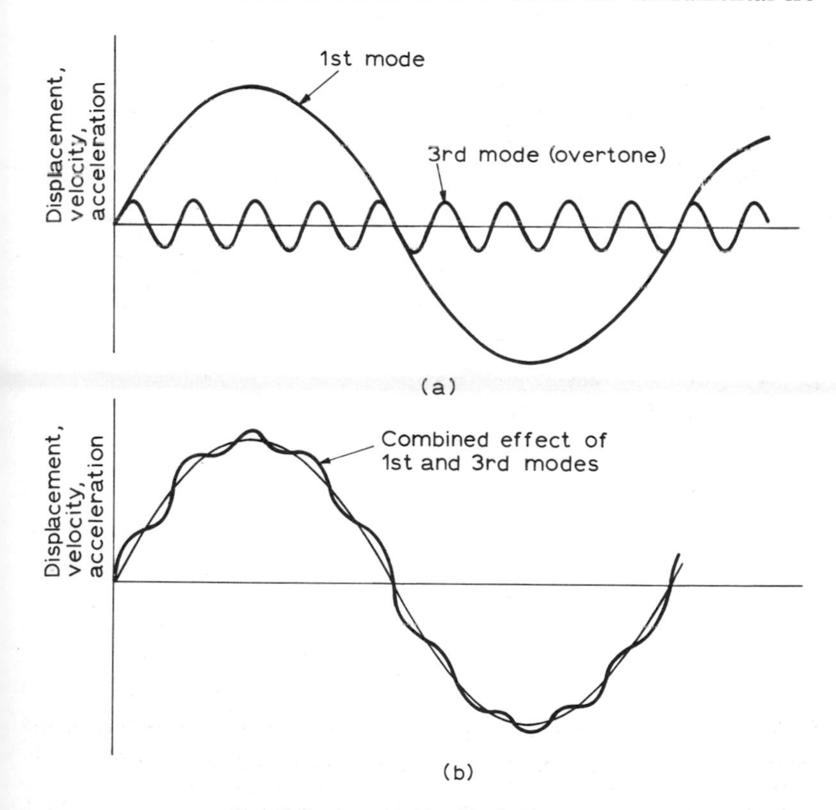

Fig. 14.3 Combined motion in two modes

quency; therefore mode 3 has a frequency 9 times that of mode 1, as shown in fig. 14.3(a). If the bar vibrated in the first *and* third modes, the resulting motion of a point on the bar would be the sum of the individual motions of that point in each mode, as shown in (b).

The exciting force applied during forced vibrations may have sinusoidal form, for example that due to the vertical component ($F_v = m\omega^2 r \sin\theta$) of an out-of-balance shaft, as shown in fig. 14.4(a). Frequently an exciting force may contain multiples of the fundamental frequency, these also being referred to as harmonics, such as those due to the reciprocating mass of a piston in an engine mechanism, resulting in a force–time relationship such as that shown in (b). Consequently, vibration-measuring instruments may measure such waveforms as these, and in the processing of readings it may be necessary to analyse the trace obtained, finding the value over a narrow frequency range at many different frequency values, using a frequency analyser.

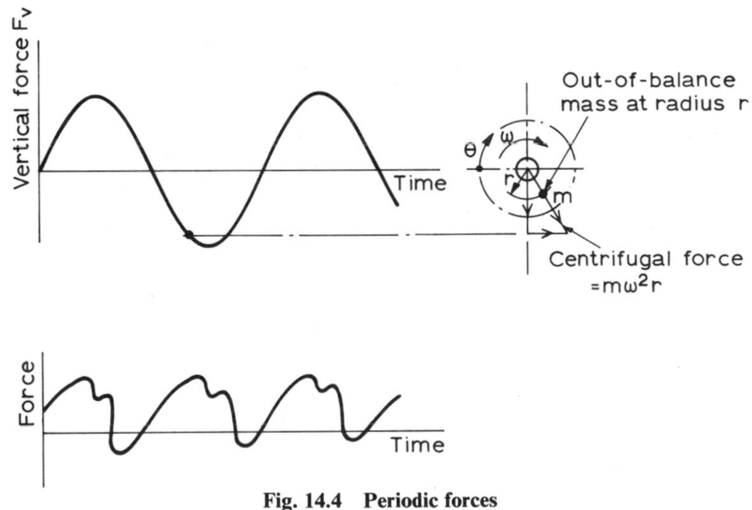

Fig. 14.4 Periodic forces

As an example of this, fig. 14.5(b) shows the steady periodic signals recorded from a no. 4 cylinder pressure transducer (top trace) and a torsional-vibration transducer attached to the nose of the crankshaft of a diesel engine running under light load, the arrangement being shown in (a). The pressure trace has a periodic time equal to two shaft revolutions, the trace being triggered and sub-marked by the half-speed crank-angle unit. The vibration trace has a periodic time equal to one crank revolution. Examination of the Polaroid picture of the oscilloscope trace (b) reveals two major pulses per revolution, one of these coinciding with the pressure trace from no. 4 cylinder. These are

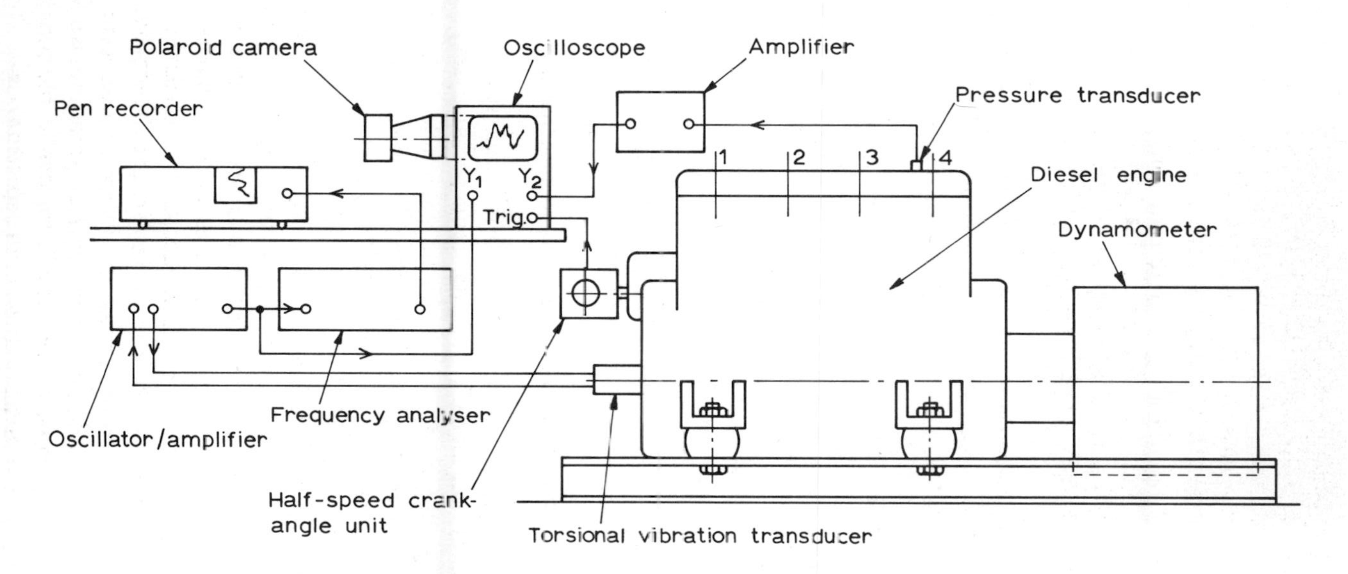

Fig. 14.5(a) Engine torsional vibration test arrangement

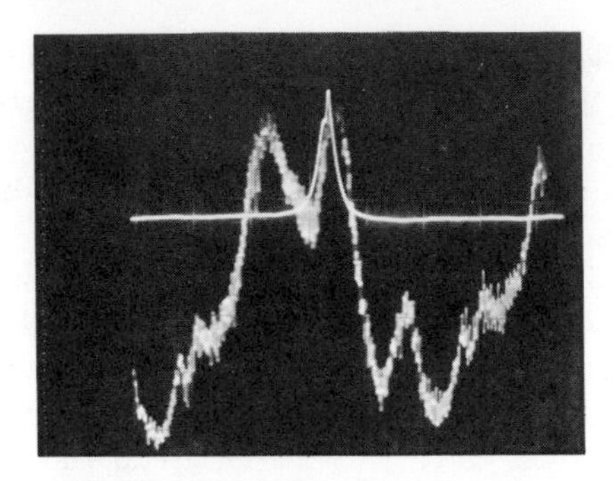

Fig. 14.5(b) Polaroid photograph of oscilloscope traces – no. 4 cylinder pressure and shaft acceleration

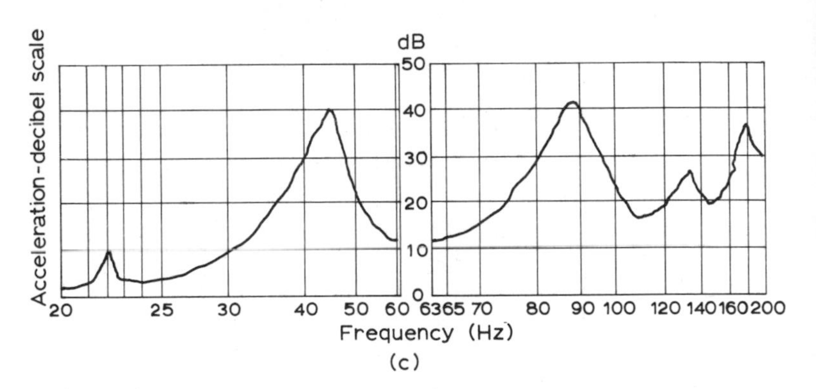

Fig. 14.5(c) Pen-recorder trace of frequency spectrum from analyser

obviously due to the two firing pulses occurring each revolution. After passing the vibration signal through a frequency analyser, the trace shown in (c) was obtained. This shows the magnitude of the measured variable against frequency on log scales. The resulting values are referred to as a *frequency spectrum*. The measured speed of the shaft was 1320 rev/min, equal to a frequency of 22 Hz.

A major peak on the spectrum is seen to be at 45 Hz, corresponding to twice the shaft speed, referred to as second engine order. It is the same as the firing frequency. A small peak is seen at the engine shaft frequency 22 Hz, probably being due to a small amount of primary unbalance of the reciprocating masses. Secondary unbalance of these, at twice engine order, is likely to be larger, and is additive to the impulses due to firing torque. Further peaks are seen to occur at 4, 6, and 8 engine orders, whilst the general level of vibration also tends to rise with frequency. The student reader will appreciate that considerable skill is necessary in the interpretation of the spectrum.

14.2 Measurement of displacement, frequency, and mode

If a vibrating object is accessible, much information can often be obtained using quite simple equipment. In the laboratory, the nodes on horizontal vibrating surfaces may be found by observing the behaviour of small particles of dry sand scattered thinly over the surface. The particles accumulate at points where the acceleration of the surface is less than $1g$, i.e. where they are staying in contact with the surface. If a wedge-type indicator as shown in fig. 14.6, usually made of paper, is fixed to a vibrating body, the distance to the dark portion of wedge where the motion overlaps is proportional to the amplitude (A) of the vibration.

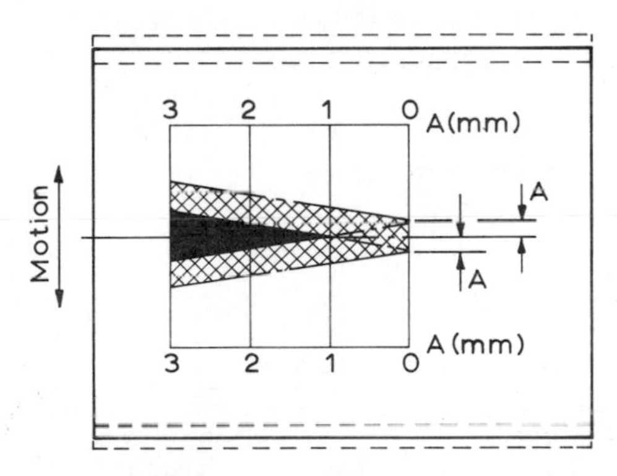

Fig. 14.6 Wedge-type vibration-amplitude indicator

The stroboscope is useful in observing the modes of vibration by adjusting the flashing frequency until slow-motion vibration is observed. The nodes may be seen, and the amplitude is sometimes large enough to be measured by a rule or suitable scale. With the usual precautions (see section 6.4.4), the frequency of the oscillation is measured directly.

Many of the displacement-measuring methods described in Chapters 10 and 11 may be applied to vibrations. However, in some of them, mass may be added to the measured system, or force may be applied to the system, leading to changes of damping ratio (ζ) and natural frequency (f_n or ω_n) of the system, leading to application error. Others, such as resistance-type displacement transducers, suffer excessive local wear when measuring oscillating displacement. Proximity-type displacement meters are more suitable for measuring dynamic displacement. They may work by the variation of reluctance or inductance due to the varying distance from a magnetic material, or the varying capacitance between a probe and an electroconductive material (sometimes non-conductive) due to a varying gap (see ref. 2). A readily available device suitable for frequency measurement only is the microphone.

Figure 14.7 illustrates an application of the Wayne Kerr capacitive-type proximity transducer, similar to those of section 11.3.2.1 and fig. 11.13, arranged to measure the central displacement of the vibrating end of an air receiver. The static distance (x_s) may be read from the distance meter, and the peak-to-peak amplitude ($= 2A$) of the displacement normal to the transducer axis may be read from the vibration meter. Typical probes have a distance and vibration range up to 2·5 mm. The meter output may be fed to an oscilloscope or recorder so that the waveform may be observed and the frequency found. Similar systems are available for inductive types.

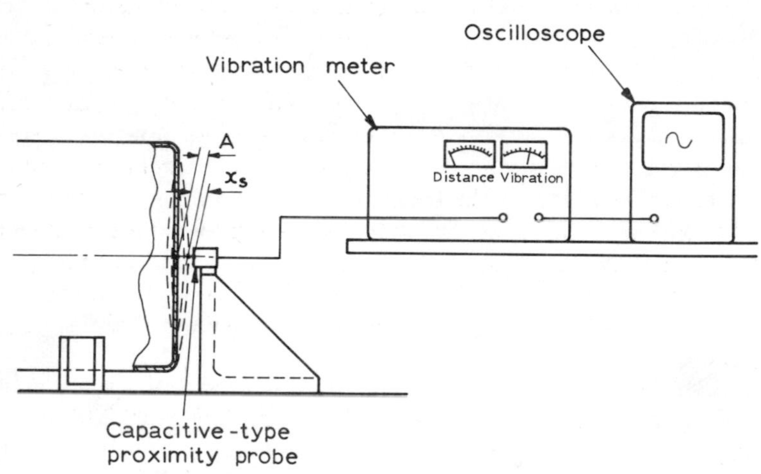

Fig. 14.7

Displacement and frequency may also be found from velocity- and acceleration-sensitive transducers, but most of these have to be mounted on the vibrating body. Their mass leads to application error, as mentioned previously, and the magnitude of this should be estimated. The only transducers presently used that have very small mass and bulk are the small electrical strain-gauges of resistance, capacitance, semiconductor, and piezoelectric-wafer type described in section 12.2.2. Of these, the resistance type in particular may be applied to components such as compressor or turbine blades with very small additional mass or interference with fluid flow round an aerofoil section. From the dynamic-strain signal, the frequency of oscillation may be found or the frequency spectrum determined as previously described. Using a number of gauges, information regarding the mode of vibration may be deduced.

14.3 Seismic instruments

Newton's laws of motion give the relationship

$$F = Kma \qquad 14.4$$

where F is the net force applied to a mass m, a is the acceleration of the mass, and K is a constant, equal to unity in the SI.

Consider the motion of a mass suspended on spring mountings in a frame, as shown in fig. 14.8. If the frame is moved by an amount θ_i, then the change of length of the springs will cause them to exert a net force (F) on the mass, causing it to accelerate in the same direction as θ_i has moved. The magnitude of the acceleration will increase as the stiffness (k) of the spring mounting increases, and will reduce as the mass increases, for a given displacement (z) of the mass *relative to the frame*. If the mass is large and the suspension stiffness low, then the acceleration is low. With conditions like this, if θ_i is a rapid oscillation of the frame, the mass cannot keep up with it. Its motion may be in the opposite direction to the frame, i.e. antiphased, as illustrated in fig. 14.15(b), and so it may stay virtually still relative to earth. Hence the motion of the mass relative to the frame may be used as a measure of the motion of the frame relative to earth. This is done in the seismograph, which uses a large mass delicately suspended. This remains

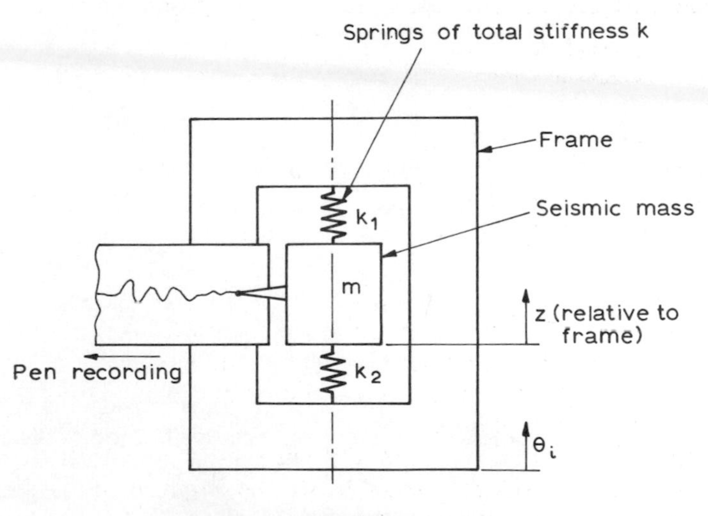

Fig. 14.8

almost stationary relative to the general earth mass, whilst the frame, fixed to the local earth mass, vibrates with locally received earth tremors. The relative motion of the mass and frame is recorded as a trace of displacement against time.

Figure 14.9 shows a vibration-sensing transducer where the spring-mounted mass is surrounded by a liquid to provide viscous damping. The displacement (z) of the mass relative to the instrument body is sensed by a secondary transducer, the one illustrated being a capacitance type, although other types such as the potentiometer or the LVDT (see section 10.4.2) are used.

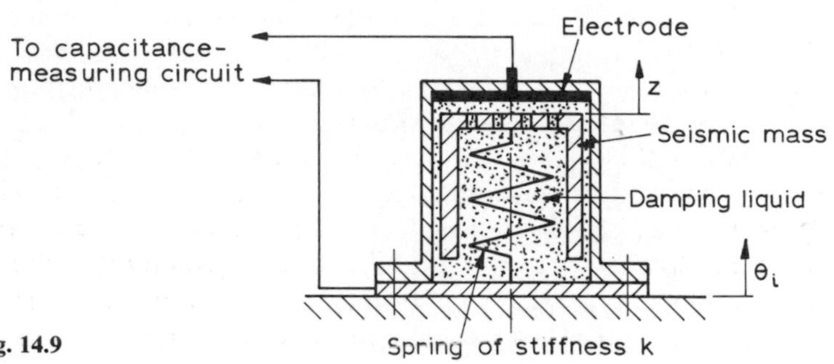

Fig. 14.9

If the motion of the transducer body is SHM, then $\theta_i = \hat{\theta}_i \cos pt$, where $\hat{\theta}_i$ is the amplitude of the displacement relative to earth and p is the circular frequency of the motion. When the mass is displaced a distance z relative to the casing, then its total displacement relative to earth is $\theta_i + z$. If the damping constant is f, and using the notation of section 3.2.2, then

$$mD^2(\theta_i + z) = -fDz - kz$$

or
$$(D^2 + (f/m)D + k/m)z = -D^2\theta_i$$

and
$$(D^2 + 2\zeta\omega_n D + \omega_n{}^2)z = -D^2\theta_i$$

But, since $\theta_i = \hat{\theta}_i \cos pt$, then

$$D\theta_i = -p\hat{\theta}_i \sin pt \quad \text{and} \quad D^2\theta_i = -p^2\hat{\theta}_i \cos pt$$

Hence
$$(D^2 + 2\zeta\omega_n D + \omega_n{}^2)z = p^2\hat{\theta}_i \cos pt \qquad 14.5$$

As discussed in section 3.2.7, the c.f. term, the transient, is not of interest, since it soon dies down. The steady-state vibration of the mass relative to the body is given by the p.i., which is (see appendix B)

$$z = [\mu^2\hat{\theta}_i/\sqrt{\{(1 - \mu^2)^2 + 4\zeta^2\mu^2\}}]\cos(pt - \alpha)$$

where μ is the frequency ratio, p/ω_n, and α is the phase-lag of z relative to θ_i.

The amplitude ratio is given by

$$\frac{Z}{\hat{\theta}_i} = \frac{\mu^2}{\sqrt{\{(1 - \mu^2)^2 + 4\zeta^2\mu^2\}}} \tag{14.6}$$

where Z is the amplitude of z.

The phase-angle is given by

$$\alpha = \arctan\{2\zeta\mu/(1 - \mu^2)\}$$

which is eqn 3.14.

14.3.1 The seismic displacement transducer

If the motion of the mass is to be a true measure of the motion of the transducer body, then the amplitude ratio must be constant. The variation of the amplitude ratio against the frequency ratio for various values of the damping ratio is shown in fig. 14.10(a), and the variation of the phase-angle is as shown for the general second-order system in fig. 3.11(b). It is seen that a damping ratio in the region of 0·6 to 0·7 gives $Z/\hat{\theta}_i$ near to unity between the frequency-ratio values 1·5 to 2, and very near above 2. Hence it is necessary for the measurement of vibrating-displacement amplitude that the natural undamped frequency (ω_n) of the vibration transducer should be low, so that the frequency range of the instrument may extend to low values. However, p should not be lower than $2\omega_n$ in general, and p/ω_n should be as large as possible, to obtain nearly constant phase-shift. This means that, when waveforms containing harmonics are measured, all the harmonic components will have the same phase-shift, and the output signal will not be distorted due to this effect (see ref. 8).

14.3.2 The seismic velocity transducer

A convenient secondary transducer is the velocity 'pick-up' of the type illustrated in fig. 2.21(a). The maximum displacement ratio $Z/\hat{\theta}_i$ is equal to the maximum velocity ratio $Zp/\hat{\theta}_i p$, and the conditions for displacement measurement discussed previously also apply to velocity measurement with this type of transducer.

14.3.3 The seismic accelerometer

To determine the values of the transducer damping ratio (ζ) and the natural circular frequency (ω_n) to use the displacement of a seismic mass as a measure of acceleration, eqn 14.6 is rewritten thus:

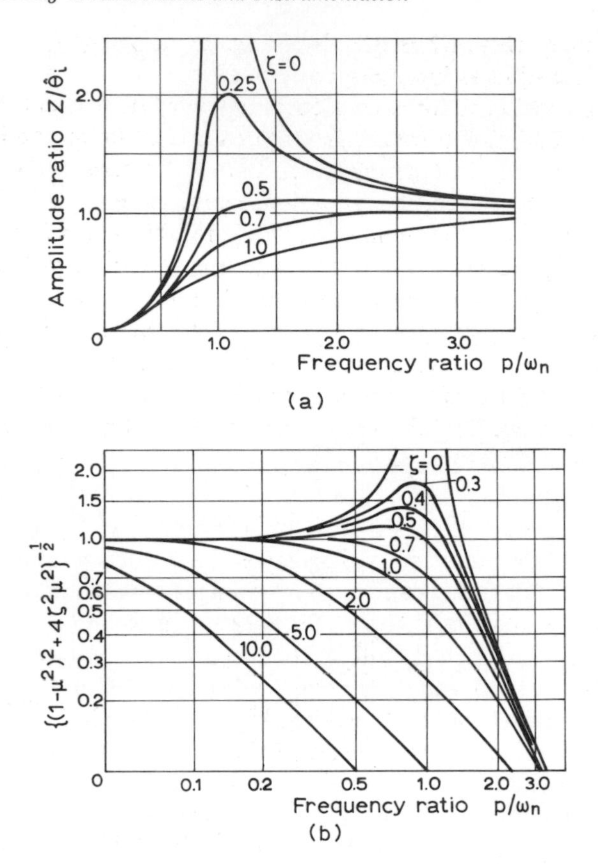

Fig. 14.10 Response of seismic vibration transducers: (a) Vibrometer response (b) Accelerometer response

$$\frac{Z}{p^2\hat{\theta}_i} = \frac{\mu^2/p^2}{\sqrt{\{(1-\mu^2)^2 + 4\zeta^2\mu^2\}}}$$

where $p^2\hat{\theta}_i$ is the amplitude ($\hat{a}$) of the measured acceleration

or
$$\frac{Z}{\hat{a}} = \frac{1/\omega_n^{\ 2}}{\sqrt{\{(1-\mu^2)^2 + 4\zeta^2\mu^2\}}} \qquad 14.7$$

Figure 14.10(b) shows $1/\sqrt{\{(1-\mu^2)^2 + 4\zeta^2\mu^2\}}$ plotted against frequency ratio. Its value must be constant for a linear relationship between Z and $\hat{a}$, and this is obtained with a damping ratio of 0·6 to 0·7 for frequency ratios up to about 0·4. Hence, to have a useful frequency range, this type of instrument must have a high stiffness (k), to give a high natural frequency; whilst the displacement and velocity types need to have a low stiffness for the opposite reason.

The piezoelectric force transducer is widely used in accelerometers, since its small compression is proportional to force applied. Figure 14.11 shows a typical arrangement. Part of the mass of the preloading nut and the piezoelectric material also contribute to the seismic mass, and a proportion of their mass is added to m. The piezoelectric material constitutes the spring mounting, which is of very high stiffness. The inherent damping is very low in piezoelectric materials, not exceeding about 0·1, and hence the usable frequency range is only about 0·2 of the natural frequency. However, since the natural frequency is high, due to the stiffness of the crystal, the usable range is also high, from 1 to 15 000 Hz in some types. Accelerometers of this type can be made with low mass and small bulk.

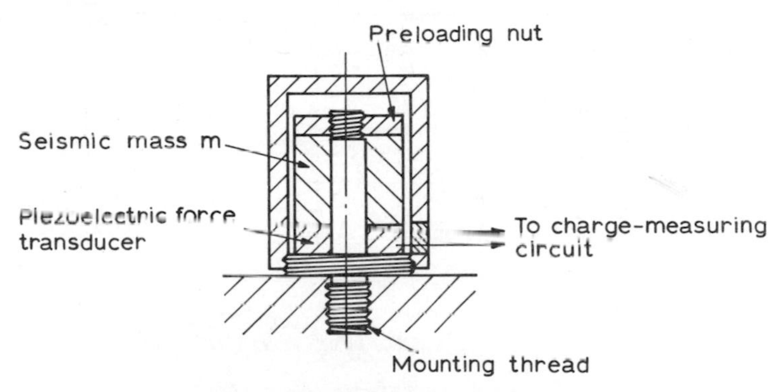

Fig. 14.11 Seismic accelerometer

Another type uses bonded or unbonded strain-gauges as a measure of displacement. The basic arrangement of such types is shown in fig. 14.12(a) and (b). In (a), the seismic mass is supported on a cantilever-spring strip to which strain-gauges are attached. The deflection at the mass end produces a proportional stress and strain at the gauges, at distance y from the mass end. In (b), the mass is supported on strain-gauge wires which constitute both the elastic mounting and the strain transducer, giving a signal proportional to the deflection of the mass.

14.4 Calibration of instruments

A variety of techniques are used in the calibration of vibrometers (i.e. instruments sensing displacement or velocity) and accelerometers, and the interested reader is referred to refs 2 and 8. Some of the methods are outlined below.

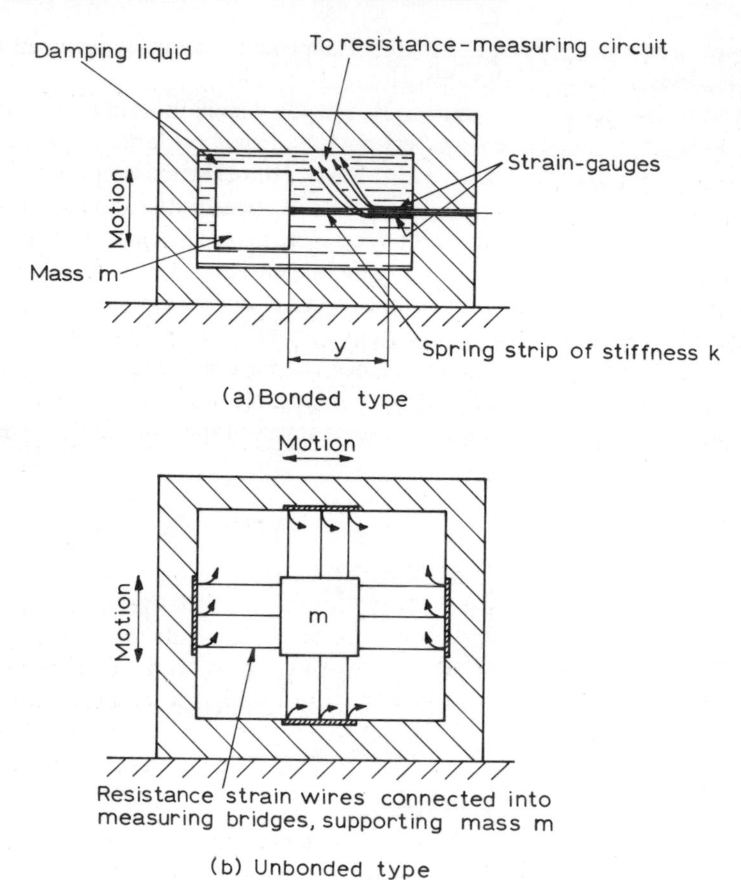

Fig. 14.12 Accelerometers with bonded and unbonded strain-gauges

14.4.1 Calibration of vibrometers

a) A simple harmonic vibration of known magnitude is applied to the transducer, and the output signal is measured. The difficulty is in knowing exactly the magnitude of the input vibration.

b) A vibration is applied as in (a), but the amplitude is measured using, as applicable, stroboscope, low-power microscope, or interferometer methods.

c) The instrument to be calibrated is subjected to the same vibration as an instrument of known accuracy, i.e. a 'standard', and the outputs of the two are compared.

14.4.2 *Calibration of accelerometers*

d) An SHM vibration of known magnitude is applied as in (a), and may be checked as in (b).
e) A steady acceleration is applied by rotating the transducer in a horizontal plane.
f) The instrument is turned over, giving it an acceleration of $2g$.
g) The instrument is rotated in a vertical plane, giving a steady acceleration value with a $\pm 1g$ fluctuation added to it.
h) The accelerometer is attached to a mass and is dropped onto a force transducer. The acceleration is calculated by $a = F/m$, and is compared with the value measured by the accelerometer.
j) For accelerometers using electrical-resistance strain-gauges as the secondary transducer, a known shunt resistance may be introduced as discussed in section 12.2.2.1.

14.5 Worked examples

14.5.1 An electric motor and pulley are bolted to a base-plate on which are mounted two identical seismic displacement pick-ups, A and B, as shown in fig. 14.13(a), these being underneath the shaft, symmetrically positioned about the bearings, and measuring vertical displacement. The outputs are fed to a displacement meter, and from that to the two beams of an oscilloscope. Deduce the out-of-balance condition of the rotating assembly if the oscilloscope trace is seen (i) as in (b), (ii) as in (c), or (iii) as in (d).

The vibrations shown in (b) are of equal magnitude and in phase, so they are likely to be due to a force system equivalent to a central unbalance force rotating with the shaft, as indicated in fig. 14.14(a), giving equal in-phase forces at the bearings. The opposite phase vibrations of equal

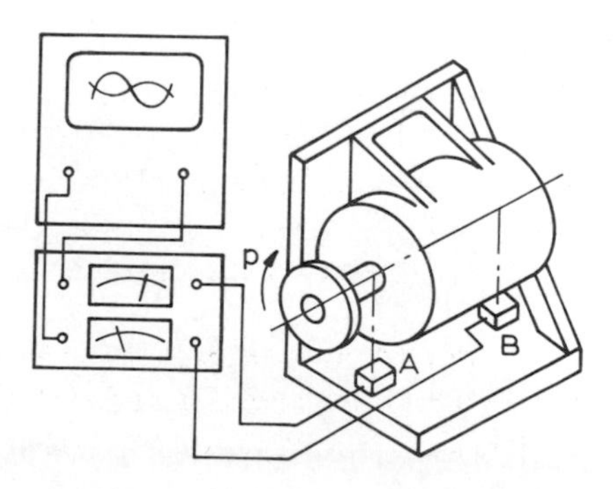

Fig. 14.13(a)

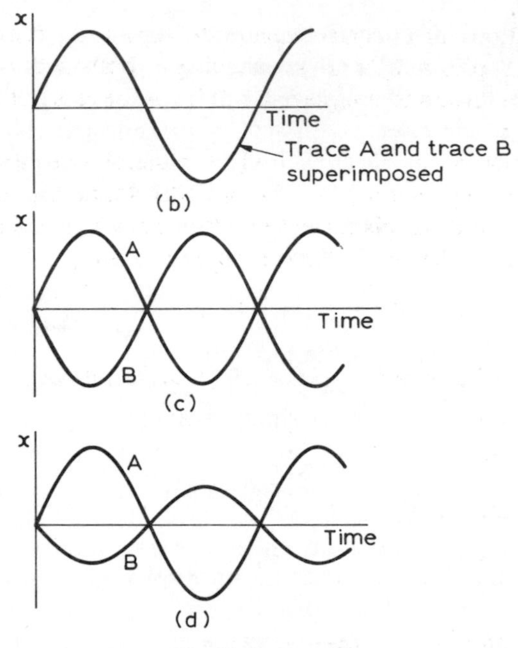

Fig. 14.13 *(cont'd)*

magnitude shown in fig. 14.13(c) are due to a force system equivalent to equal and opposite rotating forces F_1 in the same plane a distance y apart, constituting a couple F_1y_1 as shown in fig. 14.14(b). The opposite

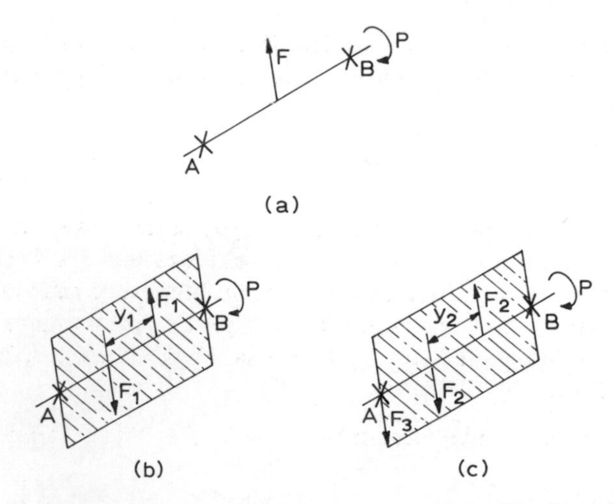

Fig. 14.14 Equivalent out-of-balance system of shaft of fig. 14.13(a)

phased vibrations of unequal magnitude shown in (c) are due to a couple as in (b) plus a force (F_3) at bearing A, both in the same plane. In general, the force and couple will not be in the same plane.

14.5.2 A point on a vibrating body has simple harmonic motion. Determine, at frequencies of 10 Hz and 10 kHz, (a) the amplitude of its acceleration if its displacement amplitude is 1 mm, (b) its displacement amplitude if the acceleration amplitude is $1g$.

a) The circular frequency $= \omega = 2\pi f$.
At 10 Hz,

$$\hat{a}_1 = \omega^2 A = 4\pi^2 \times 10^2 \times 10^{-3}\ \text{m/s}^2 = (0{\cdot}4\pi^2/9{\cdot}81)g$$
$$= 0{\cdot}40g$$

At 10 kHz,

$$\hat{a}_2 = (4\pi^2 \times 10^8 \times 10^{-3}/9{\cdot}81)g$$
$$= 400\,000g$$

b) At 10 Hz,

$$A_1 = \hat{a}/\omega^2 = 9{\cdot}81/(4\pi^2 \times 10^2)\ \text{m}$$
$$= 2{\cdot}5\ \text{mm}$$

At 10 kHz,

$$A_2 = \hat{a}/\omega^2 = 9{\cdot}81/(4\pi^2 \times 10^8)\ \text{m}$$
$$= 0{\cdot}0025\ \mu\text{m}$$

From these values, it can be seen that larger displacement amplitude at higher frequencies is necessarily coupled with very high acceleration amplitude.

14.5.3 A seismic vibrometer sensing displacement has an undamped natural frequency (f_n) of 20 Hz and a damping ratio of 0·7. Calculate (a) its damped natural frequency (f_d), (b) the amplitude ratio ($Z/\hat{\theta}_i$) and the phase-angle (α) between the motion of the seismic mass and the applied vibration if the latter is a sinusoidal displacement at a frequency (f) of (i) 30 Hz, (ii) 1 kHz.

a) The damped circular frequency $= \omega_d = \sqrt{(1 - \zeta^2)}\omega_n$ (see section 3.2.4);

hence $$2\pi f_d = \sqrt{(1 - \zeta^2)}2\pi f_n$$

$$\therefore \quad f_{\mathrm{d}} = \sqrt{(1 - 0{\cdot}7^2)} \times 20$$
$$= 14 \text{ Hz}$$

b) (i) At 30 Hz, the frequency ratio $\mu = f/f_{\mathrm{n}} = p/\omega_{\mathrm{n}} = 30/20 = 1{\cdot}5$. From eqn 14.6,

$$\frac{Z}{\hat{\theta}_{\mathrm{i}}} = \frac{\mu^2}{\sqrt{\{(1-\mu^2)^2 + 4\zeta^2\mu^2\}}} = \frac{1{\cdot}5^2}{\sqrt{\{(1-1{\cdot}5^2)^2 + 4 \times 0{\cdot}7^2 \times 1{\cdot}5^2\}}}$$
$$= 0{\cdot}92$$

Hence, an inherent output error of -8% occurs at this frequency. From eqn 3.14, the lag of z with respect to θ_{i} is

$$\alpha = \arctan\{2\zeta\mu/(1-\zeta^2)\} = \arctan\{2 \times 0{\cdot}7 \times 1{\cdot}5/(1-1{\cdot}5^2)\}$$
$$= \arctan(-1{\cdot}68)$$
$$= 120°46'$$

(ii) The frequency ratio is $\mu = 1000/20 = 50$.

$$\frac{Z}{\hat{\theta}_{\mathrm{i}}} = \frac{50^2}{\sqrt{\{(1-50^2)^2 + 4 \times 0{\cdot}7^2 \times 50^2\}}} \approx 1{\cdot}0$$

And the inherent error is virtually zero.

$$\alpha = \arctan\{2 \times 0{\cdot}7 \times 50/(1-50^2)\} = \arctan(-0{\cdot}28)$$
$$= 178°24'$$

The relative displacements of the body of the transducer and the seismic mass are shown in fig. 14.15.

14.6 Tutorial and practical work

14.6.1 How would you determine (a) the stiffness, (b) the natural frequency of vibration, of a motor-vehicle engine on its mountings? Would your method give the damped or the undamped natural frequency value? Could the other be found from this? (Refer if necessary to Chapter 3.)

14.6.2 Discuss examples of self-excited vibrations, indicating how the vibration is sustained.

14.6.3 Are the springs of the lumped systems of fig. 14.2(a) and (b) themselves continuous systems? If so, discuss the modes in which they may be expected to vibrate. Devise a method of testing and observing these.

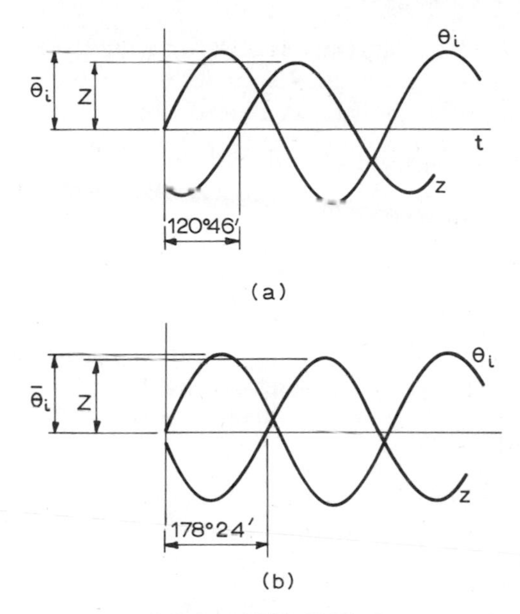

Fig. 14.15(a) Displacement and phase at 30 Hz (b) Displacement and phase at 1 kHz

14.6.4 What kind of force–time relationship would you expect to find from (a) a milling cutter in operation, (b) meshing gear teeth, (c) a turbine shaft, (d) a reciprocating pump?

14.6.5 Investigate the vibration of a circular flat plate supported at the centre, by scattering fine sand on the surface whilst it is excited by a violin bow or a vibration generator.

14.6.6 Sketch a method of mounting one or more proximity transducers to measure (a) the vibration of a milling-machine bed relative to the concrete floor, (b) the vibration of a milling-machine bed relative to its head.

Discuss sources of error in making such measurements.

14.6.7 Sketch some of the possible modes of vibration of a cantilever such as an aircraft wing or a pin-root-fixed compressor blade.

14.6.8 What relative amplitudes of vibration would you expect to read from three proximity transducers placed at $\frac{1}{4}$, $\frac{1}{2}$, and $\frac{3}{4}$ way along a bar simply supported at the ends as in fig. 14.2(c), vibrating in a combination of the first and second modes?

14.6.9 What relative amplitudes of strain would you expect to read from three strain-gauges aligned along the bar and positioned as in the previous example for the same combined vibration mode?

14.6.10 Is the application error due to the mass of a seismic vibration transducer caused by (a) the seismic mass, (b) the casing mass, or (c) both?

14.6.11 What determines the angle of the 'wedge' in the displacement indicator of fig. 14.6?

14.6.12 Sketch the arrangement of a seismic transducer for torsional vibration measurement, including a suitable secondary transducer. State what the output of the instrument is measuring and whether eqn 14.6 or eqn 14.7 applies.

14.6.13 For the vibration-testing arrangement shown in fig. 14.13, sketch the oscilloscope trace you would expect if the rotating force F_3 shown in fig. 14.14(c) were at 90° to the plane of the couple $F_2 y_2$.

14.6.14 For each of the instruments shown in fig. 14.12, draw up system block diagrams (see section 2.7) showing for each function the input, the output, and the transfer operator.

14.6.15 Discuss a system for the measurement of the out-of-balance of electric-motor rotating assemblies, suitable for production-line use.

14.6.16 Sketch the components and the resulting signal for one cycle consisting of a sinusoidal wave plus its first and second harmonics, in the relative magnitudes 4:2:1.

14.6.17 Examine vibrometers and accelerometers and manufacturers' literature on these available to you, and note their natural frequency, damping ratio, sensitivity, mass, and frequency range. Discuss their relative merits regarding range, accuracy, application, cost, etc.

14.6.18 Discuss possible optical methods of measuring accurately the amplitude of vibration of a body during a vibrometer or accelerometer test.

14.6.19 In the test of example 14.5.1, if a third pick-up were placed midway between the two shown, show the traces you would expect from this for the conditions shown in fig. 14.13(b), (c), and (d).

14.7 Exercises

14.7.1 A simple spring–mass system as shown in fig. 14.2(a) has a spring of stiffness 800 N/m supporting a block of mass 8 kg. Determine the percentage change of frequency of undamped natural vibrations in the direction of the spring axis due to a vibrometer of mass 0·3 kg being attached to the mass. [−1·5%]

14.7.2 A velocity-sensitive vibrometer has an undamped natural frequency of 10 Hz and a damping ratio of 0·7. Calculate the maximum velocity ratio ($Zp/\hat{\theta}_i p$) and the phase-angle (α) between the input and output waveforms if the input is a sinusoidal vibration of (a) 20 Hz, (b) 1 kHz. [0·98, 136°59′, 1, 179°12′]

14.7.3 A piezoelectric-type accelerometer is stated to have a natural frequency of 50 kHz and a linearity of ±2% over the frequency range of 2 Hz to 7 kHz. By calculating the value of the denominator of the right-hand side of eqn 14.7 for the ends of the range, assuming the damping ratio is 0·1, show that the linearity figure is within the inherent error for the instrument.

Appendix A
Unit Systems and Conversions

The International System of metric units (SI) has provided a coherent and elegant framework which has removed many of the artificial divisions between quantities in, for example, mechanics, electrics, and heat. Consequently, it has found wide acceptance by engineers and scientists throughout the world, and will undoubtedly be universally adopted. The system is based on seven base units and two supplementary units, from which the host of derived units is obtained. The system is described in the British Standards publication PD5686: 1972, 'The use of SI units', from which the following definitions are taken.

Definitions of the SI base units

metre

The metre is the length equal to 1 650 763·73 wavelengths in vacuum of the radiation corresponding to the transition between the levels $2p_{10}$ and $5d_5$ of the krypton-86 atom.
[11th CGPM (1960), resolution 6]

kilogram

The kilogram is the unit of mass; it is equal to the mass of the international prototype of the kilogram.
[1st CGPM (1889) and 3rd CGPM (1901)]

second

The second is the duration of 9 192 631 770 periods of the radiation corresponding to the transition between the two hyperfine levels of the ground state of the caesium-133 atom.
[13th CGPM (1967), resolution 1]

ampere

The ampere is that constant current which, if maintained in two straight parallel conductors of infinite length, of negligible circular cross-section, and placed 1 metre apart in vacuum, would produce between these conductors a force equal to 2×10^{-7} newtons per metre of length.
[CIPM (1946), resolution 2 approved by the 9th CGPM (1948)]

kelvin

The kelvin, unit of thermodynamic temperature, is the fraction 1/273·16 of the thermodynamic temperature of the triple point of water. [13th CGPM (1967), resolution 4]

Note 1. The 13th CGPM (1967, resolution 3) also decided that the unit kelvin and its symbol K should be used to express an interval or a difference of temperature.

Note 2. In addition to the thermodynamic temperature (symbol T) expressed in kelvins, use is also made of Celsius temperature (symbol θ) defined by the equation $\theta = T - T_0$, where $T_0 = 273{\cdot}15$ K by definition.

The Celsius temperature is in general expressed in degrees Celsius (symbol °C). The unit 'degree Celsius' is thus equal to the unit 'kelvin', and an interval or a difference of Celsius temperature may also be expressed in degrees Celsius.

candela

The candela is the luminous intensity, in the perpendicular direction, of a surface of 1/600 000 square metre of a black body at the temperature of freezing platinum under a pressure of 101 325 newtons per square metre.
[13th CGPM (1967), resolution 5]

mole

The mole is the amount of substance of a system which contains as many elementary entities as there are atoms in 0·012 kilogram of carbon 12.

Note. When the mole is used, the elementary entities must be specified, and may be atoms, molecules, ions, electrons, other particles, or specified groups of such particles.
[14th CGPM (1971), resolution 3]

The supplementary units, the *radian* (plane angle) and *steradian* (solid angle), may be regarded either as base units or as derived units, and may be defined thus:

radian – the plane angle subtended at the centre by an arc of unit length at unit radius;

steradian – the solid angle subtended at the centre by unit area of a spherical surface at unit radius.

The base- and supplementary-unit symbols are as follows.

	Quantity	*Name of unit*	*Symbol*
Base units	length	metre	m
	mass	kilogram	kg
	time	second	s
	electric current	ampere	A
	thermodynamic temperature	kelvin	K
	luminous intensity	candela	cd
	amount of substance	mole	mol
Supplementary units	plane angle	radian	rad
	solid angle	steradian	sr

Derived units may be expressed in the units of the base or supplementary units; for example, the unit of force may be expressed in terms of mass and acceleration, in accordance with Newton's laws, as $kg\,m/s^2$, but in the SI it is given the name '*newton*'.

Other units having special names are listed below.

Quantity	*Name of SI derived*	*Symbol*	*Expressed in terms of SI base or supplementary units or in terms of other derived units*
frequency	hertz	Hz	$1\ Hz = 1\ s^{-1}$
force	newton	N	$1\ N = 1\ kg\,m/s^2$
pressure and stress	pascal	Pa	$1\ Pa = 1\ N/m^2$
work, energy, quantity of heat	joule	J	$1\ J = 1\ N\,m$
power	watt	W	$1\ W = 1\ J/s$
quantity of electricity	coulomb	C	$1\ C = 1\ A\,s$
electrical potential, potential difference, electromotive force	volt	V	$1\ V = 1\ W/A$
electric capacitance	farad	F	$1\ F = 1\ A\,s/V$
electric resistance	ohm	Ω	$1\ \Omega = 1\ V/A$
electric conductance	siemens	S	$1\ S = 1\ \Omega^{-1}$
magnetic flux, flux of magnetic induction	weber	Wb	$1\ Wb = 1\ V\,s$
magnetic flux density, magnetic induction	tesla	T	$1\ T = 1\ Wb/m^2$
inductance	henry	H	$1\ H = 1\ V\,s/A$
luminous flux	lumen	lm	$1\ l = 1\ cd\,sr$
illuminance	lux	lx	$1\ lx = 1\ lm/m^2$

Other units, not part of the SI, are recognised by the International Committee for weights and measures (CIPM), and are listed below.

Quantity	*Name of unit*	*Unit symbol*	*Definition*
time	minute	min	1 min = 60s
	hour	h	1 h = 60 min
	day	d	1 d = 24 h
plane angle	degree	°	$1° = (\pi/180)$rad
	minute	′	$1' = (1/60)°$
	second	″	$1'' = (1/60)'$
volume	litre	l	$1 \text{ l} = 1 \text{ dm}^3$
mass	tonne	t	$1 \text{ t} = 10^3 \text{ kg}$
fluid pressure	bar	bar	$1 \text{ bar} = 10^5 \text{ Pa}$

The following prefixes are used to give decimal multiples and sub-multiples of quantities.

Factor by which unit is to be multiplied	*Prefix*	
	Name	*Symbol*
10^{12}	tera	T
10^9	giga	G
10^6	mega	M
10^3	kilo	k
10^2	hecto	h
10	deca	da
10^{-1}	deci	d
10^{-2}	centi	c
10^{-3}	milli	m
10^{-6}	micro	μ
10^{-9}	nano	n
10^{-12}	pico	p
10^{-15}	femto	f
10^{-18}	atto	a

Many existing instruments are calibrated in units of the foot–pound-force–second (f.p.s.) system, or in non-SI metric units such as kilogram-force (kgf), or kgf/cm^2 for pressure or stress. Much data is available, for example from North American sources, in f.p.s. units. For these reasons it may be necessary to convert quantities expressed in these units into SI units. The use of *Newton unity brackets* for this purpose is recommended. The equation

$$1 \text{ in} = 25{\cdot}4 \text{ mm}$$

may be written as

$$1 = \left[\frac{25{\cdot}4 \text{ mm}}{1 \text{ in}}\right]$$

where [ ] is the Newton unity bracket.

Any expression may be multiplied by an unlimited number of unity brackets. A few examples are listed below.

Quantity	*Exact unity bracket*	*Approximate unity bracket*
mass	$1 = \left[\frac{1 \text{ lb}}{0{\cdot}453\,592\,37 \text{ kg}}\right]$	
length	$1 = \left[\frac{1 \text{ microinch}}{0{\cdot}0254\ \mu\text{m}}\right]$	
volume		$1 = \left[\frac{1000 \text{ cm}^3}{1 \text{ litre}}\right]$
force	$1 = \left[\frac{1 \text{ kgf}}{9{\cdot}806\,65 \text{ N}}\right]$	$1 = \left[\frac{1 \text{ lbf}}{4{\cdot}448\,22 \text{ N}}\right]$
pressurc	$1 = \left[\frac{10^5 \text{ Pa}}{1 \text{ bar}}\right]$	
power		$1 = \left[\frac{1 \text{ h.p.}}{745{\cdot}7 \text{ W}}\right]$
temperature	$1 = \left[\frac{1{\cdot}8\,°\text{F}}{1\,°\text{C}}\right]$	

The use of unity brackets for converting an instrument setting is shown in example 2.8.4. Examples of data conversion are shown below.

Example A1 The force required for cutting in a lathe is given by $F = KA$, where F is the tangential force (lbf), K is a constant for the material used, and A is the area of the cut (in^2). Derive an expression for a constant K_{SI} in SI units in terms of K_{fps} in f.p.s. units so that if A is expressed in mm^2 then F will be given in newton units.

Rearranging the given equation to obtain the units of K_{fps},

$$K_{fps} = \frac{F}{A}\left(\frac{\text{lbf}}{\text{in}^2}\right)$$

$$K_{SI} = K_{fps}\left(\frac{\text{lbf}}{\text{in}^2}\right) \times \left[\frac{4{\cdot}448\ \text{N}}{\text{lbf}}\right] \times \left[\frac{\text{in}^2}{25{\cdot}4^2\ \text{mm}^2}\right]$$

$$= \frac{4{\cdot}448}{25{\cdot}4^2} \times K_{fps}\left(\frac{\text{N}}{\text{mm}^2}\right)$$

i.e. $K_{SI} = 0{\cdot}0069\ K_{fps}\ \text{N/mm}^2$

Example A2 Dynamic viscosity (η) is given by (shear stress)/(velocity gradient), expressed in units of lbf s/ft^2 or N s/m^2. Derive an expression for η_{SI} in terms of SI units, given values of η_{fps} in f.p.s. units.

$$\eta_{SI} = \eta_{fps}\left(\frac{\text{lbf s}}{\text{ft}^2}\right) \times \left[\frac{4{\cdot}448\ \text{N}}{\text{lbf}}\right] \times \left[\frac{\text{ft}^2}{0{\cdot}3048^2\ \text{m}^2}\right]$$

$$= 47{\cdot}9\ \eta_{fps}(\text{Ns/m}^2)$$

Appendix B
Mathematics of First- and Second-order Systems

In Chapter 3 it is shown that many instrument components and systems may be represented by first- or second-order differential equations thus:

1st order

$$\frac{\theta_0}{\theta_i} = \frac{1}{1 + \tau D} \qquad 3.4$$

2nd order

$$\frac{\theta_0}{\theta_i} = \frac{1}{1 + 2\zeta\tau D + \tau^2 D^2} \qquad 3.5$$

where ζ is the damping ratio, τ the time-constant of the system, and D is the differential operator d/dt.

The output signal (θ_0) for a given input signal (θ_i) is found by obtaining solutions of these equations when θ_i is a step input, a ramp input, or a sinusoidal input. The transient response is given by the complementary function (c.f.), and the steady-state solution by the particular integral (p.i.). The value of θ_0 is the sum of the c.f. and the p.i.

B1 First-order responses

B1.1 Step input: θ_i = *constant for* $t > 0$

The equation is rewritten in the form $\tau D\theta_0 + \theta_0 = 0$.

For the c.f.

The auxiliary equation is $\tau m + 1 = 0$,

$\therefore \quad m = -1/\tau$ and $\theta_0 = Pe^{-t/\tau}$, where P is a constant (see ref. 19).

For the p.i.

Putting $\theta_0 = Q$, a constant (since θ_i is constant), then $D\theta_0 = 0$ and, putting these values in the differential equation, $0 + Q = \theta_i$

$$\therefore \quad Q = \theta_i, \quad \text{and the p.i. is } \theta_0 = \theta_i$$

The general solution is $\theta_0 = \theta_i + Pe^{-t/\tau}$.

At the instant the step is applied, $\theta_0 = 0$ and $t = 0$. Putting these values into the solution, $0 = \theta_i + Pe^0$

$$\therefore \quad P = -\theta_i$$

Hence $\theta_0 = \theta_i - \theta_i e^{-t/\tau}$ is the response.

B1.2 *Ramp input:* $\theta_i = \Omega t$, *where* Ω *is a constant*
The c.f. is the same as in B1.1, i.e. $\theta_0 = Pe^{-t/\tau}$.

For the p.i.
Putting $\theta_0 = Qt + R$, where Q and R are constants, then $D\theta_0 = Q$, giving on substitution into the differential equation

$$\tau Q + Qt + R = \Omega t$$

Equating coefficients of t, $Q = \Omega$.
Equating constants, $\tau Q + R = 0$

$$\therefore \quad R = -\Omega\tau$$

Hence the p.i. is $\theta_0 = \Omega t - \Omega\tau$.
The general solution is $\theta_0 = Pe^{-t/\tau} + \Omega t - \Omega\tau$.
When $t = 0$, $\theta_0 = 0$,

$$\therefore \quad 0 = Pe^0 - \Omega\tau \text{ and } P = \Omega\tau$$

Hence

$$\theta_0 = \Omega\tau e^{-t/\tau} + \Omega t - \Omega\tau$$
$$= \theta_i - \Omega\tau + \Omega\tau e^{-t/\tau} \qquad 3.9$$

B1.3 *Sinusoidal input:* $\theta_i = A_i \cos pt$
The c.f. is the same as in B1.1 and B1.2, i.e. $\theta_0 = Pe^{-t/\tau}$.

For the p.i.
Putting $\theta_0 = Q \sin pt + R \cos pt$, then

$$D\theta_0 = pQ \cos pt - pR \sin pt$$

Substituting these in the differential equation,

$$p\tau Q \cos pt - p\tau R \sin pt + Q \sin pt + R \cos pt = A_i \cos pt$$

Equating coefficients of sin pt,

$$-p\tau R + Q = 0 \qquad \text{(i)}$$

Equating coefficients of cos pt,

$$p\tau Q + R = A_i \qquad \text{(ii)}$$

Substituting for Q from (i) in (ii) gives

$$R = \frac{1}{1 + (p\tau)^2} A_i$$

Substituting this value in (i) gives

$$Q = \frac{p\tau}{1 + (p\tau)^2} A_i$$

Hence the p.i. is

$$\theta_0 = \left\{ \frac{p\tau}{1 + (p\tau)^2} \cdot \sin pt + \frac{1}{1 + (p\tau)^2} \cdot \cos pt \right\} A_i$$

But $Q \sin pt + R \cos pt$ may be expressed in the form

$$\sqrt{(Q^2 + R^2)} \cos(pt - \arctan Q/R)$$

$\therefore$ the p.i. is $\theta_0 = A_i[1/\sqrt{\{1 + (p\tau)^2\}}]\cos(pt - \alpha)$
where $\alpha = \arctan(p\tau)$.

The general solution is $\theta_0 = Pe^{-t/\tau} + A_i[1/\sqrt{\{1 + (p\tau)^2\}}]\cos(pt - \alpha)$.

If at the start of thc oscillation $\theta_0 = 0$ when $t = 0$, then

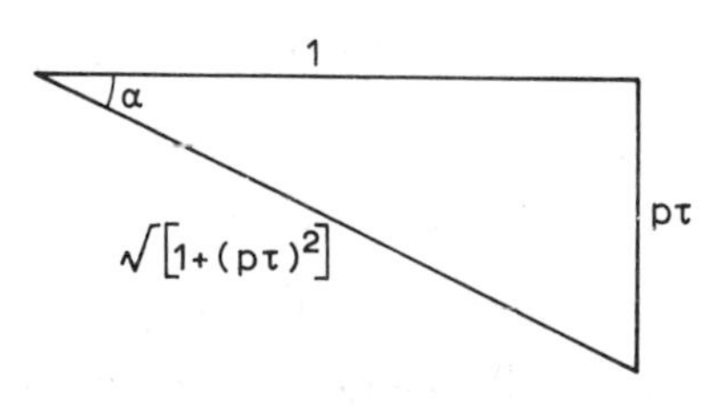

Fig. B1

$$\begin{aligned} 0 &= Pe^0 + A_i[1/\sqrt{\{1 + (p\tau)^2\}}]\cos(-\alpha) \\ &= Pe^0 + A_i[1/\sqrt{\{1 + (p\tau)^2\}}] \,.\, [1/\sqrt{\{1 + (p\tau)^2\}}] \\ &= P + A_i[1/\sqrt{\{1 + (p\tau)^2\}}] \end{aligned}$$

$\therefore \quad P = -A_i[1/\sqrt{\{1 + (p\tau)^2\}}]$

Hence

$$\theta_0 = A_i\left[\frac{-e^{-t/\tau}}{1 + (p\tau)^2} + \frac{1}{\sqrt{\{1 + (p\tau)^2\}}} \cos(pt - \alpha)\right] \qquad 3.11$$

B2 Second-order responses

Equation 3.5 is written $\tau^2 D^2\theta_0 + 2\zeta\tau D\theta_0 + \theta_0 = \theta_i$.

B2.1 The complementary function

$$\tau^2 D^2\theta_0 + 2\zeta\tau D\theta_0 + \theta_0 = 0$$

The auxiliary equation is

$$\tau^2 m^2 + 2\zeta\tau m + 1 = 0$$

and the roots of this are

$$m = -\frac{\zeta}{\tau} \pm \frac{1}{\tau}\sqrt{(\zeta^2 - 1)}$$

and may be real and different, real and equal, or imaginary when $\zeta < 1$.

Case 1: $\zeta > 1$
The roots are real, m_1 and m_2, and the c.f. is

$$\theta_0 = P_1 e^{m_1 t} + P_2 e^{m_2 t}$$

where P_1 and P_2 are constants.

Case 2: $\zeta < 1$
The roots are imaginary: $m_1 = -\zeta/\tau + (j/\tau)\sqrt{(1 - \zeta^2)}$
and $m_2 = -\zeta/\tau - (j/\tau)\sqrt{(1 - \zeta^2)}$
where $j = \sqrt{(-1)}$.

The c.f. is

$$\begin{aligned}\theta_0 &= e^{-\zeta t/\tau}\{P_1 e^{+j\sqrt{(1-\zeta^2)}t/\tau} + P_2 e^{-j\sqrt{(1-\zeta^2)}t/\tau}\} \\ &= e^{-\zeta t/\tau}\{(P_1 + P_2)\cos[(\sqrt{1-\zeta^2}/\tau)t] + (P_1 - P_2)\sin[(\sqrt{1-\zeta^2}/\tau)t]\} \\ &= e^{-\zeta t/\tau}\{P_3\cos(\omega_d t) + P_4\sin(\omega_d t)\} \quad \text{since } \omega_d = \sqrt{1-\zeta^2}/\tau \\ &= P_5 e^{-\zeta t/\tau}\cos(\omega_d t - \phi)\end{aligned}$$

where $\phi = \arctan(P_4/P_3)$.

Case 3: $\zeta = 1$
This is the condition of critical damping, and $m = -\zeta/\tau$.

The c.f. is $\theta_0 = (P + Qt)e^{-\zeta t/\tau}$ (see ref. 19).

B2.2 The particular integral

B2.2.1 Step inputs: θ_i *= constant for* $t > 0$
The p.i. is $\theta_0 = \theta_i$.

Case 1: $\zeta > 1$

$\theta_0 = P_1 e^{m_1 t} + P_2 e^{m_2 t} + \theta_i$ is the general solution

$$\therefore \quad D\theta_0 = m_1 P_1 e^{m_1 t} + m_2 P_2 e^{m_2 t}$$

When $t = 0, \theta_0 = 0, \quad \therefore \quad 0 = P_1 + P_2 + \theta_i.$
When $t = 0, D\theta_0 = 0, \quad \therefore \quad 0 = m_1 P_1 + m_2 P_2.$

Hence $\quad P_1 = \left(\dfrac{m_2}{m_1 - m_2}\right)\theta_i$ and $P_2 = \left(\dfrac{-m_1}{m_1 - m_2}\right)\theta_i.$

$$\therefore \quad \theta_0 = \left(\frac{m_2 e^{m_1 t} - m_1 e^{m_2 t}}{m_1 - m_2}\right)\theta_i + \theta_i \qquad \text{B1}$$

where m_1 and m_2 are the roots of the auxiliary equation. Since m_1 and m_2 are always negative, the first term is an exponential decay, and the steady-state output equals the input when t is large.

Case 2: $\zeta < 1$

$\theta_0 = P_5 e^{-\zeta t/\tau} \cos(\omega_d t - \phi) + \theta_i$ is the general solution

$$\therefore \quad D\theta_0 = -P_5(\zeta/\tau)e^{-\zeta t/\tau}\cos(\omega_d t - \phi) - P_5 e^{-\zeta t/\tau}\sin(\omega_d t - \phi)$$

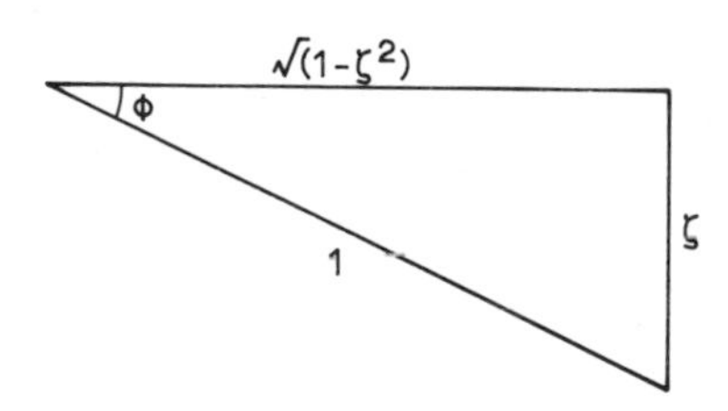

Fig. B2

When $t = 0, \theta_0 = 0$ and $D\theta_0 = 0,$

$$\therefore \quad 0 = P_5 e^0 \cos(-\phi) + \theta_i$$

and $0 = -P_5(\zeta/\tau)\cos(-\phi) - P_5\omega_d \sin(-\phi)$
Solution of these gives

$$\tan\phi = -(\zeta\tau/\omega_d) = -\zeta/\sqrt{(1 - \zeta^2)}$$

$$P_5 = -\theta_i/\cos(-\phi) = -\theta_i/\sqrt{(1 - \zeta^2)}$$

Hence $$\theta_0 = \theta_i - \theta_i \frac{e^{-\zeta t/\tau}}{\sqrt{(1 - \zeta^2)}}\cos(\omega_d t - \phi) \qquad 3.8$$

Case 3: $\zeta = 1$
$\theta_0 = (P + Qt)\mathrm{e}^{-\zeta t/\tau} + \theta_i$ is the general solution

$$\therefore \quad \mathrm{D}\theta_0 = \{(P + Qt)\zeta/\tau\}\mathrm{e}^{-\zeta t/\tau} + Q\mathrm{e}^{-\zeta t/\tau}$$

When $t = 0$, $\theta_0 = 0$ and $\mathrm{D}\theta_0 = 0$,

$$\therefore \quad 0 = P + \theta_i \qquad \therefore \quad P = -\theta_i$$

and $0 = -P\zeta/\tau + Q \qquad \therefore \quad Q = P\zeta/\tau = -\theta_i\zeta/\tau$

Hence $$\theta_0 = \theta_i - \theta_i\{1 + (\zeta t/\tau)\}\mathrm{e}^{-\zeta t/\tau} \qquad \text{B2}$$

and θ_0 will approach the step value θ_i without overshoot.

B2.2.2 Ramp inputs: $\theta_i = \Omega t$, where Ω is constant
Let the output be $\theta_0 = Rt + S$, then $\mathrm{D}\theta_0 = R$ and $\mathrm{D}^2\theta_0 = 0$.
Putting these into the differential equation,

$$0 + 2\zeta\tau R + Rt + S = \Omega t$$

Equating coefficients of t, $\quad R = \Omega$
Equating constants, $\quad 2\zeta\tau R + S = 0 \qquad \therefore \quad S = -2\zeta\tau\Omega$

$$\therefore \quad \text{the p.i. is } \theta_0 = \Omega t - 2\zeta\tau\Omega.$$

Case 1: $\zeta > 1$
$\theta_0 = P_1\mathrm{e}^{m_1 t} + P_2\mathrm{e}^{m_2 t} + \Omega t - 2\zeta\tau\Omega$ is the general solution

$$\therefore \quad \mathrm{D}\theta_0 = m_1P_1\mathrm{e}^{m_1 t} + m_2P_2\mathrm{e}^{m_2 t} + \Omega$$

When $t = 0$, $\theta_0 = 0$ and $\mathrm{D}\theta_0 = 0$,

$$\therefore \quad 0 = P_1 + P_2 - 2\zeta\tau\Omega \qquad P_2 = 2\zeta\tau\Omega - P_1$$

$$\begin{aligned} \text{and } 0 &= m_1P_1 + m_2P_2 + \Omega \\ &= m_1P_1 + m_2(2\zeta\tau\Omega - P_1) + \Omega \\ &= m_1P_1 + 2m_2\zeta\tau\Omega - m_2P_1 + \Omega \end{aligned}$$

$$\therefore \quad P_1 = \Omega(2m_2\zeta\tau + 1)/(m_2 - m_1)$$

$$\begin{aligned} P_2 &= 2\zeta\tau\Omega - P_1 \\ &= 2\zeta\tau\Omega - \Omega(2m_2\zeta\tau + 1)/(m_2 - m_1) \\ &= -\Omega(1 + 2m_1\zeta\tau)/(m_2 - m_1) \end{aligned}$$

Hence

$$\theta_0 = \Omega\left\{\frac{(2m_2\zeta\tau + 1)\mathrm{e}^{m_1 t} - (1 + 2m_1\zeta\tau)\mathrm{e}^{m_2 t}}{m_2 - m_1}\right\} + \Omega t - 2\zeta\tau\Omega \qquad \text{B3}$$

This response is seen to be of similar form to that for the first-order system to a ramp input, having a term which becomes zero when t is large, giving then a constant lag of $2\zeta\tau\Omega$.

Case 2: $\zeta < 1$

$\theta_0 = P_5 e^{-\zeta t/\tau} \cos(\omega_d t - \phi) + \Omega t - 2\zeta\tau\Omega$ is the general solution,

$$\therefore \quad D\theta_0 = P_5 e^{-\zeta t/\tau}\{-\omega_d \sin(\omega_d t - \phi)\} - P_5(\zeta/\tau)\cos(\omega_d t - \phi) + \Omega$$

When $t = 0$, $\theta_0 = 0$ and $D\theta_0 = 0$,

$$\therefore \quad 0 = P_5 \cos(-\phi) - 2\zeta\tau\Omega \qquad \therefore \quad P_5 = 2\zeta\tau\Omega/\cos(-\phi)$$

and $0 = -P_5\omega_d \sin(-\phi) - P_5(\zeta/\tau)\cos(-\phi) + \Omega$

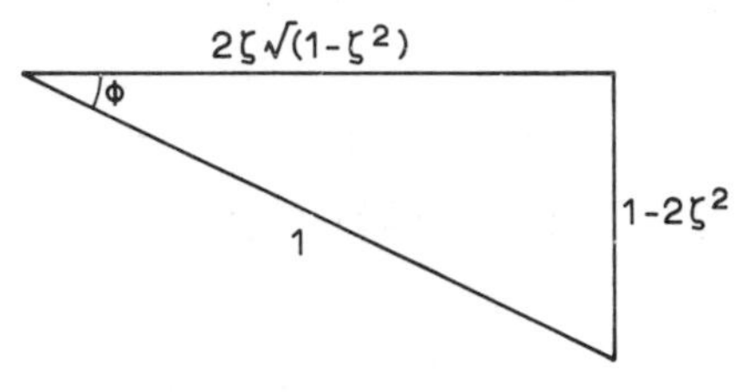

Fig. B3

Substituting for P_5 gives

$$\tan(-\phi) = (1 - 2\zeta^2)/\{2\zeta\sqrt{(1 - \zeta^2)}\}$$

and $\phi = \arcsin(2\zeta^2 - 1)$

$$\therefore \quad P_5 = 2\zeta\tau\Omega/\{2\zeta\sqrt{(1 - \zeta^2)}\}$$
$$= \Omega/\omega_d$$

Hence

$$\theta_0 = \Omega t - 2\zeta\tau\Omega + (\Omega/\omega_d)e^{-\zeta t/\tau} \cos(\omega_d t - \phi) \qquad 3.10$$

Case 3: $\zeta = 1$

$\theta_0 = (P + Qt)e^{-t/\tau} + \Omega t - 2\zeta\tau\Omega$ is the general solution,

$$\therefore \quad D\theta_0 = (P + Qt)(-1/\tau)e^{-t/\tau} + Qe^{-t/\tau} + \Omega$$

When $t = 0$, $\theta_0 = 0$ and $D\theta_0 = 0$,

$$\therefore \quad 0 = P - 2\zeta\tau\Omega \qquad P = 2\zeta\tau\Omega$$

and $0 = (P/\tau) + Q + \Omega$.

$$\therefore \quad Q = \Omega(2\zeta - 1)$$

and $\theta_0 = \Omega\{2\zeta\tau + (2\zeta - 1)t\}e^{-t/\tau} + \Omega t - 2\zeta\tau\Omega$ B4

B2.2.3 Sinusoidal inputs: $\theta_i = A_i \cos pt$, *where* A_i *is constant*

Let the output be $\theta_0 = R \sin pt + S \cos pt$, then

$$D\theta_0 = pR \cos pt - pS \sin pt$$

$$D^2\theta_0 = -p^2R \sin pt - p^2S \cos pt$$

Putting these into the differential equation,

$$-\tau^2(p^2R \sin pt + p^2S \cos pt) + 2\zeta\tau(pR \cos pt - pS \sin pt) + R \sin pt + S \cos pt = A_i \cos pt$$

Equating coefficients of sin pt,

$$-\tau^2p^2R - 2\zeta\tau pS + R = 0$$

Letting $\tau p = \mu$, then

$$-\mu^2R - 2\zeta\mu S + R = 0$$

and $$R = 2\zeta\mu S/(1 - \mu^2) \qquad \text{(i)}$$

Equating coefficients of cos pt,

$$-\tau^2p^2S + 2\zeta\tau pR + S = A_i$$

or $$-\mu^2S + 2\zeta\mu R + S = A_i \qquad \text{(ii)}$$

Substituting for R from (i) in (ii) gives

$$S(1 - \mu^2) + \frac{2\zeta\mu \,.\, 2\zeta\mu S}{1 - \mu^2} = A_i$$

$$\therefore \quad S = \left\{\frac{1 - \mu^2}{(1 - \mu^2)^2 + (2\zeta\mu)^2}\right\}A_i$$

and $R = \left\{\dfrac{2\zeta\mu}{(1 - \mu^2)^2 + (2\zeta\mu)^2}\right\}A_i$

The p.i. is

$$\theta_0 = A_i\{2\zeta\mu \sin pt + (1 - \mu^2)\cos pt\}/\{(1 - \mu^2)^2 + (2\zeta\mu)^2\}$$

$$= A_i \frac{\sqrt{\{(1 - \mu^2)^2 + (2\zeta\mu)^2\}}}{(1 - \mu^2)^2 + (2\zeta\mu)^2} \cos(pt - \alpha)$$

$$= \frac{A_i}{\sqrt{\{(1 - \mu^2)^2 + (2\zeta\mu)^2\}}} \cos(pt - \alpha)$$

where $\alpha = \arctan\{2\zeta\mu/(1 - \mu^2)\}$.

Case 2: $\zeta < 1$

$$\theta_0 = P_5 e^{-\zeta t/\tau} \cos(\omega_d t - \phi) + \frac{1}{\sqrt{\{(1 - \mu^2)^2 + (2\zeta\mu)^2\}}} A_i \cos(pt - \alpha) \qquad 3.13$$

This is the general solution. However, the c.f., representing the transient response, is not usually of interest. If the output amplitude is A_0, then, when t is large,

$$\frac{A_0}{A_i} = \frac{1}{\sqrt{\{(1 - \mu^2)^2 + (2\zeta\mu)^2\}}} \qquad 3.14$$

Equation 3.14 applies for all values of ζ. For $\zeta \geqslant 1$, the transient response will be different from that of eqn 3.13. However, for most sinusoidal signals only the steady-state relationship is of interest.

For the particular case of the seismic vibration transducer, eqn 14.5 is seen to be the equivalent of

$$(\tau^2 D^2 + 2\zeta\tau D + 1)z = \mu^2\hat{\theta}_i \cos pt$$

Comparing this with $(\tau^2 D^2 + 2\zeta\tau D + 1)\theta_0 = A_i \cos pt$, whose solutions are eqns 3.13 and 3.14, it is seen that μ^2 is an added factor on the r.h.s. Carrying this factor through the solutions, to eqns 3.13 and 3.14, gives

$$\frac{Z}{\bar{\theta}_i} = \frac{\mu^2}{\sqrt{\{(1 - \mu^2)^2 + (2\zeta\mu)^2\}}} \qquad 14.6$$

Appendix C
Mathematical Theory of Errors

C1 Binomial approximations

If a quantity u is dependent on the quantities x, y, and z such that

$$u = kx^a y^b z^c$$

where k is a constant, then, if δx is a small change in x, and δy is a small change in y, and δz a small change in z,

$$u + \delta u = k(x + \delta x)^a (y + \delta y)^b (z + \delta z)^c$$

where δu is the change in u.

Then $$u + \delta u = kx^a y^b z^c \left(1 + \frac{\delta x}{x}\right)^a \left(1 + \frac{\delta y}{y}\right)^b \left(1 + \frac{\delta z}{z}\right)^c$$

or $$1 + \frac{\delta u}{u} = \left(1 + \frac{\delta x}{x}\right)^a \left(1 + \frac{\delta y}{y}\right)^b \left(1 + \frac{\delta z}{z}\right)^c$$

If δx, δy, and δz are sufficiently small, then binomial expansion gives

$$1 + \frac{\delta u}{u} \approx 1 + a\left(\frac{\delta x}{x}\right) + b\left(\frac{\delta y}{y}\right) + c\left(\frac{\delta z}{z}\right)$$

or $$\frac{\delta u}{u} \approx a\left(\frac{\delta x}{x}\right) + b\left(\frac{\delta y}{y}\right) + c\left(\frac{\delta z}{z}\right) \qquad \text{C1}$$

If this equation is applied to a measurement of x, of y, and of z, to determine u, then the fractional error $\delta u/u$ is seen to be the sum of the fractional errors $\delta x/x$, $\delta y/y$, and $\delta z/z$ of the individual measurements, each multiplied by the power to which the variable is raised in the measurement equation.

Example C1 The measurements of a collecting tank are nominally 200 mm × 300 mm × 500 mm. If the steel tape used for measuring these dimensions can be read to an accuracy of not more than 0·5 mm error, determine the possible percentage error in the volume value due to this.

$$\frac{\delta x}{x} = \pm\frac{0{\cdot}5}{200}, \qquad \frac{\delta y}{y} = \pm\frac{0{\cdot}5}{300}, \qquad \frac{\delta z}{z} = \pm\frac{0{\cdot}5}{500}$$

$$\therefore \quad \frac{\delta u}{u} = \pm\left(\frac{0{\cdot}5}{200} + \frac{0{\cdot}5}{300} + \frac{0{\cdot}5}{500}\right) \times 100\%$$
$$= \pm 0{\cdot}25 \pm 0{\cdot}17 \pm 0{\cdot}10\%$$
$$= \pm 0{\cdot}5\%$$

Hence the total maximum error, expressed in fractional form, is seen to be the sum of the individual maximum errors, and cannot be less than the largest of these.

C2 Partial differentials

If p is a function of variables q, r, and s, i.e. $p = f(q, r, s)$, then the *total differential*, i.e. the total change in p due to the small changes δq, δr, and δs in q, r, and s, is given by

$$\delta p \approx \frac{\partial p}{\partial q}\,\delta q + \frac{\partial p}{\partial r}\,\delta r + \frac{\partial p}{\partial s}\,\delta s \qquad \text{C2}$$

where $\partial p/\partial q$ is the rate of change of p with respect to q when r and s are constant, i.e. it is a *partial derivative*, and similarly with $\partial p/\partial r$ and $\partial p/\partial s$.

If δq, δr, and δs are errors in the measurement of q, r, and s, then δp is the error in the determination of p due to these.

Example C2 The cross-section of a duct has the inside shape shown in fig. C1. Calculate the error and percentage error in the determination of the cross-sectional area if the error in measuring l is not more than $\pm 2\%$, and for $r \pm 1\%$.

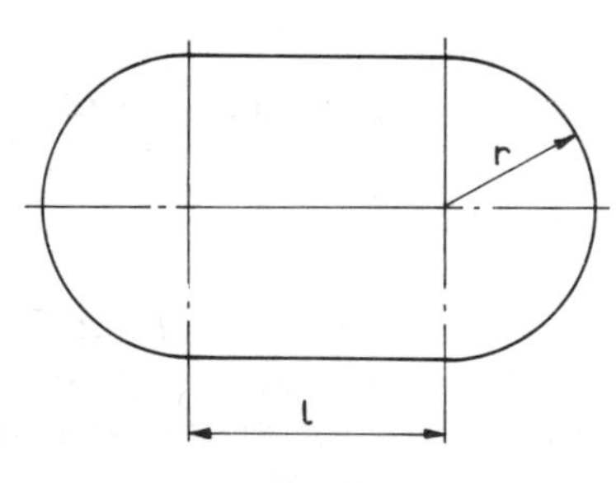

Fig. C1

Cross-sectional area $= A = \pi r^2 + 2rl$

$$\therefore \quad \frac{\partial A}{\partial r} = 2\pi r + 2l, \qquad \frac{\partial A}{\partial l} = 2r$$

Therefore, from eqn C2,

$$\begin{aligned}
\delta A &\approx 2(\pi r + l)\,\delta r + 2r\,\delta l \\
&\approx 2(\pi r + l)\left(\pm\frac{r}{100}\right) + 2r\left(\pm\frac{2l}{100}\right) \\
&\approx \pm 2\left(\frac{\pi r^2 + lr + 2rl}{100}\right) \\
&\approx \pm 2\left(\frac{\pi r^2 + 3lr}{100}\right) \\
&\approx 2\left(\frac{\pi \times 50^2 \times 3 \times 200 \times 50}{100}\right) \\
&\approx \pm 757\ \text{mm}^2 \\
A &= \pi \times 50^2 + 200 \times 100 \\
&= 27\,852\ \text{mm}^2
\end{aligned}$$

$$\therefore \quad \text{percentage error} = 757 \times 100/27\,852 = \pm 2{\cdot}7\%$$

C3 Logarithmic differentiation

If $u = kx^a y^b z^c$,

then $$\ln u = \ln k + a \ln x + b \ln y + c \ln z$$

Differentiating,

$$\frac{\mathrm{d}u}{u} = a\frac{\mathrm{d}x}{x} + b\frac{\mathrm{d}y}{y} + c\frac{\mathrm{d}z}{z}$$

If small finite changes δx, δy, and δz occur, then

$$\frac{\delta u}{u} \approx a\frac{\delta x}{x} + b\frac{\delta y}{y} + c\frac{\delta z}{z}$$

and this equation is seen to be identical with eqn C1.

Other types of equation may be solved by carrying out the differentiation and then simplifying, as in the following example.

Example C3 The angle set on a sine bar (section 11.1.3) is related to the difference in height (h) and the slant distance (l) (fig. 11.4) by the equation $\sin\theta = h/l$. Determine the error $\delta\theta$ in the angle setting if the small errors in setting h and l are δh and δl respectively.

$$\sin\theta = h/l$$

Taking logs,

$$\ln(\sin\theta) = \ln h - \ln l$$

Differentiating,

$$\frac{\cos\theta}{\sin\theta}\,\mathrm{d}\theta = \frac{\mathrm{d}h}{h} - \frac{\mathrm{d}l}{l}$$

or

$$\mathrm{d}\theta = \left(\frac{\mathrm{d}h}{h} - \frac{\mathrm{d}l}{l}\right)\tan\theta$$

$$= \left(\frac{\mathrm{d}h}{h} - \frac{\mathrm{d}l}{l}\right)\frac{h}{\sqrt{(l^2 - h^2)}}$$

$$= (\mathrm{d}h - h\,\mathrm{d}l/l)/\sqrt{(l^2 - h^2)}$$

For small finite errors, δh and δl, the error in θ is

$$\delta\theta \approx \pm(\delta h + h\,\delta l/l)/\sqrt{(l^2 - h^2)}$$

Appendix D
Formulae for Deflection and Stress of Elastic Elements

E = modulus of elasticity, G = modulus of rigidity, I = second moment of area of the section about the neutral axis.

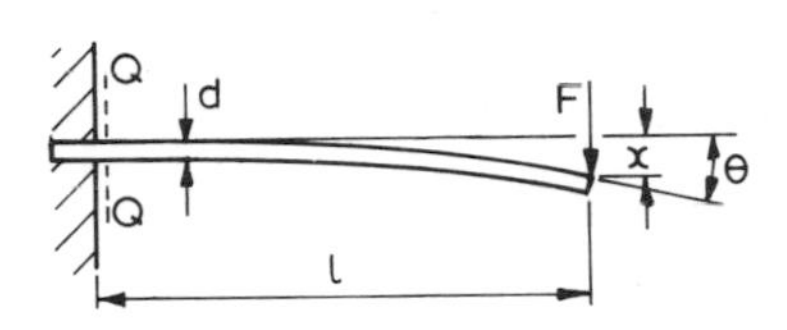

Cantilever

deflection	$x = Fl^3/3EI$
stiffness	$k = F/x = 3EI/l^3$
slope at R	$\theta = Fl^2/2EI$
skin stress at Q	$\sigma = \pm Fld/2I$

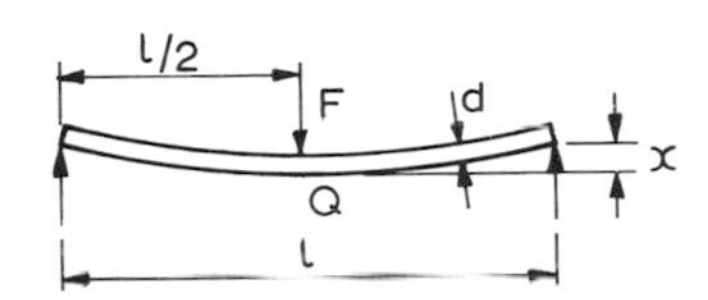

Simply supported beam

deflection	$x = Fl^3/48EI$
stiffness	$k = F/x = 48EI/l^3$
skin stress at Q	$\sigma = \pm Fld/8I$

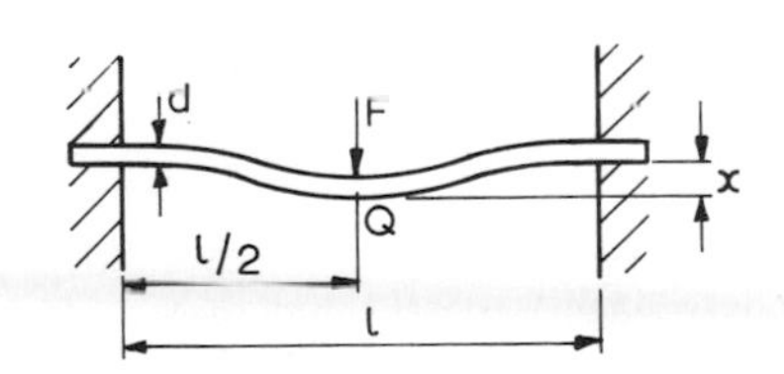

Built-in (encastré) beam

deflection	$x = Fl^3/192EI$
stiffness	$k = F/x = 192EI/l^3$
stress at Q	$\sigma = \pm Fld/32EI$

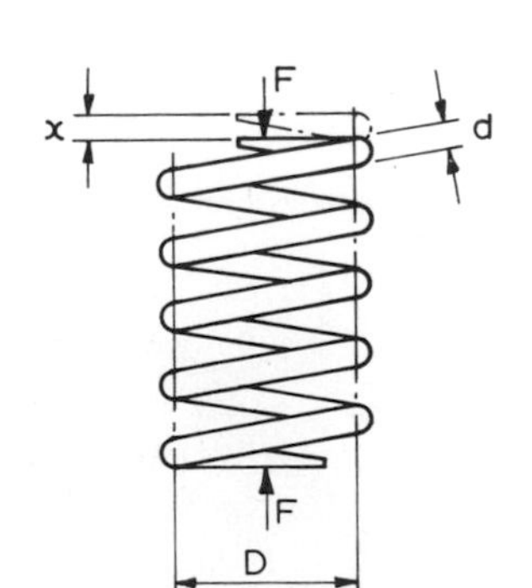

Close-coiled helical spring

stiffness	$k = F/x = Gd^4/8D^3n$
max. shear stress	$\tau = 8FD/\pi d^3$

where d = wire diameter, D = mean coil diameter, and n = number of coils.

These formulae are nearly correct only for close-coiled helical springs.

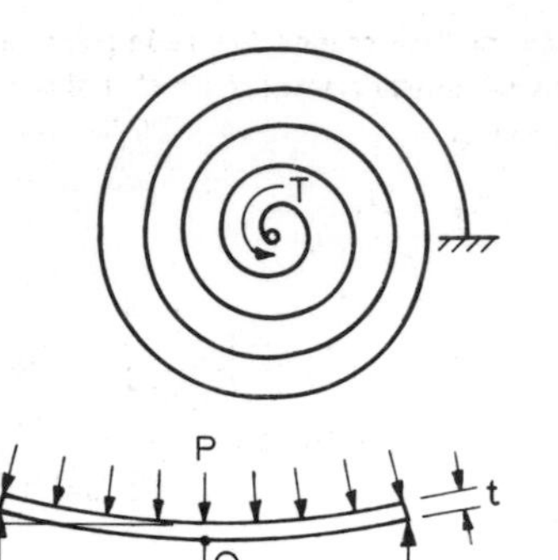

Flat spiral spring

stiffness $k = T/\theta = EI/l$

max. skin stress $\sigma = \pm 12T/bt^2$

where l = length of spiral strip, b = width of strip, t = thickness of strip, θ = rotation of centre, and T = applied torque.

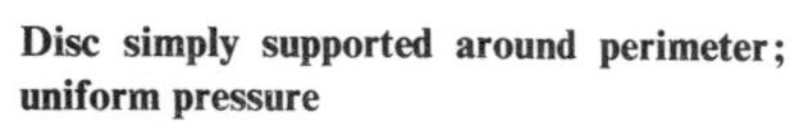

Disc simply supported around perimeter; uniform pressure

deflection

$x = 3PR^4(5 + \nu)(1 - \nu)/16Et^3$

radial stress at Q

$\sigma_r = 3PR^2(3 + \nu)/8t^2$

where P = uniform pressure and ν = Poisson's ratio

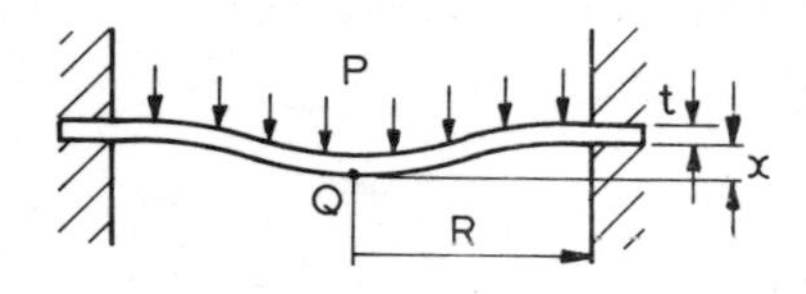

Disc built-in around the perimeter; uniform pressure applied

deflection

$x = 3PR^4(1 - \nu^2)/16Et^3$*

radial stress at Q

$\sigma_r = 3PR^2(1 + \nu)/8t^2$

where P = uniform pressure and ν = Poisson's ratio.

(* nearly linear for $x \leqslant 0{\cdot}5t$)

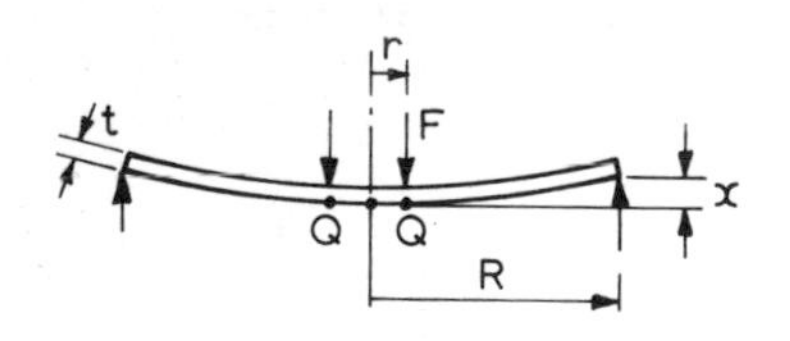

Disc simply supported around perimeter; force applied around central circle radius *r*

deflection:

$$x = \frac{3F(1 - \nu^2)}{2\pi E t^3} \times \left\{\frac{(R^2 - r^2)(3 + \nu)}{2(1 + \nu)} - r^2 \ln\left(\frac{R}{r}\right)\right\}$$

radial stress at Q (radius $\leqslant r$) = tangential stress:

$$\sigma_r = \frac{3F}{4\pi t^2}\left\{(1 + \nu)2 \ln\left(\frac{R}{r}\right) + (1 - \nu)\frac{(R^2 - r^2)}{R^2}\right\} = \sigma_t$$

where ν = Poisson's ratio and F = total force around circle of radius r.

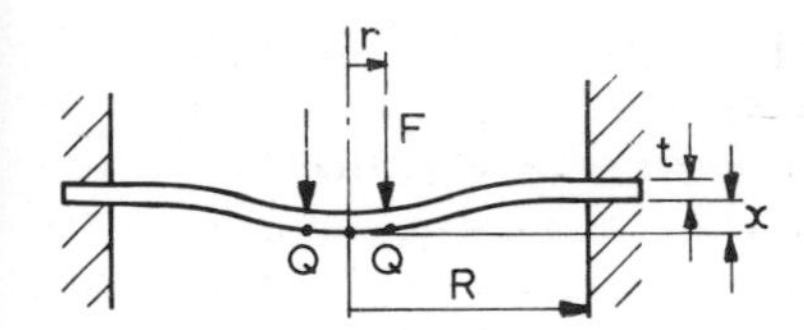

Disc built-in around the perimeter; force applied around central circle of radius *r*
deflection:

$$x = \frac{3F(1 - \nu^2)}{2\pi E t^3}$$

$$\times \left\{ \tfrac{1}{2}(R^2 - r^2) \quad r^2 \ln\left(\frac{R}{r}\right) \right\}$$

radial stress at Q (radius $\leqslant r$) = tangential stress:

$$\sigma_r = \frac{3F}{4\pi t^2}$$

$$\times \left[(1+\nu)\left\{ 2 \ln\left(\frac{R}{r}\right) + \frac{r^2}{R^2} - 1 \right\} \right]$$

(max. stress if $r < 0{\cdot}31R$)

where ν = Poisson's ratio and F = total force around circle of radius r.

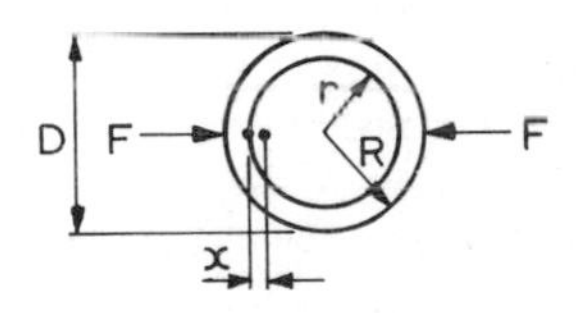

Ring
F = tensile or compressive force (radial)
x = change of radius in direction of F

$$= \frac{k}{16}\left(\frac{\pi}{2} - \frac{4}{\pi}\right)\frac{FD^3}{EI}$$

R/r	1·3	1·4	1·5	1·6
k	1·030	1·055	1·090	1·114

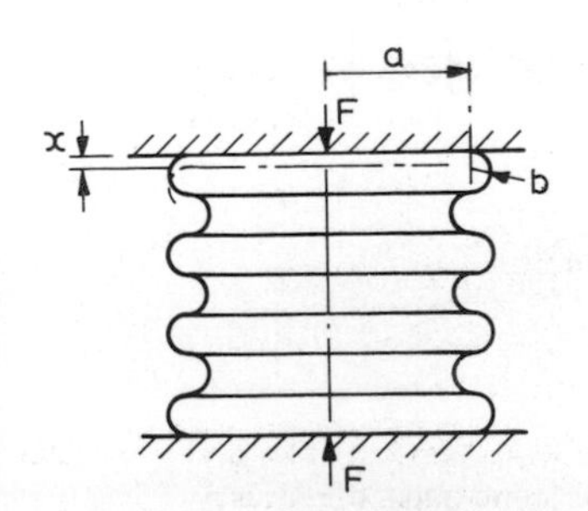

Bellows
No change of length occurs due to internal or external pressure unless acting on the end plates. Compression

$$x = 1{\cdot}813 F b n \sqrt{1 - \nu^2} / \pi E t^2$$

where a = mean radius of corrugations, b = radius of each convolution, n = number of semicircular corrugations, t = thickness of wall, and ν = Poisson's ratio.

Selection of British Standards relating to Measurement

BS 89: 1954	Electrical indicating instruments
BS 89: part 1: 1970	Single-purpose direct-acting electrical indicating instruments and their accessories
BS 188: 1957	Methods for the determination of the viscosity of liquids in c.g.s. units
BS 350	Conversion factors and tables
BS 599: 1966	Methods of testing pumps
BS 726: 1957	Measurement of air flow for compressors and exhausters
BS 848	Methods of testing fans for general purposes, including mine fans
part 1: 1963	Performance
part 2: 1966	Fan noise testing
BS 888: 1950	Slip (or block) gauges and their accessorics
BS 907: 1965	Dial gauges for linear measurement
BS 958: 1968	Spirit levels for use in precision engineering
BS 1041: 1943	Code for temperature measurement
part 2:	Expansion thermometers
section 2: 1: 1969	Liquid-in-glass expansion thermometers
part 3: 1969	Industrial resistance thermometry
part 4: 1966	Thermocouples
part 5: 1972	Radiation pyrometers
part 7: 1964	Temperature/time indicators
BS 1042	Code for flow measurement
part 1: 1964	Orifice plates, nozzles, and venturi tubes
part 3: 1965	Guide to the effects of departure from the methods in part 1
BS 1054: 1954	Engineers' comparators for external measurement
BS 1134	Method for the assessment of surface texture
part 1: 1972	Method and instrumentation
part 2: 1972	General information and guidance
BS 1610: 1964	Methods for the load verification of testing machines

BS 1780: 1960	Bourdon-tube pressure- and vacuum-gauges
part 2: 1971	Metric units
BS 1828: 1961	Reference tables for copper *v* constantan thermocouples
BS 1904: 1964	Industrial platinum resistance thermometer elements
BS 2643: 1955	Glossary of terms relating to the performance of measuring instruments
BS 3403: 1972	Indicating tachometers and speedometer systems for industrial, railway, and marine use
BS 3680	Methods of measurement of liquid flow in open channels
BS 3693	Recommendations for the design of scales and indexes
BS 3730: 1964	Methods for the assessment of departures from roundness
BS 3763: 1970	The International System of units (SI)
BS 4311: 1968	Metric gauge blocks
BS 4358: 1968	Glossary of terms used in air gauging, with notes on the technique
BS 4937	International thermocouple reference tables
part 1: 1973	Platinum–10% rhodium/platinum thermocouples
part 2: 1973	Platinum–13% rhodium/platinum thermocouples
part 3: 1973	Iron/copper–nickel thermocouples
part 4: 1973	Nickel–chromium/nickel–aluminium thermocouples
part 5: 1974	Copper/copper–nickel thermocouples

Bibliography

The author has found the following publications helpful in his teaching of measurements and in the preparation of this book.

1) Barry, B. A. *Engineering Measurements*. (John Wiley)
2) Oliver, F. J. *Practical Instrumentation Transducers*. (Pitman)
3) Cerni, R. H., and Foster, L. E. *Instrumentation for Engineering Measurement*. (John Wiley)
4) Sutcliffe, H. *Electronics for Students of Mechanical Engineering*. (Longman)
5) Neubert, H. K. P. *Instrument Transducers*. (Clarendon Press)
6) Pearson, E. B. *Technology of Instrumentation*. (The English Universities Press)
7) Bass, H. G. *Introduction to Engineering Measurements*. (McGraw-Hill)
8) Beckwith, T. G., and Buck, N. L. *Mechanical Measurements*. (Addison-Wesley)
9) Jones, E. B. *Instrument Technology*, *Vol. 1* (*Pressure, level, flow, and temperature*). (Butterworth)
10) Doyle, F. E. *Instrumentation. Pressure and Liquid Level*. (Blackie)
11) Miller, J. T. *Industrial Instrument Technology*. (United Trade Press)
12) Martin, S. J. *Numerical Control of Machine Tools*. (The English Universities Press)
13) Galyer, J. F. W., and Shotbolt, C. R. *Metrology for Engineers*. (Cassell)
14) Hume, K. J. *Engineering Metrology*. (MacDonald)
15) Lissaman A. J. *Metrology for the Technician*. (The English Universities Press)
16) Parsons, S. A. J. *Metrology and Gauging*. (MacDonald and Evans)
17) Scarr, A. J. T. *Metrology and Precision Engineering*. (McGraw-Hill)
18) Sharp, K. W. B. *Practical Engineering Metrology*. (Pitman)
19) Pedoe, J. *Advanced National Certificate Mathematics*, *Vols 1 and 2*. (The English Universities Press)
20) Warnock, F. V., and Benham, P. P. *Mechanics of Solids and Strength of Materials*. (Pitman)

21) Probert, S. D., Marsden, J. P., and Holmes, T. W. *Experimentation for Students of Engineering, Vol. 1: Experimental Method and Measurement*. (Heinemann)
22) Webb, C. R. and Luxmore, A. R. *Experimentation for Students of Engineering, Vol. 2: Applied Mechanics*. (Heinemann)
23) Yarnell, J. *Resistance Strain-gauges*. (Electronic Engineering)
24) Nottingk, B. E., McLachlan, D. F. A., Owen, C. K. V., and O'Neill, P. C. 'High-stability capacitance strain-gauge for use at extreme temperature'. *IEE Proceedings* Vol. 119. July 1972
25) Hendry, A. W. *Photoelastic Analysis*. (Pergamon)
26) Perry, C. C., and Lissner, H. R. *The Strain-gauge Primer*. (McGraw-Hill)
27) Kaye, G. W. C., and Laby, T. H. *Tables of Physical and Chemical Constants*. (Longmans)
28) Doyle, F. E., and Byrom, G. T. *Instrumentation–Temperature*. (Blackie)
29) Eckman, D. P. *Industrial Instrumentation*. (John Wiley)
30) Rogers, G. F. C., and Mayhew, Y. R. *Engineering Thermodynamics: Work and Heat Transfer*. (Longmans)
31) Condon, E. U., and Odishaw, H. *Handbook of Physics*. (McGraw-Hill)
32) NPL. *The International Practical Temperature Scale of 1968*. (HMSO)
33) Barber, C. R. (NPL). *The Calibration of Thermometers*. (HMSO)
34) Heselwood, W. C. *Instrumentation in the Metallurgical Industry*. (Institution of Metallurgists)
35) Wallace, R. H. *Understandings and Measuring Vibrations*. (Wykeham Publications)
36) Ryder, G. H. *Strength of Materials*. (Macmillan)
37) Roark, R. J. *Formulas for Stress and Strain*. (McGraw-Hill)
38) Bryan, G. T. *Control Systems for Technicians*. (The English Universities Press)
39) Hughes, E. *Electrical Technology*. (Longman)
40) Puckle, O. S., and Arrowsmith, J. R. *An Introduction to the Numerical Control of Machine Tools*. (Chapman and Hall)

Index

(Entries in **bold** type refer to figure numbers)

507